Helioseismology and Dynamics of the Solar Interior

M.J. Thompson • A.S. Brun • J.L. Culhane •
L. Gizon • M. Roth • T. Sekii
Editors

Helioseismology and Dynamics of the Solar Interior

Chapters 2–13 were previously published in *Space Science Reviews* Volume 196, Issues 1–4, 2015

Springer

Editors
M.J. Thompson
National Center for Atmospheric Research
Boulder, CO, USA

A.S. Brun
CEA/IRFU, Université Paris-Diderot
CNRS/INSU
Gif-sur-Yvette, France

J.L. Culhane
Mullard Space Science Laboratory
University College London
Dorking, UK

L. Gizon
Max-Planck-Institut fur
Sonnensystemforschung
Goettingen, Germany

M. Roth
Kiepenheuer-Institut für Sonnenphysik
Freiburg, Baden-Württemberg, Germany

T. Sekii
National Astronomical Observatory of Japan
Tokyo, Japan

ISSN 1385-7525 Space Sciences Series of ISSI
ISBN 978-94-024-1033-4

Library of Congress Control Number: 2017932284

Cover Image: 3-D rendering of the radial normalized velocity in a global simulation of the Sun from $r = 0$ to $r = 0.99\ R_S$. A quadrant has been omitted in order to better see the meridional and equatorial planes and the interplay between the surface turbulent convective motions and the excitation and propagation of internal waves in the deep radiative interior. Redish tones correspond to upflows. Visualization performed with SDvision software.
Copyrights: A.S. Brun, CEA-Saclay

Printed on acid-free paper

This Springer imprint is published by Springer Nature
The registered company is Springer Science+Business Media B.V.
The registered company address is: Van Godewijckstraat 30, 3311 GX Dordrecht, The Netherlands

Contents

Helioseismology with *Solar Orbiter*
B. Löptien, A.C. Birch, L. Gizon, J. Schou, T. Appourchaux, J. Blanco Rodríguez, P.S. Cally, C. Dominguez-Tagle, A. Gandorfer, F. Hill, J. Hirzberger, P.H. Scherrer and S.K. Solanki: Space Science Reviews 2014, 2015:251–283 (15 July 2014) DOI:10.1007/s11214-014-0065-3

Future Mission Concepts for Helioseismology
T. Sekii, T. Appourchaux, B. Fleck and S. Turck-Chièze: Space Science Reviews 2015, 2015:285–302 (17 March 2015) DOI:10.1007/s11214-015-0142-2

The Solar-Stellar Connection
A.S. Brun, R.A. García, G. Houdek, D. Nandy and M. Pinsonneault: Space Science Reviews 2014, 2015:303–356 (26 November 2014) DOI:10.1007/s11214-014-0117-8

Preface: Helioseismology and Dynamics of the Solar Interior

Laurent Gizon[1,2] · **Michael J. Thompson**[3] · **A. Sacha Brun**[4] · **J. Len Culhane**[5] · **Markus Roth**[6] · **Takashi Sekii**[7]

Keywords Sun · Helioseismology · Solar convection zone dynamics · Space missions

1 Motivation

It has long been the practice to study the properties of the Earth by examining the seismic waves propagating in its interior. Helioseismology, the analogous study of the Sun's interior from observations of the vibrations of its surface, has become a highly productive technique. Unlike earthquakes, solar acoustic oscillations are continuously excited by turbulence throughout the Sun's convection zone with periods in the range ~2–10 minutes. Long observation times of months to years are essential to measure the frequencies of the modes of oscillations with high precision. Ground-based observations have been undertaken with worldwide networks, including the Global Oscillations Network Group (GONG) and the Birmingham Solar Oscillations Network (BiSON). The value of observations from space was successfully demonstrated by the VIRGO, LOI, GOLF and MDI instruments on the Solar and Heliospheric Observatory (SOHO), in operations since 1996. Since 2010, the Helioseismic and Magnetic Imager (HMI) onboard the Solar Dynamics Observatory (SDO) has been providing nearly-continuous helioseismic observations of the full disk of the Sun with unprecedented spatial resolution (images have 16 million pixels).

A workshop was held during September 2012 under the auspices of the International Space Science Institute (ISSI) to review the status of helioseismology and its contribution to

1 Max-Planck-Institut für Sonnensystemforschung, 37077 Göttingen, Germany

2 Institut für Astrophysik, Georg-August-Universität Göttingen, 37077 Göttingen, Germany

3 High Altitude Observatory, NCAR, Boulder, CO, USA

4 CEA/Service d'Astrophysique, Gif-sur-Yvette, France

5 International Space Science Institute, Bern, Switzerland

6 Kiepenheuer-Institut für Sonnenphysik, 79104 Freiburg i. Br., Germany

7 National Astronomical Observatory of Japan, Mitaka, Japan

our understanding the solar internal dynamics. The authors of this preface were the conveners of the workshop. Information about the programme and the participants can be accessed at http://www.issibern.ch/workshops/helioseismology/. The workshop focussed on the following topics in helioseismology:

- Achievements of helioseismology, summarising in particular recent results from HMI.
- New mission concepts, assessing advances likely to arise from farside and out-of-ecliptic missions such as *Solar Orbiter*.
- Future ground-based technologies, including multi-line studies of solar oscillations at different heights in the atmosphere.
- Long-term synoptic observations.
- Improved and/or new data analysis techniques.

In addition, outstanding problems in solar convection zone dynamics were discussed in light of the observational constraints from helioseismology:

- Large-scale dynamics: rotation (near-surface shear layer, tachocline).
- Large-scale dynamics: meridional circulation.
- Convective velocities and the maintenance of mean flows.
- The solar dynamo; flux transport; flux emergence.
- Synergies between solar and stellar physics.

2 Organization of the Book

The present book covers most of the topics that were discussed at the conference and is organized according to the following twelve chapters. Each chapter has already appeared as a peer-reviewed article in *Space Science Reviews*, Volume 196, Issue 1–4 (December 2015):

- Prospects and challenges for helioseismology (Toomre and Thompson 2015).
- Some glimpses from helioseismology at the dynamics of the deep solar interior (Gough 2015).
- Understanding the internal chemical composition and physical processes of the solar interior (Basu et al. 2015).
- Solar dynamics, rotation, convection and overshoot (Hanasoge et al. 2015).
- Recent advances on solar global magnetism and variability (Brun et al. 2015a).
- The importance of long-term synoptic observations and data sets for solar physics and helioseismology (Elsworth et al. 2015).
- Sub-photosphere to solar atmosphere connection (Komm et al. 2015).
- Interpretation of helioseismic travel times (Burston et al. 2015).
- Recent developments in helioseismic analysis methods and solar data assimilation (Schad et al. 2015).
- Helioseismology with *Solar Orbiter* (Löptien et al. 2015).
- Future mission concepts for helioseismology (Sekii et al. 2015).
- The solar-stellar connection (Brun et al. 2015b).

3 Background Material, Important Numbers, Vocabulary

The following books, review papers, conference proceedings and topical issues provide an introduction to helioseismology and solar interior dynamics:

- Lectures on solar physics (Antia et al. 2003).
- The internal rotation of the Sun (Thompson et al. 2003)
- The Sun: an introduction (Stix 2004).
- Local helioseismology (Gizon and Birch 2005).
- Helioseismology and solar abundances (Basu and Antia 2008).
- Helioseismology, asteroseismology, and MHD connections (Gizon et al. 2008).
- The origin and dynamics of solar magnetism (Thompson et al. 2009).
- Physics, formation and evolution of rotating stars (Maeder 2009).
- Evolution and seismic tools for stellar astrophysics (Monteiro 2009).
- Asteroseismology (Aerts et al. 2010).
- Synergies between solar and stellar modelling (di Mauro et al. 2010).
- Three-dimensional imaging of the solar interior (Gizon et al. 2010).
- The solar tachocline (Hughes et al. 2012).
- Solar dynamics and magnetism from the interior to the atmosphere (Mansour et al. 2013).
- The solar activity cycle: physical causes and consequences (Balogh et al. 2015).
- Imaging convection and magnetism in the Sun (Hanasoge 2015).
- Extraterrestrial seismology (Tong and García 2015).
- Seismic probing of convection (Hanasoge et al. 2016)

The dynamics of the interior of the Sun is set by a number of parameters which describe the Sun and its motions. One of the most important of these is the radius of the Sun, which has been determined using observations of Mercury's transit to be $R_\odot = 695.74 \pm 0.0011$ Mm (Kuhn et al. 2004), which is consistent with the helioseismically determined radius of 695.68 ± 0.0003 Mm (Schou et al. 1997). The total mass of the Sun, $M_\odot$, is approximately 1.9884×10^{30} kg (Folkner et al. 2009). The gravitational acceleration at the solar surface (based on $M_\odot$ and $R_\odot$) is 274 m/s^2. The solar irradiance is about 1366 W m^{-2} (Froelich 1980), which implies a solar luminosity $L_\odot = 3.8416 \times 10^{33}$ ergs/s. Several models of the solar interior have been constructed (Monteiro 2009), for example the commonly used Model S (Christensen-Dalsgaard et al. 1996), which can be downloaded at http://astro.phys.au.dk/~jcd/solar_models/. This model is consistent with helioseismic inferences in placing the base of the convection zone at $r_{cz} = 0.713 \pm 0.003 R_\odot$ (Christensen-Dalsgaard et al. 1991). However the model and helioseismic inferences differ slightly (but with high statistical significance) in the surface layers and near the base of the convection zone (Ball et al. 2016). The sound speed, c, scales like the square of temperature, it is ≈7 km/s in the solar photosphere (the solar surface), ≈200 km/s at the base of the convection zone, and ≈500 km/s at the center of the Sun.

The modes of oscillation observed on the Sun are p modes and the f mode. The p modes are essentially acoustic waves, modified by stratification, rotation and in active regions by magnetism. They are excited by turbulent convection in the near-surface layers and are sensitive to conditions between the Sun's surface and their lower turning point, where the horizontal wave speed of the waves matches the local adiabatic sound speed. Thus p modes with larger values of $\nu/(l + 1/2)$ penetrate more deeply into the Sun, where ν is frequency of the mode and l is the degree of the spherical harmonic that describes the mode's horizontal structure: for example, for $\nu = 3$ mHz, modes of degree greater than about 40 are confined to the convection zone, while modes with degrees smaller than 40 penetrate into the radiative interior beneath it. At high degree the f mode has the character of a surface gravity wave. Internal gravity-wave modes (g modes) have been reported in the Sun but have not been unambiguously detected.

The Sun is differentially rotating, with the usually quoted value for the rotation period being that of Carrington at sunspot latitudes, which is 27.2753 days (synodic, as viewed from Earth) or 25.38 days (sidereal). This yields a solar rotational frequency of $\Omega_{\text{Carr}}/2\pi =$ 0.46 μHz and implies a surface velocity $V = R\Omega_{\text{Carr}} = 1.99$ km/s. In the convection zone, the equator rotates faster than the poles by approximately 2.8 deg/day (Schou et al. 1998). The near surface shear layer covers the top $\approx$35 Mm below the surface. In this layer, the rotation rate decreases outward fast, such that $d \ln \Omega / d \ln r \simeq -1$ independently of latitude up to 60° (Barekat et al. 2014).

Solar convection manifests itself at the solar surface in the form of granules, with a lifetime of about 10 min and a characteristic length scale of about 1.7 Mm (Rieutord et al. 2010), and supergranules, with a lifetime of $\tau \simeq 2$ days and a scale of about 35 Mm. (Rieutord and Rincon 2010). The convective motions at scales larger than supergranulation have smaller velocity amplitudes and are constrained by rotation. It is generally believed that the effect of the Coriolis force on the convective motions deep in the convection zone sets up anisotropic turbulent stresses that maintain the differential rotation and a meridional flow (Kitchatinov and Rüdiger 2005). The meridional flow is poleward at the surface with a maximum (in latitude) flow speed of about 20 m/s (Giles et al. 1997). Measuring the meridional flow at depth is an important goal of helioseismology and a matter of debate (Zhao et al. 2013; Rajaguru and Antia 2015).

These large-scale flows in combination with the convective flows and the Coriolis force, are thought to drive a large-scale $\alpha\Omega$ dynamo (Krause and Rädler 1980). In particular the Ω effect is the winding up of poloidal field by the differential rotation to produce toroidal magnetic field, and the α effect corresponds to the systematic effect of the Coriolis force on either convective flows or on flows associated with rising magnetic flux tubes. One special example of an $\alpha\Omega$ dynamo is the Babcock-Leighton dynamo, where the α effect is associated with the systematic tendency of the leading sunspot in a bipolar active region to be slightly closer to the equator than the following sunspot (Leighton 1969). Within the mean field (and therefore the Babcock-Leighton) framework, the observed equatorial propagation of the activity belt during a cycle can be explained in terms of a dynamo wave (Parker 1955) or advection by the (return) meridional flow (Wang and Sheeley 1991). For those theories that place the flux tube formation at the base of the convection zone, the magnetic field there needs to be about 10^5 Gauss for the emergence to not occur at too high latitudes (Choudhuri and Gilman 1987), which matches the additional constraint of Joys law (D'Silva and Choudhuri 1993). Exactly how and where magnetic flux concentrations form and amplify in the convection zone is not a solved problem. Another question of importance is how the toroidal magnetic field in emerging active regions leads to the correct poloidal component to 'close the dynamo'. Also the large-scale dynamo has a clear back-reaction on the large-scale flows, with temporal changes in the zonal flows and the meridional flow (e.g. Gizon et al. 2010). The changes in the longitudinally average meridional flow have been attributed to the local inflows into active regions measured by local helioseismology (Gizon 2004). These inflows into active regions provide a nonlinearity in the framework of the Babcock-Leighton model that may control the variability of solar activity from cycle to cycle (Cameron et al. 2014).

Acknowledgements We thank the International Space Science Institute for hosting the Workshop in Bern and for providing financial support to the Workshop participants. Special thanks are due to the ISSI Science Committee and to the Director of ISSI, Roger Bonnet, for approving a topic on helioseismology and the solar interior.

References

C. Aerts, J. Christensen-Dalsgaard, D.W. Kurtz, *Asteroseismology* (Springer, Berlin, 2010)

H.M. Antia, A. Bhatnagar, P. Ulmschneider (eds.), *Lectures on Solar Physics*. Lecture Notes in Physics, vol. 619, (Springer, Berlin, 2003), p. 1
W. Ball, B. Beeck, R. Cameron, L. Gizon Astron. Astrophys. **592**, A159 (2016)
A. Balogh, H. Hudson, K. Petrovay, R. von Steiger (eds.), *The Solar Activity Cycle* (Springer, Berlin, 2015)
A. Barekat, J. Schou, L. Gizon, Astron. Astrophys. **570**, L12 (2014)
S. Basu, H.M. Antia, Phys. Rep. **457**, 217 (2008)
S. Basu, N. Grevesse, S. Mathis, S. Turck-Chièze, Space Sci. Rev. **196**, 49 (2015)
A.S. Brun, M.K. Browning, M. Dikpati, H. Hotta, A. Strugarek, Space Sci. Rev. **196**, 101 (2015a)
A.S. Brun, R.A. García, G. Houdek, D. Nandy, M. Pinsonneault, Space Sci. Rev. **196**, 303 (2015b)
R. Burston, L. Gizon, A.C. Birch, Space Sci. Rev. **196**, 201 (2015)
R.H. Cameron, J. Jiang, M. Schüssler, L. Gizon, J. Geophys. Res. Space Phys. **119**, 680 (2014)
A.R. Choudhuri, P.A. Gilman, Astrophys. J. **316**, 788 (1987)
J. Christensen-Dalsgaard, W. Däppen, S.V. Ajukov et al., Science **272**, 1286 (1996)
J. Christensen-Dalsgaard, D.O. Gough, M.J. Thompson, Astrophys. J. **378**, 413 (1991)
M.P. di Mauro, D. Cardini, M. Marconi (eds.), *Synergies Between Solar and Stellar Modelling* (Springer, Berlin, 2010)
S. D'Silva, A.R. Choudhuri, Astron. Astrophys. **272**, 621 (1993)
Y. Elsworth, A.-M. Broomhall, S. Gosain et al., Space Sci. Rev. **196**, 137 (2015)
W.M. Folkner, J.G. Williams, D.H. Boggs, Interplanet. Netw. Prog. Rep. **178**, 1 (2009)
C. Froelich, in *ESA SP 150: On Climatology, Hydrology, Atmospheric Research and Meteorology (CHARM) from Space*, ed. by T. Guyenne (ESA, Nordwijk, 1980), pp. 91–94
P.M. Giles, T.L. Duvall, P.H. Scherrer, R.S. Bogart, Nature **390**, 52 (1997)
L. Gizon, Sol. Phys. **224**, 217 (2004)
L. Gizon, A.C. Birch, Living Rev. Sol. Phys. **2**, 6 (2005)
L. Gizon, A.C. Birch, H.C. Spruit, Annu. Rev. Astron. Astrophys. **48**, 289 (2010)
L. Gizon, P. Cally, J. Leibacher (eds.), *Helioseismology, Asteroseismology, and MHD Connections* (Springer, Berlin, 2008)
D.O. Gough, Space Sci. Rev. **196**, 15 (2015)
S. Hanasoge, L. Gizon, K.R. Sreenivasan, Annu. Rev. Fluid Mech. **48**, 191 (2016)
S. Hanasoge, M.S. Miesch, M. Roth et al., Space Sci. Rev. **196**, 79 (2015)
S.M. Hanasoge, *Imaging Convection and Magnetism in the Sun*. SpringerBriefs in Mathematics (2015)
D.W. Hughes, R. Rosner, N.O. Weiss (eds.), *The Solar Tachocline* (Cambridge University Press, Cambridge, 2012)
L.L. Kitchatinov, G. Rüdiger, Astron. Nachr. **326**, 379 (2005)
R. Komm, I. De Moortel, Y. Fan, S. Ilonidis, O. Steiner, Space Sci. Rev. **196**, 167 (2015)
F. Krause, K.H. Rädler, *Mean-Field Magnetohydrodynamics and Dynamo Theory* (Pergamon, Oxford, 1980)
J.R. Kuhn, R.I. Bush, M. Emilio, P.H. Scherrer, Astrophys. J. **613**, 1241 (2004)
R.B. Leighton, Astrophys. J. **156**, 1 (1969)
B. Löptien, A.C. Birch, L. Gizon et al., Space Sci. Rev. **196**, 251 (2015)
A. Maeder, *Physics, Formation and Evolution of Rotating Stars* (Springer, Berlin, 2009)
N.N. Mansour, A.G. Kosovichev, R. Komm, D. Longcope (eds.), *Solar Dynamics and Magnetism from the Interior to the Atmosphere* (Springer, Berlin, 2013)
M.J.P.F.G. Monteiro, *Evolution and Seismic Tools for Stellar Astrophysics*, vol. 316 (Springer, Berlin, 2009), reprinted from Astrophys. Space Sci.
E.N. Parker, Astrophys. J. **122**, 293 (1955)
S.P. Rajaguru, H.M. Antia, Astrophys. J. **813**, 114 (2015)
M. Rieutord, F. Rincon Living Rev. Sol. Phys. **7**, 2 (2010)
M. Rieutord, T. Roudier, F. Rincon et al., Astron. Astrophys. **512**, A4 (2010)
A. Schad, L. Jouve, T.L. Duvall, M. Roth, S. Vorontsov, Space Sci. Rev. **196**, 221 (2015)
J. Schou, H.M. Antia, S. Basu et al., Astrophys. J. **505**, 390 (1998)
J. Schou, A.G. Kosovichev, P.R. Goode, W.A. Dziembowski, Astrophys. J. **489**, L197 (1997)
T. Sekii, T. Appourchaux, B. Fleck, S. Turck-Chièze, Space Sci. Rev. **196**, 285 (2015)
M. Stix, *The Sun: An Introduction* (Springer, Berlin, 2004)
M.J. Thompson, A. Balogh, J.L. Culhane et al. (eds.), *The Origin and Dynamics of Solar Magnetism*. Space Sciences Series of ISSI, vol. 32 (Springer, Berlin, 2009)
M.J. Thompson, J. Christensen-Dalsgaard, M.S. Miesch, J. Toomre, Annu. Rev. Astron. Astrophys. **41**, 599 (2003)
V.C.H. Tong, R.A. García, *Extraterrestrial Seismology* (Cambridge University Press, Cambridge, 2015)
J. Toomre, M.J. Thompson, Space Sci. Rev. **196**, 1 (2015)
Y.-M. Wang, N.R. Sheeley Jr., Astrophys. J. **375**, 761 (1991)
J. Zhao, R.S. Bogart, A.G. Kosovichev, T.L. Duvall Jr., T. Hartlep, Astrophys. J. **774**, L29 (2013)

Space Sci Rev (2015) 196:1–14
DOI 10.1007/s11214-015-0147-x

Prospects and Challenges for Helioseismology

J. Toomre[1] · **M.J. Thompson**[2]

Received: 3 December 2014 / Accepted: 24 March 2015 / Published online: 9 April 2015

Abstract Helioseismology has advanced considerably our knowledge of the interior of the Sun over the past three decades. Our understanding of the Sun's internal structure, its dynamics, rotation, convection and magnetism, have all been advanced. Yet there are challenges, areas where the results from helioseismology are tantalizing but inconclusive, and aspects where the interpretation of the data has still to be put on a firm footing. In this paper we shall focus on a number of those challenges and give our assessment of where progress needs to be made in the next decade.

Keywords Sun · Helioseismology

1 Introduction

The field of helioseismology is now over thirty years old. Helioseismology has led to some remarkable advances in our knowledge and understanding of the solar interior. The run of sound speed with depth in the Sun has been measured from the surface into the core (Christensen-Dalsgaard et al. 1985; Basu et al. 1997). The depth of the convection zone has been measured (Christensen-Dalsgaard et al. 1991), as has the abundance of helium in the convection zone (Däppen et al. 1988; Kosovichev et al. 1992; Richard et al. 1998). The internal rotation has also been mapped over most of the interior, excluding the inner core and a region around the rotation axis (Duvall et al. 1984; Tomczyk et al. 1995; Thompson et al. 1996; Schou et al. 1998; Thompson et al. 2003). This has led to the discovery of the solar tachocline (e.g. Christensen-Dalsgaard and Thompson 2007) and the subsurface shear layer (Corbard and Thompson 2002), and a re-evaluation of the preconception that rotation in the convection zone would be so organized to be constant on cylinders (e.g. Gilman and Howe 2003). For more details of the methods and the developing results in helioseismology, see

✉ M.J. Thompson
mjt@ucar.edu

1 JILA and Department of Astrophysical and Planetary Sciences, University of Colorado, Boulder, CO 80309-0440, USA

2 High Altitude Observatory, National Center for Atmospheric Research, Boulder, CO 80301, USA

e.g. the reviews by Deubner and Gough (1984), Gough and Toomre (1991), Christensen-Dalsgaard (2002), Gizon and Birch (2005), Basu and Antia (2008), Howe (2009), Gizon et al. (2010). In this article, we shall focus not on the well established results and successes of helioseismology, but on the future prospects and challenges. In particular we shall focus on scientific issues that are unresolved, helioseismic results that have raised more questions than answers, and aspects of helioseismology itself that are not yet well understood. We shall conclude with some thoughts about what may be required to resolve these issues and advance the subject.

2 Prospects and Challenges in Global Helioseismology

Global helioseismology is the study of the Sun using the properties of global resonant solar oscillations, in particular the mode frequencies. This is a mature part of the subject: the forward problem of how the Sun's physical properties influence the frequencies is relatively well understood, and the successful inferences from global helioseismology have for the most part been achieved with a toolbox of well established linear inversion techniques. There are a number of open questions in the physics of the solar interior that are being addressed with global helioseismology. In the spirit of looking forward we focus on some of these in the following three subsections, and address some needed thrusts in global helioseismology in Sect. 2.4.

2.1 On Abundances and Equation of State

The frequencies of global p modes depend primarily on the adiabatic sound speed c, given by

$$c^2 \equiv \left(\frac{\partial p}{\partial \rho}\right)_s \equiv \Gamma_1 p/\rho \tag{1}$$

where p is pressure, ρ is density, s is specific entropy and Γ_1 is the first adiabatic exponent. To an excellent approximation, the plasma in the solar interior is a perfect gas, so that $c^2 \propto \Gamma_1 T/\mu$ where T is temperature and μ is the mean molecular weight. The run of sound speed with depth in the Sun is dominated by the variation of temperature with depth, which in turn depends on the heat transport and hence on the opacity in the radiative interior and the transition from convective to radiative heat transport at the base of the convective envelope. The opacity in turn is sensitive to the abundance of different elements, and in particular the heavy-element abundances; and so the sound speed in the radiative zone is sensitive to these chemical abundances. The sound speed's sensitivity to Γ_1 manifests itself in the ionization zones of different elements, where Γ_1 departs from 5/3; and its sensitivity to μ is most marked in the core, where fusion has converted hydrogren to helium over the lifetime of the Sun and hence the mean molecular weight has been increased. Thus there are three ways in which some information about chemical abundances can be inferred seismologically from the sound speed: from the general run of sound speed in the interior due to the sensitivity of the opacity; in the ionization zones of specific elements; and, for abundant elements, from their effect on the mean molecular weight. For a review of the role of chemical abundances in helioseismology, see Basu and Antia (2008).

Solar models in the 1990s, which generally used the solar chemical abundances of Grevesse and Noels (1993), were in pretty good agreement with the helioseismically inferred sound speed in the Sun, as a function of depth. This was particularly true when the gravitational settling of elements heavier than hydrogen was taken into account in the models (Christensen-Dalsgaard et al. 1993). Early in 2000's, Martin Asplund and colleagues revised the spectroscopic determinations of the photospheric abundances (Asplund et al. 2004, 2005), in particular revising sharply downwards the oxygen abundance from the earlier abundances of Grevesse and Noels (1993). This greatly increased the disagreement between the models and the helioseismically inferred sound speed in the Sun, at the base of the convection zone and over a broad extent in the radiative interior (Bahcall et al. 2005a,b,c).

The location and extent of the new discrepancy between models and Sun suggests that the opacity may be largely responsible, though other "nonstandard" solutions such as mass loss and dark matter have been explored, as well as the effects of electron screening on nuclear reaction rates: see the review by Guzik (2014). There have also been more recent revisions of the solar abundances (Asplund et al. 2009; Caffau et al. 2010, 2011; Lodders 2010) which have metallicities intermediate between the abundances of the 1990s and those of the mid-2000s. Nonetheless, these still leave a significant discrepancy between solar models and the helioseismically inferred structure of the Sun (Basu and Antia 2013).

The above efforts to reconcile modern solar abundances and helioseismology have involved 1-D solar modeling. The residual discrepancies between models and the Sun have led some to consider macrophysical effects in 2- and 3-D solar and stellar models that can variously redistribute chemical abundances and angular momentum within the solar interior. The additional macrophysics currently under consideration includes rotation, magnetic fields, and internal gravity waves (e.g. Talon and Charbonnel 2008; Eggenberger et al. 2010; Mathis 2010, 2013).

Inferences about the abundances based on the sound speed and density profiles in the radiative interior inevitably involve assumptions and uncertainties about the opacities and other physics. In principle, a cleaner way to investigate the abundances would be to estimate them from their effects on the adiabatic exponent Γ_1 (and hence on the sound speed) in the ionization zones of oxygen and carbon in the convection zone, where the stratification is fairly simple. This indeed is the way in which the Sun's envelope helium abundance has been measured (Däppen and Gough 1984, 1986; Däppen et al. 1988). Such an approach would mean that one could check whether the abundances in the solar convection zone were consistent with those measured at the surface by Asplund and others. If they were different, this could suggest other ways to reconcile the surface abundances and helioseismology. Such an approach to determining the abundances in the solar interior has been proposed (Mussack and Gough 2009), and the actual investigation is in progress. A related approach has been used by Antia and Basu (2006), estimating the heavy element abundances in the solar convection zone by using a mean value of the function W (which depends on the radial derivative of the sound speed) in the lower convection zone. This therefore averages over the ionization zones of several elements. A possible concern with this approach is that other, intrinsic, equation-of-state effects might also potentially affect this measurement (c.f. Baturin et al. 2000).

2.2 Macrophysics: Mixing, Magnetic Fields and Solar Activity

A common feature of many sound-speed inversions in the deep interior is that the sound speed at a fractional radius of about $r = 0.2R$ in the Sun appears to be lower than that in "standard" solar models. Perhaps this is due to an error in the opacities used to construct the models. A more intriguing possibility is that this may be evidence for some macroscopic mixing that has taken place in the solar core, such as might arise from some instability. This is faintly reminiscent of the "solar spoon" mixing proposed by Dilke and Gough (1972), which came about from an instability caused by build-up of He-3. A similar signature might arise in the sound speed if the central sound speed in the Sun were higher than in models: one might imagine that, after the model had been properly calibrated to the present radius and luminosity, this might then give rise to a lower sound speed in the outer core. What might cause such a higher central sound speed is a matter of speculation.

Baldner and Basu (2008) have reported an intriguing variation in the inferred sound speed over the solar cycle. The immediate interpretation is that this might be the signature of a magnetic field, which would appear to the modes as an apparent sound-speed enhancement. Certainly, various solar dynamo models envisage that magnetic flux is stored and built up at the base of the convection zone before rising buoyantly and emerging at the solar surface as sunspots. Interestingly, Chou and Serebryanskiy (2002) also found some anomaly in the travel time corresponding to rays that just reach from the surface to the base of the convection zone, and attributed this to the possible effect of magnetic field stored there. See also Chou and Serebryanskiy (2005).

Finally, in addition to the well known modulation of mode frequencies with the solar cycle (Woodard and Noyes 1985), there are also shorter period variations in the frequencies, on a timescale of about two years It has been speculated that these are also magnetic in origin, and some have further speculated that the Sun has a dynamo that operates on that timescale as well as the better known 11-year one (Fletcher et al. 2010).

2.3 Dynamics of the Convection Zone and Tachocline

Clearly one of the major triumphs of global helioseismology has been the ability to map the angular velocity $\Omega(r,\theta)$ profile with radius and latitude for almost all of the solar interior, with only the central core region being less certain due to the few p modes that penetrate so deeply and the absence of g mode data. With such differential rotation mapping came several surprises: the tachocline of strong shear at the base of the convection zone (CZ) had not been anticipated, nor the distinctive near-surface shear layer (NSSL), both of which are striking boundary layers that may have substantial roles in how the Sun achieves the dynamo action responsible for building its magnetic fields. With the discovery of a tachocline came the possibility that an interface dynamo may be centered there to help build the cycling global fields (e.g. Charbonneau 2010; Miesch 2012; Charbonneau 2014), whereas the NSSL is a region of intensely turbulent multi-scale convection that likely builds much of the small-scale magnetic fields visible at the surface. It is a significant theoretical challenge to understand how these two boundary layers are achieved and maintained (e.g. Miesch and Toomre 2009).

The extended and continuous global p mode data from both GONG and MDI, and now also from SDO, has allowed study of temporal variations in the zonal flows, or thus the differential rotation, as the solar cycle proceeds, both within the convection zone proper and in the tachocline (e.g. Howe 2009). By their nature, the splitting of the global mode frequencies sample only the symmetric part of Ω with latitude about the equator, and so too any temporal changes. There was an interval of possible oscillations in the zonal flow within the tachocline with a 1.3 yr period, and a phasing difference in the radiative region just below, but that behavior appears to have faded away (Howe et al. 2000, 2011). In contrast, there are regular though modest speedups in the zonal flows (of order 10 nHz changes in Ω), called torsional oscillations, that propagate equatorward in advance of the activity belts, and more substantial poleward propagating speedups as the solar cycles proceed (e.g. Howe et al. 2013; Komm et al. 2014). These temporal adjustments to the zonal flows extend from the near-surface to at least the mid-CZ depth (Vorontsov et al. 2002). There is keen interest to continue to pursue such evidence for changes in the differential rotation as the solar cycles advance, for it places constraints and provides guidance to the global dynamo models based on 3-D simulations that are becoming available.

The solar tachocline must be an intricate boundary layer of strong rotational shear that is laced with substantial magnetic fields and filled with convective plumes overshooting into the stable radiative interior. These serve to define its structure and yet keep it localized and apparently not spreading, variously presenting real challenges to theory. The global p mode data has provided estimates of the shape of the tachocline and its variation with latitude (e.g. Thompson et al. 2003; Howe 2009), though the inversions can only place upper limits on its thickness. An important continuing goal of helioseismology would be to improve the precision of these estimates of tachocline properties, for here too the Sun serves as a unique laboratory for defining a complex boundary layer that must be present in most other stars with convective envelopes and dynamos.

2.4 Needed Thrusts in Global Helioseismology

The measurement of the frequencies themselves from the observations, via some kind of fitting ("peakbagging") in a power spectrum or Fourier transform of the observed oscillations (e.g. Schou 1992; Hill et al. 1996), is still an area of active research and advancement. In part because we do not observe the whole of the Sun's surface at once, there is leakage of different modes into the power spectrum, and handling that leakage in the fitting is one challenge. Better modeling of the leakage, including taking into account such aspects as the line asymmetry and the horizontal component of the eigenfunctions, is leading to improvement to standard fitting pipelines (Larson and Schou 2008). Another radical approach combines careful modeling of the leakage with fitting all modes (or a large number of modes) simultaneously (Jefferies and Vorontsov 2004; Vorontsov and Jefferies 2005, 2013). Advances in this area will reduce systematic errors in the mode frequencies which in many instances, as Vorontsov and collaborators discuss, now dominate the effect of random errors in the results of helioseismic inversions.

A mostly neglected area of global helioseismology to date is the coupling of different modes, which affects the frequencies of both. An example of a physical effect that can couple modes is bulk flow (Lavely and Ritzwoller 1992). In particular, meridional flows, which to first order do not manifest themselves in the global mode frequencies, can in principle be detected in a second-order perturbation of the frequencies due to mode coupling (to use the language of geophysics). This has been researched by Roth and Stix (2008) and Schad et al. (2013).

Another challenging area is the detection of solar gravity modes (g modes). These would be extremely valuable for improving inferences about the core, because the g modes are sensitive to conditions in the core. The challenge to detecting g modes is that they are evanescent throughout the Sun's convective envelope, so even if the modes have substantial amplitude in the radiative interior, their amplitude at the photosphere will be very small. Estimates of their amplitude in velocity are 10^{-2} cm/sec or smaller (Gough 1985; Kumar et al. 1996). Early reports of g-mode detections were probably in error: modern detections have greater credibility (e.g. Appourchaux et al. 2010, and references therein). There are ongoing efforts and proposals to build instrumentation with the necessary stability and sensitivity to try to observe solar g modes (Turck-Chièze et al. 2006). An alternate approach is to try to detect g modes through their perturbative effect on the observed p modes (Kennedy et al. 1993; Lou 2001). Because of their great potential for probing the solar core, the search for g modes will no doubt continue.

3 Prospects and Challenges in Local Helioseismology

The advent of nearly continuous full-disk Doppler imaging with GONG starting in 1995 and with MDI on SOHO in 1996 made it possible to consider alternative helioseismic analyses of the acoustic wave fields in addition to the global mode studies. Thereby began the rapidly developing field of local helioseismology that encompasses a variety of different techniques, such as ring-diagram analysis, time-distance helioseismology, and helioseismic holography. These are most recently discussed and reviewed in excellent detail in Gizon et al. (2010). Unlike global helioseismology for which most of the analysis elements are quite mature, local helioseismology involves very active ongoing work to refine and understand the sensitivities of the different approaches. Local helioseismology is clearly at the forefront of research discoveries now coming forth from using the refined 4096^2 high-resolution Doppler and magnetic imaging with HMI on SDO. Thus we entered a new phase in seismic probing with the launch of SDO in 2010.

3.1 Solar Subsurface Flows and Structures

Local helioseismology provides the means to probe a wide range of 3-D subsurface flows and associated structures, at least within the upper reaches of the solar convection zone that extend to about 30 Mm depths. Unlike global helioseismology which utilizes the low- to medium-degree modes that have long lifetimes and appear discretely in power spectra, the high-degree ($\ell \geq 120$) acoustic waves appear as continuous ridges in power distinguished by their radial order. Ring-diagram analyses study the Doppler displacements of these ridges when the wave fields are studied over a succession of localized small domains that tile the surface, and then inversions are carried out to determine the horizontal flows with depth below each tile. Time-distance methods similarly measure the cross-covariance function of these wave fields separated by a range of angular distances, determining time lags that are attributable in part to advection by the underlying flows. These several approaches have revealed the richness of streaming and meandering flows below the surface, not unlike jet streams, many of which clearly are influenced by active magnetic complexes (e.g. Hindman et al. 2004; Gizon et al. 2010; Kosovichev et al. 2011; Baldner et al. 2013; Seligman et al. 2014). The spatial resolution at the shallower depths also shows distinctive outflow cellular structures that are comparable to supergranular scales. Having access to both horizontal velocity components may admit estimating transport properties by some of the Reynolds stresses of the convection, though measuring the vertical component is problematic.

Significant findings from both ring-diagram and time-distance analyses are that large active regions typically exhibit broad surrounding inflows at 2 to 5 Mm depths, and outflows at about 10 Mm depths. This may impact how magnetic field is able to escape from active regions, with poleward drift of fields of a particular sense being a crucial element in flux-transport dynamo models. Time-distance probing has also had sufficient spatial resolution to be able to detect radial moat outflows close to the surface from individual sunspots. However, there is no clear agreement about the nature of flows at greater depths below sunspots using different implementations of time-distance analyses. Comparisons with realistic 3-D near-surface convection and sunspot simulations are now helping to clarify the sensitivity of such time-distance probing (e.g. Braun et al. 2012; DeGrave et al. 2014a,b). When suitably averaged in longitude and time, the local helioseismic zonal flow probing also captures the evolving torsional oscillation signal deduced from the global studies (e.g. Komm et al. 2014). Major efforts are under way to sharpen the spatial resolution of ring-diagram studies, either involving multiple tile sizes or using the larger tile sizes with repeated small lateral displacements, now offering 0.5° horizontal resolution over about the upper 30 Mm. Multi-ridge fitting is now being combined with 3-D inversions for the flows using new tools such as AARDI (Featherstone 2011; Featherstone et al. 2011; Greer et al. 2014).

Determining the amplitude and structure of large-scale flows deep within the convection zone, which appear as giant-cells within 3-D global convection simulations and are responsible for both transporting the heat flux and maintaining the differential rotation, is another major goal for local helioseismology. Time-distance assessments suggest far smaller amplitudes for such convective flows than reported from global modeling (e.g. Hanasoge et al. 2012; Gizon and Birch 2012; Miesch et al. 2008, 2012), thus raising very basic questions about how the Sun variously achieves its heat transport and continually redistributes its angular momentum. The very low flow estimates from time distance are however not in accord with those from ring-diagram measurements (Greer et al. 2015) that are closer to values realized with global convection simulations. A related issue concerns why neither direct Doppler imaging nor local helioseismic probing is able to detect any clear signature of the global scales of convection in the surface layers. The depth dependence of supergranulation is also continuing to be a target of interest, with some time-distance studies reporting that these striking cellular flows at the surface may be confined to rather shallow depths, becoming difficult to detect at depths greater than about 3 Mm (e.g. Duvall and Hanasoge 2013; Duvall et al. 2014).

3.2 Meridional Circulation

Meridional circulation—that is, the component of the mean flow that is in meridional planes—is a key component of some contemporary dynamo models, particularly flux-transport ones (e.g. Choudhuri et al. 1995; Dikpati and Gilman 2009; Charbonneau 2010). In contrast with the Sun's internal rotation, which is relatively well determined from helioseismology, the meridional circulation is still very uncertain. This is principally because, unlike the rotation, meridional flows have no effect on the global mode frequencies, to first order. Most of the helioseismic investigations of the meridional circulation have therefore used local helioseismic techniques (for a fuller discussion of these and global-mode approaches, see Hanasoge et al. (2015, this issue). In the very near surface layers, at least down to 10 Mm depth, it is well established that the flow is poleward in both hemispheres at low and mid latitudes, though there is evidence for some temporal variation and developing equatorward counter-flows beneath the immediate surface layers at mid and higher latitudes (Haber et al. 2002; Komm et al. 2013).

Since the pioneering work of Giles (2000), there have been a number of efforts to determine the meridional flow deeper in the convection zone, and in particular to determine the depth at which a return (i.e. equatorward) flow occurs, as by mass conservation it must (e.g. Mitra-Kraev and Thompson 2007). Perhaps the most convincing result on this to date is that of Zhao et al. (2012). These authors find a multi-cellular meridional flow pattern in depth, with poleward flow down to almost $0.90R$, equatorward flow between about $0.83R$ and $0.90R$, and poleward flow again beneath that depth. The implications of this finding for the flux-transport dynamo hypothesis are currently discussed (e.g. Hazra et al. 2014).

The results of Zhao et al. seem to depend crucially on their correction of a center-to-limb systematic effect in the travel-time measurements, which they do by assuming that the East-West and North-South systematics are numerically the same. Whether the systematic is purely circularly symmetric is not known, nor just what is its underlying cause. Until this is much better understood, the robustness of the result must remain an open question.

3.3 Magnetic-Field Effects

A major goal of local helioseismology is to help unravel the coupling of magnetic fields and subsurface flows both within individual sunspots and within broader active magnetic complexes. Such efforts are made challenging by the acoustic waves changing their character as they encounter strong magnetic fields at what would be their upper reflection points. Mode conversion can convert an acoustic wave into magnetoacoustic fast and slow waves and into Alfven waves, each with differing wave speeds and trajectories and depending both upon magnetic field strengths and orientations. Upon emerging from the region of magnetic influence, the magnetoacoustic fast wave emerges as the simpler but somewhat changed acoustic wave. Thus acoustic waves traversing regions of magnetism can have their amplitudes diminished and their phasing modified. Since these are the waves currently used for all local helioseismic analyses, such modifications can lead to substantial uncertainties and differences in the various helioseismic interpretations of flows and structures within sunspots and their surroundings (e.g. Cally 2009; Birch et al. 2009; Gizon et al. 2010; Moradi et al. 2010; Hanasoge et al. 2011; Braun et al. 2012; Moradi and Cally 2013; DeGrave et al. 2014a). The helioseismic implications of such mode conversion are complex and thus are a clear focus of directed research. Having access to 3-D vector magnetic field data as now being provided by HMI, along with the Doppler velocity sampling of the wave fields, may provide new constraints to be incorporated into the helioseismic analyses.

3.4 Predicting Space Weather

Predicting space weather is one of the most compelling societally relevant goals of solar physics. This includes understanding the causes and triggers of flares and coronal mass ejections (CMEs) and being able to predict when they will occur or at least when conditions are such that their occurrence is most likely. To date, helioseismology has most obviously played a role in advancing this area in three ways. Firstly, measurements of the subsurface flows and derivation of subsurface kinetic helicity is suggestive of possible causal links between the bulk motions under magnetic complexes and the occurrence of flares (Reinard et al. 2010). Secondly, far-side imaging of large, newly emergent active regions on the far side of the Sun now provides the potential to give a week's or more advance notice of a potentially flare- and CME-producing active region rotating onto the Earthward face of the Sun (Lindsey and Braun 2000; González Hernández et al. 2013). Thirdly, and most recently, work by Ilonidis et al. (2011) has suggested that emerging active regions can be detected up to two days before they become visible at the surface of the Sun. This last result is one of the most intriguing from local helioseismology in the era of SDO/HMI (though the initial findings used SOHO/MDI data). The result and its interpretation has been questioned (Braun 2012), though the topic is complex (Birch et al. 2013). This important result therefore needs further investigation. The choice of filters applied to the acoustic power spectrum as part of the analysis seems crucial. It is possible that movement of power into or out of the range of the filter could have a bearing on the origin of the signal (P. Scherrer, private communication).

3.5 Needed Thrusts in Local Helioseismology

A central issue for all local helioseismic analyses is to understand how to better take into account the effects of strong magnetic fields upon the acoustic waves that are being used to probe subsurface flows and structures close to and possibly within active regions and sunspots. Mode conversion can change the wave fields that enter and leave such regions, leading to substantial uncertainties in the wave travel times, amplitudes and phases being used at least implicitly in all the local helioseismic analyses. Since HMI on SDO enables detailed vector magnetic field reconstructions in the photosphere in addition to the Doppler velocity imaging, this may be able to provide a path to introduce some constraints into the helioseismic inversions being used to assess the subsurface flow fields. A related goal is whether helioseismic probing can be refined to detect rising magnetic structures well before they emerge through the solar surface. There are hints that this may have been realized, and if borne out by other studies, may provide an important focus for helioseismology in space weather efforts, in addition to the far-side imaging now developed and operational. Another topic calling for continuing advances concerns improving the probing of large-scale dynamics deep within the convection zone, including the structure and amplitude of its meridional circulation and the nature of the giant cells that are responsible for angular momentum transport to maintain the differential rotation profile.

4 Conclusions

Helioseismology has made some great achievements. But further opportunities remain: probing the structure and dynamics of the Sun's inner core; constraining the abundances in the solar interior of elements heavier than helium; providing tools for detecting the sub-surface emergence of magnetic flux; constraining large-scale flows deep in the solar interior, particularly near the base of the solar convection zone. These opportunities require investment of efforts into improving the analysis of helioseismic data, notably global-mode peakbagging; and disentangling the effects of strong magnetic fields on solar oscillations especially for local helioseismology but also for global-mode helioseismic analyses. They call for continuing and improved helioseismic observations, including observations from vantage points away from the Sun-Earth line. The challenges are considerable, but so are the gains in understanding the dynamics of the Sun. Helioseismology has achieved insights that may once have seemed impossible. It can do so again, but getting there requires focused attention on several major fronts.

References

H.M. Antia, S. Basu, Determining solar abundances using helioseismology. Astrophys. J. **644**, 1292–1298 (2006). doi:10.1086/503707

T. Appourchaux, K. Belkacem, A.-M. Broomhall, W.J. Chaplin, D.O. Gough, G. Houdek, J. Provost, F. Baudin, P. Boumier, Y. Elsworth, R.A. García, B.N. Andersen, W. Finsterle, C. Fröhlich, A. Gabriel, G. Grec, A. Jiménez, A. Kosovichev, T. Sekii, T. Toutain, S. Turck-Chièze, The quest for the solar G modes. Astron. Astrophys. Rev. **18**, 197–277 (2010). doi:10.1007/s00159-009-0027-z

M. Asplund, N. Grevesse, A.J. Sauval, The solar chemical composition, in *Cosmic Abundances as Records of Stellar Evolution and Nucleosynthesis*, ed. by T.G. Barnes III, F.N. Bash. Astronomical Society of the Pacific Conference Series, vol. 336 (2005), p. 25

M. Asplund, N. Grevesse, A.J. Sauval, C. Allende Prieto, D. Kiselman, Line formation in solar granulation. IV. [O I], O I and OH lines and the photospheric O abundance. Astron. Astrophys. **417**, 751–768 (2004). doi:10.1051/0004-6361:20034328

M. Asplund, N. Grevesse, A.J. Sauval, P. Scott, The chemical composition of the Sun. Annu. Rev. Astron. Astrophys. **47**, 481–522 (2009). doi:10.1146/annurev.astro.46.060407.145222

J.N. Bahcall, S. Basu, A.M. Serenelli, What is the neon abundance of the Sun? Astrophys. J. **631**, 1281–1285 (2005a). doi:10.1086/431926

J.N. Bahcall, A.M. Serenelli, S. Basu, New solar opacities, abundances, helioseismology, and neutrino fluxes. Astrophys. J. Lett. **621**, 85–88 (2005b). doi:10.1086/428929

J.N. Bahcall, S. Basu, M. Pinsonneault, A.M. Serenelli, Helioseismological implications of recent solar abundance determinations. Astrophys. J. **618**, 1049–1056 (2005c). doi:10.1086/426070

C.S. Baldner, S. Basu, Solar cycle related changes at the base of the convection zone. Astrophys. J. **686**, 1349–1361 (2008). doi:10.1086/591514

C.S. Baldner, S. Basu, R.S. Bogart, O. Burtseva, I. González Hernández, D. Haber, F. Hill, R. Howe, K. Jain, R.W. Komm, M.C. Rabello-Soares, S. Tripathy, Latest results found with ring-diagram analysis. Sol. Phys. **287**, 57–69 (2013). doi:10.1007/s11207-012-0171-x

S. Basu, H.M. Antia, Helioseismology and solar abundances. Phys. Rep. **457**, 217–283 (2008). doi:10.1016/j.physrep.2007.12.002

S. Basu, H.M. Antia, Revisiting the issue of solar abundances. J. Phys. Conf. Ser. **440**(1), 012017 (2013). doi:10.1088/1742-6596/440/1/012017

S. Basu, J. Christensen-Dalsgaard, W.J. Chaplin, Y. Elsworth, G.R. Isaak, R. New, J. Schou, M.J. Thompson, S. Tomczyk, Solar internal sound speed as inferred from combined BiSON and LOWL oscillation frequencies. Mon. Not. R. Astron. Soc. **292**, 243 (1997)

V.A. Baturin, W. Däppen, D.O. Gough, S.V. Vorontsov, Seismology of the solar envelope: sound-speed gradient in the convection zone and its diagnosis of the equation of state. Mon. Not. R. Astron. Soc. **316**, 71–83 (2000). doi:10.1046/j.1365-8711.2000.03459.x

A.C. Birch, D.C. Braun, S.M. Hanasoge, R. Cameron, Surface-focused seismic holography of sunspots: II. Expectations from numerical simulations using sound-speed perturbations. Sol. Phys. **254**, 17–27 (2009). doi:10.1007/s11207-008-9282-9

A.C. Birch, D.C. Braun, K.D. Leka, G. Barnes, B. Javornik, Helioseismology of pre-emerging active regions. II. Average emergence properties. Astrophys. J. **762**, 131 (2013). doi:10.1088/0004-637X/762/2/131

D.C. Braun, Comment on 'Detection of emerging sunspot regions in the solar interior'. Science **336**, 296 (2012). doi:10.1126/science.1215425

D.C. Braun, A.C. Birch, M. Rempel, T.L. Duvall, Helioseismology of a realistic magnetoconvective sunspot simulation. Astrophys. J. **744**, 77 (2012). doi:10.1088/0004-637X/744/1/77

E. Caffau, H.-G. Ludwig, P. Bonifacio, R. Faraggiana, M. Steffen, B. Freytag, I. Kamp, T.R. Ayres, The solar photospheric abundance of carbon. Analysis of atomic carbon lines with the CO5BOLD solar model. Astron. Astrophys. **514**, 92 (2010). doi:10.1051/0004-6361/200912227

E. Caffau, H.-G. Ludwig, M. Steffen, B. Freytag, P. Bonifacio, Solar chemical abundances determined with a CO5BOLD 3D model atmosphere. Sol. Phys. **268**, 255–269 (2011). doi:10.1007/s11207-010-9541-4

P.S. Cally, Magnetic and thermal phase shifts in the local helioseismology of sunspots. Mon. Not. R. Astron. Soc. **395**, 1309–1318 (2009). doi:10.1111/j.1365-2966.2009.14708.x

P. Charbonneau, Dynamo models of the solar cycle. Living Rev. Sol. Phys. **7**, 3 (2010). doi:10.12942/lrsp-2010-3

P. Charbonneau, Solar dynamo theory. Annu. Rev. Astron. Astrophys. **52**, 251–290 (2014). doi:10.1146/annurev-astro-081913-040012

D.-Y. Chou, A. Serebryanskiy, Searching for the signature of the magnetic fields at the base of the solar convection zone with solar cycle variations of p-mode travel time. Astrophys. J. Lett. **578**, 157–160 (2002). doi:10.1086/344635

D.-Y. Chou, A. Serebryanskiy, In search of the solar cycle variations of p-mode frequencies generated by perturbations in the solar interior. Astrophys. J. **624**, 420–427 (2005). doi:10.1086/428925

A.R. Choudhuri, M. Schussler, M. Dikpati, The solar dynamo with meridional circulation. Astron. Astrophys. **303**, 29 (1995)

J. Christensen-Dalsgaard, Helioseismology. Rev. Mod. Phys. **74**, 1073–1129 (2002). doi:10.1103/RevModPhys.74.1073

J. Christensen-Dalsgaard, M.J. Thompson, Observational results and issues concerning the tachocline, in *The Solar Tachocline*, ed. by D.W. Hughes, R. Rosner, N.O. Weiss (2007), p. 53

J. Christensen-Dalsgaard, D.O. Gough, M.J. Thompson, The depth of the solar convection zone. Astrophys. J. **378**, 413–437 (1991). doi:10.1086/170441

J. Christensen-Dalsgaard, C.R. Proffitt, M.J. Thompson, Effects of diffusion on solar models and their oscillation frequencies. Astrophys. J. Lett. **403**, 75–78 (1993). doi:10.1086/186725

J. Christensen-Dalsgaard, T.L. Duvall Jr., D.O. Gough, J.W. Harvey, E.J. Rhodes Jr., Speed of sound in the solar interior. Nature **315**, 378–382 (1985). doi:10.1038/315378a0

T. Corbard, M.J. Thompson, The subsurface radial gradient of solar angular velocity from MDI f-mode observations. Sol. Phys. **205**, 211–229 (2002). doi:10.1023/A:1014224523374

W. Däppen, D.O. Gough, On the determination of the helium abundance of the solar convection zone, in *Liege International Astrophysical Colloquia*, vol. 25 (1984), pp. 264–268

W. Däppen, D.O. Gough, Progress report on helium abundance determination, in *NATO ASIC Proc. 169: Seismology of the Sun and the Distant Stars*, ed. by D.O. Gough (1986), pp. 275–280

W. Däppen, D.O. Gough, M.J. Thompson, Further progress on the helium abundance determination, in *Seismology of the Sun and Sun-Like Stars*, ed. by E.J. Rolfe. ESA Special Publication, vol. 286 (1988), pp. 505–510

K. DeGrave, J. Jackiewicz, M. Rempel, Time-distance helioseismology of two realistic sunspot simulations. Astrophys. J. **794**, 18 (2014a). doi:10.1088/0004-637X/794/1/18

K. DeGrave, J. Jackiewicz, M. Rempel, Validating time-distance helioseismology with realistic quiet-Sun simulations. Astrophys. J. **788**, 127 (2014b). doi:10.1088/0004-637X/788/2/127

F.-L. Deubner, D. Gough, Helioseismology: oscillations as a diagnostic of the solar interior. Annu. Rev. Astron. Astrophys. **22**, 593–619 (1984). doi:10.1146/annurev.aa.22.090184.003113

M. Dikpati, P.A. Gilman, Flux-transport solar dynamos. Space Sci. Rev. **144**, 67–75 (2009). doi:10.1007/s11214-008-9484-3

F.W.W. Dilke, D.O. Gough, The solar spoon. Nature **240**, 262–294 (1972). doi:10.1038/240262a0

T.L. Duvall, S.M. Hanasoge, Subsurface supergranular vertical flows as measured using large distance separations in time-distance helioseismology. Sol. Phys. **287**, 71–83 (2013). doi:10.1007/s11207-012-0010-0

T.L. Duvall, S.M. Hanasoge, S. Chakraborty, Additional evidence supporting a model of shallow, high-speed supergranulation. Sol. Phys. **289**, 3421–3433 (2014). doi:10.1007/s11207-014-0537-3

T.L. Duvall Jr., W.A. Dziembowski, P.R. Goode, D.O. Gough, J.W. Harvey, J.W. Leibacher, Internal rotation of the Sun. Nature **310**, 22–25 (1984). doi:10.1038/310022a0

P. Eggenberger, G. Meynet, A. Maeder, A. Miglio, J. Montalban, F. Carrier, S. Mathis, C. Charbonnel, S. Talon, Effects of rotational mixing on the asteroseismic properties of solar-type stars. Astron. Astrophys. **519**, 116 (2010). doi:10.1051/0004-6361/201014713

N.A. Featherstone, Exploring convection and dynamos in the cores and envelopes of stars. PhD thesis, University of Colorado at Boulder, 2011

N.A. Featherstone, B.W. Hindman, M.J. Thompson, Ring-analysis flow measurements of sunspot outflows. J. Phys. Conf. Ser. **271**(1), 012002 (2011). doi:10.1088/1742-6596/271/1/012002

S.T. Fletcher, A.-M. Broomhall, D. Salabert, S. Basu, W.J. Chaplin, Y. Elsworth, R.A. Garcia, R. New, A seismic signature of a second dynamo? Astrophys. J. Lett. **718**, 19–22 (2010). doi:10.1088/2041-8205/718/1/L19

P.M. Giles, Time-distance measurements of large-scale flows in the solar convection zone. PhD thesis, Stanford University, 2000

P.A. Gilman, R. Howe, Meridional motion and the slope of isorotation contours, in *GONG+ 2002. Local and Global Helioseismology: The Present and Future*, ed. by H. Sawaya-Lacoste. ESA Special Publication, vol. 517 (2003), pp. 283–285

L. Gizon, A.C. Birch, Local helioseismology. Living Rev. Sol. Phys. **2**, 6 (2005). doi:10.12942/lrsp-2005-6

L. Gizon, A.C. Birch, Helioseismology challenges models of solar convection. Proc. Natl. Acad. Sci. **109**, 11896–11897 (2012). doi:10.1073/pnas.1208875109

L. Gizon, A.C. Birch, H.C. Spruit, Local helioseismology: three-dimensional imaging of the solar interior. Annu. Rev. Astron. Astrophys. **48**, 289–338 (2010). doi:10.1146/annurev-astro-082708-101722

I. González Hernández, C. Lindsey, D.C. Braun, R.S. Bogart, P.H. Scherrer, F. Hill, Far-side helioseismic maps: the next generation. J. Phys. Conf. Ser. **440**(1), 012029 (2013). doi:10.1088/1742-6596/440/1/012029

D. Gough, J. Toomre, Seismic observations of the solar interior. Annu. Rev. Astron. Astrophys. **29**, 627–685 (1991). doi:10.1146/annurev.aa.29.090191.003211

D.O. Gough, Theory of solar oscillations, in *Future Missions in Solar, Heliospheric & Space Plasma Physics*, ed. by E. Rolfe, B. Battrick. ESA Special Publication, vol. 235 (1985), pp. 183–197

B.J. Greer, B.W. Hindman, J. Toomre, Multi-ridge fitting for ring-diagram helioseismology. Sol. Phys. **289**, 2823–2843 (2014). doi:10.1007/s11207-014-0514-x

B.J. Greer, B.W. Hindman, N.A. Featherstone, J. Toomre, Helioseismic imaging of fast convective flows throughout the near-surface shear layer. Astrophys. J. Lett. (2015, in press). arXiv:1504.00699

N. Grevesse, A. Noels, Cosmic abundances of the elements, in *Origin and Evolution of the Elements*, ed. by N. Prantzos, E. Vangioni-Flam, M. Casse (1993), pp. 15–25

J.A. Guzik, Challenges in solar and stellar model physics (2014). arXiv:1401.0547

D.A. Haber, B.W. Hindman, J. Toomre, R.S. Bogart, R.M. Larsen, F. Hill, Evolving submerged meridional circulation cells within the upper convection zone revealed by ring-diagram analysis. Astrophys. J. **570**, 855–864 (2002). doi:10.1086/339631

S.M. Hanasoge, A. Birch, L. Gizon, J. Tromp, The adjoint method applied to time-distance helioseismology. Astrophys. J. **738**, 100 (2011). doi:10.1088/0004-637X/738/1/100

S.M. Hanasoge, T.L. Duvall, K.R. Sreenivasan, Anomalously weak solar convection. Proc. Natl. Acad. Sci. **109**, 11928–11932 (2012). doi:10.1073/pnas.1206570109

S. Hanasoge, M.S. Miesch, M. Roth, J. Schou, M. Schüssler, M.J. Thompson, Solar dynamics, rotation, convection and overshoot. Space Sci. Rev. (2015, this issue). doi:10.1007/s11214-015-0144-0

G. Hazra, B.B. Karak, A.R. Choudhuri, Is a deep one-cell meridional circulation essential for the flux transport solar dynamo? Astrophys. J. **782**, 93 (2014). doi:10.1088/0004-637X/782/2/93

F. Hill, P.B. Stark, R.T. Stebbins, E.R. Anderson, H.M. Antia, T.M. Brown, T.L. Duvall Jr., D.A. Haber, J.W. Harvey, D.H. Hathaway, R. Howe, R.P. Hubbard, H.P. Jones, J.R. Kennedy, S.G. Korzennik, A.G. Kosovichev, J.W. Leibacher, K.G. Libbrecht, J.A. Pintar, E.J. Rhodes Jr., J. Schou, M.J. Thompson, S. Tomczyk, C.G. Toner, R. Toussaint, W.E. Williams, The solar acoustic spectrum and eigenmode parameters. Science **272**, 1292–1295 (1996). doi:10.1126/science.272.5266.1292

B.W. Hindman, L. Gizon, T.L. Duvall Jr., D.A. Haber, J. Toomre, Comparison of solar subsurface flows assessed by ring and time-distance analyses. Astrophys. J. **613**, 1253–1262 (2004). doi:10.1086/423263

R. Howe, Solar interior rotation and its variation. Living Rev. Sol. Phys. **6**, 1 (2009). doi:10.12942/lrsp-2009-1

R. Howe, J. Christensen-Dalsgaard, F. Hill, R.W. Komm, R.M. Larsen, J. Schou, M.J. Thompson, J. Toomre, Dynamic variations at the base of the solar convection zone. Science **287**, 2456–2460 (2000). doi:10.1126/science.287.5462.2456

R. Howe, R. Komm, F. Hill, J. Christensen-Dalsgaard, T.P. Larson, J. Schou, M.J. Thompson, J. Toomre, Rotation-rate variations at the tachocline: an update. J. Phys. Conf. Ser. **271**(1), 012075 (2011). doi:10.1088/1742-6596/271/1/012075

R. Howe, J. Christensen-Dalsgaard, F. Hill, R. Komm, T.P. Larson, M. Rempel, J. Schou, M.J. Thompson, The high-latitude branch of the solar torsional oscillation in the rising phase of cycle 24. Astrophys. J. Lett. **767**, 20 (2013). doi:10.1088/2041-8205/767/1/L20

S. Ilonidis, J. Zhao, A. Kosovichev, Detection of emerging sunspot regions in the solar interior. Science **333**, 993 (2011). doi:10.1126/science.1206253

S.M. Jefferies, S.V. Vorontsov, A new way to model the solar oscillation ℓ-ν power spectrum. Sol. Phys. **220**, 347–359 (2004). doi:10.1023/B:SOLA.0000031404.13076.73

J.R. Kennedy, S.M. Jefferies, F. Hill, Solar G-mode signatures in P-mode signals, in *GONG 1992. Seismic Investigation of the Sun and Stars*, ed. by T.M. Brown. Astronomical Society of the Pacific Conference Series, vol. 42 (1993), p. 273

R. Komm, I. González Hernández, F. Hill, R. Bogart, M.C. Rabello-Soares, D. Haber, Subsurface meridional flow from HMI using the ring-diagram pipeline. Sol. Phys. **287**, 85–106 (2013). doi:10.1007/s11207-012-0073-y

R. Komm, R. Howe, I. González Hernández, F. Hill, Solar-cycle variation of subsurface zonal flow. Sol. Phys. **289**, 3435–3455 (2014). doi:10.1007/s11207-014-0490-1

A.G. Kosovichev, J. Christensen-Dalsgaard, W. Daeppen, W.A. Dziembowski, D.O. Gough, M.J. Thompson, Sources of uncertainty in direct seismological measurements of the solar helium abundance. Mon. Not. R. Astron. Soc. **259**, 536–558 (1992)

A.G. Kosovichev, S. Basu, R. Bogart, T.L. Duvall Jr., I. Gonzalez-Hernandez, D. Haber, T. Hartlep, R. Howe, R. Komm, S. Kholikov, K.V. Parchevsky, S. Tripathy, J. Zhao, Local helioseismology of sunspot regions: comparison of ring-diagram and time-distance results. J. Phys. Conf. Ser. **271**(1), 012005 (2011). doi:10.1088/1742-6596/271/1/012005

P. Kumar, E.J. Quataert, J.N. Bahcall, Observational searches for solar G-modes: some theoretical considerations. Astrophys. J. Lett. **458**, 83 (1996). doi:10.1086/309926

T.P. Larson, J. Schou, Improvements in global mode analysis. J. Phys. Conf. Ser. **118**(1), 012083 (2008). doi:10.1088/1742-6596/118/1/012083

E.M. Lavely, M.H. Ritzwoller, The effect of global-scale, steady-state convection and elastic-gravitational asphericities on helioseismic oscillations. Philos. Trans. R. Soc. Lond. Ser. A **339**, 431–496 (1992). doi:10.1098/rsta.1992.0048

C. Lindsey, D.C. Braun, Seismic images of the far side of the Sun. Science **287**, 1799–1801 (2000). doi:10.1126/science.287.5459.1799

K. Lodders, Solar system abundances of the elements, in *Principles and Perspectives in Cosmochemistry*, ed. by A. Goswami, B.E. Reddy (2010), p. 379. doi:10.1007/978-3-642-10352-0_8

Y.-Q. Lou, Solar P-modes modulated by interior G-modes. Astrophys. J. Lett. **556**, 121–125 (2001). doi:10.1086/322270

S. Mathis, The interior of the Sun in 3-D: beyond the spherical Sun picture. Astron. Nachr. **331**, 883 (2010). doi:10.1002/asna.201011419

S. Mathis, Transport processes in stellar interiors, in *Lecture Notes in Physics*, ed. by M. Goupil, K. Belkacem, C. Neiner, F. Lignières, J.J. Green. Lecture Notes in Physics, vol. 865 (Springer, Berlin, 2013), p. 23. doi:10.1007/978-3-642-33380-4_2

M.S. Miesch, The solar dynamo. Philos. Trans. R. Soc. Lond. Ser. A **370**, 3049–3069 (2012). doi:10.1098/rsta.2011.0507

M.S. Miesch, J. Toomre, Turbulence, magnetism, and shear in stellar interiors. Annu. Rev. Fluid Mech. **41**, 317–345 (2009). doi:10.1146/annurev.fluid.010908.165215
M.S. Miesch, A.S. Brun, M.L. De Rosa, J. Toomre, Structure and evolution of giant cells in global models of solar convection. Astrophys. J. **673**, 557–575 (2008). doi:10.1086/523838
M.S. Miesch, N.A. Featherstone, M. Rempel, R. Trampedach, On the amplitude of convective velocities in the deep solar interior. Astrophys. J. **757**, 128 (2012). doi:10.1088/0004-637X/757/2/128
U. Mitra-Kraev, M.J. Thompson, Meridional flow profile measurements with SOHO/MDI. Astron. Nachr. **328**, 1009–1012 (2007). doi:10.1002/asna.200710873
H. Moradi, P.S. Cally, Helioseismic implications of mode conversion, in *Astronomical Society of the Pacific Conference Series*, ed. by K. Jain, S.C. Tripathy, F. Hill, J.W. Leibacher, A.A. Pevtsov. Astronomical Society of the Pacific Conference Series, vol. 478 (2013), p. 263
H. Moradi, C. Baldner, A.C. Birch, D.C. Braun, R.H. Cameron, T.L. Duvall, L. Gizon, D. Haber, S.M. Hanasoge, B.W. Hindman, J. Jackiewicz, E. Khomenko, R. Komm, P. Rajaguru, M. Rempel, M. Roth, R. Schlichenmaier, H. Schunker, H.C. Spruit, K.G. Strassmeier, M.J. Thompson, S. Zharkov, Modeling the subsurface structure of sunspots. Sol. Phys. **267**, 1–62 (2010). doi:10.1007/s11207-010-9630-4
K. Mussack, D. Gough, Measuring solar abundances with seismology, in *Solar-Stellar Dynamos as Revealed by Helio- and Asteroseismology: GONG 2008/SOHO 21*, ed. by M. Dikpati, T. Arentoft, I. González Hernández, C. Lindsey, F. Hill. Astronomical Society of the Pacific Conference Series, vol. 416 (2009), p. 203
A.A. Reinard, J. Henthorn, R. Komm, F. Hill, Evidence that temporal changes in solar subsurface helicity precede active region flaring. Astrophys. J. Lett. **710**, 121–125 (2010). doi:10.1088/2041-8205/710/2/L121
O. Richard, W.A. Dziembowski, R. Sienkiewicz, P.R. Goode, On the accuracy of helioseismic determination of solar helium abundance. Astron. Astrophys. **338**, 756–760 (1998)
M. Roth, M. Stix, Meridional circulation and global solar oscillations. Sol. Phys. **251**, 77–89 (2008). doi:10.1007/s11207-008-9232-6
A. Schad, J. Timmer, M. Roth, Global helioseismic evidence for a deeply penetrating solar meridional flow consisting of multiple flow cells. Astrophys. J. Lett. **778**, 38 (2013). doi:10.1088/2041-8205/778/2/L38
J. Schou, On the analysis of helioseismic data. PhD thesis, Aarhus University, 1992
J. Schou, H.M. Antia, S. Basu, R.S. Bogart, R.I. Bush, S.M. Chitre, J. Christensen-Dalsgaard, M.P. Di Mauro, W.A. Dziembowski, A. Eff-Darwich, D.O. Gough, D.A. Haber, J.T. Hoeksema, R. Howe, S.G. Korzennik, A.G. Kosovichev, R.M. Larsen, F.P. Pijpers, P.H. Scherrer, T. Sekii, T.D. Tarbell, A.M. Title, M.J. Thompson, J. Toomre, Helioseismic studies of differential rotation in the solar envelope by the solar oscillations investigation using the Michelson Doppler Imager. Astrophys. J. **505**, 390–417 (1998). doi:10.1086/306146
D. Seligman, G.J.D. Petrie, R. Komm, A combined study of photospheric magnetic and current helicities and subsurface kinetic helicities of solar active regions during 2006–2013. Astrophys. J. **795**, 113 (2014). doi:10.1088/0004-637X/795/2/113
S. Talon, C. Charbonnel, Angular momentum transport by internal gravity waves. IV. Wave generation by surface convection zone, from the pre-main sequence to the early-AGB in intermediate mass stars. Astron. Astrophys. **482**, 597–605 (2008). doi:10.1051/0004-6361:20078620
M.J. Thompson, J. Toomre, E.R. Anderson, H.M. Antia, G. Berthomieu, D. Burtonclay, S.M. Chitre, J. Christensen-Dalsgaard, T. Corbard, M. De Rosa, C.R. Genovese, D.O. Gough, D.A. Haber, J.W. Harvey, F. Hill, R. Howe, S.G. Korzennik, A.G. Kosovichev, J.W. Leibacher, F.P. Pijpers, J. Provost, E.J. Rhodes Jr., J. Schou, T. Sekii, P.B. Stark, P.R. Wilson, Differential rotation and dynamics of the solar interior. Science **272**, 1300–1305 (1996). doi:10.1126/science.272.5266.1300
M.J. Thompson, J. Christensen-Dalsgaard, M.S. Miesch, J. Toomre, The internal rotation of the Sun. Annu. Rev. Astron. Astrophys. **41**, 599–643 (2003). doi:10.1146/annurev.astro.41.011802.094848
S. Tomczyk, J. Schou, M.J. Thompson, Measurement of the rotation rate in the deep solar interior. Astrophys. J. Lett. **448**, 57 (1995). doi:10.1086/309598
S. Turck-Chièze, P.-H. Carton, J. Ballot, J.-C. Barrière, P. Daniel-Thomas, A. Delbart, D. Desforges, R.A. Garcia, R. Granelli, S. Mathur, F. Nunio, Y. Piret, P.L. Pallé, A.J. Jiménez, S.J. Jiménez-Reyes, J.M. Robillot, E. Fossat, A.M. Eff-Darwich, B. Gelly, GOLF-NG spectrometer, a space prototype for studying the dynamics of the deep solar interior. Adv. Space Res. **38**, 1812–1818 (2006). doi:10.1016/j.asr.2005.09.033
S.V. Vorontsov, S.M. Jefferies, Modeling solar oscillation power spectra. I. Adaptive response function for Doppler velocity measurements. Astrophys. J. **623**, 1202–1214 (2005). doi:10.1086/427913
S.V. Vorontsov, S.M. Jefferies, Modeling solar oscillation power spectra. II. Parametric model of spectral lines observed in Doppler-velocity measurements. Astrophys. J. **778**, 75 (2013). doi:10.1088/0004-637X/778/1/75
S.V. Vorontsov, J. Christensen-Dalsgaard, J. Schou, V.N. Strakhov, M.J. Thompson, Helioseismic measurement of solar torsional oscillations. Science **296**, 101–103 (2002). doi:10.1126/science.1069190

M.F. Woodard, R.W. Noyes, Change of solar oscillation eigenfrequencies with the solar cycle. Nature **318**, 449–450 (1985). doi:10.1038/318449a0

J. Zhao, K. Nagashima, R.S. Bogart, A.G. Kosovichev, T.L. Duvall Jr., Systematic center-to-limb variation in measured helioseismic travel times and its effect on inferences of solar interior meridional flows. Astrophys. J. Lett. **749**, 5 (2012). doi:10.1088/2041-8205/749/1/L5

Space Sci Rev (2015) 196:15–47
DOI 10.1007/s11214-015-0159-6

Some Glimpses from Helioseismology at the Dynamics of the Deep Solar Interior

D.O. Gough[1]

Received: 18 January 2015 / Accepted: 9 May 2015 / Published online: 26 May 2015

Abstract Helioseismology has taught us a great deal about the stratification and kinematics of the solar interior, sufficient for us to embark upon dynamical studies more detailed than have been possible before. The most sophisticated studies to date have been the very impressive numerical simulations of the convection zone, from which, especially in recent years, a great deal has been learnt. Those simulations, and the seismological evidence with which they are being confronted, are reviewed elsewhere in this volume. Our understanding of the global dynamics of the radiative interior of the Sun is in a much more primitive state. Nevertheless, some progress has been made, and seismological inference has provided us with evidence of more to come. Some of that I summarize here, mentioning in passing hints that are pointing the way to the future.

Keywords Sun · Helioseismology · Dynamics

1 Introduction

The most studied dynamics of the Sun occurs in the convective envelope, where the timescales are human. The large-scale convective flow and its associated angular-momentum transport is to some degree accessible to seismological probing. However, I shall pay little attention to that here, because it is reviewed elsewhere in this issue by Brun et al. (2015) and Hanasoge et al. (2015), who discuss principally the results of the impressive modern simulations and the physics that has been gleaned from them. Instead, I shall try to reflect on the global behaviour of the Sun, recalling some of the issues that have concerned heliophysicists in the past—issues that to some degree have been thought by commentators not in the field to have been resolved, yet in the light of our increasingly sophisticated helioseismological findings have left some room for more than a modicum of uncertainty.

✉ D.O. Gough
douglas@ast.cam.ac.uk

1 Institute of Astronomy & Department of Applied Mathematics and Theoretical Physics, Madingley Road, Cambridge, CB3 0HA, UK

In order to establish the what-one-might-call standard view of the Sun, I shall first describe briefly what we believe to have been the principal dynamical processes that have influenced the evolution of the Sun to its present state. Only then do we have a basis for discussing the extent to which those beliefs should really be relied upon. Quite naturally, I first adopt the simplest of descriptions, consistent with our general understanding of stellar evolution; that seems to provide a gratifyingly accurate first-order class of models. There have been attempts to standardize some versions of those models, principally by John Bahcall. However, a universal standard has never been produced, because heliophysicists have not agreed on the manner in which the physics should be simplified. The discord has been aired many times, such as during IAU Colloquium 121 (Berthomieu and Cribier 1990), at which Evry Schatzman (reported by Gough 1990b) favoured the inclusion, by some means or other, of all the pertinent generally accepted physics, including macroscopic dynamics (represented in some precisely stated, albeit approximate, manner), in order to obtain the most realistic model possible. By contrast, John argued for extreme dynamical simplicity, ignoring all the effects of macroscopic motion except the heat flux by thermal convection (yet excluding the momentum and kinetic-energy fluxes), parametrized by a local mixing-length formula, in order to produce a very straightforward model that can be used as a standard benchmark, although in practice that standard has evolved with time, even for John. Here, I shall adopt Model S of Christensen-Dalsgaard et al. (1996) as my reference, because the assumptions upon which it was built have been stated clearly, and because, unlike many other solar models, it has been produced with sufficient care for reasonably precise, well defined (adiabatic) oscillations to be computed from it.

I do not go into the details of the early, pre-main-sequence, dynamics. That remains a complicated area of (active) research which continues to benefit from advances in (principally numerical) simulation. Suffice it to say that, because the age of the Sun is very much greater than any of the dynamical timescales, its present structure is hardly likely to have any recall of the precise manner in which the Sun condensed from the interstellar medium, save for the relic chemical composition, and the total mass and angular momentum. It is normally presumed that the chemical composition was either initially uniform, or that mixing during Hayashi contraction homogenized any initial internal inhomogeneity well before the star settled onto the main sequence. However, that assumption is open to question. It is also normally presumed that there was no substantial mass loss nor accretion, although that presumption is also questioned from time to time, with good reason. It is generally believed that, in common with other stars, there has been some loss of angular momentum, particularly in the immediately pre-main-sequence and early main-sequence stages of evolution (and even more before), as is evinced by the decrease with age of the rotational velocities of the photospheres of stars in young clusters (e.g. Skumanich 1972). However, that process is unlikely to have had a great effect on the state of the Sun today (in other words, it is of little concern that the angular momentum in the very early days might have been rather greater than it is today), and is accordingly rarely discussed. Nevertheless, a word or two about it later in my discussion may not be out of place. It may have been responsible for a degree of mixing of the products of nuclear reactions in the core, especially in early times. Be that as it may, the essentially static effect of rotation, via the centrifugal force, is very much smaller than the pressure gradient and the force of gravity (as also are Maxwell stresses and the momentum flux due to large-scale convection, except in the very outer, near-surface, layers) so to a first approximation the Sun can well be regarded as being spherically symmetric. Oblateness of the figure resulting from rotation can be treated subsequently as a small perturbation, as I discuss later. So can the effect of mass loss.

2 Standard Main-Sequence Evolution

The basic principle behind the evolution is quite straightforward and very robust. Nuclear reactions in the core convert hydrogen into helium, reducing the number of particles—nuclei, ions and electrons—per unit mass of the stellar material, reducing the pressure at a given temperature, permitting the star to contract under gravity and thereby raising the temperature in the innermost regions to maintain hydrostatic balance. The temperature- and density-sensitive nuclear reaction rates are augmented with this contraction by more than the reduction due to the loss of hydrogen fuel, essential to maintain global stability, so the luminosity of the star increases. This basic result is an inevitable consequence of thermonuclear physics and the conditions for hydrostatic balance, and is beyond challenge (provided, of course, that standard nuclear and gravitational physics are accepted).

The theoretical variation of the luminosity L with time t is given approximately by

$$L(t) \simeq \frac{L_\odot}{1+\beta(1-t/t_\odot)} \tag{1}$$

(Gough 1977b) with $\beta = L(0)/L_\odot - 1 \simeq 2/5$, where the origin of time t is at the start of the evolution on the main sequence. It is robust against minor variations in the assumptions behind the model (e.g. Gough 1988; Bahcall et al. 2001). A 'derivation' of this formula is presented by Gough (1988, 1990) based on homology scaling, although the constant $2/5$ was obtained by adjusting an initial theoretical estimate to account for the deviation from homologous variation due to nuclear transmutation such as to render Eq. (1) consistent with numerical stellar-evolution computation, so the relation should perhaps be regarded more as a physically motivated interpolation formula. The formula has been generalized to take account of a putative temporal variation of the gravitational constant, G, and has also been adapted to accommodate main-sequence mass loss (Gough 1990), thereby summarizing the numerical computations that have been carried out both before and since.

One of the reasons for considering a temporal variation in the gravitational constant was to obviate what has been called the faint-Sun problem (some have even called it a paradox). It was posed by Sagan and Mullen (1972), who argued from simple equilibrium energy balance that early in the Sun's main-sequence evolution the Earth would have been completely glaciated had the Sun evolved essentially according to Eq. (1), which is counter to geological finding. The solar luminosity is a steeply increasing function of G (Teller 1948): $L \propto G^{7.8}$ (Gamow 1967). Therefore adopting an appropriate small decline of G with time can almost annul the rise in L described by Eq. (1): if, for example, $G(t')/G(t_u) = (t'/t_u)^{-q}$ (where t' is time measured from the Big Bang and t_u is the age of the Universe), then the irradiance on Earth would have been very nearly constant[1] if $-G/\dot{G}|_{t'=t_u} \simeq 1.2 \times 10^{11}$ y, almost irrespective of the value of q (e.g. Gough 1988, 1990). However, this is not strong evidence in favour of there being a temporal variation in G, because it is unlikely that the climatological equilibrium energy-balance assumption is correct. Apparently more sophisticated meteorological arguments have even been perceived to exacerbate the issue. For example, in the summer of 1973, motivated by a prediction that the Sun's luminosity might have declined suddenly by a few per cent some million or so years ago (Dilke and Gough 1972), Tzvi Gal-Chen and I (unpublished) engineered a 5-per-cent reduction in solar irradiance in the Global Circulation Model (GCM) of the Earth's atmosphere (Kasahara and Washington 1971) at the US National Center for Atmospheric Research. As expected, the Earth became completely glaciated; but what was not expected is that when the irradiance was subsequently

[1] Having deviated from its initial value by no more than 0.5 per cent.

restored to its present value, the ice that had just been produced would not melt. That result is evidently contrary to observation, because the Earth is not completely glaciated today. The moral, for me at least, was to confirm that one must always beware of the products of complicated computer programmes, particularly when they are state of the art yet depend, as does the GCM, on rather primitive physics. The many physical processes that the GCM had to account for were ill understood—as indeed many are still—and their influence on the atmospheric dynamics had to be parametrized; the controlling parameters whose values were not previously known were measured, where possible, and then held constant in the model. That was the most obvious flaw. We know that the stability and evolution of any dynamical system can depend quite sensitively, and sometimes critically, on the nature of the constraints that are imposed; evidently a meteorological model, however sophisticated, with processes whose controlling parameters do not change appreciably on a timescale of days or weeks could not possibly be expected to apply to the climate on timescales of millions of years. Indeed, Dilke and Gough had presumed that the response of the Earth's climate to irradiance variation was quite different on a timescale of 10^6 y from that on a timescale of 10^8 y. The processes responsible for such timescale dependence have been studied subsequently by, amongst many others, Margulis and Lovelock (1974), Lovelock and Margulis (1974) and Lovelock and Watson (1982), who likened the long-term stability of the climate to the work of Gaia, the ancient Greek goddess of Earth. The cause of the suspected sudden relatively recent decline in L was itself a dynamical process, considered to have resulted from a nonlinear instability of g modes in the Sun's core, the theory of whose development also depending critically on what is held constant, a matter to which I shall return later.

The computation of the structure and evolution of (standard) solar models is straightforward once the equation of state, the nuclear reaction rates and the opacity of the stellar material have been specified. There is also a matter of determining the mean stratification of the convection zone: to that end some form of mixing-length formalism is normally used to specify the energy (usually just heat) flux and, sometimes, the Reynolds stress.[2] For most purposes the deficiencies in such a procedure are of little concern to the overall structure of the star, because throughout almost all of the convection zone the stratification is adiabatic and the Reynolds stress is negligible; it is only in the outer boundary layer where the details of the formalism matter. In practice the mixing-length formalism contains at least one explicitly adjustable constant which can be calibrated to determine the value of the adiabatic constant deep in the convection zone that is required to reproduce the observed radius of the Sun; calibrated in that way, the outcome is a function of chemical composition, and it formally determines the depth of the convection zone. Of course, that is not to say that in reality the dynamics of the upper convective boundary layer depends explicitly on the chemical composition in the manner determined by the solar calibration.

The chemical composition is normally specified by two of three parameters: the relative abundances X, Y, Z by mass of H, ^{4}He and of all other elements combined;[3] they satisfy $X + Y + Z = 1$. (The relative abundances of the elements incorporated into Z are provided principally, but not completely, by spectroscopic analyses of the Sun's atmosphere (Asplund et al. 2009; Caffau et al. 2009).) Stellar evolution theory provides a further relation between them from the requirement that the current luminosity $L(t_\odot)$ agrees with the value measured (assuming the age $t_\odot$ to be known). That leaves a single infinity of apparently acceptable models, which can be labelled by any one of X, Y and Z. I record that along the sequence

[2]Yet hardly ever the flux of kinetic energy.

[3]Usually referred to as 'heavy elements', even though they include ^{3}He.

of these models, both Y and the depth d_c of the convection zone increase monotonically with Z.

Before helioseismology, there was just one measured quantity associated with the Sun that in principle could be used to select the most appropriate model, namely the neutrino 'luminosity' L_ν (usually expressed as an associated flux F_ν at 1 AU computed under the assumption that neutrinos do not decay, which itself is expressed as a capture rate by whatever neutrino detector is being considered). Theoretically, L_ν is a monotonic increasing function of Z. Therefore a unique model can thereby be chosen. However, that model left many people uncomfortable, and was deemed unacceptable by most astronomers and astrophysicists because the value of Y in that model was much lower than that observed in open stellar clusters containing apparently Sun-like stars; moreover, it was lower than the amount of helium created in generally accepted models of Big-Bang nucleosynthesis. Astronomical opinion prevailed, and the discrepancy posed what was accordingly named the solar neutrino problem.[4]

Perhaps the first genuinely seismological inference to be drawn about the structure of the Sun was from a crude calibration of the depth d_c of the convection zone (Gough 1977a). It implied a value of Y even greater than those preferred at the time, exacerbating the neutrino problem. The inference was soon confirmed with more detailed, numerical, calculations by Rhodes et al. (1977), and later by Berthomieu et al. (1980) and Lubow et al. (1980), who addressed the sensitivity of the eigenfrequencies to uncertainties in the background model envelope. Subsequently, a helioseismological inference of the sound speed throughout most of the solar interior (Christensen-Dalsgaard et al. 1985) not only provided a consistent direct measurement of d_c but also ruled out the F_ν-reproducing low-Y model. A more recent, and cleaner, demonstration of that inference is reproduced in Fig. 1. It convincingly demonstrated that the resolution of the solar neutrino problem must lie in nuclear or particle physics, and not in stellar physics, although that conclusion was not accepted immediately by most of the heliophysics community who did not yet appreciate the power of helioseismology (e.g. Bahcall and Ulrich 1988). I shall say no more about that here because it is only indirectly associated with interior dynamics (but see Cumming and Haxton 1996).

3 Angular Velocity and the Solar Oblateness

The most obvious indicator of global solar dynamics is the angular velocity $\Omega(r, \theta, t)$ (I adopt spherical polar coordinates (r, θ, ϕ)). A seismological determination of Ω in the Sun's envelope by Schou et al. (1998) is illustrated in Fig. 4, and a schematic representation by Chaplin et al. (1999) extending to the centre is plotted in Fig. 2. Roughly speaking, the angular velocity is independent of radius and increases with colatitude θ in the convective envelope ($r > r_e$), and it is approximately uniform in the radiative envelope beneath, except possibly in the energy-generating core ($r < r_c$) where there is (slight) evidence[5] that the spherically averaged Ω is lower than Ω in the radiative envelope (Elsworth et al. 1995). The transition at the base of the convection zone, called the tachocline (Spiegel and Zahn 1992; see also Spiegel 1972), is too abrupt to be properly resolved seismologically, as is the transition, if there is one, at the edge of the core.

[4]Rather than, for example, the solar abundance problem.

[5]Which is commonly doubted.

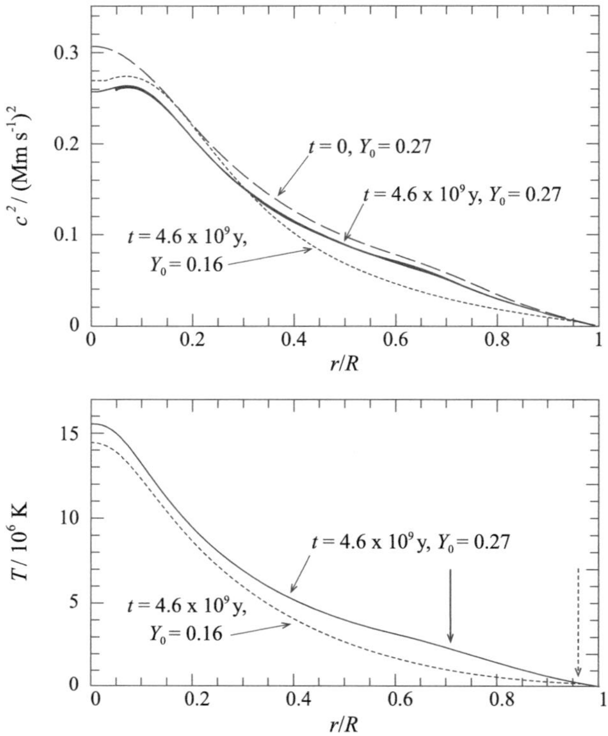

Fig. 1 Square of the sound speed, c^2, and temperature, T, in two solar models computed with a local mixing-length formalism and having different initial heavy-element abundances Z_0: one (*continuous curves*) is the standard Models S of Christensen-Dalsgaard et al. (1996), having $Z_0 = 0.020$ and a corresponding initial helium abundance $Y_0 = 0.27$; the other (*short-dashed curves*) has $Z_0 = 0.001$, $Y_0 = 0.16$ and has been continuously contaminated with heavy elements at the surface at such a constant rate as to have a surface abundance $Z_s = 0.020$ today (Christensen-Dalsgaard et al. 1979a,b). The depression in c^2 (relative to T) in the central regions of the cores of the evolved models is a result of the augmentation of the mean molecular mass in the core produced by the transmutation of hydrogen into helium. Included, for comparison, as a *long-dashed curve*, is the square of the sound speed of the higher-Y model at zero main-sequence age, which, of course, has no such depression. The *vertical arrows* (with line styles corresponding to the models to which they refer) mark the bases of the convection zones of the two present-day models, where the second derivatives of c^2 and T are discontinuous. The square of the sound speed in the Sun, inferred seismologically, is drawn also as a continuous curve, not quite reaching the centre $r/R = 0$ of the star; it is barely distinguishable from c^2 in the higher-Y model (after Gough 1999)

It is common practice to expand the latitudinal dependence of Ω in even powers of $\mu = \cos\theta$:

$$\Omega(r,\theta,t) = \sum_{k\geq 0} \Omega_k(r,t)\mu^{2k}, \tag{2}$$

where, on the whole, Ω_k varies weakly with r and t, except in the tachocline.[6] It is interesting and, so far as I am aware, unexplained, that for most values of r the magnitude of Ω_k is particularly small for all $k > 2$. The coefficients of the terms of lower degree are ob-

[6] Often, for computational convenience, Ω is expanded in orthogonal polynomials, such as Clebsch-Gordon coefficients (Ritzwoller and Lavely 1991).

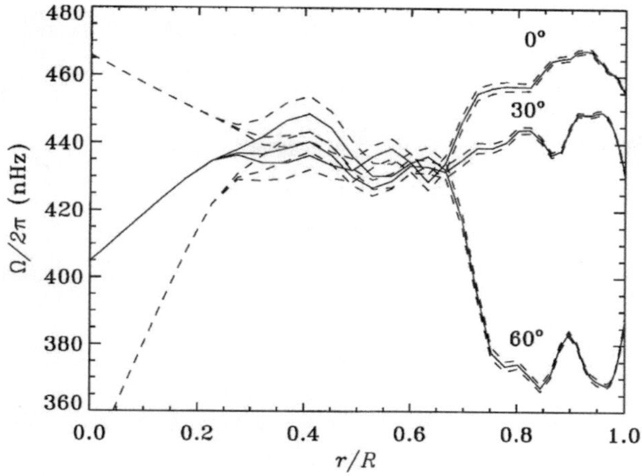

Fig. 2 Angular velocity through the Sun, plotted against radius at three different latitudes. The *continuous curves* are the expectations (averages), and are flanked by *dashed curves* with added or subtracted standard errors propagated from the estimated errors in the data. The tachocline is not well resolved; nevertheless its base appears to be spherical, as one would expect, and penetration of the shear into the convection zone preferentially at high latitudes, rendering the overall shear layer prolate, is discernible. The angular velocity in the radiative envelope (outside the energy-generating core) is essentially uniform, within the uncertainty, and there is a hint that the core is rotating more slowly (from Chaplin et al. 1999). The confluence at $r/R \simeq 0.25$ of the values of Ω at the three different latitudes and the almost linear dependence beneath are both consequences of the regularization, which, because the seismic data are too scant to resolve the angular velocity well, dominates the inversion procedure. All that can be inferred is that an equatorially biassed latitudinal average of the angular velocity in the core appears likely to be lower than the angular velocity of the radiative envelope (cf. Elsworth et al. 1995); therefore the apparent uniformity of Ω with respect to latitude in the core cannot be regarded as evidence for an absence of latitudinal variation

served to vary somewhat with the sunspot cycle, but that variation has been convincingly detected only in the convection zone, so I refrain from discussing it in any detail here. There is also a variation with a characteristic period of about 2 years (Broomhall et al. 2012), which has been potentially misleadingly called the quasi-biennial oscillation—it can hardly be compared with the terrestrial disturbance with the same name, having the acronym QBO, because the latter is driven by differential gravity-wave dissipation, a process which I explain in Sect. 10, and which can hardly be sustained in the convection zone of the Sun, where gravity waves cannot propagate.

The first conclusion to be drawn from the earliest determination of Ω was an estimate of the quadrupole moment J_2 of the external gravitational field (Duvall et al. 1984). It is induced by the centrifugal force acting on the stellar material, causing the mass distribution to be oblate. Therefore J_2 can be represented as an appropriately weighted average of Ω^2 over the volume of the star (Gough 1981; Pijpers 1998), the weight function F being determined by linear perturbation about the nonrotating state. The radial dependence of the component of F pertaining to the spherically averaged Ω is plotted in Fig. 3. Knowledge of J_2 is essential for testing theories of gravity from measurements of planetary precession and spacecraft orbits, for non-Newtonian theories produce a perturbation to Newtonian orbits that at present are observationally indistinguishable from that which is produced by a deviation from spherical symmetry induced by an appropriate internal centrifugal force.

Early attempts to measure J_2 were made by Auwers (1891), Ambronn and Schur (1905) and Poor (1905a,b) from direct observations of the presumedly centrifugally induced visual oblateness, defined as $\Delta_v = (R_e - R_p)/R$, where R_e, R_p and R are respectively the

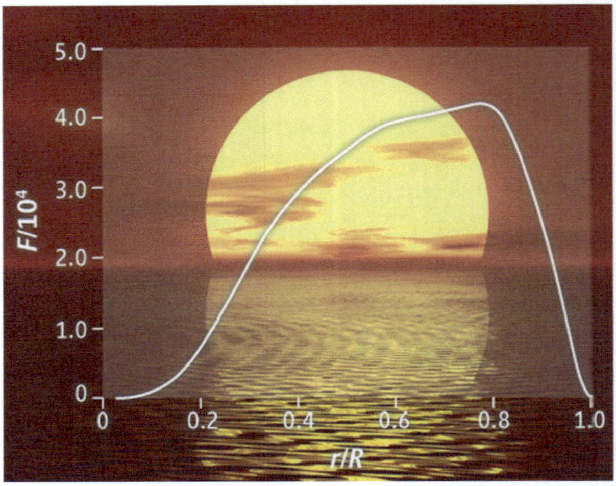

Fig. 3 Superposed on the setting Sun is a plot of the weighting function $F(x)$ in the approximate formula $J_2 \simeq \int F\overline{\Omega}^2 \mathrm{d}x$, where $x = r/R$ and $\overline{\Omega}$ is the spherical average of the (latitudinally mildly varying) internal angular velocity of the Sun, in units of s^{-1} (after Gough 2012a). The function F is small near the centre of the Sun, where the ratio of centrifugal force to angular velocity is low, and near the surface where the low density, however oblate, makes only a minor imprint on the gravitational potential

equatorial and polar radii and their average;[7] in retrospect those observations were inconclusive. More modern measurements by Dicke and Goldenberg (1967, 1974) suggested that Δ_v is about 5 times greater than what one would expect from a distortion by a uniform rotation of the Sun's interior, consistent with the Brans and Dicke (1961) theory of gravity, appropriately calibrated. However, that was challenged by Hill and Stebbins (1975), whose measured value was consistent with uniform rotation. These and subsequent measurements have been reviewed by, for example, Damiani et al. (2011). The shape measurements were very difficult, partly because Δ_v is very small, of the order of the centrifugal parameter $\Lambda = R^3\Omega^2/GM \simeq 2 \times 10^{-5}$, where M is the mass of the Sun, and partly because we now know that the oblateness Δ_v of the visible solar disc is dominated not by the oblateness Δ_Φ of the gravitational field caused by the action of the centrifugal force on the dense interior material, but principally by the oblateness Δ_Ω caused by the direct effect of the centrifugal force on the diffuse visible surface layers. Therefore Δ_Φ would need to be determined as the relatively small difference between Δ_v and Δ_Ω, which is an intrinsically unreliable procedure. The difficulties in the entire measurement procedure have been compounded by the fact that the relation between the oblateness of the observed radiative intensity and the oblateness of, say, the surfaces of constant pressure in the photosphere is contaminated with brightness variations due to sunspots and magnetically generated excess emission, such as faculae. Accounting for these is an uncertain process, as is evident from even the most recent investigations (Lefebvre et al. 2007; Fivian et al. 2008; Kuhn et al. 2012).

The advent of helioseismology changed the situation dramatically, because the seismic signature pertinent to determining J_2 can be measured much more accurately: that signature is the rotational frequency splitting, whose magnitude is of order $2m\Omega$, where m is the magnitude of the azimuthal order of a seismic mode; modern measurement error of mean dipole-mode splitting, for example, is approximately 1 % (Howe 2009), which offers some idea of the precision to which Ω can be inferred. Furthermore, the dynamical relation between Ω, which is inferred directly from the frequency splitting, and J_2 is not influenced significantly by non-dynamical variables such as radiative intensity. From even the first primitive measurement of the interior angular velocity (Duvall et al. 1984) it was evident

[7] The oblateness Δ_v depends also on J_4 and the higher moments, but the additional contributions appear to be less than the observational uncertainty, so for clarity I do not take them explicitly into account here.

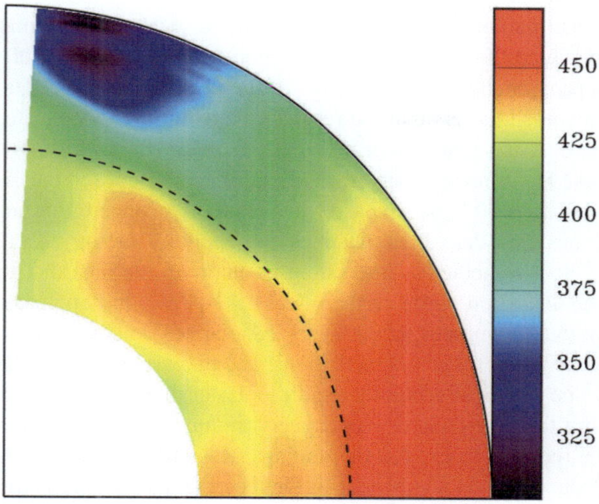

Fig. 4 Colour rendering of the rotation rate of a meridional quadrant of the Sun. The scale is in nHz. The continuous *quarter circle* is located in the photosphere, the *dashed quarter circle* at the base of the convection zone. The *white area* denotes the region in which the rotational splitting data provide only scant indication of the angular velocity (from the MDI Image Gallery: http://soi.stanford.edu/results/Triana/modslice.ps, constructed from the work of Schou et al. 1998)

that the prediction of the non-Newtonian component of the precession of the perihelion of the orbit of Mercury is consistent with General Relativity. Subsequent measurements, such as those reported by Schou et al. (1998), tightened that conclusion, and will in future provide more stringent tests of theories of gravitation once finer orbital measurements become available.

It should be noted that the only unambiguous seismic signature of rotation comes from inertial (such as Coriolis) effects on the nonaxisymmetric ($m > 0$) modes, which are tiny in the evanescent region of those modes: namely, near the axis and, doubly, near the centre (because the degree l cannot be less than m), which explains the white region in Fig. 4. That region coincides, accidently, with one of the regions in which the oblateness kernel (whose spherical-average component is illustrated in Fig. 3) is small, because the centrifugal force is small. Therefore the principal uncertainty in the inference of Ω contributes little to the uncertainty in the value of Δ_Φ that is derived.

4 A Dynamical Issue Raised by Recent Oblateness Measurements

Taken at face value the most recent direct measurements raise an interesting dynamical question. After doing their best to remove facular contamination by rejecting excessively bright regions at the solar limb, which are equatorially concentrated and therefore contribute positively to the oblateness of the intensity distribution, Kuhn et al. (2012) found a residual visual oblateness that is actually less than Δ_Ω by about 8×10^{-7}. The observations by Fivian et al. (2008) led to a similar, although less extreme, result.[8] So is the Sun gravitationally prolate? The only way that that could be is probably for the radiative interior to be constricted by a predominantly toroidal belt of magnetic field (with an associated poloidal component, for stability) encircling the rotation axis, of intensity that would need to be 10^4 T or more. Is that possible? What are the alternatives?

Of course there is always the possibility that the observations were inadequately analyzed. Kuhn et al. explicitly rejected bright facular regions. Did they take adequate care to

[8]In their paper, Fivian et al. (2008) actually claim a value that exceeds Δ_Ω by about 2×10^{-7}, which they offer as an estimate of Δ_Φ, but their result was obtained with Dicke's (1970) outdated underestimate of Δ_Ω.

reject darkened, presumably penumbral regions whose inappropriate inclusion could have rendered the radiative intensity more prolate? Fivian et al. accounted for faculae by using a proxy indicator for rejecting data and extrapolating Δ_v to 100 % rejection; there is a fear in some quarters that the resulting statistics are inadequate for rendering the outcome reliable. A naïve theorist insufficiently familiar with the details of the observations might simply take the difference between the two inferred values of the uncontaminated visual oblateness, about 8×10^{-7}, as a plausible estimate of the uncertainty, and, noticing that it is nearly three times the value of Δ_Φ obtained from helioseismology, set the matter aside.

It is nonetheless interesting to entertain the idea that there is a prolate contribution to the distortion of the surface. As Dicke (1970) has emphasized, that must necessarily occur in the seen photospheric layers. A solar physicist's immediate reaction is likely to be to invoke a superficial magnetic field. The photospheric mean (spatially smoothed) field appears to be principally poloidal and dipolar at most times, and of insufficient intensity to maintain the pole-equator photospheric radius difference of the order of the $10^{-6}R$ that would be required to explain the findings of Kuhn et al. (2012). Moreover, it would enhance, not reduce, the visual oblateness. On the other hand, a preferential suppression of polar convection by only a mere 10^{-3} per cent, which is arguably more plausible, would be sufficient to induce the required reduction in oblateness if the local static adjustment of the star in only the vertical direction were to matter. What would be the dynamical consequence? The poles would be elevated relative to a surface that is everywhere perpendicular to the combined gravitational and centrifugal forces, and if convective Reynolds stresses acted like a scalar viscosity, which is not uncommonly assumed yet which is almost certainly incorrect, photospheric matter would flow downhill towards the equator, which is contrary to the direction observed. The dynamical problem is evidently not straightforward. Yet it is an interesting and evidently important problem, for, irrespective of the oblateness issue, a cogent explanation of the poleward meridional flow in the outermost layers of the convection zone is lacking (e.g. Toomre and Thompson 2015, this issue).

5 Spin-Down

The solar wind is rotationally coupled to the Sun via a large-scale magnetic field. It is caused to rotate roughly at the solar rate out to about 5 solar radii, thereby removing angular momentum from the Sun; beyond, angular momentum in the wind is more-or-less conserved. Thus the rotation of the outer layers of the Sun must be decelerating, in common with inferences for other stars drawn from observations of rotational spectral-line broadening in open clusters of different ages (Skumanich 1972). An interesting dynamical question that arises from this process is how far into the Sun this deceleration penetrates. Of course we now know the answer from seismological measurement, but the dynamical issues, to some extent, remain.

The matter was debated in the late 1960s after (Dicke 1964) had tried to maintain that the Sun's core is rotating so rapidly (with a period of a little over a day) as to induce a gravitational oblateness of magnitude sufficient to sever the agreement between the observed residual precession of the orbit of Mercury and the prediction of General Relativity. Dicke pointed out that the global viscous diffusion time $\tau_v = R^2/\nu$ (where ν is a mean kinematic viscosity) exceeds even the age of the Universe, and that therefore the Sun's core is rotating essentially at its pre-main-sequence rate. He argued that the shear could be stable to the Richardson criterion, so shear turbulence would not add to the viscous stress, and that Maxwell stresses could be insignificant too. Howard et al. (1967) and Bretherton and

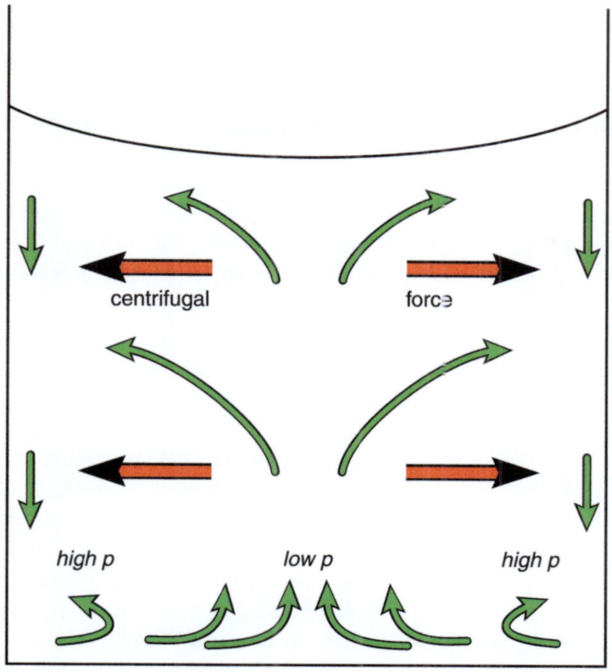

Fig. 5 Spin-down in a stirred cup of tea. The free surface of the fluid is denoted by the *curved black line*, the cylindrical vessel containing the fluid by the *vertical* and *horizontal lines*. The *thicker arrows* (*red*) *with black heads* indicate the centrifugal force acting on the body of the rotating fluid; that force is essentially absent from the Ekman boundary layer near the base of the container, which hardly rotates. The *thinner* (*green*) *arrows* indicate the direction of the induced meridional flow, which, in the main body of the fluid, is essentially inviscid and therefore angular-momentum conserving. Angular momentum is lost from the fluid principally by viscous transfer in the thin Ekman boundary layer. Such a boundary layer is not essential to the global mechanism of spin-down, however: as Bretherton and Spiegel (1968) have argued, penetration of the meridional flow into the Sun's convection zone could provide an (even more) effective mechanism for divesting angular momentum. That process can be modelled on Earth by spreading a layer of heavy beads over the bottom of the teacup

Spiegel (1968) pointed out a fundamental flaw in Dicke's claim, using simple dynamical models to demonstrate that meridional advection, and not viscous stress, is likely to be the dominant agent transporting angular momentum through the body of the Sun, in a process called spin-down (Greenspan and Howard 1963), as in a stirred cup of tea.

Interestingly, the spin-down process had been discussed qualitatively long before by Einstein (1926) as an explanation of the meanders of rivers (unwittingly defending his theory of General Relativity), and quantitatively by Bondi and Lyttleton (1948) in a discussion of the deceleration of the Earth's core. I describe the broad principles briefly here, because, as I shall mention later, the general conclusion may find astrophysical application elsewhere.

Consider a stationary cylindrical vessel of radius R containing water initially rotating (approximately) uniformly with angular velocity Ω about its vertical axis, a situation discussed, for example, by Greenspan (1969). The situation is illustrated in Fig. 5. The fatter (red), horizontal, double arrows represent the centrifugal force, which is balanced by a pressure gradient directed away from the axis. The excess pressure Δp near the outside wall of the container supports a greater head of water, and therefore the upper, free, surface of the water is concave upwards. Surfaces of constant pressure are not normal to gravity, a situation known as baroclinicity. Viscous stresses slow the rotation of the water near where it touches the container, particularly at the bottom, thereby reducing the centrifugal force. That leaves an unbalanced component of the pressure gradient which drives a spiralling inward flow with characteristic speed u, indicated in Fig. 5 by the thinner (green) arrows, in a thin boundary layer (of characteristic thickness $\delta \ll R$), now called an Ekman layer (after Ekman 1905), near the bottom of the container. It is there that angular momentum is removed from the water.[9] In the boundary layer the Coriolis force (in a frame rotating with the bulk of the fluid) is balanced by the viscous stress in the shear: $\Omega u \simeq \nu\delta^{-2}u$. The radial pressure gradient, of order $\Delta p/R$, is also balanced by viscous stress, which determines the boundary-layer velocity: $\Delta p/R \simeq R\Omega^2 \simeq \nu u/\delta^2$. The return flow—upwards and outwards in the bulk of the fluid where viscous stresses are negligible, with characteristic velocity $w \simeq u\delta/R$—slows the rotation in the bulk of the fluid mainly by angular-momentum conservation, on the timescale $\tau \simeq R/w$. This spin-down (equilibration) time is thus $\tau \simeq R^2/\delta u \simeq R/\sqrt{\nu\Omega}$; it is the geometric mean of the global viscous diffusion time $\tau_v = R^2/\nu$ and the characteristic dynamical time Ω^{-1}, which is much shorter than the viscous time. More accurate formulae for δ and τ can be obtained by a local analysis of the boundary layer, which relates the vertical shear in the angular velocity to the upflow velocity out of the Ekman layer (Greenspan 1969) consequent on what is now known as gyroscopic pumping.

An important observation of the process is that the flow near the bottom of the container is inwards, towards the central axis. That is why loose tealeaves in a stirred cup of tea migrate to the middle of the cup, and silt at the bottom of a slowly flowing, meandering, river is transported from the outer to the inner bank of a bend, accentuating the meander (Einstein 1926). It is evident from the balance of forces that if instead the container were caused to rotate more rapidly than the fluid, a flow with the same geometry would be induced, but with its direction reversed.

The principal importance of this process for spin-down is not in the details of the Ekman layer, but simply in the fact that the layer extracts angular momentum from the fluid, leaving an unbalanced pressure gradient within the layer to drive a large-scale flow towards (or away from, were the teacup to be rotating faster than the tea) the axis. I have included the account of the boundary layer here, however, simply to complete my discussion of some interesting physics.

Returning now to the discussion of the Sun, one must first beware, quite generally, of attributing the properties of oversimplified models to complicated situations without careful consideration of the implications of those simplifications, especially when dynamics is involved. Good models might exhibit some of the physical processes in operation, but it must always be appreciated that they may be no more than merely illustrative. It is therefore not uncommon for arguments based on the understanding gleaned from them to be challenged, in disbelief of the extension of the domain of applicability necessarily required for addressing the matter in hand.

For example, Dicke (1967) pointed out that the Sun is no cup of tea, and that the base of the convection zone does not exert a stress in the manner of the base (or top) of a rigid container to produce a diffusive Ekman layer. Bretherton and Spiegel (1968) countered by

[9] Viscous stress operates also on the side walls, but there the boundary layer is not as thin as that at the bottom of the container, and removes negligible angular momentum (just as Bondi and Lyttleton had found, in the case of spin-down in a sphere, that negligible angular momentum is removed near the equator). The bottom boundary layer is thinner as a result of the vertical 'rigidity' imparted on the fluid immediately above by the vortex stretching (which is intimately related to a tendency towards local angular-momentum conservation) produced by the shear, and which is also responsible for the better-known Taylor-Proudman theorem for steady incompressible inviscid flow.

explaining that an Ekman layer is not essential: the pertinent agent is the induced meridional flow which advects angular momentum essentially inviscidly throughout the body of the fluid; that flow can penetrate into the convection zone where it can divest its angular momentum even more efficiently than in a viscous boundary layer. They illustrated the process by modelling the convection zone as a rigid porous medium.[10] They modelled the radiative zone as an incompressible fluid, since the details of how exactly angular momentum is transported in the essentially inviscid radiative interior, be it compressible or not, is of relatively minor importance.

Dicke also tried arguing that the stable stratification of the radiative interior performs the crucial role of inhibiting large-scale, angular-momentum-advecting flow—as indeed had Howard et al. (1967) recognized already—confining it to a shallow Holton (1965a,b) layer immediately beneath the convection zone, thereby insulating the core from the deceleration of the surface (a conclusion which is valid only on timescales shorter than the shortest diffusion—here thermal—timescale); McDonald and Dicke (1967) illustrated the process experimentally, concluding that the stratification of the Sun should preclude core spin-down. Clark et al. (1969) and Modisette and Novotny (1969) concurred. Apparently oblivious of the arguments that had been presented by Howard et al. (1967), Dicke had failed to recognize that actually the impediment is thermally moderated—negative buoyancy generated by vertical adiabatic motion can be annulled by thermal diffusion—so one must estimate the buoyancy annihilation time in order to assess the validity of the simplified model. Roxburgh (1964) had already pointed out the importance of thermal diffusion, positing that the diffusively moderated spin-down time is the (thermally driven) Eddington-Sweet timescale, which is greater than the age of the Sun. But Howard, Moore and Spiegel considered, more realistically, the thermal control of the (dynamically driven) flow, estimating the spin-down time to be at most of order only 10^9 yr. To fluid dynamicists at the time of the 1960s debate, the balance of the evidence seemed to favour substantial global spindown,[11] but the case had not yet been proven. Subsequent analysis by Spiegel and Zahn (1992) of a similar, though not identical, situation relating to the seismologically observed near-uniform rotation of the radiative interior in the face of a differentially rotating convection zone, to which I turn my attention in the next section, has added substantial dynamical support to the evidence. And, of course, the seismological findings themselves negate the view that purported weakness of spin-down has enabled the Sun to have sustained a high gravitational oblateness, although they do not tell us how.

The situation is changed dramatically once it is recognized that the radiative envelope might be pervaded by a large-scale magnetic field. Mestel and Weiss (1987; see also Mestel 1953, 2012 and Cowling 1976) have argued that a quite modest poloidal field (about 5×10^{-6} T) is sufficient to connect the core to the convection zone in the lifetime of the Sun, and have advanced arguments to suggest that the actual field could be of order 10^{-2} T. That field is substantially weaker than the only *ab initio*, albeit poor, estimate—of order 30 T—(Gough 1990a) that I know. The magnetic spin-down process was subsequently studied numerically by Charbonneau and MacGregor (1992, 1993a,b) with an idealized model having a rigidly rotating convection zone. That study confirmed that magnetic spin-down is efficient, but the model was too simplistic to explain why the radiative interior rotates uniformly, an issue to which I now turn.

[10] And subsequently demonstrated it in the laboratory (unpublished) in a rotating beaker of water containing several layers of glass beads.

[11] Perhaps even in the face of the steep gradient of molecular weight (Huppert and Spiegel 1977) which had previously been regarded as isolating the angular momentum of the energy-generating core (Mestel 1953; Mestel 1957).

6 The Steady Laminar Tachocline

The angular velocity of the convection zone varies with latitude in an almost depth-independent manner, and interfaces with a nearly uniformly rotating radiative interior via the thin tachocline. How is that achieved? The matter was first seriously considered by Spiegel and Zahn (1992; see also Spiegel 1972). They, like I here, did not discuss the cause of differential rotation of the convection zone: it is evidently the result of a balance between the angular-momentum-transporting Reynolds stress, Maxwell stress and advection by large-scale meridional flow, a matter which is reviewed briefly by Hanasoge et al. (2015) in this issue. Instead, recognizing that the global equilibration timescale of convection is no doubt much shorter than the dynamical timescales of processes operating beneath (even if they are related to the solar cycle), Spiegel and Zahn considered the convection-zone shear to be given, and ignored any back-reaction of the tachocline dynamics on the convection zone. They then asked why the bulk of the fluid below rotates uniformly.

Spiegel and Zahn first established that although the radiative interior is very highly stratified, thermal diffusion is sufficiently efficacious to permit baroclinically driven flow (induced in a manner similar to that in my spin-down discussion of the previous section) to advect the required angular momentum essentially throughout the Sun in its lifetime. Therefore, the (imposed) differential rotation of the convection zone has to be insulated from the radiative interior to allow the latter to rotate uniformly. To achieve that, Spiegel and Zahn invoked a tachocline pervaded by turbulence generated by the tachocline shear itself, suppressed vertically by the stable stratification and therefore being layerwise two-dimensional. Moreover, they implicitly assumed the turbulence to be horizontally isotropic, notwithstanding the fact that the rms vorticity of the turbulence is likely to be comparable with, or perhaps even less than, the angular velocity of the Sun, a situation which one would naturally expect to lead to substantial anisotropy. Thus they introduced a (turbulent) viscous stress tending to force the rotational flow towards being horizontally shear-free, and were thus able to achieve a steady state with a thin tachocline abutting a uniformly rotating interior, having angular velocity Ω_c. In the tachocline itself there was a flow similar to that in the Ekman layer in Fig. 5: towards the axis of rotation—i.e. poleward, on essentially horizontal surfaces—in an equatorial region where the tachocline rotation exceeds that of the interior, and equatorward in polar regions. Spiegel and Zahn estimated that the two regions meet at latitude 42°. That implied that Ω_c is essentially the value of the photospheric angular velocity at that latitude, namely $0.90\Omega_e$, where Ω_e is the photospheric angular velocity at the equator.

The study stimulated a potentially interesting numerical simulation by Miesch (2003), who set out to determine, amongst other matters, whether the shear turbulence would actually be likely to induce the rigidity produced by a scalar viscosity. He generated turbulence by imposing a grid of point sources of (gravity waves) in a stably stratified fluid under gravity. The grid was forced to rotate rigidly, and therefore the resulting mean flow inevitably tended towards rigid rotation too, in just the same way that gravity waves generated by wind over mountains exerts a drag on the wind (in the frame of reference in which the mountains are stationary). The conclusion that the turbulence leads to rigid rotation was therefore unjustified. What would be much more revealing is to simulate a situation having the sources move with the flow, so that no external torque is applied to the fluid.

Spiegel and Zahn's study has no doubt introduced much of the pertinent dynamics of the tachocline. However, the details—indeed even the basic principle—have come under criticism. For example, it is unheard of that continuously maintained shear-generated turbulence can completely annul the shear that drives it. Indeed, an investigation by Elliott (1997) demonstrated that even the magnitude of the turbulent stress generated would be insufficient for requirements. Moreover, in the observed natural environment, predominantly the Earth's (rotating) atmosphere, layerwise two-dimensional turbulence leads to augmentation, rather than suppression, of larger-scale shear, partly through angular-momentum transport by waves (Haynes et al. 1991). Such considerations led Gough and McIntyre (1998) to conclude that the angular velocity of the radiative interior could never by kept uniform by fluid-dynamical processes alone, and that the interior must necessarily be held rigid by a large-scale magnetic field, presumably primordial. They outlined a nonlinear dynamical balance in what they regarded as the simplest model of the tachocline; it again led to a meridional flow, both geometrically and dynamically similar to that inferred by Spiegel and Zahn, associated with which is downflow from the convection zone in both polar and equatorial regions, and upflow between, near the latitude of zero shear, observed seismologically to be about 30° (see Fig. 4), and implying that $\Omega_c \simeq 0.93\Omega_e$ (cf. Fig. 2). Advection of magnetic flux by the downflow counters the tendency for the field to expand by diffusion, yielding a steady-state balance which determines the thickness of the tachocline. Near the shear-free latitude the outflow might lift the primordial field into the convection zone, anchoring the angular velocity in the convection zone to that in the radiative interior, and possibly fuelling the magnetohydrodynamical processes responsible for the solar cycle (Byington et al. 2014). An obvious consequence of this picture is that in this region the shear might actually be quenched completely by the penetrating magnetic field, a feature that in principle is testable seismologically. It is interesting to note that the latitude of this region is the same as that at which sunspots first appear at the start of a solar cycle. It would be an unlikely coincidence if that had no dynamical significance.

Of course, the potential role of a large-scale magnetic field holding the radiative interior rigid had been discussed before. I have already mentioned the work of Mestel and Weiss (1987) and Charbonneau and MacGregor (1992, 1993a,b) in connection with spin-down. Additionally, Rüdiger and Kitchatinov (1997) imagined the presence of such a field, without careful regard to the direct action on it of the differentially rotation convection zone, and proposed the tachocline to be just a simple Hartmann layer (e.g. Hartmann 1937; Shercliff 1965; Roberts 1967) as an explanation of why it is so thin. What Gough and McIntyre argued is that the magnetic field with a poloidal component must necessarily be present in the radiative zone, and they provided an albeit approximate analysis of how it is prevented from crossing (most of) the tachocline by gyroscopic pumping from the convection zone, thereby disenabling it from imparting on the radiative interior the latitudinal rotational shear in the convection zone. Some aspects of the analysis were subsequently supported by numerical simulations carried out by Kitchatinov and Rüdiger (2006) and Rüdiger and Kitchatinov (2007), in particular the advection domination by the meridional flow, although the models differ in their details. Rüdiger and Kitchatinov assumed a meridional flow much more rapid than the value derived by Gough and McIntyre, and produced a tachocline above an essentially uniformly rotating interior with a less intense field. An essential feature of both descriptions is the absence of Maxwell stress at the base of the convection zone. There are more complicated pictures in which the tachocline is magnetized and turbulent (e.g. McIntyre 2007; Diamond et al. 2007), a result of the nonlinear breakup of the flow resulting from the magnetorotational instability (Velikhov 1959; Chandrasekhar 1960; Balbus and Hawley 1991) or the Tayler instability (Tayler 1957, 1973; Markey and Tayler 1973)—see also Spruit (2002), Arlt et al. (2007), Kitchatinov and Rüdiger (2007), Rüdiger and Kitchatinov (2007)—and these are perhaps more representative of reality.

Baroclinic meridional flow in the tachocline is an inevitable consequence of the basic steady-state dynamics of any model in which Maxwell stresses play only a minor role within the body of the tachocline itself. Therefore a prediction of the latitude of the confluence of the poleward and equatorward flows provides an easily testable consequence of the theory, because it determines Ω_c, which has been measured seismologically. That latitude in the Spiegel-Zahn theory is obtained from a linear equation, and is well determined once the latitudinal dependence of the turbulent stress-strain relation is specified (Spiegel and Zahn assume it to be constant, without comment). By contrast, assuming a local linear stress-strain relation across the tachocline, with a coefficient of proportionality that is independent of latitude, yields $\Omega_c \simeq 0.96\Omega_e$ (Gough 1985). The Gough-McIntyre theory is fundamentally nonlinear, and has not yet been worked out in sufficient detail for a prediction to be made. Some simpler toy models have been looked at too (e.g. Garaud and Guervilly 2009), which at least shed some light on the potentially relevant processes: vertical shear, horizontal advection, geometry of Maxwell and Reynolds stresses.

One of the issues that needs to be addressed in the Gough-McIntyre picture is how the configuration could have been established. Numerical simulations by Brun and Zahn (2006) and Strugarek et al. (2011a,b), for example, have failed to reproduce an appropriately confined interior magnetic field, leading the authors to conclude that the theory must be wrong. But the establishment depends critically on the dominant role of field advection, which could not be reproduced by the simulations because, as in all such computations, realistically small diffusion coefficients could not be achieved. Greater success has been achieved in a programme led by Garaud, who was able to reduce diffusion by restricting attention to axisymmetric configurations (a restriction also adopted originally in the discussion by Gough and McIntyre). One of the consequent difficulties that she found initially is that any field penetrating the tachocline at the poles could not readily be advected equatorwards (e.g. Garaud 2002), leaving a large stress on the axis which locked the rotation of the convection zone and the radiative interior, even though an unlocked steady-state solution was subsequently shown to exist (Wood and McIntyre 2011). However, the most recent work (Acevedo-Arreguin et al. 2013) has succeeded for the first time to reproduce a steady two-dimensional numerical configuration consistent with the Gough-McIntyre theory, and provide a simple explanation of why previous numerical investigations had failed.

The assumption of axisymmetry, adopted originally for superficial simplicity, is perhaps also an extremely inhibiting impediment to the success of many of the simulations. Provided other completing influences cannot dominate, an axisymmetric magnetic field, for example, whose axis of symmetry is initially nearly but not exactly aligned with the rotation axis, could perhaps be advected away from the rotation axis by the equatorward meridional flow in the tachocline, because it is on the magnetic axis that the field strength is greatest. The field would be left, maybe, with its axis of symmetry intersecting the tachocline at the latitude of the poleward-equatorward confluence: the latitude of zero shear. There it penetrates into the convection zone, where perhaps it suppresses the tachocline shear in a non-zero range of latitude. A seismological test for such an outcome is currently being developed. Indeed, it is not wholly out of the question that a similar process is at least partially responsible for the obliquity of the orientation of large-scale magnetic fields in earlier-type stars (Gough 2012b). More complicated asymmetric configurations can be envisaged. The outcome would be essentially a steady magnetic field configuration rotating with angular velocity Ω_c, which overall leads to a nonaxisymmetric Sun, even if the field itself, in its appropriate frame of reference, were axisymmetric. Could this be the root of an explanation for the existence of active longitudes? The asymmetry would rotate with the Sun, maintaining its phase. However, long-term phase stability of the active longitudes is not obviously evident in the observations (e.g. Gyenge et al. 2014). Nevertheless, other indicators could be more telling: it is interesting that the sector structure of the solar wind, which might mirror the magnetic asymmetry within the Sun, has maintained its phase for at least four sunspot cycles prior to the mid seventies (Svalgaard and Wilcox 1975). Further, similar, analysis

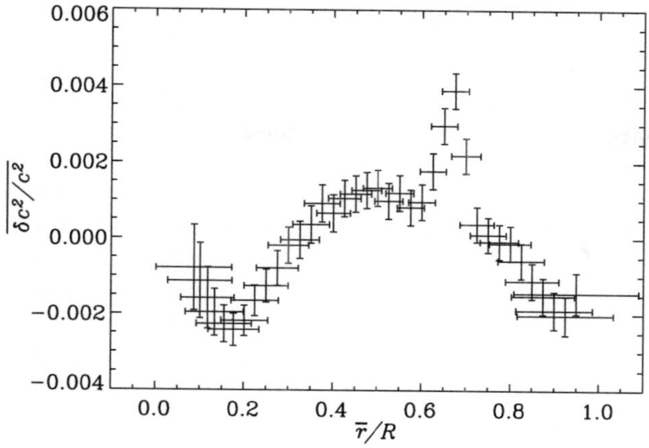

Fig. 6 Optimally localized averages of relative differences between the squared sound speed in the Sun and in Model S of Christensen-Dalsgaard et al. (1996), computed by M. Takata from MDI 360-day data and plotted against the centres $\bar{x} = \bar{r}/R$ of the averaging kernels $A(x; \bar{x})$—which here resemble Gaussian functions—and are defined by $\bar{x} = \int x A^2 \mathrm{d}x / \int A^2 \mathrm{d}x$. The length of each horizontal bar is *twice* the spread s of the corresponding averaging kernel, defined as $s = 12 \int (x - \bar{x})^2 A^2 \mathrm{d}x$; an averaging kernel A that is well represented by a Gaussian function of variance Δ^2 has spread approximately $1.7\Delta \simeq 0.72 FWHM$; were it to be a top-hat function, its spread would be the full width, which is why s is so defined. The *vertical bars* extend to ± one standard deviation of the errors, computed from the frequency errors quoted by the observers assuming them to be statistically independent with zero mean. The sharp positive anomaly centred at $\bar{x} \simeq 0.67$ immediately beneath the convection zone is no doubt the consequence of chemical homogenization of the tachocline with the convection zone. The outward decline in $\overline{\delta c^2/c^2}$ in the convection zone is a result of having underestimated the seismic radius of the Sun. The convex variation about $\bar{x} \simeq 0.15$ provides a hint of there having been some large-scale meridional flow in the core (cf. Gough and Kosovichev 1990), which may also be responsible for the low equatorially biassed angular velocity evident in Fig. 2. The broad positive discrepancy in the radiative envelope is not understood

of solar-wind data up to the present day would therefore be very welcome, and could add (or subtract) credence to the hypothesis.

Descriptions of the genre that I have discussed have all led to a ventilation timescale by the baroclinic meridional flow of order 10^6 y. That is long enough for thermal diffusion to establish a radiatively balanced thermal stratification, yet too short to be countered by microscopic diffusion, or gravitational settling of chemical species. Therefore the tachocline is chemically homogeneous with the convection zone, and thus has lower mean molecular mass than it would have had gravitational settling been unopposed. Therefore the sound speed is higher than it would otherwise be, creating the positive anomaly prominent in Fig. 6. Elliott and Gough (1999) used that property to calibrate the thickness Δ of the tachocline,[12] yielding $\Delta \simeq 2.0 \times 10^{-2} R$, which is more precise (and smaller) than seismological measurements of shear, because hydrostatic stratification can be measured more precisely than rotation; however, the outcome depends on the reliability of the value of the diffusion coefficients required for evaluating the extent of gravitational settling, so the estimate may not be as accurate. Indeed, a yet unpublished investigation by Christensen-Dalsgaard and myself has established that the precise form of the tachocline anomaly cannot easily be reproduced by standard spherically symmetrical solar-structure theory, suggesting, perhaps, the pres-

[12] Here I adopt the original Spiegel and Zahn (1992) definition of the tachocline: the gyroscopically pumped shear layer confined to only the stable region beneath the convection zone, despite the etymology of the appellation.

ence of a degree of asphericity of the tachocline, in contradiction to an earlier finding of Basu and Antia (2001). From an assessment of the horizontal balance of forces, it is inconceivable that the base of the tachocline is both steady and aspherical, although the transition between the stably stratified tachocline and the unstably stratified convection zone, aside from the inevitable convective buffeting, could be.

Finally, a word about the overall shear near the base of the convection zone. Most seismological investigations of the tachocline have—quite naturally, given its appellation—considered only the shear itself, and have tried to characterize its thickness by fitting to the ill resolved inversions for Ω a chosen functional form with a quantifiable width (e.g. Kosovichev 1996). Not surprisingly, the values found tended to exceed that determined from the sound-speed anomaly (Kosovichev 1996; Basu 1997; Antia et al. 1998; Corbard et al. 1998; Charbonneau et al. 1999), because, in accord with the original definition, the true tachocline shear exists only in the stably stratified interior, although Corbard et al. (1999) suggested subsequently that Δ might be as low as $0.01R$. Almost all theoretical studies to date have ignored the reaction of the convection zone to the tachocline shear, which must penetrate to some degree into the convection zone, especially at high latitudes where vortex stretching is the greatest. Of course, some upward penetration of the vertical shear must occur, and is indeed clearly visible in the seismological inferences (Fig. 2), especially in the polar regions where vertical shear is resisted the most strongly by vortex stretching. It is that penetration that has led to the conclusion by Kosovichev (1996), Antia et al. (1998), Charbonneau et al. (1999) and Basu and Antia (2003) that the tachocline is thicker at the poles. A recent detailed helioseismic study of the shear, including its temporal variations, has been presented by Antia and Basu (2011).

7 The Temporally Varying Tachocline

The description of the tachocline outlined in the previous section is dominated by essentially laminar dynamics, superposed upon which there might be some weak small-scale turbulence providing additional, diffusive, transport. Moreover, the bulk of the tachocline is usually considered to be free of magnetic field, except in the upwelling region of near-zero shear. But there are more complicated situations that have been considered, and which suffer temporal variation on a timescale much shorter than the 10^6-year ventilation time mentioned above.

The most obvious time variation to consider is the buffeting by large-scale plumes from the convection zone. This is likely to cause the stability boundary—i.e. the surface across which the (local) convective stability changes—to undulate, in a manner similar to the tops (and bottoms) of terrestrial clouds (which, it should be pointed out, are not demarcated precisely by the visible vapour interface). Beneath that interface are tight ripples—gravity waves predominantly with timescales and horizontal lengthscales comparable with those of the buffeting—because the stable stratification is very much more intense than the unstable stratification above, with vertical lengthscales very much less than the horizontal scales. The group velocity is very nearly horizontal, so the ripples penetrate only a very short vertical distance before they dissipate (e.g. Gough 1977a). The undulations are usually called overshooting, but they do not necessarily induce as much mixing as is often presumed for causing the region of nearly adiabatic stratification to penetrate downwards and terminate in a yet sharper interface (e.g. Christensen-Dalsgaard et al. 2011). Instead they induce a smoothing of the horizontally averaged stratification (to which global seismology is sensitive). There is, however, a small degree of material mixing resulting from minor disruptions to the interface. The gravity waves themselves also contribute to material transport by (nonlinear)

Taylor (1953, 1954) dispersion, though on a much smaller vertical scale (e.g. Press and Rybicki 1981; Knobloch and Merryfield 1992). It is likely that the pressure perturbations associated with the convective plumes are transmitted to the tachopause—the base of the tachocline—causing undulation in the boundary beneath. A consequence of the smoothing of the horizontally averaged stratification near the base of the convection zone is a reduction in the magnitude of the (negative) leading constant in the asymptotic formula for the periods of high-order g modes of low degree (Ellis 1984), which could be detectable should the modes be observed. At present there is no hope of a direct observation. However, indirect methods, such as seeking g-mode-induced undulations in p modes, the dynamics of which I refrain from discussing here, may one day be successful.

The simple Gough-McIntyre description of the tachocline is magnetic-field free, except where it interfaces with the primordial field in the interior and also near the almost shear-free region where the tachocline flow is upwelling. There is undoubtedly also a magnetic field in the convection zone above, vacillating with the solar cycle with a characteristic 22-year period. If the temporally averaged field were strictly zero, the field would not penetrate far into the tachocline (e.g. Garaud 1999), which is why in their most elementary description the possible existence of such a field was ignored by Gough and McIntyre. But isn't it more likely that there is a randomness in the vacillation that leaves a nonzero residue? The residual field crossing the tachocline would be wound up by the tachocline shear. The situation would be ripe for Tayler-type instability, as Spruit (1999, 2002) has advocated (see also Diamond et al. 2007). It would lead to additional transport by the ensuing turbulence, and it would modify the relation between the tachocline thickness and the large-scale interior magnetic field (McIntyre 2007). I think it is fair to say that this very complex subject is still ill understood. It remains an area of active research, attracting not only those who regard themselves as solar physicists. For more information I refer the reader especially to the book on the tachocline edited by Hughes et al. (2007) and to subsequent publications by the contributing authors.

It has even been proposed that the residual vacillating field leaking from the convection zone is alone sufficient to render the rotation of the radiative interior uniform (Forgács-Dajka and Petrovay 2001, 2002), without recourse to requiring the rigidity of a large-scale field. The idea seems to demand that the leaking field that is generated by dynamic action in the differentially rotating convection zone somehow takes on a rigid configuration. What must surely be the case is that the field takes on at least part of the convection-zone shear, which is then transmitted through the tachocline. So far as I am aware, not even a highly idealized model of the pertinent dynamics has been fully investigated. However, if one imagines a toy model in which a periodically oscillating source of field with zero mean generated in a convection zone with high, turbulent, scalar magnetic diffusivity rotating at the observed differential rate, and diffusing into the radiative interior below, where the diffusivity is very much lower, then the temporally averaged Maxwell stress on the interior does not vanish, and is such as to generate differential rotation in the radiative envelope of the same form, although not necessarily with the same amplitude, as that in the convection zone. Solar-cycle-related time dependence is also not out of the question. Antia et al. (2013) have embarked upon a programme to study potential dynamical consequences of the seismologically determined variations of the angular velocity in the convection zone. It is the only truly dynamical study that stems directly from helioseismological inference. The magnetic field required to produce the seismically observed angular-velocity variations was obtained under the assumption that the only azimuthal force is solely a Maxwell stress, and in this initial study only the azimuthal balance of forces was considered. The analysis is therefore far from complete. However, a magnetic field was obtained which oscillates not only in the body of the convection zone but also in the tachocline. That result therefore reinforces the opinion that the early, steady, tachocline studies were grossly oversimplified.

I conclude this section by mentioning the so-called tachocline oscillation, an apparently wave-like oscillation in the angular velocity near the equatorial plane immediately above and below the base of the convection zone, with a period of about 1.3 years. It was discovered seismologically by Howe et al. (2000), with opposite phase in the convection zone from that at the top of the radiative interior, thus exhibiting a maximum of the shear amplitude in the tachocline. Soon afterwards the oscillation almost disappeared (Howe et al. 2011); the eye of faith of an imaginative observer might perceive hints of its return (see Howe 2009), although the evidence is not statistically significant in any plausible sense. What is the restoring force? Howe et al. suggested a magnetic field, which is quite plausible. Indeed, there is (unpublished) evidence that the oscillation extends more deeply than has been reported in the literature, with a further maximum in the shear at about $r = 0.55R$, but with rather lower amplitude, as one would expect if the field does not increase with depth faster than $\sqrt{\rho}$. It is interesting, although perhaps merely coincidental, that the intensity of the vertical component of the field required to produce such an oscillation would need to be about 1 T, which is similar to, yet rather less than, the rough *ab initio* estimate of the global field in the radiative zone.[13] Howe and her colleagues (personal communication) consider the evidence for the deep penetration of the oscillation to be too insecure for them to report, but in trying to understand the workings of the Sun it is useful for theorists at least to bear in mind the possibility. What might drive the oscillation? One possibility is its influence on the anisotropic dissipation of gravity waves, a process which drives the terrestrial QBO, to which I turn my attention later.

8 Excitation of Seismic Modes

It is generally agreed that in the Sun acoustic oscillation modes are intrinsically stable, their energy being absorbed into the background configuration of the star by the combined action of appropriately phased heat and momentum transport by convection and, to a lesser degree, radiative heat transfer in the convectively stably stratified interior. The modes are generated stochastically by the turbulence in the upper layers of the convection zone, by both random impulses from the turbulent motion and by the associated fluctuations in buoyancy. There has been a sequence of more-and-more sophisticated analyses of the processes involved (e.g. Stein 1967; Goldreich and Keeley 1977; Balmforth and Gough 1990; Balmforth 1992; Belkacem et al. 2009; Chaplin et al. 2005). They derive initially from the work of Batchelor (1953) and Lighthill (1952, 1954), who considered respectively the stochastic excitation of a (general) simple harmonic oscillator and the mechanism of turbulent acoustic wave generation by momentum transfer—also of Gough (1965, 1977a), Unno (1967), Xiong (1989), Gabriel (1996), who addressed the role of convection in determining the linear growth and damping rates of stellar oscillations (see also Dupret et al. 2005, 2006, 2009). The absolute intensity of the nonlinear excitation is extremely difficult to quantify, because it is extremely sensitive to details of the turbulence, which are not well defined by the mixing-length formulations that are adopted to represent the solar convection zone (Gough 1977a, 2002). Therefore the absolute amplitudes of the seismic waves are very uncertain. Nevertheless,

[13] And similar also to field intensities commonly discussed elsewhere in connection with the tachocline (e.g. Spruit 2002; Gilman and Cally 2007), although others (e.g. Gough and McIntyre 1998; Kitchatinov and Rüdiger 2007) have entertained weaker fields.

an appropriate combination of the uncertain factors in the theory can be adjusted at least to harmonize with seismic observations; that leads to a functional form of amplitude against frequency—a true prediction of the theory—which is in reasonable accord with observation (Houdek et al. 1999; Chaplin et al. 2005). It is interesting, and perhaps somewhat surprising, that, once the adjustment has been made, the remaining weakest part of the theory, at least when applied to the Sun, then appears to be the prediction of the linear damping rates of the modes; that inference was obtained by replacing the theoretical damping rates by observed acoustic power-spectrum line widths, and finding substantial amelioration (Chaplin et al. 2005).

Further work on the convection-pulsation interaction is therefore called for, not merely to make minor improvements to derivations of mode amplitudes in the Sun, but, more importantly, to improve our understanding of the underlying dynamics, both for its own sake and for application to other stars for which the gross solar adjustment of the theory may no longer be appropriate.

Observations of solar-like observations in other stars add new data with which to compare the theory. Particularly useful are stars rather different from the Sun. An observation of the mode amplitudes in the (coolish) more evolved star ξ Hydrae, for example, has been well predicted (Houdek and Gough 2002), although subsequent direct estimates of the damping rates (Stello et al. 2006) were not in accord with the theory, suggesting, perhaps, the presence of cancelling errors in the theoretical growth and excitation rates. Before jumping to that conclusion, however, one must recognize that the damping rates were obtained from the observations by equating them with the line widths in the power spectrum, which is a correct procedure only if it is known that the background state of the star is truly invariant over the duration of the observations (Batchelor 1953); otherwise non-dissipative phase (frequency) wandering can broaden the spectral lines and thereby falsify the conclusion. Indeed, subsequent analysis of high-quality data from the space missions CoRoT and Kepler (Huber et al. 2010; Baudin et al. 2011) yielded values in accord with the theoretical predictions.

The situation is much worse with the hot star Procyon, whose oscillation amplitudes are grossly overestimated (Houdek et al. 1999; Matthews et al. 2004a,b). In this case there must be something fundamentally wrong with how one describes either how the turbulence properties scale with variations of stellar structure, or, worse still, the manner in which they combine to quantify the excitation; alternatively, there might be some missing ingredient in the calculation of the damping rates, such as wave scattering or some process unaccounted for in the time-dependent convection theory.

9 On the Dynamics of Gravity Modes

In an attempt to resolve the solar neutrino problem before helioseismology, Dilke and Gough (1972) suggested that intermittent nonlinear breakdown of a grave unstable gravity mode would trigger finite-amplitude convection that almost completely mixes the core. That suggestion is now known not to be correct, because the form of the depression of the sound-speed in the central regions of the Sun, evident in Fig. 1, incontrovertibly demonstrates the existence of a substantial gradient of helium abundance (Gough and Kosovichev 1990). Moreover, an important analysis by Dziembowski (1982) of triad coupling to a resonating pair of otherwise damped daughter modes suggested that it is unlikely that the principal unstable mode would ever achieve an amplitude high enough to overcome the barrier to the nonlinear onset of convection. Nevertheless, the dynamical processes that were thought to operate are not necessarily irrelevant today, and can perhaps be profitably rehearsed. The driving of grave solar g modes is principally by the modulation of the nuclear reactions, a mechanism understood first by Eddington (1926), and now called the ϵ mechanism. Extreme sensitivity of the nuclear reaction rates to temperature causes (local) temperature excesses to be augmented in the face of (global) losses by radiative diffusion. Key to the process in cool stars like the Sun, which are powered by the relatively insensitive p-p reaction chain, is the destruction of nuclear balance by the oscillations, which vary on the timescale of an hour, much shorter than the 10^5-year nuclear equilibration time, exposing the variation of the energy generated by the temperature-sensitive ^{3}He–^{3}He, and the ^{3}He–^{4}He, nuclear reactions (cf. Ledoux and Sauvenier-Goffin 1950). Also important is that the g-mode-depressing convection zone is deep compared with those in hotter stars, preventing the mode from achieving so great an amplitude in the outer stellar envelope as to suffer sufficient damping to overcome the nuclear driving. Consistent calculations by Christensen-Dalsgaard et al. (1974), Shibahashi et al. (1975), Boury et al. (1975) and Saio (1980) confirmed the original estimate that some g modes were indeed unstable.

The analysis by Dziembowski (1982) is particularly interesting dynamically, because the outcome is superficially counterintuitive. It is well known that a pair of daughter gravity modes can sap energy from their parent, provided that they resonate precisely enough. What is less well appreciated are the conditions required for that process to be severely debilitating. When I was a student at the Program of Geophysical Fluid Dynamics (GFD) at the Woods Hole Oceanographic Institution in 1965 my attention was caught by the daily water being heated for coffee in an aluminium pan on a flat electric hob. The bottom of the pan was rounded with age, and the pan rocked on the heated plate. If the quantity of water was right, there was a resonance between the rocking frequency and the frequency of surface gravity modes in the water, and the differently located bursts of heat through the bottom of the pan as different parts of it touched the hot plate caused the oscillation to be sustained. Eddington's ϵ mechanism was operating at home. The oscillation amplitude was not tiny, yet surely the intrinsic dissipation—measured by the inverse Reynolds number, for example—was much greater than it is in the Sun. So shouldn't gravity modes in the Sun attain even greater amplitudes? The answer is negative: oddly enough, it is actually because the propensity for dissipation in the GFD pot was so high that the oscillation could be sustained. To understand that one must first realize that to be an effective energy drain the difference between the frequencies of the two daughters must match the frequency of the parent within an inverse intrinsic growth time, a condition which in general becomes harder and harder as the thermal diffusion coefficient, and hence the nonadiabaticity, diminishes. This requires finding daughter pairs amongst modes with higher and higher characteristic wave numbers, which actually leads to higher and higher damping rates, notwithstanding the smaller diffusion coefficient, therefore lower and lower filial amplitudes and consequently an even lesser and lesser ability of the daughters to extract energy from their parent. Only if the dissipation rate (diffusion coefficient) is low enough can the daughters attain high-enough amplitudes for them to be able to limit the growth of the parent effectively.

Dziembowski assumed the background stratification of the Sun would not be changing with time. He found that it was not possible always to find daughter pairs sufficiently close to resonance, yet with adequately low damping rates to permit them to grow to amplitudes high enough to severely inhibit the parent. So he carried out a statistical calculation to estimate the expected amplitude of the parent, and found the amplitude to be a mere $10\ \mathrm{cm\,s^{-1}}$ in the early stages of the Sun's evolution on the main sequence (Dziembowski 1983). That is surely insufficient to cause serious disruption to the structure of the core. Nevertheless, it appeared to permit accidentally much higher amplitudes. What is the possible outcome? That question is answerable once it is realized that the Sun is evolving, and that resonance with particular daughters cannot be maintained. In fact the passage from one resonating pair to another as the Sun evolves typically occurs on a timescale shorter than the growth time of the daughters. Moreover, it turns out that strict formal phase coherence over the time it takes for the parent to reach its limiting amplitude is not required: the parent's phase drifts as the Sun evolves, and as new energy-sapping daughters are encountered their phases are established by the triad interaction itself. Therefore, as reported briefly by Jordinson and Gough (2000), the limiting amplitude of the parent, as its frequency drifts amongst the family of resonances, is simply the expectation value calculated by Dziembowski. A simplifying assumption of the original calculation was that interaction with only a single daughter pair was considered; in practice, however, there are interactions at any time with an entire hierarchy, limiting the amplitude of the parent yet more severely. However, in general it is the gravest pair who sap by far the most energy, and it is likely that the extra loss to all the other daughter pairs is actually quite small, leaving Dziembowski's estimate more-or-less intact.

It appears from this discussion that the structure of the Sun, even the temperature-sensitive ^{8}B neutrino production, is unlikely to have been modified seriously by gravity modes. Indeed, an end to this matter has been called by many astrophysicists who have grappled with neutrinos. However, anyone interested in dynamics might still ask whether there is any small remnant consequence. There has for a long time (cf. Gough and Kosovichev 1990; Gough et al. 1995) been a hint, albeit rather weak, that some degree of material redistribution, possibly laminar, has taken place in the inhomogeneous energy-generating core: this is evinced by the generally negative slope of $\overline{\delta c^2/c^2}$ inwards of $r \simeq 0.2R$, visible in Fig. 6, which appears to imply a radial gradient in the spherically averaged Y that is somewhat lower than theoretical expectation, and which, both in the very early days of helioseismology and also more recently (Basu et al. 2009), has been recognized as being suggestive of mild mixing. Moreover, a general redistribution of angular momentum within the core from an initially uniformly rotating state could make the core appear to be rotating more slowly, as is suggested by Fig. 2, because on the whole the angular velocity near the edge of the core, especially near the equatorial plane, where rotational splitting is most prone to be affected, is likely to be diminished. Could these hints be manifestations of a soliton in the core, controlled either by gravity-wave-like dynamics, or by some more slowly operating dynamics that might be influencing reported variations in the neutrino luminosity (e.g. Sturrock 2008)? A possibly simpler, but less interesting, solution might be that mixing in the early stages of the main-sequence evolution was much more extensive than it is now, because the angular velocity (and especially its deceleration) was much greater then than it is today, possibly causing adequate baroclinicity to drive a laminar circulation through the core when the impedance by chemical inhomogeneity was quite small.

Could there have been significant Taylor dispersion by daughter g modes? These are matters that appear not to have been addressed. One might note that the grave g modes have similar amplitudes both in the core and near the surface, and that Dziembowski's 10 $\mathrm{cm\,s^{-1}}$ estimate is high enough to be observed at the surface today. However, the structure of the Sun today differs substantially from its early main-sequence state, and some calculations suggest that no g mode is now unstable (Boury et al. 1975; Shibahashi et al. 1975), although the matter is a delicate balance between driving and damping processes, leaving open the possibility that some g modes are today intrinsically unstable (Noels et al. 1975; Saio 1980). There have been various claims of direct detection, but these are not universally accepted. I think it is fair to say that there is yet no convincing evidence for their existence at observable amplitudes (Appourchaux et al. 2010). Finding gravity modes would be extremely important for seismology; as has long been appreciated (e.g. Gough 1978; Gabriel et al. 1995), the additional seismological resolving power that the g modes would provide would be quite substantial. Are there modes that are unstable, or are they all intrinsically stable and excited stochastically as pressure fluctuations associated with the undulating tachocline or by the same pressure fluctuations that excite acoustic modes in the outer layers of the convection zone? Estimates of the latter are summarized by Appourchaux et al. (2010). They vary quite widely, although amplitudes are typically lower than the observational claims. A great deal of uncertainty remains.

10 Gravity-Wave Transport

In an early discussion of gravity-wave transport between core and convection zone (Gough 1977a) it was estimated that grave modes of the kind that might be self-excited in the core to amplitudes comparable with those discussed in the previous section are too weak to carry a substantial amount of heat, and too weak also to transfer a substantial proportion of the Sun's angular momentum in its lifetime. Moreover, stable gravity waves that resonate in space and time with the undulating lower boundary of the convection zone penetrate only about 1 km into the radiative interior. A more recent estimate of the latter (Gough 2002) suggests that penetration is about 100 times deeper, but that is still too shallow to have a material impact on the overall structure. It was estimated that modes generated off-resonance would have too low an amplitude to transport a substantial amount of angular momentum. It was also pointed out that any group of gravity waves that do penetrate must dissipate asymmetrically, and are expected to enhance, rather than suppress, differential rotation.

Subsequently Press (1981) and Garcia Lopez and Spruit (1991) estimated the wave amplitudes by balancing inertially generated pressure fluctuations near the base of the convection zone against corresponding resonant gravity-wave fluctuations in the radiative zone beneath. As with the estimates of acoustic wave generation (e.g. Gough 1977a), there is substantial uncertainty arising from uncertainties in the properties of the convective turbulence.

An important consequence of the enhancement of rotational shear by gravity-wave transport was illustrated with a pioneering experiment carried out by Plumb and McEwan (1978), who demonstrated that the necessarily anisotropic wave dissipation can induce global oscillations of the shear, with characteristic period inversely proportional to the fourth power of the amplitude of the gravity waves (and proportional to the sixth power of the horizontal phase speed). The experiment was carried out to confirm a theory of Lindzen and Holton (1968) and Holton and Lindzen (1972) explaining the mechanism of the terrestrial QBO. It is of some interest to summarize the basic process here, because there is misinformation in the literature.

Low-amplitude gravity waves generated with a broad spectrum of frequencies at an essentially horizontal surface propagate upwards (in the case of the Earth's atmosphere) into the sheared stratosphere. The large-scale flow in the stratosphere is essentially horizontal, its variation with time and horizontal distance being much weaker than the frequency and horizontal wavenumber of the waves (and was therefore ignored in discussing the linearized local wave dynamics). Frequency and horizontal wavenumber in an inertial frame of reference are therefore conserved. Given that the gravity-wave dispersion relation is approximately $\omega^2 = N^2 k_h^2/(k_v^2 + k_h^2)$, where ω is the wave frequency in the local frame of the fluid, N is the buoyancy (Brunt-Väïsäla) frequency of the stratification, and k_h and k_v are respectively the horizontal and vertical components of the wave number, it follows that the horizontal component of the group velocity, and the (angular) momentum flux, are in the same direction as the horizontal phase velocity.[14] Also prograde waves, having negatively

[14]The vertical component of the group velocity is directed oppositely to the phase velocity.

Doppler-shifted frequencies ω, have higher k_v than do the corresponding retrograde waves, and therefore dissipate faster, causing the momentum that is imparted on the mean flow to be preferentially in the prograde direction. If prograde and retrograde waves are excited to the same amplitude, the shear is therefore at first enhanced, but further from the excitation layer, where the remnant prograde waves have lower amplitude as a result of earlier higher dissipation, the direction of the wave stress acting on the mean flow is reversed.

An initially static atmosphere into which gravity waves are propagating is (then) susceptible to shear enhancement. If initially there is no mean shear, a random fluctuation in the symmetry of the wave excitation will first produce an initial weak shear which is then amplified by the process described above. The increased shear near the excitation layer eventually loses momentum to the excitation layer, and the whole pattern descends, producing a temporal oscillation in the mean horizontal flow at any given altitude. Evidently the amplitude and period of that oscillation must depend on the excitation rate and the dissipation rate, the latter, in view of the dispersion relation, depending on the horizontal phase velocity. The existence of the oscillation requires the fluid not to be too diffusive, for otherwise the oscillation is damped.

Is there an analogous phenomenon in the Sun? Likely amplitude estimates lead to a characteristic period of about a month (Gough 2002), in a very thin layer within the tachocline into which the waves penetrate before being dissipated. It is most unlikely that any significance can be attributed to the coincidence of that period with the characteristic large-eddy turnover time at the base of the convection zone or with the rotation period of the Sun. In fact there may be no such coincidence, because the amplitudes of the waves generated, and hence the period of the oscillation in the shear generated, are so uncertain. A characteristic period even a hundred times longer is not out of the question. Indeed, Gough (1998) wondered whether Plumb and McEwan's (non-magnetic) mechanism could be responsible for the 22-year solar cycle, leaving the magnetic field to play only a secondary role in the dynamics. The process could also induce meridional flow (Fritts et al. 1998), which might conceivably provide a significant transport mechanism for chemical species, and possibly explain the Li deficiency in the Sun's photosphere. An example of a rather different claim comes from an investigation by Kumar et al. (1999), who estimated an enormously greater gravity-wave amplitude for waves that penetrate more deeply.[15] They too commented that Plumb and McEwan's (non-magnetic) mechanism might be responsible for the 22-year solar cycle. In a subsequent investigation, Talon et al. (2002) argued that shear instability would soften the oscillations. The effect of the instability was modelled with a scalar viscosity which increased the QBO-like oscillation period to about 300 y, and the spectrum of the waves had enough power at low order to permit angular momentum to be redistributed throughout the radiative interior, again supposed to establish an almost uniform rotation profile. How the waves would suppress the imprint of the differential rotation in the convection zone discussed by Spiegel and Zahn (1992) remained unexplained.

The simplest discussions of this process are based on the action of an almost monochromatic group of waves (e.g. Kim and MacGregor 2001, 2003). A more realistic broad spectrum of waves such as would be expected in the Earth's atmosphere (Baldwin et al. 2001) and the Sun can lead to much more complicated behaviour, as Mathis et al. (2008) portend. From a series of numerical simulations, Rogers et al. (2008), Rogers and MacGregor (2010, 2011) and MacGregor and Rogers (2011) have concluded that simple linearized wave

[15] As had both Kumar and Quataert (1997) and Zahn et al. (1997) earlier, in the mistaken belief that angular momentum transport by the waves would be in the reverse sense and would thus supply the reason for the observed uniform rotation of the radiative envelope.

dynamics is inadequate for computing the mean Reynolds stress because (nonlinear) wave-wave interactions are important at the amplitudes they expected to be generated near the lower boundary of the solar convection zone. They argued that oscillatory shear flow of the QBO type is much less likely to be produced—although rigid rotation is not to be expected either, as McIntyre (e.g. 1994, 2003) has often stressed. One might wonder whether in the numerical simulations excessive diffusion requires unrealistically large wave amplitudes, as is possibly the case also of convective motion (Toomre and Thompson 2015, this issue), and that the quasi-linear dynamics of the kind described by Plumb and McEwan (1978) really is a good guide. We await with interest the further, more realistic, simulations that are promised.

References

L.A. Acevedo-Arreguin, P. Garaud, T.S. Wood, Dynamics of the solar tachocline—III. Numerical solutions of the Gough and McIntyre model. Mon. Not. R. Astron. Soc. **434**, 720–741 (2013). doi:10.1093/mnras/stt1065

L. Ambronn, A.C.W. Schur, Die Messungen des Sonnendurchmessers an dem Repsold'schen 6-zoelligen Heliometer der Sternwarte zu Göttingen. Astron. Mitt. Koenig. Sternwarte Göttingen, 7:T (1905)

H.M. Antia, S. Basu, Revisiting the solar tachocline: average properties and temporal variations. Astrophys. J. Lett. **735**, 45 (2011). doi:10.1088/2041-8205/735/2/L45

H.M. Antia, S. Basu, S.M. Chitre, Solar internal rotation rate and the latitudinal variation of the tachocline. Mon. Not. R. Astron. Soc. **298**, 543–556 (1998). doi:10.1046/j.1365-8711.1998.01635.x

H.M. Antia, S.M. Chitre, D.O. Gough, On the magnetic field required for driving the observed angular-velocity variations in the solar convection zone. Mon. Not. R. Astron. Soc. **428**, 470–475 (2013). doi:10.1093/mnras/sts040

T. Appourchaux, K. Belkacem, A.-M. Broomhall, W.J. Chaplin, D.O. Gough, G. Houdek, J. Provost, F. Baudin, P. Boumier, Y. Elsworth, R.A. García, B.N. Andersen, W. Finsterle, C. Fröhlich, A. Gabriel, G. Grec, A. Jiménez, A. Kosovichev, T. Sekii, T. Toutain, S. Turck-Chièze, The quest for the solar g modes. Astron. Astrophys. Rev. **18**, 197–277 (2010). doi:10.1007/s00159-009-0027-z

R. Arlt, A. Sule, G. Rüdiger, Stability of toroidal magnetic fields in the solar tachocline. Astron. Astrophys. **461**, 295–301 (2007). doi:10.1051/0004-6361:20065192

M. Asplund, N. Grevesse, A.J. Sauval, P. Scott, The chemical composition of the Sun. Annu. Rev. Astron. Astrophys. **47**, 481–522 (2009). doi:10.1146/annurev.astro.46.060407.145222

A. Auwers, Der Sonnendurchmesser und der Venusdurchmesser nach den Beobachtungen an den Heliometern der deutschen Venus-Expeditionen. Astron. Nachr. **128**, 361 (1891). doi:10.1002/asna.18911282002

J.N. Bahcall, R.K. Ulrich, Solar models, neutrino experiments, and helioseismology. Rev. Mod. Phys. **60**, 297–372 (1988). doi:10.1103/RevModPhys.60.297

J.N. Bahcall, M.H. Pinsonneault, S. Basu, Solar models: current epoch and time dependences, neutrinos, and helioseismological properties. Astrophys. J. **555**, 990–1012 (2001). doi:10.1086/321493

S.A. Balbus, J.F. Hawley, A powerful local shear instability in weakly magnetized disks. I—Linear analysis. II—Nonlinear evolution. Astrophys. J. **376**, 214–233 (1991). doi:10.1086/170270

M.P. Baldwin, L.J. Gray, T.J. Dunkerton, K. Hamilton, P.H. Haynes, W.J. Randel, J.R. Holton, M.J. Alexander, I. Hirota, T. Horinouchi, D.B.A. Jones, J.S. Kinnersley, C. Marquardt, K. Sato, M. Takahashi, The quasi-biennial oscillation. Rev. Geophys. **39**, 179–229 (2001). doi:10.1029/1999RG000073

N.J. Balmforth, Solar pulsational stability. Part Three—Acoustical excitation by turbulent convection. Mon. Not. R. Astron. Soc. **255**, 639 (1992)

N.J. Balmforth, D.O. Gough, Mixing-length theory and the excitation of solar acoustic oscillations. Sol. Phys. **128**, 161–193 (1990). doi:10.1007/BF00154155

S. Basu, Seismology of the base of the solar convection zone. Mon. Not. R. Astron. Soc. **288**, 572–584 (1997)

S. Basu, H.M. Antia, A study of possible temporal and latitudinal variations in the properties of the solar tachocline. Mon. Not. R. Astron. Soc. **324**, 498–508 (2001). doi:10.1046/j.1365-8711.2001.04364.x

S. Basu, H.M. Antia, Changes in solar dynamics from 1995 to 2002. Astrophys. J. **585**, 553–565 (2003). doi:10.1086/346020

S. Basu, W.J. Chaplin, Y. Elsworth, R. New, A.M. Serenelli, Fresh insights on the structure of the solar core. Astrophys. J. **699**, 1403–1417 (2009). doi:10.1088/0004-637X/699/2/1403

G.K. Batchelor, *The Theory of Homogeneous Turbulence* (Cambridge University Press, Cambridge, 1953)
F. Baudin, C. Barban, K. Belkacem, S. Hekker, T. Morel, R. Samadi, O. Benomar, M.-J. Goupil, F. Carrier, J. Ballot, S. Deheuvels, J. De Ridder, A.P. Hatzes, T. Kallinger, W.W. Weiss, Amplitudes and lifetimes of solar-like oscillations observed by CoRoT (Corrigendum). Red-giant versus main-sequence stars. Astron. Astrophys. **535**, 1 (2011). doi:10.1051/0004-6361/201014037e
K. Belkacem, R. Samadi, M.J. Goupil, M.A. Dupret, A.S. Brun, F. Baudin, Stochastic excitation of nonradial modes. II. Are solar asymptotic gravity modes detectable? Astron. Astrophys. **494**, 191–204 (2009). doi:10.1051/0004-6361:200810827
G. Berthomieu, M. Cribier (eds.), Inside the Sun, in *IAU Colloq. 121: Inside the Sun*, Astrophysics and Space Science Library, vol. 159 (1990)
G. Berthomieu, J. Provost, A. Rocca, A.J. Cooper, D.O. Gough, Y. Osaki, Sensitivity of five minute eigenfrequencies to the structure of the Sun, in *Nonradial and Nonlinear Stellar Pulsation*, ed. by H.A. Hill, W.A. Dziembowski Lecture Notes in Physics, vol. 125 (Springer, Berlin, 1980), pp. 307–312. doi:10.1007/3-540-09994-8-29
H. Bondi, R.A. Lyttleton, On the dynamical theory of the rotation of the Earth. Proc. Camb. Philos. Soc. **44**, 345 (1948). doi:10.1017/S0305004100024361
A. Boury, M. Gabriel, A. Noels, R. Scuflaire, P. Ledoux, Vibrational instability of a 1 solar mass star towards non-radial oscillations. Astron. Astrophys. **41**, 279–285 (1975)
C. Brans, R.H. Dicke, Mach's principle and a relativistic theory of gravitation. Phys. Rev. **124**, 925–935 (1961). doi:10.1103/PhysRev.124.925
F.P. Bretherton, A.E. Spiegel, The effect of the convection zone on solar spin-down. Astrophys. J. Lett. **153**, 77 (1968). doi:10.1086/180224
A.-M. Broomhall, W.J. Chaplin, Y. Elsworth, R. Simoniello, Quasi-biennial variations in helioseismic frequencies: can the source of the variation be localized? Mon. Not. R. Astron. Soc. **420**, 1405–1414 (2012). doi:10.1111/j.1365-2966.2011.20123.x
A.S. Brun, J.-P. Zahn, Magnetic confinement of the solar tachocline. Astron. Astrophys. **457**, 665–674 (2006). doi:10.1051/0004-6361:20053908
A.S. Brun, M.K. Browning, M. Dikpati, H. Hotta, A. Strugarek, Recent advances on solar global magnetism and variability. Space Sci. Rev. (2015). doi:10.1007/s11214-013-0028-0
B.M. Byington, N.H. Brummell, J.M. Stone, D.O. Gough, Stoked nondynamos: sustaining field in magnetically non-closed systems. New J. Phys. **16**(8), 083002 (2014). doi:10.1088/1367-2630/16/8/083002
E. Caffau, E. Maiorca, P. Bonifacio, R. Faraggiana, M. Steffen, H.-G. Ludwig, I. Kamp, M. Busso, The solar photospheric nitrogen abundance. Analysis of atomic transitions with 3D and 1D model atmospheres. Astron. Astrophys. **498**, 877–884 (2009). doi:10.1051/0004-6361/200810859
S. Chandrasekhar, The stability of non-dissipative Couette flow in hydromagnetics. Proc. Natl. Acad. Sci. **46**, 253–257 (1960). doi:10.1073/pnas.46.2.253
W.J. Chaplin, J. Christensen-Dalsgaard, Y. Elsworth, R. Howe, G.R. Isaak, R.M. Larsen, R. New, J. Schou, M.J. Thompson, S. Tomczyk, Rotation of the solar core from BiSON and LOWL frequency observations. Mon. Not. R. Astron. Soc. **308**, 405–414 (1999). doi:10.1046/j.1365-8711.1999.02691.x
W.J. Chaplin, G. Houdek, Y. Elsworth, D.O. Gough, G.R. Isaak, R. New, On model predictions of the power spectral density of radial solar p modes. Mon. Not. R. Astron. Soc. **360**, 859–868 (2005). doi:10.1111/j.1365-2966.2005.09041.x
P. Charbonneau, K.B. MacGregor, Solar spin-down with internal magnetic fields. Astrophys. J. Lett. **397**, 63–66 (1992). doi:10.1086/186545
P. Charbonneau, K.B. MacGregor, Angular momentum transport in magnetized stellar radiative zones. II. The solar spin-down. Astrophys. J. **417**, 762 (1993a). doi:10.1086/173357
P. Charbonneau, K.B. MacGregor, Solar spin-down with internal magnetic fields: erratum. Astrophys. J. Lett. **403**, 87 (1993b). doi:10.1086/186728
P. Charbonneau, J. Christensen-Dalsgaard, R. Henning, R.M. Larsen, J. Schou, M.J. Thompson, S. Tomczyk, Helioseismic constraints on the structure of the solar tachocline. Astrophys. J. **527**, 445–460 (1999). doi:10.1086/308050
J. Christensen-Dalsgaard, F.W.W. Dilke, D.O. Gough, The stability of a solar model to non-radial oscillations. Mon. Not. R. Astron. Soc. **169**, 429–445 (1974)
J. Christensen-Dalsgaard, D.O. Gough, J.G. Morgan, Dirty solar models. Astron. Astrophys. **73**, 121–128 (1979a)
J. Christensen-Dalsgaard, D.O. Gough, J.G. Morgan, Erratum: "Dirty solar models" [Astron. Astrophys. **79**, 121–128 (1979)]. Astron. Astrophys. **79**, 260 (1979b)
J. Christensen-Dalsgaard, T.L. Duvall Jr., D.O. Gough, J.W. Harvey, E.J. Rhodes Jr., Speed of sound in the solar interior. Nature **315**, 378–382 (1985). doi:10.1038/315378a0
J. Christensen-Dalsgaard, W. Däppen, S.V. Ajukov, E.R. Anderson, H.M. Antia, S. Basu, V.A. Baturin, G. Berthomieu, B. Chaboyer, S.M. Chitre, A.N. Cox, P. Demarque, J. Donatowicz, W.A. Dziembowski, M.

Gabriel, D.O. Gough, D.B. Guenther, J.A. Guzik, J.W. Harvey, F. Hill, G. Houdek, C.A. Iglesias, A.G. Kosovichev, J.W. Leibacher, P. Morel, C.R. Proffitt, J. Provost, J. Reiter, E.J. Rhodes Jr., F.J. Rogers, I.W. Roxburgh, M.J. Thompson, R.K. Ulrich, The current state of solar modeling. Science **272**, 1286–1292 (1996). doi:10.1126/science.272.5266.1286

J. Christensen-Dalsgaard, M.J.P.F.G. Monteiro, M. Rempel, M.J. Thompson, A more realistic representation of overshoot at the base of the solar convective envelope as seen by helioseismology. Mon. Not. R. Astron. Soc. **414**, 1158–1174 (2011). doi:10.1111/j.1365-2966.2011.18460.x

A. Clark Jr., J.H. Thomas, P.A. Clark, Solar differential rotation and oblateness. Science **164**, 290–291 (1969). doi:10.1126/science.164.3877.290

T. Corbard, G. Berthomieu, J. Provost, P. Morel, Inferring the equatorial solar tachocline from frequency splittings. Astron. Astrophys. **330**, 1149–1159 (1998)

T. Corbard, L. Blanc-Féraud, G. Berthomieu, J. Provost, Non linear regularization for helioseismic inversions. Application for the study of the solar tachocline. Astron. Astrophys. **344**, 696–708 (1999)

T.G. Cowling, *Magnetohydrodynamics* (Hilger, Bristol, 1976)

A. Cumming, W.C. Haxton, ^{3}He transport in the Sun and the solar neutrino problem. Phys. Rev. Lett. **77**, 4286–4289 (1996). doi:10.1103/PhysRevLett.77.4286

C. Damiani, J.P. Rozelot, S. Lefebvre, A. Kilcik, A.G. Kosovichev, A brief history of the solar oblateness. A review. J. Atmos. Sol.-Terr. Phys. **73**, 241–250 (2011). doi:10.1016/j.jastp.2010.02.021

P.H. Diamond, S.-I. Itoh, K. Itoh, L.J. Silvers, β-Plane Mhd turbulence and dissipation in the solar tachocline, in *The Solar Tachocline*, ed. by D.W. Hughes, R. Rosner, N.O. Weiss (2007), p. 213

R.H. Dicke, The Sun's rotation and relativity. Nature **202**, 432–435 (1964). doi:10.1038/202432a0

R.H. Dicke, The solar spin-down problem. Astrophys. J. Lett. **149**, 121 (1967). doi:10.1086/180072

R.H. Dicke, The solar oblateness and the gravitational quadrupole moment. Astrophys. J. **159**, 1 (1970). doi:10.1086/150286

R.H. Dicke, H.M. Goldenberg, Solar oblateness and general relativity. Phys. Rev. Lett. **18**, 313–316 (1967). doi:10.1103/PhysRevLett.18.313

R.H. Dicke, H.M. Goldenberg, The oblateness of the Sun. Astrophys. J. Suppl. **27**, 131 (1974). doi:10.1086/190292

F.W.W. Dilke, D.O. Gough, The solar spoon. Nature **240**, 262–294 (1972). doi:10.1038/240262a0

M.-A. Dupret, A. Grigahcène, R. Garrido, M. Gabriel, R. Scuflaire, Convection-pulsation coupling. II. Excitation and stabilization mechanisms in δ Sct and γ Dor stars. Astron. Astrophys. **435**, 927–939 (2005). doi:10.1051/0004-6361:20041817

M.A. Dupret, C. Barban, M.-J. Goupil, R. Samadi, A. Grigahcène, M. Gabriel, Theoretical damping rates and phase-lags for solar-like oscillations, in *Proceedings of SOHO 18/GONG 2006/HELAS I, Beyond the Spherical Sun*. ESA Special Publication, vol. 624 (2006), p. 97

M.-A. Dupret, K. Belkacem, R. Samadi, J. Montalban, O. Moreira, A. Miglio, M. Godart, P. Ventura, H.-G. Ludwig, A. Grigahcène, M.-J. Goupil, A. Noels, E. Caffau, Theoretical amplitudes and lifetimes of non-radial solar-like oscillations in red giants. Astron. Astrophys. **506**, 57–67 (2009). doi:10.1051/0004-6361/200911713

T.L. Duvall Jr., W.A. Dziembowski, P.R. Goode, D.O. Gough, J.W. Harvey, J.W. Leibacher, Internal rotation of the Sun. Nature **310**, 22–25 (1984). doi:10.1038/310022a0

W. Dziembowski, Nonlinear mode coupling in oscillating stars. I—Second order theory of the coherent mode coupling. Acta Astron. **32**, 147–171 (1982)

W. Dziembowski, Resonant coupling between solar gravity modes. Sol. Phys. **82**, 259–266 (1983). doi:10.1007/BF00145568

A.S. Eddington, *The Internal Constitution of the Stars* (Cambridge University Press, Cambridge, 1926)

A. Einstein, Die Ursache der Mäanderbildung der Flußläufe und des sogenannten Baerschen Gesetzes. Naturwissenschaften **14**, 223–224 (1926). doi:10.1007/BF01510300

V.W. Ekman, Ark. Mat. Astron. Fys. **2**(11), 1–52 (1905)

J.R. Elliott, Aspects of the solar tachocline. Astron. Astrophys. **327**, 1222–1229 (1997)

J.R. Elliott, D.O. Gough, Calibration of the thickness of the solar tachocline. Astrophys. J. **516**, 475–481 (1999). doi:10.1086/307092

A.N. Ellis, The base of the solar convection zone, in *Theoretical Problems in Stellar Stability and Oscillations*. Liege International Astrophysical Colloquia, vol. 25 (1984), pp. 290–292

Y. Elsworth, R. Howe, G.R. Isaak, C.P. McLeod, B.A. Miller, R. New, S.J. Wheeler, D.O. Gough, Slow rotation of the Sun's interior. Nature **376**, 669–672 (1995). doi:10.1038/376669a0

M.D. Fivian, H.S. Hudson, R.P. Lin, H.J. Zahid, A large excess in apparent solar oblateness due to surface magnetism. Science **322**, 560 (2008). doi:10.1126/science.1160863

E. Forgács-Dajka, K. Petrovay, Tachocline confinement by an oscillatory magnetic field. Sol. Phys. **203**, 195–210 (2001). doi:10.1023/A:1013389631585

E. Forgács-Dajka, K. Petrovay, Dynamics of the fast solar tachocline. I. Dipolar field. Astron. Astrophys. **389**, 629–640 (2002). doi:10.1051/0004-6361:20020586

D.C. Fritts, S.L. Vadas, O. Andreassen, Gravity wave excitation and momentum transport in the solar interior: implications for a residual circulation and lithium depletion. Astron. Astrophys. **333**, 343–361 (1998)

M. Gabriel, Solar oscillations: theory. Bull. Astron. Soc. India **24**, 233 (1996)

A.H. Gabriel, G. Grec, J. Charra, J.-M. Robillot, T. Roca Cortés, S. Turck-Chièze, R. Bocchia, P. Boumier, M. Cantin, E. Cespédes, B. Cougrand, J. Crétolle, L. Damé, M. Decaudin, P. Delache, N. Denis, R. Duc, H. Dzitko, E. Fossat, J.-J. Fourmond, R.A. García, D.O. Gough, C. Grivel, J.M. Herreros, H. Lagardère, J.-P. Moalic, P.L. Pallé, N. Pétrou, M. Sanchez, R.K. Ulrich, H.B. van der Raay, Global Oscillations at Low Frequency from the SOHO Mission (GOLF). Sol. Phys. **162**, 61–99 (1995). doi:10.1007/BF00733427

G. Gamow, Does gravity change with time? Proc. Natl. Acad. Sci. **57**, 187–193 (1967). doi:10.1073/pnas.57.2.187

P. Garaud, Propagation of a dynamo field in the radiative interior of the Sun. Mon. Not. R. Astron. Soc. **304**, 583–588 (1999). doi:10.1046/j.1365-8711.1999.02338.x

P. Garaud, Dynamics of the solar tachocline—I. An incompressible study. Mon. Not. R. Astron. Soc. **329**, 1–17 (2002). doi:10.1046/j.1365-8711.2002.04961.x

P. Garaud, C. Guervilly, The rotation rate of the solar radiative zone. Astrophys. J. **695**, 799–808 (2009). doi:10.1088/0004-637X/695/2/799

R.J. Garcia Lopez, H.C. Spruit, Li depletion in F stars by internal gravity waves. Astrophys. J. **377**, 268–277 (1991). doi:10.1086/170356

P.A. Gilman, P.S. Cally, Global Mhd instabilities of the tachocline, in *The Solar Tachocline*, ed. by D.W. Hughes, R. Rosner, N.O. Weiss (2007), p. 243

P. Goldreich, D.A. Keeley, Solar seismology. II—The stochastic excitation of the solar p modes by turbulent convection. Astrophys. J. **212**, 243–251 (1977). doi:10.1086/155043

D.O. Gough, The internal structure of late-type main-sequence stars, in *Astrophysics: Recent Progress and Future Possibilities*, ed. by B. Gustafsson, P.E. Nissen (1990), pp. 13–50. K. Dan. Vidensk. Selsk., Mat-fys Medd, Mat-fys Medd **42**(4)

D.O. Gough, The pulsational stability of a convective atmosphere, in *Geophys. Fluid Dynamics II*. Woods Hole Oceanographic Institution (1965), pp. 49–84

D.O. Gough, Random remarks on solar hydrodynamics, in *IAU Colloq. 36: The Energy Balance and Hydrodynamics of the Solar Chromosphere and Corona*, ed. by R.M. Bonnet, P. Delache, G. de Bussac (de Bussac, Clermont-Ferrand, 1977a), pp. 3–36

D.O. Gough, Theoretical predictions of variations in the solar output, in *The Solar Output and Its Variation*, ed. by O.R. White (1977b), p. 451

D.O. Gough, The significance of solar oscillations, in *Pleins Feux sur la Physique Solaire*, ed. by S. Dumont, J. Roesch (1978), p. 81

D.O. Gough, A new measure of the solar rotation. Mon. Not. R. Astron. Soc. **196**, 731–745 (1981)

D.O. Gough, Theory of solar oscillations, in *Future Missions in Solar, Heliospheric & Space Plasma Physics*, ed. by E. Rolfe, B. Battrick ESA Special Publication, vol. 235 (1985), pp. 183–197

D.O. Gough, Theory of solar variation, in *Solar-Terrestrial Relationships and the Earth Environment in the Last Millennia*, ed. by G. Cini Castagnoli (Soc. Italiana di Fisica, Bologna, 1988), pp. 90–132

D.O. Gough, On Possible Origins of Relatively Short-Term Variations in the Solar Structure. Philos. Trans. R. Soc. Lond. Ser. A **330**, 627–640 (1990a). doi:10.1098/rsta.1990.0043

D.O. Gough, Open questions, in *IAU Colloq. 121: Inside the Sun*, ed. by G. Berthomieu, M. Cribier Astrophysics and Space Science Library, vol. 159 (1990b), p. 451

D.O. Gough, On the composition of the solar interior: rapporteur paper 1. Space Sci. Rev. **85**, 141–158 (1998). doi:10.1023/A:1005105224272

D.O. Gough, Helioseismology and solar neutrinos. Nucl. Phys. B, Proc. Suppl. **77**, 81–88 (1999). doi:10.1016/S0920-5632(99)00401-6

D.O. Gough, How is solar activity influencing the structure of the Sun, in *From Solar Min to Max: Half a Solar Cycle with SOHO*, ed. by A. Wilson ESA Special Publication, vol. 508 (2002), pp. 577–592

D.O. Gough, How oblate is the Sun? Science **337**, 1611–1612 (2012a). doi:10.1126/science.1226988

D.O. Gough, Pattern formation in rapidly oscillating peculiar A stars. Geophys. Astrophys. Fluid Dyn. **106**, 429–449 (2012b)

D.O. Gough, A.G. Kosovichev, Using helioseismic data to probe the hydrogen abundance in the solar core, in *IAU Colloq. 121: Inside the Sun*, ed. by G. Berthomieu, M. Cribier Astrophysics and Space Science Library, vol. 159 (1990), p. 327

D.O. Gough, M.E. McIntyre, Inevitability of a magnetic field in the Sun's radiative interior. Nature **394**, 755–757 (1998). doi:10.1038/29472

D.O. Gough, A.G. Kosovichev, T. Toutain, Constrained estimates of low-degree mode frequencies and the determination of the interior structure of the Sun. Sol. Phys. **157**, 1–15 (1995). doi:10.1007/BF00680605

H.P. Greenspan, *The Theory of Rotating Fluids* (Cambridge University Press, Cambridge, 1969)
H.P. Greenspan, L.N. Howard, On a time-dependent motion of a rotating fluid. J. Fluid Mech. **17**, 385–404 (1963). doi:10.1017/S0022112063001415
N. Gyenge, T. Baranyi, A. Ludmány, Migration and extension of solar active longitudinal zones. Sol. Phys. **289**, 579–591 (2014). doi:10.1007/s11207-013-0424-3
S. Hanasoge, M.S. Miesch, M. Roth, J. Schou, M. Schüssler, M.J. Thompson, Solar dynamics, rotation, convection and overshoot. Space Sci. Rev. (2015). doi:10.1007/s11214-015
J. Hartmann, Theory of the laminar flow of an electrically conductive liquid in a homogeneous magnetic field. K. Dan. Vidensk. Selsk., Mat-fys Medd **15**(6), 1–28 (1937)
P.H. Haynes, M.E. McIntyre, T.G. Shepherd, C.J. Marks, K.P. Shine, On the 'Downward Control' of extratropical diabatic circulations by Eddy-induced mean zonal forces. J. Atmos. Sci. **48**, 651–680 (1991). doi:10.1175/1520-0469(1991)048<0651:OTCOED>2.0.CO;2
H.A. Hill, R.T. Stebbins, The intrinsic visual oblateness of the Sun. Astrophys. J. **200**, 471–475 (1975). doi:10.1086/153813
J.R. Holton, The influence of viscous boundary layers on transient motions in a stratified rotating fluid. Part I. J. Atmos. Sci. **22**, 402–411 (1965a). doi:10.1175/1520-0469(1965)022<0402:TIOVBL>2.0.CO;2
J.R. Holton, The influence of viscous boundary layers on transient motions in a stratified rotating fluid. Part II. J. Atmos. Sci. **22**, 535–540 (1965b). doi:10.1175/1520-0469(1965)022<0535:TIOVBL>2.0.CO;2
J.R. Holton, R.S. Lindzen, An updated theory for the quasi-biennial cycle of the tropical stratosphere. J. Atmos. Sci. **29**, 1076–1080 (1972). doi:10.1175/1520-0469(1972)029<1076:AUTFTQ>2.0.CO;2
G. Houdek, D.O. Gough, Modelling pulsation amplitudes of ξ Hydrae. Mon. Not. R. Astron. Soc. **336**, 65–69 (2002). doi:10.1046/j.1365-8711.2002.06024.x
G. Houdek, N.J. Balmforth, J. Christensen-Dalsgaard, D.O. Gough, Amplitudes of stochastically excited oscillations in main-sequence stars. Astron. Astrophys. **351**, 582–596 (1999)
L.N. Howard, D.W. Moore, E.A. Spiegel, Solar spin-down problem. Nature **214**, 1297 (1967)
R. Howe, Solar interior rotation and its variation. Living Rev. Sol. Phys. **6**, 1 (2009). doi:10.12942/lrsp-2009-1
R. Howe, J. Christensen-Dalsgaard, F. Hill, R.W. Komm, R.M. Larsen, J. Schou, M.J. Thompson, J. Toomre, Dynamic variations at the base of the solar convection zone. Science **287**, 2456–2460 (2000). doi:10.1126/science.287.5462.2456
R. Howe, R. Komm, F. Hill, J. Christensen-Dalsgaard, T.P. Larson, J. Schou, M.J. Thompson, J. Toomre, Rotation-rate variations at the tachocline: an update. J. Phys. Conf. Ser. **271**(1), 012075 (2011). doi:10.1088/1742-6596/271/1/012075
D. Huber, T.R. Bedding, D. Stello, B. Mosser, S. Mathur, T. Kallinger, S. Hekker, Y.P. Elsworth, D.L. Buzasi, J. De Ridder, R.L. Gilliland, H. Kjeldsen, W.J. Chaplin, R.A. García, S.J. Hale, H.L. Preston, T.R. White, W.J. Borucki, J. Christensen-Dalsgaard, B.D. Clarke, J.M. Jenkins, D. Koch, Asteroseismology of red giants from the first four months of Kepler data: global oscillation parameters for 800 stars. Astrophys. J. **723**, 1607–1617 (2010). doi:10.1088/0004-637X/723/2/1607
D.W. Hughes, R. Rosner, N.O. Weiss (eds.), *The Solar Tachocline* 2007
H.E. Huppert, E.A. Spiegel, Penetration of the molecular-weight barrier. Astrophys. J. **213**, 157–164 (1977). doi:10.1086/155140
C. Jordinson, D.O. Gough, The effect of the solar cycle on the resonant coupling of G modes, in *IAU Colloq. 176: The Impact of Large-Scale Surveys on Pulsating Star Research*, ed. by L. Szabados, D. Kurtz Astronomical Society of the Pacific Conference Series, vol. 203 (2000), p. 390
A. Kasahara, W.M. Washington, General circulation experiments with a six-layer NCAR model, including orography, cloudiness and surface temperature calculations. J. Atmos. Sci. **28**, 657–701 (1971). doi:10.1175/1520-0469(1971)028<0657:GCEWAS>2.0.CO;2
E.-J. Kim, K.B. MacGregor, Gravity wave-driven flows in the solar tachocline. Astrophys. J. Lett. **556**, 117–120 (2001). doi:10.1086/322973
E.-J. Kim, K.B. MacGregor, Gravity wave-driven flows in the solar tachocline. II. Stationary flows. Astrophys. J. **588**, 645–654 (2003). doi:10.1086/373943
L.L. Kitchatinov, G. Rüdiger, Magnetic field confinement by meridional flow and the solar tachocline. Astron. Astrophys. **453**, 329–333 (2006). doi:10.1051/0004-6361:20064867
L.L. Kitchatinov, G. Rüdiger, Stability of toroidal magnetic fields in the solar tachocline and beneath. Astron. Nachr. **328**, 1150 (2007). doi:10.1002/asna.200710859
E. Knobloch, W.J. Merryfield, Enhancement of diffusive transport in oscillatory flows. Astrophys. J. **401**, 196–205 (1992). doi:10.1086/172052
A.G. Kosovichev, Helioseismic constraints on the gradient of angular velocity at the base of the solar convection zone. Astrophys. J. Lett. **469**, 61 (1996). doi:10.1086/310253
J.R. Kuhn, R. Bush, M. Emilio, I.F. Scholl, The precise solar shape and its variability. Science **337**, 1638–1640 (2012). doi:10.1126/science.1223231

P. Kumar, E.J. Quataert, Angular momentum transport by gravity waves and its effect on the rotation of the solar interior. Astrophys. J. Lett. **475**, 143–146 (1997). doi:10.1086/310477

P. Kumar, S. Talon, J.-P. Zahn, Angular momentum redistribution by waves in the Sun. Astrophys. J. **520**, 859–870 (1999). doi:10.1086/307464

P.J. Ledoux, E. Sauvenier-Goffin, The vibrational stability of white dwarfs. Astrophys. J. **111**, 611 (1950). doi:10.1086/145305

S. Lefebvre, A.G. Kosovichev, J.P. Rozelot, Helioseismic test of nonhomologous solar radius changes with the 11 year activity cycle. Astrophys. J. Lett. **658**, 135–138 (2007). doi:10.1086/515394

M.J. Lighthill, On sound generated aerodynamically. I. General theory. Proc. R. Soc. Lond. Ser. A **211**, 564–587 (1952). doi:10.1098/rspa.1952.0060

M.J. Lighthill, On sound generated aerodynamically. II. Turbulence as a source of sound. Proc. R. Soc. Lond. Ser. A **222**, 1–32 (1954). doi:10.1098/rspa.1954.0049

R.S. Lindzen, J.R. Holton, A theory of the quasi-biennial oscillation. J. Atmos. Sci. **25**, 1095–1107 (1968). doi:10.1175/1520-0469(1968)025<1095:ATOTQB>2.0.CO;2

J.E. Lovelock, L. Margulis, Homeostatic tendencies of the Earth's atmosphere. Orig. Life **5**, 93–103 (1974). doi:10.1007/BF00927016

J.E. Lovelock, A.J. Watson, The regulation of carbon dioxide and climate—Gaia or geochemistry. Planet. Space Sci. **30**, 795–802 (1982). doi:10.1016/0032-0633(82)90112-X

S.H. Lubow, E.J. Rhodes Jr., R.K. Ulrich, Five minute oscillations as a probe of the solar interior, in *Nonradial and Nonlinear Stellar Pulsation*, ed. by H.A. Hill, W.A. Dziembowski Lecture Notes in Physics, vol. 125 (Springer, Berlin, 1980), pp. 300–306

K.B. MacGregor, T.M. Rogers, Reflection and ducting of gravity waves inside the Sun. Sol. Phys. **270**, 417–436 (2011). doi:10.1007/s11207-011-9771-0

L. Margulis, J.E. Lovelock, Biological modulation of the Earth's atmosphere. Icarus **21**, 471 (1974). doi:10.1016/0019-1035(74)90150-X

P. Markey, R.J. Tayler, The adiabatic stability of stars containing magnetic fields. II. Poloidal fields. Mon. Not. R. Astron. Soc. **163**, 77–91 (1973). doi:10.1093/mnras/163.1.77

S. Mathis, S. Talon, F.-P. Pantillon, J.-P. Zahn, Angular momentum transport in the Sun's radiative zone by gravito-inertial waves. Sol. Phys. **251**, 101–118 (2008). doi:10.1007/s11207-008-9157-0

J.M. Matthews, R. Kuschnig, D.B. Guenther, G.A.H. Walker, A.F.J. Moffat, S.M. Rucinski, D. Sasselov, W.W. Weiss, Erratum: No stellar p-mode oscillations in space-based photometry of Procyon. Nature **430**, 921 (2004a). doi:10.1038/nature02812

J.M. Matthews, R. Kuschnig, D.B. Guenther, G.A.H. Walker, A.F.J. Moffat, S.M. Rucinski, D. Sasselov, W.W. Weiss, No stellar p-mode oscillations in space-based photometry of Procyon. Nature **430**, 51–53 (2004b). doi:10.1038/nature02671

B.E. McDonald, R.H. Dicke, Solar oblateness and fluid spin-down. Science **158**, 1562–1564 (1967). doi:10.1126/science.158.3808.1562

M.E. McIntyre, Solar tachocline dynamics: Eddy viscosity, anti-friction, or something in between? in *Stellar Astrophysical Fluid Dynamics*, ed. by M.J. Thompson, J. Christensen-Dalsgaard (2003), p. 111

M.E. McIntyre, The Quasi-Biennial Oscillation (QBO): some points about the terrestrial QBO and the possibility of related phenomena in the solar interior, in *The Solar Engine and its Influence on Terrestrial Atmosphere and Climate*, ed. by E. Nesme-Ribes (1994), p. 293

M.E. McIntyre, Magnetic confinement and the sharp tachopause, in *The Solar Tachocline*, ed. by D.W. Hughes, R. Rosner, N.O. Weiss (2007), p. 183

L. Mestel, *Stellar Magnetism*, 2nd edn. (Oxford University Press, London, 2012)

L. Mestel, Rotation and stellar evolution. Mon. Not. R. Astron. Soc. **113**, 716 (1953)

L. Mestel, Meridional circulation in shell-source stars. Astrophys. J. **126**, 550 (1957). doi:10.1086/146429

L. Mestel, N.O. Weiss, Magnetic fields and non-uniform rotation in stellar radiative zones. Mon. Not. R. Astron. Soc. **226**, 123–135 (1987)

M.S. Miesch, Numerical modeling of the solar tachocline. II. Forced turbulence with imposed shear. Astrophys. J. **586**, 663–684 (2003). doi:10.1086/367616

J.L. Modisette, J.E. Novotny, Buoyancy and solar spin-down. Science **166**, 872–874 (1969). doi:10.1126/science.166.3907.872

A. Noels, M. Gabriel, A. Boury, R. Scuflaire, P. Ledoux, Vibrational stability of g^+ modes towards non-radial oscillations for solar models. Mem. Soc. R. Sci. Liege **8**, 317–323 (1975)

F.P. Pijpers, Helioseismic determination of the solar gravitational quadrupole moment. Mon. Not. R. Astron. Soc. **297**, 76–80 (1998). doi:10.1046/j.1365-8711.1998.01801.x

R.A. Plumb, A.D. McEwan, The instability of a forced standing wave in a viscous stratified fluid: a laboratory analogue of the quasi-biennial oscillation. J. Atmos. Sci. **35**, 1827–1839 (1978). doi:10.1175/1520-0469(1978)035<1827:TIOAFS>2.0.CO;2

C.L. Poor, The figure of the Sun. Astrophys. J. **22**, 103 (1905a). doi:10.1086/141245

C.L. Poor, The figure of the Sun. II. Astrophys. J. **22**, 305 (1905b). doi:10.1086/141285
W.H. Press, Radiative and other effects from internal waves in solar and stellar interiors. Astrophys. J. **245**, 286–303 (1981). doi:10.1086/158809
W.H. Press, G.B. Rybicki, Enhancement of passive diffusion and suppression of heat flux in a fluid with time varying shear. Astrophys. J. **248**, 751 (1981). doi:10.1086/159199
E.J. Rhodes Jr., R.K. Ulrich, G.W. Simon, Observations of nonradial p-mode oscillations on the Sun. Astrophys. J. **218**, 901–919 (1977). doi:10.1086/155745
M.H. Ritzwoller, E.M. Lavely, A unified approach to the helioseismic forward and inverse problems of differential rotation. Astrophys. J. **369**, 557–566 (1991). doi:10.1086/169785
P.H. Roberts, *An Introduction to Magnetohydrodynamics* (Longmans, Green, New York, 1967)
T.M. Rogers, K.B. MacGregor, On the interaction of internal gravity waves with a magnetic field—I. Artificial wave forcing. Mon. Not. R. Astron. Soc. **401**, 191–196 (2010). doi:10.1111/j.1365-2966.2009.15618.x
T.M. Rogers, K.B. MacGregor, On the interaction of internal gravity waves with a magnetic field—II. Convective forcing. Mon. Not. R. Astron. Soc. **410**, 946–962 (2011). doi:10.1111/j.1365-2966.2010.17493.x
T.M. Rogers, K.B. MacGregor, G.A. Glatzmaier, Non-linear dynamics of gravity wave driven flows in the solar radiative interior. Mon. Not. R. Astron. Soc. **387**, 616–630 (2008). doi:10.1111/j.1365-2966.2008.13289.x
I.W. Roxburgh, Solar rotation and the perihelion advance of the planets. Icarus **3**, 92–97 (1964). doi:10.1016/0019-1035(64)90049-1
G. Rüdiger, L.L. Kitchatinov, The slender solar tachocline: a magnetic model. Astron. Nachr. **318**, 273 (1997)
G. Rüdiger, L.L. Kitchatinov, Structure and stability of the magnetic solar tachocline. New J. Phys. **9**, 302 (2007). doi:10.1088/1367-2630/9/8/302
C. Sagan, G. Mullen, Earth and Mars: evolution of atmospheres and surface temperatures. Science **177**, 52–56 (1972). doi:10.1126/science.177.4043.52
H. Saio, Stability of nonradial g^+-mode pulsations in 1 solar mass models. Astrophys. J. **240**, 685–692 (1980). doi:10.1086/158275
J. Schou, H.M. Antia, S. Basu, R.S. Bogart, R.I. Bush, S.M. Chitre, J. Christensen-Dalsgaard, M.P. di Mauro, W.A. Dziembowski, A. Eff-Darwich, D.O. Gough, D.A. Haber, J.T. Hoeksema, R. Howe, S.G. Korzennik, A.G. Kosovichev, R.M. Larsen, F.P. Pijpers, P.H. Scherrer, T. Sekii, T.D. Tarbell, A.M. Title, M.J. Thompson, J. Toomre, Helioseismic studies of differential rotation in the solar envelope by the solar oscillations investigation using the Michelson Doppler imager. Astrophys. J. **505**, 390–417 (1998). doi:10.1086/306146
J.A. Shercliff, *A Textbook of Magnetohydrodynamics* (Pergamon, Elmsford, 1965)
H. Shibahashi, Y. Osaki, W. Unno, Nonradial g-mode oscillations and the stability of the Sun. Publ. Astron. Soc. Jpn. **27**, 401–410 (1975)
A. Skumanich, Time scales for CA II emission decay, rotational braking, and lithium depletion. Astrophys. J. **171**, 565 (1972). doi:10.1086/151310
E.A. Spiegel, A history of solar rotation. NASA Spec. Publ. **300**, 61 (1972)
E.A. Spiegel, J.-P. Zahn, The solar tachocline. Astron. Astrophys. **265**, 106–114 (1992)
H.C. Spruit, Differential rotation and magnetic fields in stellar interiors. Astron. Astrophys. **349**, 189–202 (1999)
H.C. Spruit, Dynamo action by differential rotation in a stably stratified stellar interior. Astron. Astrophys. **381**, 923–932 (2002). doi:10.1051/0004-6361:20011465
R.F. Stein, Generation of acoustic and gravity waves by turbulence in an isothermal stratified atmosphere. Sol. Phys. **2**, 385–432 (1967). doi:10.1007/BF00146490
D. Stello, H. Kjeldsen, T.R. Bedding, D. Buzasi, Oscillation mode lifetimes in ξ Hydrae: will strong mode damping limit asteroseismology of red giant stars? Astron. Astrophys. **448**, 709–715 (2006). doi:10.1051/0004-6361:20054016
A. Strugarek, A.S. Brun, J.-P. Zahn, Magnetic confinement of the solar tachocline: II. Coupling to a convection zone. Astron. Astrophys. **532**, 34 (2011a). doi:10.1051/0004-6361/201116518
A. Strugarek, A.S. Brun, J.-P. Zahn, Magnetic confinement of the solar tachocline: the oblique dipole. Astron. Nachr. **332**, 891 (2011b). doi:10.1002/asna.201111613
P.A. Sturrock, Solar neutrino variability and its implications for solar physics and neutrino physics. Astrophys. J. Lett. **688**, 53–56 (2008). doi:10.1086/594993
L. Svalgaard, J.M. Wilcox, Long-term evolution of solar sector structure. Sol. Phys. **41**, 461–475 (1975). doi:10.1007/BF00154083
S. Talon, P. Kumar, J.-P. Zahn, Angular momentum extraction by gravity waves in the Sun. Astrophys. J. Lett. **574**, 175–178 (2002). doi:10.1086/342526
R.J. Tayler, Hydromagnetic instabilities of an ideally conducting fluid. Proc. Phys. Soc. B **70**, 31–48 (1957). doi:10.1088/0370-1301/70/1/306

R.J. Tayler, The adiabatic stability of stars containing magnetic fields—I. Toroidal fields. Mon. Not. R. Astron. Soc. **161**, 365 (1973)

G.I. Taylor, Dispersion of soluble matter in solvent flowing slowly through a tube. Proc. R. Soc. Lond. Ser. A **219**, 186–203 (1953). doi:10.1098/rspa.1953.0139

G.I. Taylor, The dispersion of matter in turbulent flow through a pipe. Proc. R. Soc. Lond. Ser. A **223**, 446–468 (1954). doi:10.1098/rspa.1954.0130

E. Teller, On the change of physical constants. Phys. Rev. **73**, 801–802 (1948). doi:10.1103/PhysRev.73.801

J. Toomre, M.J. Thompson, Prospects and challenges for helioseismology. Space Sci. Rev. (2015, this issue). doi:10.1007/s11214-015-0147-x

W. Unno, Stellar radial pulsation coupled with the convection. Publ. Astron. Soc. Jpn. **19**, 140 (1967)

E.P. Velikhov, Stability of an ideally conducting liquid flowing between cylinders rotating in a magnetic field. Sov. Phys. JETP **36**, 1398–1404 (1959)

T.S. Wood, M.E. McIntyre, Polar confinement of the Sun's interior magnetic field by laminar magnetostrophic flow. J. Fluid Mech. **677**, 445–482 (2011). doi:10.1017/jfm.2011.93

D.-R. Xiong, Radiation-hydrodynamic equations for stellar oscillations. Astron. Astrophys. **209**, 126–134 (1989)

J.-P. Zahn, S. Talon, J. Matias, Angular momentum transport by internal waves in the solar interior. Astron. Astrophys. **322**, 320–328 (1997)

Space Sci Rev (2015) 196:49–77
DOI 10.1007/s11214-014-0035-9

Understanding the Internal Chemical Composition and Physical Processes of the Solar Interior

Sarbani Basu · Nicolas Grevesse · Stephane Mathis · Sylvaine Turck-Chièze

Received: 24 May 2013 / Accepted: 20 January 2014 / Published online: 20 February 2014

Abstract The Sun, the closest and most well studied of stars, is generally used as a standard that other stars are compared to. Models of the Sun are constantly tested with helioseismic data. These data allow us to probe the internal structure and dynamics of the Sun. Among the main sources of the data is the SOHO spacecraft that has been continuously observing the Sun for more than a solar cycle. Current solar models, although good, do not include all the physical processes that are present in the Sun. In this chapter we focus on specific inputs to solar models and discuss generally neglected dynamical physical processes whose inclusion could result in models that are much better representatives of the Sun.

Keywords Sun: interior · Sun: helioseismology · Sun: evolution · Sun: abundances

1 Introduction

The Sun is the closest star and hence used as a benchmark to study other stars. In contrast with other stars, its radius, luminosity and mass are known with great accuracy, allowing us

S. Basu (✉)
Department of Astronomy, Yale University, PO Box 208101, New Haven, CT 06520-8101, USA
e-mail: sarbani.basu@yale.edu

N. Grevesse
Centre Spatial de Liège and Institut d'Astrophysique et de Géophysique, Université de Liège, 4031 Angleur-Liège, Belgium
e-mail: Nicolas.Grevesse@ulg.ac.be

S. Mathis
Laboratoire AIM, CEA/DSM-CNRS-Université Paris Diderot, IRFU/SAp Centre de Saclay, 91191 Gif-sur-Yvette, France
e-mail: stephane.mathis@cea.fr

S. Turck-Chièze
SAp/IRFU/CEA CE Saclay, 91191 Gif sur Yvette, France
e-mail: turck-chieze@cea.fr

to make very precise models of the Sun. The current mass of the Sun ($1M_\odot$) is estimated to be $1.98892(1 \pm 0.00013) \times 10^{33}$ g (Cohen and Taylor 1987), the radius ($1R_\odot$) is $6.9599(1 \pm 0.0001) \times 10^{10}$ cm (Allen 1973), the luminosity ($1L_\odot$) is $3.8418(1 \pm 0.004) \times 10^{33}$ ergs s^{-1} (Fröhlich and Lean 1998; Bahcall et al. 1995). The age of the Sun as determined from radioactive dating of meteorites is $4.57(1 \pm 0.0044)$ Gyr (Bahcall et al. 1995).

One of the more important concepts in modelling the Sun is that of the Standard Solar Model (SSM). The Sun is modelled in a manner somewhat different from other stars since its age, luminosity and radius are known independently. To be called a solar model, a $1M_\odot$ model must have a luminosity of $1L_\odot$ and a radius of $1R_\odot$ at the solar age. The stellar structure equations are solved to satisfy the mass, radius and luminosity constraints. This is done by varying the mixing-length parameter α and the initial helium abundance Y_0 until we get a model with the required characteristics. Mathematically speaking, we have two unknown parameters (α and Y_0) and two constraints (radius and luminosity) at the solar age, and hence this is a well defined problem. The definition of an SSM is somewhat more stringent, with the exception of the mixing-length parameter, the construction of an SSM cannot include any physical quantity or effect that has free parameters in its implementation. Thus for instance SSMs do not include any effects that solar dynamics (rotation, meridional and zonal flows, etc.) may have on solar structure. SSMs are sometimes modified to include a specified amount of convective overshoot, however, since there is no consensus on how overshoot is implemented, this is not common. Just like any other stellar model, an SSM depends on a number of inputs such as radiative opacities, equation of state, nuclear reaction rates, etc. These ingredients have been refined with time as observations from the solar interior (acoustic modes, solar neutrino fluxes) have become more and more accurate. A description of these and other inputs can be found in the review of Turck-Chièze et al. (1993). The observed value of the solar metallicity is also used as a constraint.

SSMs model solar structure very well by the standards of astronomy. For instance, the discrepancy in the sound-speed profile is within a few percent. As helioseismic analyses have shown (see e.g. Christensen-Dalsgaard 2002), the structure of an SSM is in remarkable agreement with the Sun, although the degree of agreement does depend on what we assume the solar metallicity to be. In this chapter we discuss two sources of uncertainty in SSMs—the metallicity, and the equation of state.

The standard solar model, while extremely useful and popular, does not represent all aspects of the Sun. For example, it does not include differential rotation, nor does it include magnetic fields that cause the solar activity cycle and its associated irradiance change (the Schwabe cycle; Schwabe 1844) of about 0.1 % (Fröhlich and Lean 1998). While a 0.1 % change is small, it is still a factor of about 10^5 larger than the change in the luminosity of standard solar models over the same period (Turck-Chièze and Lambert 2007). Without dynamical effects and the effects of magnetic fields one could not hope to model these subtle effects.

In parallel with SSMs, there has also been the development of so-called seismic models of the Sun (see e.g., Antia 1996; Takata and Shibahashi 1998; Turck-Chièze et al. 2001). These are models of the present-day Sun that are forced to satisfy the helioseismic constraints. Some of these are evolutionary models, some others not. For instance the Antia (1996) model is not an evolutionary model, but those of Turck-Chièze et al. (2001) and Couvidat et al. (2003) are evolutionary models obtained by changing opacity and/or reaction rates within their uncertainties. The aim of these models is to predict the thermodynamic properties of the solar core and to estimate neutrino fluxes more accurately. An example of the agreement between the Sun and solar models, one an up-to-date SSM and one seismic, is shown in Fig. 1.

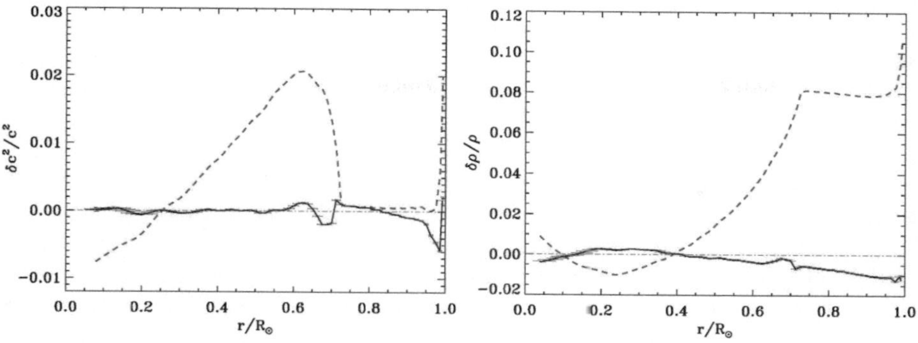

Fig. 1 The difference between the squared sound speed and density profiles of the Sun and two models, one an SSM constructed with Asplund et al. (2009) composition (*red dashed lines*) and the other the seismic model of Turck-Chièze et al. (2011) (*black continuous lines*). The errorbars are from the inversion described in Turck-Chièze et al. (2001). See Turck-Chièze et al. (2012) for the list of the values

We have organised this chapter as follows: We first describe the reliability of the data used in determining properties of the solar interior and discuss what we have learned of solar structure thus far. We then discuss the question of solar abundances and describe how the more-recent set of abundances were derived. In the same section we look at the effect of the abundances on SSMs and how they fare in helioseismic tests; we also discuss several attempts that have been made to determine solar metallicity through helioseismic analyses. Next we examine the solar equation of state and review tests of the different equations of state. We then turn our attention to the details of the solar radiative zone and core to discuss what other physics may be needed to model these regions properly. And finally, we discuss angular-momentum transport which is perhaps the most neglected process in solar models.

2 Helioseismic Insights into the Solar Radiative Zone

It is generally believed that SSMs should describe the radiative zone of the Sun reasonably well since, unlike in the convection zone (CZ), dynamical motions are supposed to be negligible in that region. One of the successes of helioseismology has been the ability to determine the structure of the Sun right to the core. MDI (Michelson Doppler Image; Scherrer et al. 1995) and GOLF (Global Oscillations at Low Frequency; Gabriel et al. 1995) are two instruments on board the Solar and Heliospheric Observatory (SoHO) that complement each other. MDI observations have been used to determine solar acoustic-mode frequencies of degree ℓ of 0 to 300. GOLF was built specifically to obtain high-precision data on the solar low-degree modes that allow us to probe the solar core. Combining MDI and GOLF data allows one to determine the sound-speed and density profile of the Sun down to about 0.08 (Basu et al. 2000; Turck-Chièze et al. 2001).

A precise helioseismic description of the solar radiative zone in its entirety presupposes that solar acoustic waves are sensitive to all regions. This however, is not the case. Acoustic modes are by their very nature sensitive to the outer layers of a star because of the lower sound speed (and hence a longer time spent there by the modes) than in the core. They also have higher amplitudes there. Thus even the frequencies of low-degree modes that penetrate deep inside are largely influenced by the outer layers. Easily detectable low-degree modes (i.e., those with frequencies $\geq$2 mHz) contain information of the core (their lower turning points are in that region), however, these modes are also affected by solar-cycle related changes in the outer layers of the Sun (Gelly et al. 2002; Salabert et al. 2009; Basu et al. 2012; Simoniello et al. 2013). This variability is frequency-dependent, and the change of frequency (up to 0.5 μHz at the maximum of the cycle) is of the same order than the information coming from the core, and thus one needs to correct for this effect or avoid it.

The advantage of GOLF is its very low intrinsic instrumental noise which allows a rapid detection of the lower-frequency part of the spectrum (<1.6 mHz). An extensive discussion of this, as well as a list of the modes, can be found in Turck-Chièze and Lopes (2012). These low-frequency low-degree modes are not visibly affected by magnetic variability in the outer parts of the Sun since the upper turning points of these modes are located below the region that is most affected by magnetic fields. Combining low-degree modes from GOLF with higher-degree MDI frequencies obtained during the solar minimum has led to the determination of sound-speed and density profile in the solar radiative zone down to about 0.08. Results can be found in Basu et al. (2000), where the frequencies that were used were determined using asymmetric Lorentzian profiles for the modes, and Turck-Chièze et al. (2001) where the frequencies had been corrected for magnetic bias. Low frequency acoustic modes are also detectable in ground-based observations collected over decades. Low-degree mode frequencies measured with the Birmingham Solar Oscillation Network (BiSON) when combined with MDI data result in a solar sound-speed profile (see Basu et al. 2009) that agrees with that obtained with the GOLF-MDI combination of data.

The solar sound speed and density profiles are extracted first by determining the sound-speed and density differences between a solar model (the 'reference model') and the Sun and then using the known sound speed and density of the model to reconstruct solar values. The sound-speed and density differences are determined from the frequency differences between the model and the Sun, though a correction needs to be applied, commonly known as the 'surface term', that accounts for the fact that solar models do not describe the solar near-surface layers correctly. Details of the procedure can be found in Christensen-Dalsgaard (2003). The reference model is usually an SSM. Of course SSMs have evolved because of improvements in input physics, as well as changes in solar composition. Some of these changes have improved the match between SSMs and the Sun, while others have degraded the match. Several groups have regularly constructed and compared SSMs to the Sun. Published SSMs include those of Christensen-Dalsgaard et al. (1996), Bahcall et al. (2001), Turck-Chièze et al. (2001), Christensen-Dalsgaard (2002), Turck-Chièze et al. (2004a), Bahcall et al. (2005), Guzik and Mussack (2010), Turck-Chièze and Couvidat (2011), Turck-Chièze and Lopes (2012), etc.

In Fig. 2, we show a detailed view of sound-speed and density in the solar core. The results, with corresponding uncertainties, can be found in Turck-Chièze and Lopes (2012). The details of the difference between the Sun and the models depend on the data used and of course the model, but most groups agree that there are statistically significant difference between the Sun and models throughout the radiative zone. Thus it can be reasonably assumed that SSMs match the Sun only to a few percent. The precision of the seismic probes is such that it impels us to determine the origin of these discrepancies that are small, but statistically significant.

3 The Issue of Solar Abundances

The heavy-element mass fraction (i.e. mass fraction of elements heavier than helium), Z, of the Sun is one of the fundamental inputs to solar models. Z affects energy transport

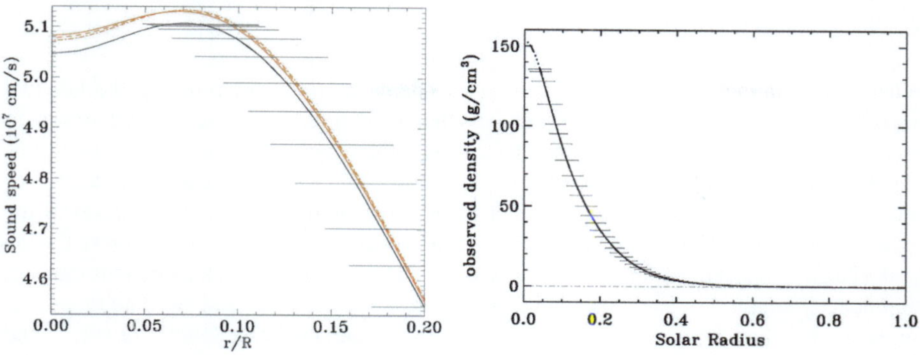

Fig. 2 The sound speed in the solar core, and density in the radiative zone. Note the better agreement with a seismic solar model in *black* (the *vertical errorbars* on the points are too small to be visible) than with the SSM or with the solar model with modified energy production (*red*), see Turck-Chièze et al. (2011) for details. The *horizontal errorbars* are a measure of the resolution of the inversions and denote the width of the resolution kernels

through its effect on radiative opacities. The abundance of some specific elements, such as oxygen, carbon, and nitrogen, can also affect the energy generation rates through the CNO cycle. The effect of Z on opacities changes the boundary between the radiative and convective zones, as well as the structure of radiative region; the effect of Z on energy generation rates can change the structure of the core. Although, the main effect of heavy element abundances is through opacity, these abundances also affect the equation of state. In particular, the adiabatic index Γ_1 is affected in regions where these elements undergo ionisation. The importance of the solar heavy-element abundance does not merely lie in being able to model the Sun correctly, it is often used as the standard against which heavy-element abundances of other stars are measured. Thus the predicted structure of those stars too become uncertain if the solar heavy-element abundance is uncertain. As analysis techniques have changed, there have been related changes in the solar heavy element abundance.

A decrease by 30 % of the photospheric iron abundance that was announced in the 1990s had deteriorated, slightly, the agreement between the sound-speed profile obtained from helioseismic analyses and that of solar models (Turck-Chieze and Lopes 1993). Indeed until about 2002, the reported abundances were such that solar models constructed with those abundances matched the structure of the Sun to an amazing degree (see e.g., Turck-Chièze et al. 2001; Christensen-Dalsgaard 2002; Basu and Antia 2008, and references therein). Since then however, there has been a steady decrease in the reported value of solar metallicity, and this has resulted in models that do not agree with the Sun (see e.g., Basu and Antia 2004, 2008; Bahcall et al. 2005; Turck-Chièze et al. 2004a, and references therein). While lower abundances, particularly lower oxygen abundance, have led to models that do not agree with the Sun, the lower solar abundances, if correct, also mean that the solar oxygen abundance is no longer higher than that of other nearby stars (Turck-Chièze et al. 2004a). Some of the authors of this chapter have been involved on opposite sides of this debate; thus we present below a detailed view of both why the new abundances are considered better than the old ones (Sect. 3.1) and what helioseismic analysis tells us about the newer heavy-element abundance determinations and that helioseismic analyses generally yield higher Z (Sect. 3.2).

3.1 Determining Solar Abundances

When one examines the evolution of the solar chemical composition during the last two decades, looking at different reviews, one sees that the progress has, essentially, been due to progress in the accuracy of the atomic data, mostly the transition probabilities. The resulting changes in Z were small. The situation has, however, changed recently with the rather severe downward revisions of the abundances of a few elements. The new analyses of the solar chemical composition are summarised in the reviews by Asplund et al. (2009), Caffau et al. (2011, see also the references in those two reviews). The new photospheric results are significantly smaller for the most abundant elements like C, N, O, and Ne than those recommended in the widely used compilations of Anders and Grevesse (1989), Grevesse and Noels (1993) and Grevesse and Sauval (1998). They are generally only somewhat smaller for the other elements. As the solar metallicity is essentially dominated by oxygen (43 %), carbon (18 %), neon (10 %), and iron (10 %), the older metallicities from these three compilations, from $Z = 0.02$ to $Z = 0.017$ (iron reduction), decrease to $Z = 0.0134$ with the new solar mixture of Asplund et al. (2009).

We briefly review the changes of the last 15 years and discuss the main reasons for the downward revisions of the abundances of the important elements like O, C and Ne.

3.1.1 The New Analysis

Asplund et al. (2009) re-determined the abundances of nearly all available elements. The authors used a new 3D hydro-dynamical solar model atmosphere instead of the classical 1D models of the photosphere used for many decades. They did a careful and demanding selection of the spectral lines and when possible, replaced the often used LTE hypothesis by non-LTE analyses. In the cases of C, N and O, they used many different indicators of the abundances, atoms as well as molecules. From the available atomic and molecular data, they only used those that they considered to be the most reliable. These points are discussed below, along with the solar spectra that are the basis of all the analyses.

3.1.2 Solar Spectra

The most commonly used optical disk-center intensity spectra of the quiet Sun are the so-called Jungfraujoch or Liège atlas (Delbouille et al. 1973, after its locations of observation and production, respectively) and the Kitt Peak solar atlas (Neckel and Labs 1984). The spectral resolution of the latter is slightly higher, whereas the former is less affected by telluric absorption owing to the much higher observing altitude. There are also corresponding IR disk-center intensity atlases observed from Kitt Peak (Delbouille et al. 1981) and from the Atmospheric Trace Molecule Spectroscopy (ATMOS) experiment flown on the space shuttle (Farmer and Norton 1989; Abrams et al. 1996), as well as from the recent ACE spacecraft experiment (Hase et al. 2010). In the new analyses, although excellent quality solar flux spectra were available, the authors avoided using such spectra because the lines are formed higher in the atmosphere, in regions that are more sensitive to departures from LTE that are still difficult to estimate in many cases. When possible, they also avoided using lines in the blue or near UV, because the density of spectral lines per unit wavelength in these spectral regions is very large, and the probability of unknown blends therefore very high. All solar atlases agree very well with each other except for spectral regions afflicted by telluric features. Thus, the quality of observations is in general not a source of significant error in solar abundance analyses.

3.1.3 Solar Photospheric Models

The new analyses of the solar chemical composition (Asplund et al. 2009; Caffau et al. 2011, see also the references in these two reviews) were driven by the availability of three-dimensional (3D), time-dependent, hydrodynamical models of the solar atmosphere that were successfully applied to the solar spectrum line formation. Various classical 1D models of the photosphere are also used for comparison. The empirical model of Holweger and Mueller (1974) has been used in a large number of solar abundance analyses since more than 30 years. In addition, Asplund et al. (2009) applied theoretical models like MARCS (Gustafsson et al. 2008) and Kurucz (see http://kurucz.harvard.edu). In stellar abundance analyses, such models are often used. When comparing stellar and solar results, data have to be obtained with the same type of model since abundances might be very model-dependent. They also used a mean 3D model, 1DAV (1D averaged) that is a result of temporarily and horizontally averaging the 3D model mentioned on surfaces of equal optical depths. Comparing 3D and 1DAV results, actually shows the role played by the heterogeneities.

Details of the main characteristics of the 3D models can be found in Stein and Nordlund (1998) and Nordlund et al. (2009). To summarise, the models successfully reproduce a wide range of observational constraints. The observed heterogeneous and ever changing nature of the photospheric granulation is successfully modelled. These 3D models have enough realistic physics that allow one to reproduce the solar line profiles as they are really observed in the solar spectrum: all the line profiles are asymmetric, the bisectors of the unblended lines show a delicate C-shape, and the wavelength of the centres of the lines are shifted to the blue. These subtle characteristics in the shapes of the lines result from the motions of matter in the photosphere caused by convective-overshoot from the solar convection-zone into the photosphere. 1D models cannot reproduce these subtle but important properties of the line profiles, leading to unshifted, symmetric lines. Furthermore, the widths of the spectral lines are predicted naturally with the 3D models without invoking any fudge parameters like micro- and macro-turbulence needed with 1D models. The new generation of 3D models reproduce in detail the observed centre-to-limb variation of the continuous intensity and the absolute intensity.

3.1.4 Selection of the Lines, Non LTE-Molecules

Asplund et al. (2009) selected lines keeping in mind that inclusion of blended lines increases the abundance scatter and skews the results towards erroneous larger abundances. Thus the shape of each line profile was compared to the ideal shape computed with the 3D model. This resulted in discarding a number of good candidates that did not show obvious traces of blends, but that were probably slightly blended since their shape do not obey the C-shape rule. It was also found that many lines do not obey the local thermodynamic equilibrium (LTE). Non-LTE analyses were done when possible i.e. when the atomic data were available. Generally missing are accurate values for the cross-sections of the collisions with the neutral hydrogen atoms. This is the main source of uncertainty for lines that show large non-LTE corrections. To determine the abundances of C, N and O, molecular lines of CH, C_2, CO, CN, NH and OH were used. Many of these lines (except for C_2) are in the IR where they are not blended at all. For these elements, the molecular lines are better indicators of the abundances, than the high-excitation lower levels of the atoms, since there is more C, N and O in molecules.

3.1.5 Differences Between New and Old Solar Mixtures

The largest downward revisions that occur for the most important elements, O and C are discussed below. Also discussed is the abundance of Ne, since the solar Ne abundance is derived from the Ne/O ratio that can only be measured in the corona.

Both O and C have low excitation forbidden lines and very high excitation permitted lines (O I and C I) as well as a large number of very good molecular lines. In the old solar mixtures of Anders and Grevesse (1989), Grevesse and Noels (1993) and Grevesse and Sauval (1998), as well as in the new one of Asplund et al. (2009) the abundances of O and C are determined from these indicators. However, in the old analyses, the solar photospheric model was the popularly used 1D model of Holweger and Mueller (1974), and the LTE hypothesis was also adopted. The solar data were essentially the same as the ones used nowadays. The main characteristics of the results of the three older analyses are the following: the results from the various indicators, permitted and forbidden atomic lines and molecular lines, agree within about 0.1 dex, leading to abundances of O between 8.8 and 8.9 (the abundances are given in the usual logarithmic astronomical scale relative to hydrogen, where $\log N(\mathrm{H}) = 12$) and the evolution from 1989 to 1998 is essentially due to progress in the atomic and molecular data used to derive the abundances. The differences between the new, 8.69 (Asplund et al. 2009) and old oxygen abundance, 8.83 (Grevesse and Sauval 1998, as mentioned in the review presented in Vol. 5 of the Space Sciences Series of ISSI) can be explained as follows. The decrease of the abundances derived from the three forbidden lines is due to the presence of blends unknown 15 years ago. The authors estimated the importance of these blends in a purely empirical way independent of any model atmosphere. The results from these forbidden lines are not sensitive to any non-LTE effects nor to the model. The decrease of the abundance derived from the very high excitation permitted O I lines is due to departures from LTE affecting those lines. These non-LTE effects could not be computed 15 years ago because of the lack of the required atomic data. The decrease of the abundance from the molecular lines is the result of the new 3D model—the mean temperature of the 1D model used 15 years ago (Holweger and Mueller 1974), is a bit too high in the somewhat higher layers where the molecular lines are formed, leading to anomalously large abundances. The new final abundances from the 3 different indicators are in agreement, the differences being less than 0.01 dex. The old values also agreed but with a larger scatter, of order 0.1 dex, between the various indicators. The situation for C is very similar to that of O. For the same reasons as for O, the abundance of C has also decreased by about 25 %.

The abundance of neon, derived from the ratio Ne/O, also decreased accordingly. The coronal Ne/O ratio, derived from far UV and X-ray lines, has been measured by different authors (Young 2005; Schmelz et al. 2005; Robrade et al. 2008) and has values between 0.15 and 0.20.

Only minor changes are observed between old and new values for heavier elements.

3.1.6 Differences Between Recent Solar Mixtures

Caffau et al. (2011, and references therein) have also recently revisited the solar abundances of a few elements using their own 3D model of the solar photosphere. The results from these authors are slightly larger than those of Asplund et al. (2009): for oxygen, 8.76 compared with 8.69, and for carbon, 8.50, compared with 8.43. The main reasons for these differences are not related to the use of different 3D models. Their somewhat larger O value comes from the forbidden lines and should decrease if the blends were more accurately estimated, the value from the permitted lines could also somewhat decrease if their solar data as well as the non-LTE results were updated. For carbon, the dispersion in their results is clearly produced by the use of many strong C I lines that are slightly blended leading to a somewhat too large abundance estimate.

In conclusion, Asplund et al. (2009) recommend to use their new solar mixture and discard the old solar mixture of Grevesse and Sauval (1998) that had been used with so much success in the past. Although helium is an important ingredient, it cannot be measured directly in the solar atmosphere and is usually determined helioseismically (e.g. Dappen and Gough 1986; Dziembowski et al. 1991; Kosovichev et al. 1992; Antia and Basu 1994; Richard et al. 1998; Basu and Antia 2004). Assuming a helioseismically determined value of the helium abundance, the new abundance yields a present day solar metallicity of $Z = 0.0134$ with an uncertainty of about 12 %. This uncertainty has been estimated, for the main contributors to Z, by taking into account the combined uncertainties due to the model, to the heterogeneities and to the departures from LTE as recommended by Asplund et al.

Very recently, new 3D models have been built that take solar photospheric magnetic fields into account (Fabbian et al. 2010, 2012; Thaler 2012). These new models might have an influence on the solar abundance results. First estimates of their impact on the O abundance for example, shows a very small increase for the results from the atomic lines and a somewhat larger increase on the abundances derived from the molecular lines. We however need to wait for more severe tests of these new 3D models before being able to really quantify their actual impacts on solar abundances

The problems caused by the lower abundances for solar models and attempts to determine solar abundances using helioseismic data are discussed next.

3.2 Solar Abundances and Helioseismology

Heavy-element abundances affect solar models in a number of ways although the effects are indirect. The main effect is on radiative opacities—higher metallicity implies higher opacities. Opacity in turn affects the position of the base of the outer convection zone and the structure of the radiative zones. Effects of metallicity are also visible in thermodynamic quantities, particularly in the ionisation zones. C, N and O abundances also affect the flux of neutrinos emitted by CNO reactions in the solar core, though the effect is small since CNO reactions account for <2 % of the energy emitted in the Sun. In the case of calibrated SSMs, heavy-element abundances also change the amount of helium required to model the Sun. The changes in the initial helium abundances in SSMs, and the reason, for the change, can be found in Table 6 of Turck-Chièze and Couvidat (2011). Serenelli and Basu (2010) recently determined what the solar initial helium abundance should be by first determining the dependence of the current and initial helium abundance on input parameters and marginalising over them. Assuming the helioseismic $Y_{\rm surf} = 0.2485 \pm 0.0035$ (Antia and Basu 1994) and a 20 % uncertainty in the diffusion rate, they find $Y_{\rm initial} = 0.273 \pm 0.006$ independent of the reference model used. Of course, the analysis was limited to SSMs, and non-standard physics could change the results.

The most dramatic manifestation of the change of metallicities is the change in the position of the convection-zone base, which changes the sound-speed difference between solar models and the Sun. In Fig. 3 we show the relative sound-speed and density differences between several solar models of the Sun. The models were constructed with different solar metallicities, but all other inputs were the same. We compare models with the composition of Grevesse and Sauval (1998; henceforth GS98), Asplund et al. (2005; henceforth AGS05),

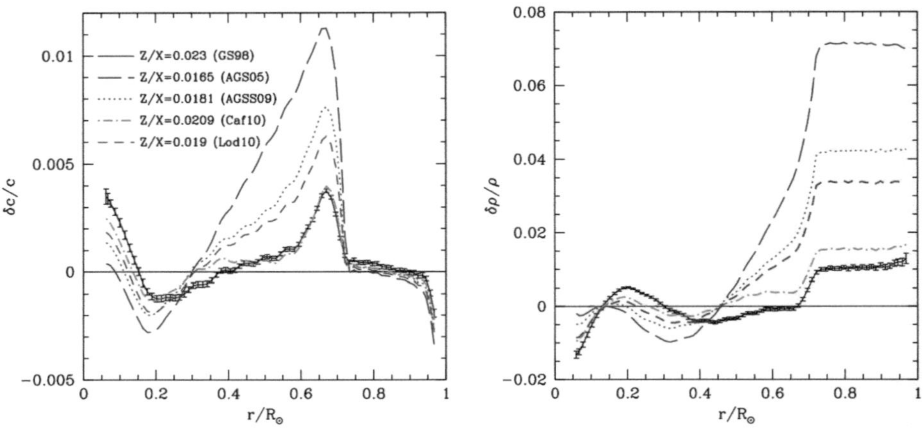

Fig. 3 The sound-speed and density differences between the Sun and standard solar models constructed with different heavy-element abundances. The differences are shown by the *lines* and the *ticks* show 1σ errorbars. For the sake of clarity, errorbars are shown only for one result

Table 1 Properties of models with different heavy-element mixtures

Mixture	Z/X	R_{CZ}	Y_{CZ}	Y_0
Helioseismic	–	0.713 ± 0.001[a]	0.2485 ± 0.0034[b]	0.273 ± 0.006[c]
GS98	0.023	0.7139	0.2456	0.2755
AGS05	0.0165	0.7259	0.2286	0.2586
AGSS09	0.018	0.7205	0.2352	0.2650
CAF10	0.0209	0.7150	0.2415	0.2711
LOD10	0.019	0.7136	0.2412	0.2665

[a]Basu and Antia (1997)

[b]Basu and Antia (2004)

[c]Serenelli and Basu (2010)

Asplund et al. (2009; AGSS09), Caffau et al. (2010, 2011, CAF10) and the compilation of Lodders (2010, Lod10). As can be seen, the AGS05 abundances result in the worst model in terms of sound-speed and density profiles. The newer AGSS09 abundances fare better, but are still less satisfactory compared with the older GS98 abundances or the Caffau et al. (2011) abundances. It is interesting to notice that the Caffau et al. abundances are lower than the GS98 one; however, the sound-speed and density profiles are almost as good because the relative abundances are such that the total opacity is about the same.

The difference in the sound-speed and density is mainly caused by differences in the convection-zone depth. The position of the base of the solar convection zone is known precisely from helioseismology ($0.713 \pm 0.001\ R_\odot$; Basu and Antia 1997 and references therein), and the lower-Z models do not match that. Also different are the current and initial helium abundances. These are listed in Table 1. As can be seen, the low-Z models also have low helium, lower than what has been determined through helioseismic analyses (see, e.g., Kosovichev et al. 1992; Antia and Basu 1994; Basu and Antia 2004, etc.).

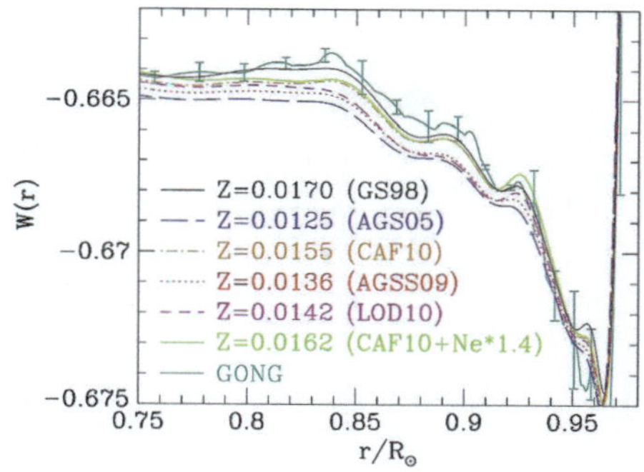

Fig. 4 The dimensionless sound-speed gradient of models constructed with different compositions compared with the solar value as determined using data obtained by GONG, the Global Oscillation Network Group. The model CAF10+Ne·1.4 is a model constructed with CAF10 metallicities where the neon abundance has been increased by a factor of 1.4

There are also, more subtle, signatures in the ionisation zone. In Fig. 4, we show the dimensionless sound-speed gradient of the Sun and the models constructed with different metallicities. This quantity, $W(r)$ defined as

$$W(r) = \frac{r^2}{Gm}\frac{dc^2}{dr}, \tag{1}$$

is sensitive to the changes in the adiabatic index in ionisation zones of various elements. As can be seen, the low-Z models do not agree with the Sun even in the ionisation zones.

As described in Basu and Antia (2008), many attempts have been made to reconcile the new abundances and resulting solar models with the structure of the Sun as determined with helioseismology. Many solution have been suggested, such as modifying opacities, increasing the rate of gravitational settling that increases the heavy-element abundance at the convection-zone base thereby increasing opacities locally, accretion of low-Z material so that the CZ has lower Z than the interior, etc. None of these steps solve all issues. For instance increasing opacities can bring the CZ base to normal, but that does not change $W(r)$; additionally the opacity changes needed have to be fine-tuned. Increasing the rate of diffusion decreases the convection-zone helium abundance even more. As a result, attempts have been made to determine the solar abundance through helioseismic analyses. This is not simple, since solar oscillation equations do not directly involve abundances, and one has assume that the inputs to solar models, such as opacity and equation of state, are correct.

Delahaye and Pinsonneault (2006) using helioseismic constraints on the position of the solar convection-zone base and the convection-zone helium abundance first demonstrated that the position of the CZ and the CZ helium abundance of an SSM have different sensitivities to different elements. They used this difference to find that the logarithmic number density of $\mathrm{Fe/H} = 7.50 \pm 0.0045 \pm 0.003$ and $\mathrm{O/H} = 8.86 \pm 0.041 \pm 0.025$, where the second error term is the uncertainty in the overall abundance scale from errors in the C, N, and Ne. This work assumes that the uncertainties in opacities are given by the difference between OPAL and OP opacities.

To avoid the issue of uncertainties in opacities, Antia and Basu (2006) avoided using constraints sensitive to opacities, and looked to the EOS instead. They assumed that the EOS is correct and calibrated the difference $W(r)$ as a function of a model's metallicity and determined the metallicity for which the $W(r)$ is the same as that of the Sun. They found $Z = 0.0172 \pm 0.002$.

Basu et al. (2007) had shown that the so-called small frequency spacing and frequency ratios between $\ell = 0$ and $\ell = 2$, and $\ell = 1$ and $\ell = 3$ modes are sensitive to the metallicity of the solar core. The frequency ratios are determined by the sound-speed gradient in the core, which in turn is determined by the mean molecular weight and its gradient. Chaplin

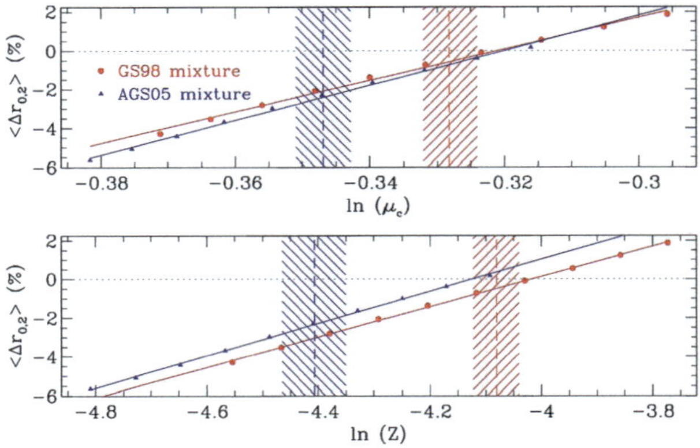

Fig. 5 The relation between the averaged difference of the frequency separation ratios between a solar model and the Sun and the average mean molecular weight in the core of solar models as well as Z for the models. Only results for the $(0, 2)$ separation ratios are shown, results for the $(1, 3)$ ratios are similar. The best-fit *straight lines* are also shown. The *hashed region* around these lines represent uncertainties in μ_c and Z. These were obtained from the Monte Carlo simulations of Bahcall et al. (2006). Errors in $\langle \Delta r_{0,2} \rangle$ are of the size of the points

et al. (2007) expanded on this work to use this to determine solar metallicity. They compared the average frequency separation ratios of models with those of the Sun as determined with BiSON data. Chaplin et al. (2007) constructed two sets of test models. The models in each set had different values of Z/X, but one sequence of models was constructed with the relative heavy element abundances of GS98, while the second sequence was made with the relative heavy-element abundances of AGS05. To fix the Z/X of a given model in either sequence, the individual relative heavy element abundances of GS98 (or AGS05) were multiplied by the same constant factor. They found that the mean difference of the separation ratios between the Sun and the models was a linear function of $\ln \mu_c$ and $\ln Z$, where μ_c is the mean molecular weight in the inner 20 % (by radius) of the Sun, and Z is the metallicity in the convection zone. This relation can be seen in Fig. 5. This monotonic relationship led the authors to argue that $\ln \mu_c$ and $\ln Z$ for the model that lead to perfect match between the separation ratios of the model and the Sun would be a good estimate of $\ln \mu_c$ and $\ln Z$ of the Sun. They found that the relative abundances of GS98 yielded a solar Z of 0.0178 and the AGS mixture a solar Z of 0.0161.

Thus most helioseismic analyses thus far have yielded high solar metallicities, bringing them into conflict with some of the new spectroscopic measurements that have been discussed earlier. It should be noted though that some analyses yield a low metallicity (see e.g., Vorontsov et al. 2013).

As mentioned earlier, changing the most obvious inputs does not bring models with low abundances back in concordance with the Sun. There are ongoing efforts that assume that the new abundances are correct and look to neglected physics, such as those discussed in Sects. 5 and 6 to bring solar models back in agreement with the Sun.

4 The Solar Equation of State

The ability of helioseismology to probe the solar interior in such detail has allowed us not just to determine solar structure and dynamics, but also to use the Sun as a laboratory to test

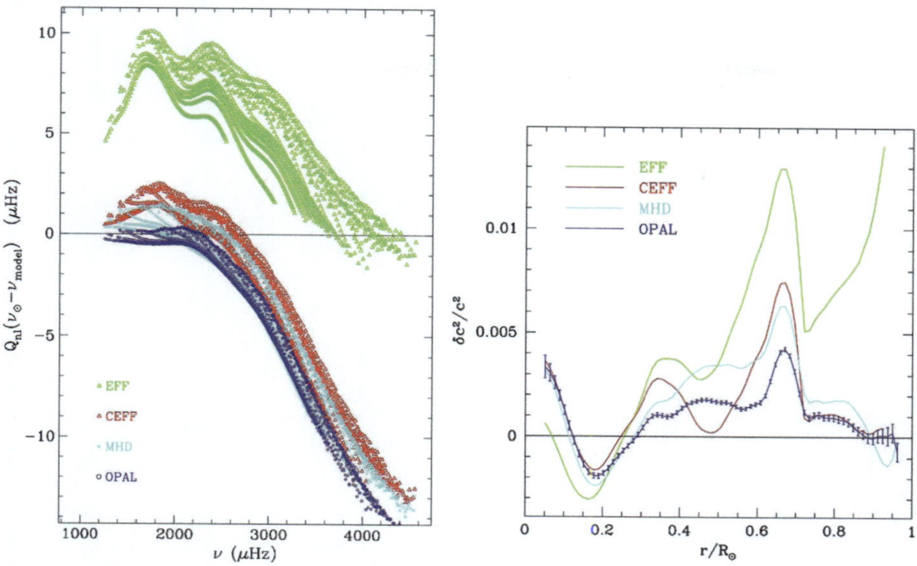

Fig. 6 *Left:* The scaled frequency differences between the Sun and SSMs constructed with different EOSs. The factor Q_{nl} corrects for the fact that frequencies of modes with low inertia change more for the same perturbation than frequencies of modes with high inertia. *Right:* The relative sound speed differences between the Sun and the models. Note that the models were constructed with the Grevesse and Noels (1993) abundances

different inputs that are used to construct solar models. One of these inputs is the equation of state (EOS). The EOS provides a relation between pressure, temperature, composition and abundances. It is a description of the fundamental properties of matter. The equation of state determines the structure of the ionization zones and the convection zone and as a result, determines the structure of the model.

Equations of state that are used as inputs in stellar models are results of extremely complicated theoretical calculations. Thus there is no certainty that they are correct, and hence they need to be tested. An indirect way to examine whether an EOS is correct is to construct an SSM with the EOS and examine its frequencies and sound-speed profile. An example of this is shown in Fig. 6. In these figures we show the frequency difference between the Sun and the models constructed with different EOSs but otherwise identical inputs. These are the Eggleton, Faulker and Flanerry (EFF; Eggleton et al. 1973) EOS, a relatively simple EOS that is still used by many stellar modellers, the Coulomb Corrected EFF (CEFF; Christensen-Dalsgaard and Daeppen 1992) EOS which is basically the EFF EOS but with Debye-Hückel Coulomb corrections that rectifies its absence in EFF EOS, the so-called MHD equation of state (Dappen et al. 1987; Hummer and Mihalas 1988; Mihalas et al. 1988) and the OPAL equation of state (Rogers et al. 1996; Rogers and Nayfonov 2002). The last two are results of complex theoretical calculations.

As is clear from Fig. 6 (left) we can see that the model with EFF EOS fares very badly. The other EOSs are better, the differences between them appear smaller, and the frequency differences alone are not enough to indicate whether the EOS are good since the frequency differences are dominated by differences in the near surface layers. Hence in Fig. 6 (right) we show the sound-speed differences between the models and the Sun. Note that one can see that OPAL models do better than MHD and CEFF. CEFF looks marginally better. The

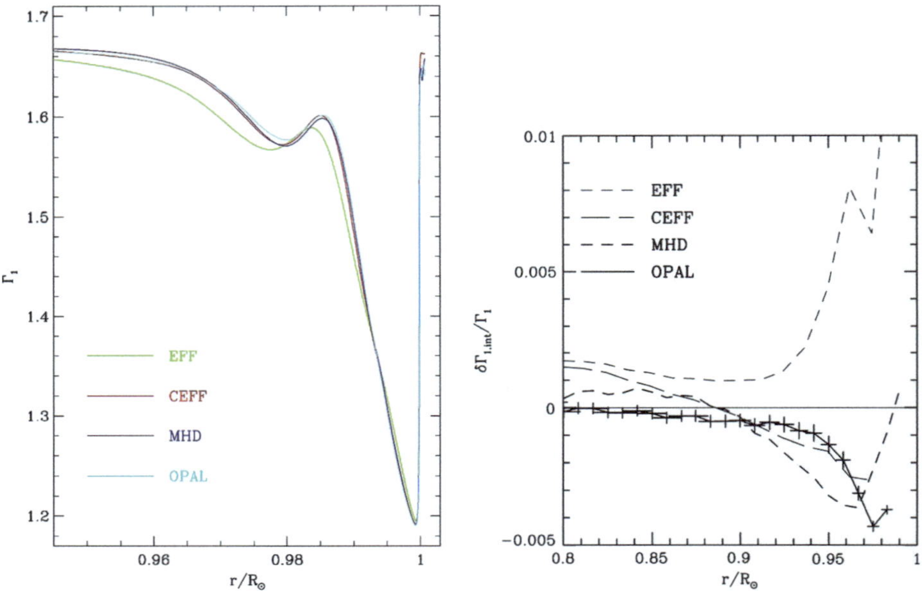

Fig. 7 *Left:* The Γ_1 profiles of models constructed with different EOSs. *Right:* The intrinsic Γ_1 differences between the Sun and models constructed with different EOSs

CEFF EOS is however not fully thermodynamically consistent (see Christensen-Dalsgaard and Daeppen 1992).

The sound-speed profile is an indirect way to test the EOS. A somewhat more direct way is to examine the profile of the adiabatic index Γ_1. The adiabatic index was used by Elliott and Kosovichev (1998) to show that the then available versions of OPAL and MHD were deficient under conditions present in the solar core. It was found that Γ_1 for the solar core was less than $5/3$ while that of the models was $5/3$. The decrease in Γ_1 in the solar core was attributed to the relativistic tail of the velocity distribution of electrons. The deficiency in the input EOSs has since been rectified (Turck-Chièze et al. 2001; Gong et al. 2001; Rogers and Nayfonov 2002).

The ionisation zones are good regions to study the equation of state since the process of ionisation changes Γ_1. This is particularly true for the He II ionisation zone which leaves a marked dip in Γ_1 (see Fig. 7[left]). The He I ionisation zone is less useful since the H I ionisation zone merges with it. However, Γ_1 also depends on the helium abundance of the models (which we cannot specify), and the ionisation zone also depends on other structural quantities. Basu and Christensen-Dalsgaard (1997) proposed a way of studying the part of Γ_1 (what they called the "intrinsic" Γ_1) that is independent of structure and helium abundance. The intrinsic Γ_1 differences between the Sun and several EOS are shown in Fig. 7(right). Note that we find again that EFF is very deficient, that OPAL does better than MHD in the deeper layers, however, closer to the surface MHD does better. Basu et al. (1999) confirmed that MHD indeed does better than OPAL in the outermost layers and could attribute that to the treatment of H II.

As we can see above, none of the input equations of state describe the EOS of the solar material perfectly and more work needs to be done in this regard. One of the biggest drawbacks of both MHD and OPAL EOSs is that they are both constructed with a fixed relative-abundance of heavy metals. This makes testing new abundances difficult.

CEFF allows the flexibility to changing relative-abundance, however, it is not a completely thermodynamically consistent EOS and this shows up as discrepancies in the deeper layers of the Sun. Fortunately for abundance tests, the outer layers of the Sun are described well by the CEFF EOS. There are some analytic EOS that have been developed in recent years and these are not-only thermodynamically consistent, but also allow the Z mixture to be changed. These include the SIREFF EOS (Guzik and Swenson 1997; Guzik et al. 2005). The use of this EOS shows that changing the relative abundance in the EOS does not change the sound-speed profile of the models enough to alleviate the problem with the low abundances (Guzik and Mussack 2010).

5 Towards a Better Description of the Solar Radiative Zone

Helioseismology has given us new, detailed information about the solar radiative zone that exposes the limits of SSMs. In this section we discuss some of the differences between the Sun and SSMs which could be connected to the limitations of these standard models and the equations used to construct them. In particular we focus on questions related to energy transport. The next section is dedicated to dynamical processes.

5.1 The Energy Balance, Neutrino Fluxes and the Central Composition

It is instructive to compare the central conditions obtained in SSMs with those of seismic models that reproduce the solar sound speed and density profiles. Seismic models suggests that the central temperature of the Sun is slightly greater than the central temperature of current SSMs and that the Sun also has a slightly greater central density (Turck-Chièze et al. 2011). The 1.5 % difference in temperature has important consequences for the prediction of the Boron neutrino flux which has an extremely high temperature-dependence (a factor of about 20). This flux has been measured by the SNO detector and is directly comparable to predictions unlike other neutrino fluxes that require neutrino flavour-transformations to be taken into account. The prediction of the seismic model is in perfect agreement with this detection: 5.31 ± 0.6 for a detected value of $5.045 \pm 0.18 \times 10^6\ \mathrm{cm}^{-2}\,\mathrm{s}^{-1}$ but the SSM prediction is a bit too low 4.5 ± 0.5 (Turck-Chièze and Couvidat 2011). This difference could be interpreted as an energy-generation related problem in SSMs that neglect any motion in the whole radiative zone; this energy difference is easy to quantify. Since p-p reactions dominate solar nuclear burning, one can estimate a maximal difference of 5–6 % between the energy production and its release at the surface between the Sun and SSMs. SSMs assume that the produced energy is transported only by photons. Assuming this is correct, the discrepancy between the sound speed and density profiles between the radiative zones of the Sun and models can be explained as being partly due to a redistribution of the produced energy through other energetic phenomena that are not included in SSMs. Possible candidates include magnetic energy, kinetic energy, meridional circulation energy in the radiative zone, gravity wave energy, and dark matter energy that can be additional channels for the transport of energy.

Making progress on these points is difficult, but this may be possible in the near future thanks to some complementary observables: a better detection of gravity modes (see the description of the instrument called GOLD in the chapter dedicated to future projects) for a much improved determination of the dynamics of the core and the use of the neutrino spectroscopy to determine the electron density totally independently (Lopes and Turck-Chièze 2013). The transport of energy by photons and microscopic diffusion in the radiative zone will be checked by the ^{13}N and ^{15}O neutrino fluxes; the accurate measurement of the *pep* neutrino flux will constrain the energy produced by the p-p chain.

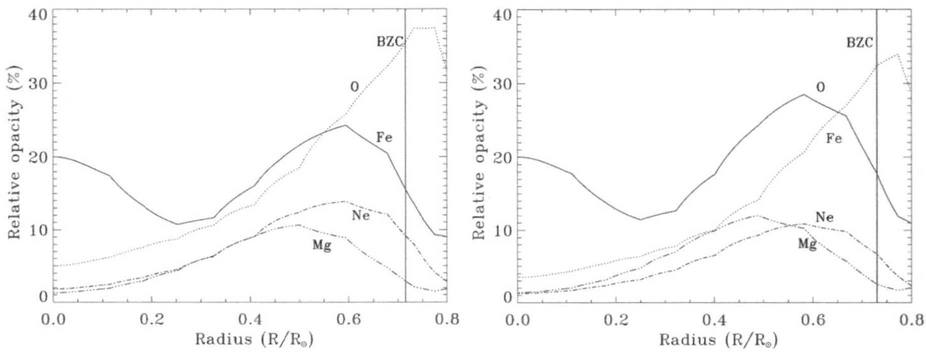

Fig. 8 Main contributions of heavy elements to the total opacity for: *Left:* the GS98 composition, and *Right:* the AGS05 composition. 'BCZ' marks the base of the convection zone. From Turck-Chièze et al. (2009)

5.2 Energy Balance and Dark Matter

Twenty years ago, the idea that weakly interactive massive particles (WIMPs) could slightly lower the solar central temperature had been explored to help solve the solar neutrino problem. Today these particles remain very good candidates for dark matter, but the problem is reversed. Since the core of the Sun shows no evidence of cooling, the absence of WIMPs' signatures put some limits on their mass: masses $\leq$12 GeV are excluded for currently accepted interaction cross sections (Turck-Chièze et al. 2012; Turck-Chièze and Lopes 2012). This result is also supported by other studies and the mass of these particles is mainly assumed around 100 GeV, a value that is not easy to detect.

Other particles, like axions and sterile neutrinos, are also considered to be potential candidates for dark matter. If these particles are present in the Sun, they will act very differently on the solar interior. They will indeed transport energy in the whole radiative zone but their very low mass means that they will not migrate to the most central regions; however, they could modify the gravitational field in the radiative zone. This justifies a dedicated study to determine their specific density and interactions.

5.3 The Transport of Energy by Photons

The equations governing SSMs only consider the production of energy by nuclear reaction and the instantaneous energy loss by neutrinos. The energy is transported by photons in the radiative zone of the Sun. This transport is calculated using Rosseland mean opacity tables which result from calculations using the local mixture of elements. Current SSMs use the work derived by Iglesias and Rogers (1996) or those of the OP consortium (Badnell et al. 2005). These calculations represent the best estimates of such complex calculations. The recent revisions of solar metallicity have deteriorated the earlier good agreement between the sound-speed and density profiles of the Sun, mainly through modification of opacity because of changes in O and Fe (Christensen-Dalsgaard et al. 2009; Turck-Chièze et al. 2009).

Figure 8 shows how the changes in relative abundances modifies the role of the main contributors to the total opacity. The contribution of oxygen was at the level of iron for the GS98 composition in a large part of the radiative zone; when the oxygen abundance is reduced, the iron contribution becomes dominant. The newer AGS09 composition gives rise to an intermediate situation. The shape of the oxygen contribution has the opposite shape of the change in the sound-speed differences between the Sun and models with GS98 and AGS05 compositions (see Fig. 3), this is not completely surprising, since ultimately the change in the O abundance is the main reason for the difference in sound speed between the two models.

The opacity calculations have never been validated through measurements made for solar conditions. New possibilities for making such measurement exist today using the Z machine (Bailey et al. 2007) and large laser facilities (Turck-Chièze et al. 2009). In parallel, new detailed calculations of opacity are being performed to better describe the transport of energy and for the microscopic diffusion of elements that can change local opacities (Blancard et al. 2012; Turck-Chièze and the OPAL consortium 2013). Another effort is underway to improve the grids of calculated opacities for conditions near the base of the convective zone, as well as improve interpolation between the grids.

6 A Dynamical View of the Inner Sun

SSMs and most stellar models ignore dynamical processes and hence avoid issues such as the transport of angular momentum from one part of the star to another. However, the Sun and solar-type stars are dynamical and magnetic objects, and hence, rotation and magnetic fields could modify their evolution as well as their interaction with their external environment. We know, for example, that differential rotation induces "non-standard" mixing processes, which modify their lifetime, their late stages of evolution and their nucleosynthetic properties (e.g., Maeder 2009). Since we now have strong constraints on the Sun's internal rotation (e.g. Schou et al. 1998; García et al. 2007, 2011a; Turck-Chièze and Lopes 2012) and on more evolved low-mass stars (e.g. Beck et al. 2012; Deheuvels et al. 2012; Mosser et al. 2012), it is time to construct solar and stellar models that account for magneto-hydrodynamic effects on both dynamical and secular time-scales. In this section, we describe recent progress that has been made in the detection and modelling of secular exchanges of angular momentum in the Sun and low-mass stars interiors focusing on mechanisms that act in the radiative core. Readers are referred to Brun (2011) for processes in the convective region.

6.1 The Role of the Turbulence

While dynamical processes are generally not included in SSMs, one process, namely convection, has to be included, and this is usually done using the "mixing length theory," or MLT (Böhm-Vitense 1958). MLT assumes that the full range of turbulent eddies can be parametrized with a single parameter, the mixing length parameter α. In the case of the Sun, this parameter can be calibrated by requiring the Sun to have its observed radius and luminosity at it current age. The solar calibrated value of α is normally used for all other stars, although there is increasing evidence that the solar α should not be used in other stars (see Bonaca et al. 2012, and references therein). In addition to the large-scale convective motions, turbulence is present in two regions: the base of the convection zone and the subsurface layers. The high-precision of the available seismic data justifies paying more attention to these regions.

6.1.1 Turbulence Below the Convective Envelope

There are two possible sources of turbulence in this region:

- Dynamical shear-induced turbulence caused by the horizontal shear imposed at the top of the stably stratified radiative zone by turbulent flows in the surrounding convective envelope (Zahn 1992; Spiegel and Zahn 1992), this case is discussed in Sect. 6.2.
- Turbulence due to the penetration of convective flows into the radiation zone because of their inertia, commonly referred to as convective overshoot (e.g. Zahn 1991; Brun et al. 2011).

An interesting point is that both sources of turbulence may lead to efficient mixing at the base of the convective envelope that reduces the differences between the sound-speed profile of a solar model and the Sun (e.g. Richard et al. 1996; Brun et al. 1999; Zhang et al. 2012).

6.1.2 Turbulence in the Subsurface Layers

Treating this region correctly is particularly important in improving predictions of acoustic-mode frequencies and also in understanding the 2-year modulation (Fletcher et al. 2010) and 11-year cycles (Elsworth et al. 1990; Libbrecht and Woodard 1990, etc.) seen in solar acoustic mode frequencies. SSMs do not include the effects of turbulence on solar structure, and this leads to the large frequency-dependent discrepancy between the solar frequencies and those of the models (see Fig. 6[Left]). Realistic 3D simulations of the convection zone show that turbulent pressure plays a role in an extended region just below and just above the surface (e.g., Nordlund et al. 2009) with a consequent modification of density and thermal structure. This leads to a direct effect on the predicted radius and in the predicted frequencies. This significantly reduces the frequency differences between models and the Sun (Rosenthal et al. 1999; Li et al. 2002, etc.). Additionally one can introduce magnetic field in the 3D code to study how the solar radius evolves with activity. First results, using the strength of the toroidal field as determined by Baldner et al. (2009) from seismic observations, can be found in Piau et al. (2014).

6.2 The Solar Rotation Profile

Early helioseismic results had shown that the surface latitudinal differential rotation (25 days around the equator up to 32 days at 75 degrees) is maintained throughout the convective zone but disappears in the turbulent layers called "tachocline" located at the transition between radiative and convective zone, A more or less rigid rotation in the radiative zone is observed down to the limit of the core (Eff-Darwich and Korzennik 2012).

It is well known that acoustic modes contain very little information about the rotation of the solar core. This is because they have their largest amplitudes close to the surface and very few modes penetrate to the core. By contrast, gravity (or g) modes are the natural way to probe the solar core but due to their evanescence in the outer convective zone, they are particularly difficult to detect. Their predicted amplitude is at the level of mm/s at the surface (Belkacem 2011).

The unique performance of the GOLF instrument at low frequencies (see Turck-Chièze and Lopes 2012) and its exceptionally long stretch of observations has made GOLF observations the prime dataset to look for g-modes. Some candidates have been found in the mixed mode range after integration over more than 4 years of observations (Gabriel et al. 2002; Turck-Chièze et al. 2004b). The most probable detections correspond to mixed modes of $\ell = 2$, $n = -3$ that are believed to be the most easily excited in that range of frequencies (Cox and Guzik 2004). The search was conducted again with more than 10 years of GOLF data (García et al. 2008). Another search was done after adding the mode power spectra in the asymptotic range where the modes are nearly equidistant in period (García et al. 2007)

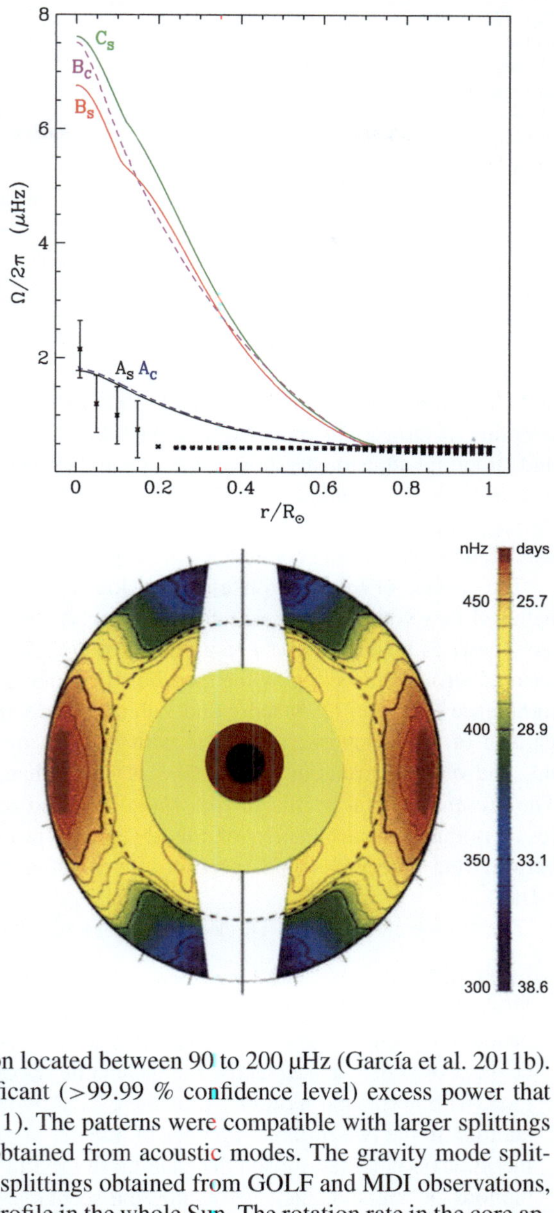

Fig. 9 *Top:* Comparison of the solar rotation profile as obtained with a combination of MDI and GOLF data with different predictions made by including transport of momentum by rotation. The data points with errorbars are results obtained using a combination of GOLF and MDI data. The label S stands for the STAREVOL code and C for CESAM code. 'A' corresponds to a very slow initial rotation, 'B' and 'C' correspond respectively to initial rotation rates of 20 km/s and 50 km/s; see Turck-Chièze et al. (2010a) for details. At the scale of the axes, the latitudinal differential rotation of the convective zone is not visible clearly. *Bottom:* 2D schematic of the solar internal rotation profile obtained with SoHO data

and then looking directly to the region located between 90 to 200 μHz (García et al. 2011b). García et al. (2011b) claimed significant (>99.99 % confidence level) excess power that they attributed to dipole modes ($\ell = 1$). The patterns were compatible with larger splittings in the solar core region than those obtained from acoustic modes. The gravity mode splittings, combined with acoustic mode splittings obtained from GOLF and MDI observations, lead to a first insight on the rotation profile in the whole Sun. The rotation rate in the core appears a factor between 5 and 8, depending on the detailed shape in the core, greater than the rest of the radiative zone (see Fig. 9 top). Such results are also seen with asteroseismology for other stars as mentioned earlier and motivate the construction more complete models of the Sun that include the transport of momentum by rotational mechanisms as suggested by Zahn (1992) and Mathis and Zahn (2004).

6.3 The Transport of Momentum by Rotation During the Sun's Lifetime

One first needs to understand differential rotation and the associated large-scale meridional circulation, and shear-induced turbulence (e.g. Zahn 1992; Maeder and Zahn 1998; Mathis

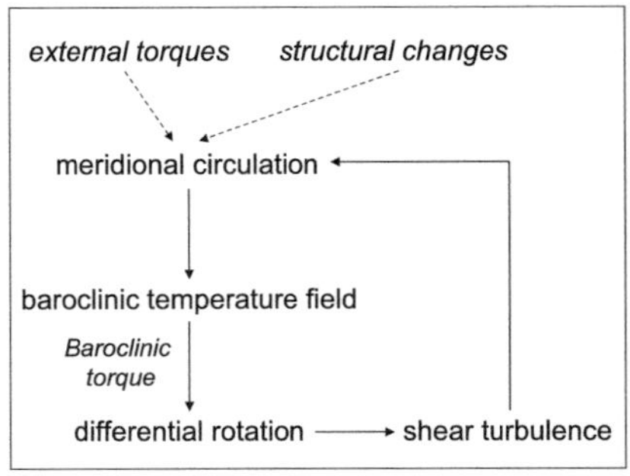

Fig. 10 The transport loop in stellar radiation zones when differential rotation, the associated large-scale meridional circulation and shear-induced turbulence are taken into account. From Decressin et al. (2009)

and Zahn 2004). Rieutord (2006) and Decressin et al. (2009) give a global view of the coupling of the various processes involved. First, viscous turbulent transport, stellar winds and structural adjustments induce meridional currents. These currents advect heat and that in turn leads to a latitudinal gradient of temperature. Because of the associated baroclinic torque, a new differential rotation profile is built which can be understood looking at the so-called thermal-wind equation, and this closes the transport loop (see Fig. 10).

The surface of the Sun rotates today slowly at about 2 km/s. However, it is generally believed that young stars rotate fast, slowing down after several millions of years. This slow-down is believed to be a result of magnetic braking and angular momentum is redistributed through advection by the meridional circulation and diffusion by the shear-induced turbulence (Zahn 1992; Maeder and Zahn 1998; Mathis and Zahn 2004). The first comparison of the solar rotation profile with detailed models that included rotation showed that the observed rotation in the core is lower than thought theoretically (see Fig. 9). The present rotation profile of the whole Sun also remains difficult to explain. Another result is that the introduction of this dynamical process does not reduce the difference between the observed sound speed profile and that of such models (Turck-Chièze et al. 2010a).

The main conclusions of Turck-Chièze et al. (2010a) can be summarized as follows:

- In the different cases that have been studied, and that vary in their initial rotation rate (50 km/s, 20 km/s and extremely small), the meridional circulation of the radiative zone becomes extremely slow in comparison with the observed meridional circulation in the convective zone. Consequently, the tachocline appears naturally as a very turbulent region, with horizontal motions. Thus there exist two very different dynamical regimes in addition to the two different regimes of transfer of energy. The calculated radiative-zone meridional velocities are smaller than those used in existing 3D simulations; these simulations of course cannot yet handle the secular evolution of the transport of angular momentum over Gyrs.
- In all cases, the transport of angular momentum was efficient mainly during the contraction phase (where core rotation increases rapidly) and during the pre-main-sequence. The transport of momentum by rotation is largely reduced during the main sequence.
- When the initial rotation rate is high (20–50 km/s), the current core rotation is too high compared with the observed value, so one possibility is that the Sun was a rather slow rotator (5–10 km/s) and the other is that the rotation rate has been reduced by some other process. Moreover, none of the calculations lead to the observed rigid rotation out of the solar core. These two facts suggest that some additional transport of momentum had been effective in the past and may be effective even today.

Thus the present solar core rotation profile appears to be a relic of the first million years of the Sun's life and its knowledge requires a better description of the young Sun, in particular, of the phases I and II of evolution as described by Lada (1987), Andre et al. (2000).

Differential rotation and associated meridional currents and shear-induced turbulence, if applied to the Sun and low-mass stars, are unable to reproduce the angular-velocity profile of the radiative core of the present Sun (Pinsonneault et al. 1989; Turck-Chièze et al. 2010b). They cannot reproduce light-element mixing in solar-type stars on the main sequence (e.g. Talon and Charbonnel 2005) either. Nor can they reproduce the rotation profiles of the central radiation zone of low-mass subgiant and red giant stars (Eggenberger et al. 2012; Ceillier et al. 2012; Marques et al. 2013). Thus other processes, such as magnetic fields (e.g. Rudiger and Kitchatinov 1997; Gough and McIntyre 1998; Garaud and Garaud 2008; Wood et al. 2011) or internal waves excited by penetrative convection (e.g. Schatzman 1993a; Zahn et al. 1997) must be studied.

6.4 The Transport of Momentum by Magnetic Fields

Another possible mechanism of angular momentum transport is the magnetic field that could have been built when the Sun was young. This was an epoch when the Sun was certainly much more active than it is today; the Sun at that epoch was fully convective as well. During this phase, a dynamo effect could have amplified an initial field in the proto-stellar cloud. At that time, when the star is a few Myrs old, mass loss could become largely greater than accretion leading to a loss of angular momentum. Spectropolarimetry today shows complex magnetic field configuration for such stars (Donati and Landstreet 2009). Eggenberger et al. (2005) have also suggested that a fossil field in the Sun can lead to a strictly uniform core rotation that matches with helioseismology in contrast to angular momentum extraction by gravity waves. Of course, these two processes can act simultaneously in the present Sun (Mathis 2013).

A reasonable model of the young Sun that contains all the phenomena observed in the other young stars is not yet available. Such a model must contain not only an appropriate, realistic, magnetic field configuration and its effect on the solar rotation, but also mass loss. Such magnetic fields, if they subsist in time, will be subject to Ohmic diffusion and to advection by the shear of differential rotation and meridional currents, and will transport angular momentum to damp differential rotation in the whole radiative zone. Some preliminary studies can be found in Duez et al. (2010b), Turck-Chièze et al. (2011). Indeed this is the most accepted hypothesis for the origin of magnetic fields in the cores of solar-type stars (Mestel 1965). One first needs to understand the field geometry in the radiation zone since it affects angular momentum transport. This problem is related to magneto-hydrodynamic relaxation processes, wherein an initial turbulent magnetic field is converted into a large-scale configuration thanks to a selective decay of the ideal magneto-hydrodynamic invariants. The case of the high-β stellar radiation zone plasma, where gas pressure dominates, has been studied both theoretically (Duez and Mathis 2010) and numerically (Braithwaite and Spruit 2004; Braithwaite and Nordlund 2006; Braithwaite 2008). Roughly axisymmetric dipolar twisted configurations are obtained if the initial magnetic energy is confined near the centre (Braithwaite and Nordlund 2006; Duez and Mathis 2010), while one gets non-axisymmetric fields in the case where the energy is distributed in the whole radiation zone (Braithwaite 2008). The field is then organised on large scales, mixed (poloidal and toroidal), non force-free configurations, which are ini-

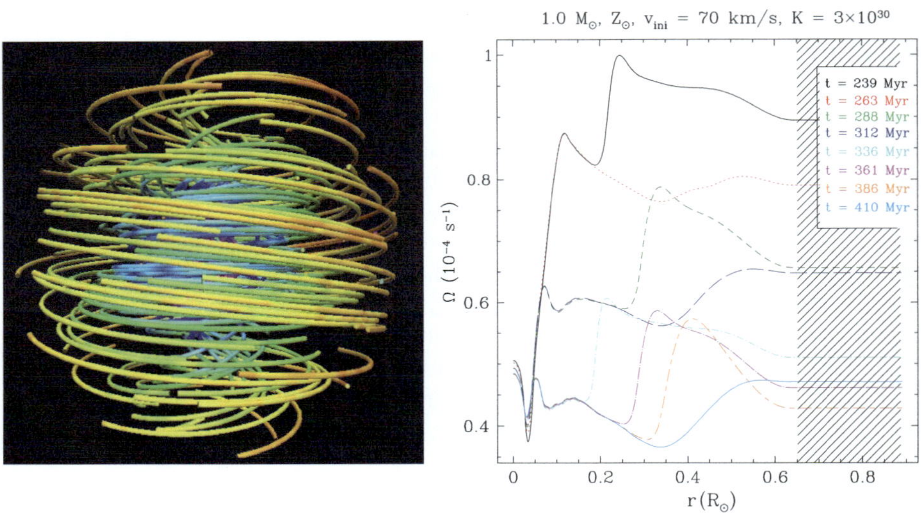

Fig. 11 *Left*: 3-D topology of an axisymmetric relaxed fossil magnetic field (from Duez et al. 2010a, 2010b). In the case of such configuration, the angular velocity will be frozen along the lines of the poloidal (i.e. meridional) component of the field. *Right*: Net extraction of angular momentum by internal waves in a solar-type star. The braking applied by the wind induces the extraction of angular momentum along a front created because of the propagation of retrograde waves deep in the core. *Hatched areas* indicate the convective envelope. from Alvan et al. (2013)

tially stable as has been demonstrated by Braithwaite (2009) and Duez et al. (2010a) (see Fig. 4 of Duez et al. 2010a).

Once the initial relaxed magnetic configuration has been established (see left panel of Fig. 11), it interacts with differential rotation. In the case of axisymmetric configurations, the rotation becomes uniform on poloidal (i.e. meridional) magnetic surfaces; this is the so-called Ferraro-state (e.g. Brun and Zahn 2006; Spada et al. 2010; Strugarek et al. 2011a). In the case of a non-axisymmetric field, the rotation will become uniform if the field is strong (e.g. the oblique rotators case; see Moss 1992) while the case of a weak field is being explored (see Spruit 1999; Strugarek et al. 2011b). This picture could be modified by magnetic instabilities if during the first phase of evolution the residual differential rotation on each magnetic surface is able to generate a strong toroidal component of the field that becomes unstable (e.g. Tayler 1973), and if this instability is able to trigger a dynamo action through an α-effect. This question is strongly debated (Spruit 2002; Zahn et al. 2007; Rüdiger et al. 2012).

Finally, let us examine the interaction of fossil fields with meridional currents. Meridional flows in radiation zones are driven by internal viscous and magnetic stresses, by external torques such as those due to stellar winds, and by structural adjustments during stellar evolution. In the case where all these sources vanish, the meridional circulation dies after an Eddington-Sweet time and the star settles in a baroclinic state described by the magnetic thermal wind equation. When applied to magnetised radiation zones of a star without structural adjustments and external torques, the meridional circulation will thus be mainly driven by the residual Lorentz torque until the star reaches a torque-free state and the advection of angular momentum by meridional circulation balances the residual magnetic torque (see Mestel et al. 1988; Mathis and Zahn 2005).

6.5 Internal Waves

The last transport mechanism that we consider here is the action of internal waves propagating in the stably stratified radiative cores of the Sun and solar-type stars. These are excited by turbulent convective flows at the radiation/convection border (e.g. Garcia Lopez and Spruit 1991; Dintrans et al. 2005; Rogers and Glatzmaier 2005; Belkacem et al. 2009). These waves transport angular momentum, which is deposited where waves are damped because of thermal diffusion (e.g. Zahn et al. 1997) or where they meet their co-rotation layers (Goldreich and Nicholson 1989; Alvan and Mathis 2011). In case of prograde waves, the momentum flux is of the same sign as that of energy flux leading to a deposition of angular momentum. For retrograde waves, the momentum and energy fluxes are of opposite signs leading to an extraction of angular momentum. Then, if a negative gradient of differential rotation is initially present, a net extraction of angular momentum is obtained because of the Doppler effect that leads to a stronger damping of prograde waves which are thus damped closer to the excitation region. Then, different angular momentum extraction fronts propagate from the interior to the exterior during the evolution (see Fig. 11 [right]). This extraction, combined with the strong horizontal shear-induced turbulence due to the stable stratification, is therefore a possible explanation of the observed quasi-uniform rotation rate in the inner $0.2\,R_{\odot}$ of the solar radiative zone, as well as an explanation of mixing in low-mass stars (Charbonnel and Talon 2005; Talon and Charbonnel 2005). Additionally, when the action of internal waves is efficient, meridional circulation is strongly affected and becomes highly multi-cellular because of Reynolds stresses.

Progress still needs to be made in the description of wave-excitation by turbulent convection (Brun et al. 2011; Alvan et al. 2012; Lecoanet and Quataert 2013) and on the impact of rotation and magnetic fields on their propagation, dissipation and on the induced transport of angular momentum. Indeed, the young Sun was rotating faster than the current Sun with an internal differential rotation caused by contraction and braking. The tachocline is believed to be the place of the large-scale toroidal magnetic field storage in the Sun (Browning et al. 2006). This is why the changes in the structure of the waves and their dissipation by differential rotation and by magnetic fields have been studied (Mathis et al. 2008; Mathis 2009; Mathis and de Brye 2011, 2012). The waves become Magneto-Gravito-Inertial waves because of the three restoring forces: i.e. the buoyancy force, the Coriolis acceleration, and the Lorentz force. Then, depending on the ratio between excited frequencies (σ_c), the inertial frequency (2Ω, where Ω is the angular velocity) and the Alfvén frequency (ω_{A}), the propagation of these waves can be strongly modified. First, in the case of an axisymmetric toroidal field (e.g. Schatzman 1993b; Kim and MacGregor 2003; Rogers and MacGregor 2011; Mathis and de Brye 2011), waves become vertically trapped as soon as $1 - m^2\omega_{\mathrm{A}}^2/\sigma_c^2 < 0$ (where m is the azimuthal order). Next, for such a field geometry, low-frequency waves (for example sub-inertial ones for which $\sigma_c < 2\Omega$) become trapped in an equatorial belt below a given critical latitude with a stronger trapping for prograde waves when the magnetic tension force is taken into account. This modifies the transmission of energy from turbulent convection to waves which is reduced in the case of trapped waves. Moreover, the thermal diffusion of waves is enhanced as soon as the rotation and the field amplitude is increased leading to a damping closer to the excitation region. The efficiency of the induced transport of angular momentum is thus a function of the rotation and of the field amplitude. It decreases as soon as vertical and horizontal trapping of wave modify wave-dynamics, with a net bias in favor of retrograde waves (see details given in Mathis and de Brye 2012).

6.6 The Coupling of the Sun and Other Solar-Type Stars to Their Surroundings

As has been emphasised in previous sections, torques applied on a star's outer envelope modify internal transport processes: for instance they can drive the large-scale meridional circulation (e.g. Zahn 1992). Therefore, one must get a coherent physical model of the interaction of the Sun and solar-type stars with their environment when their rotational evolution is studied (Bouvier 2008). First, stellar winds must be carefully examined as a function of a star's rotation rate and of its magnetic field topology and amplitude (e.g. Kawaler 1988; Pinto et al. 2011; Matt et al. 2012). Furthermore, if the star under scrutiny hosts a planetary system, the coupling with the proto-planetary disk (Matt and Pudritz 2005; Zanni and Ferreira 2013) and tidal interactions should be taken into account. For the latter, the equilibrium tide applied to the hydrostatic adjustment of the star to the tidal excitation, leads to a net torque applied on the outer convective envelope of solar-type stars (Zahn 1966; Remus et al. 2012) and this modifies their rotational evolution (Zahn and Bouchet 1989). Moreover, internal waves are also excited by the tidal potential at the radiation/convection boundary and are able to transport angular momentum in the same way that those excited by the turbulent penetrative convection: this is the so-called dynamical tide (see e.g. Zahn 1975; Ogilvie and Lin 2007). Therefore, we find that one needs a detailed picture of the secular exchanges of angular momentum to get a precise and exact picture of the structure and evolution of the Sun and solar-type stars. Constraints from helio- and asteroseismology will probably be our best guide.

7 Concluding Thoughts

The solar model is a work in progress. In this article we have tried to give a flavour of some of the uncertainties in the model ingredients, such as abundances and the equation of state, and discuss one of the most neglected physical properties, namely angular momentum transport.

Abundances and equation of state are important ingredients for all solar models. We have described the pain-staking analysis needed to measure solar abundances from solar spectra. We have also discussed the current discrepancy between abundance estimates from helioseismic analyses and spectroscopic analyses. We have also shown how equations of state can be tested and the state of the currently used equations of state.

An examination of what seismic data reveals about the radiative zone and comparing the results models reveals the limitations of standard solar models. The equations of standard solar models need to be supplemented with equations that describe dynamical evolutions and angular momentum transport. We have described some of the ways angular momentum may be transported in the Sun. Seismic data also point the way to investigations into other interesting processes that may play a role in transporting energy from the solar core, including gravity mode propagation, neutrinos, plasma processes and perhaps even dark matter. We except the next generation of observations and experiments to provide better constraints on these ingredients.

References

M.C. Abrams, A. Goldman, M.R. Gunson, C.P. Rinsland, R. Zander, Appl. Opt. **35**, 2747–2751 (1996)

C.W. Allen, *Astrophysical Quantities* (1973)

L. Alvan, S. Mathis, Critical layers for internal waves in stellar radiation zones, in *SF2A-2011: Proceedings of the Annual Meeting of the French Society of Astronomy and Astrophysics*, ed. by G. Alecian, K. Belkacem, R. Samadi, D. Valls-Gabaud (2011), pp. 443–447

L. Alvan, A.S. Brun, S. Mathis, 3D simulations of internal gravity waves in stellar interiors, in *SF2A-2012: Proceedings of the Annual Meeting of the French Society of Astronomy and Astrophysics*, ed. by S. Boissier, P. de Laverny, N. Nardetto, R. Samadi, D. Valls-Gabaud, H. Wozniak (2012), pp. 289–293
L. Alvan, S. Mathis, T. Decressin, Astron. Astrophys. **553**, A86 (2013). doi:10.1051/0004-6361/201321210
E. Anders, N. Grevesse, Geochim. Cosmochim. Acta **53**, 197–214 (1989)
P. Andre, D. Ward-Thompson, M. Barsony, *Protostars and Planets IV*, p. 59 (2000)
H.M. Antia, Astron. Astrophys. **307**, 609–623 (1996)
H.M. Antia, S. Basu, Astrophys. J. **426**, 801–811 (1994)
H.M. Antia, S. Basu, Astrophys. J. **644**, 1292–1298 (2006)
M. Asplund, N. Grevesse, A.J. Sauval, in *Cosmic Abundances as Records of Stellar Evolution and Nucleosynthesis*, ed. by T.G. Barnes III, F.N. Bash. Astronomical Society of the Pacific Conference Series, vol. 336 (2005), p. 25
M. Asplund, N. Grevesse, A.J. Sauval, P. Scott, Annu. Rev. Astron. Astrophys. **47**, 481–522 (2009)
N.R. Badnell, M.A. Bautista, K. Butler, F. Delahaye, C. Mendoza, P. Palmeri, C.J. Zeippen, M.J. Seaton, Mon. Not. R. Astron. Soc. **360**, 458–464 (2005)
J.N. Bahcall, M.H. Pinsonneault, G.J. Wasserburg, Rev. Mod. Phys. **67**, 781–808 (1995). doi:10.1103/RevModPhys.67.781
J.N. Bahcall, M.H. Pinsonneault, S. Basu, Astrophys. J. **555**, 990–1012 (2001)
J.N. Bahcall, A.M. Serenelli, S. Basu, Astrophys. J. Lett. **621**, 85–88 (2005)
J.N. Bahcall, A.M. Serenelli, Sarbani Basu, Astrophys. J. Suppl. Ser. **165**, 400 (2006). doi:10.1086/504043
J.E. Bailey, G.A. Rochau, C.A. Iglesias, J. Abdallah Jr., J.J. Macfarlane, I. Golovkin, P. Wang, R.C. Mancini, P.W. Lake, T.C. Moore, M. Bump, O. Garcia, S. Mazevet, Phys. Rev. Lett. **99**(26), 265002 (2007)
C.S. Baldner, H.M. Antia, S. Basu, T.P. Larson, Astrophys. J. **705**, 1704–1713 (2009)
S. Basu, H.M. Antia, Mon. Not. R. Astron. Soc. **287**, 189–198 (1997)
S. Basu, H.M. Antia, Astrophys. J. Lett. **606**, 85–88 (2004)
S. Basu, H.M. Antia, Phys. Rep. **457**, 217–283 (2008)
S. Basu, J. Christensen-Dalsgaard, Astron. Astrophys. **322**, 5–8 (1997)
S. Basu, W. Däppen, A. Nayfonov, Astrophys. J. **518**, 985–993 (1999)
S. Basu, S. Turck-Chièze, G. Berthomieu, A.S. Brun, T. Corbard, G. Gonczi, J. Christensen-Dalsgaard, J. Provost, S. Thiery, A.H. Gabriel, P. Boumier, Astrophys. J. **535**, 1078–1084 (2000)
S. Basu, W.J. Chaplin, Y. Elsworth, R. New, A.M. Serenelli, G.A. Verner, Astrophys. J. **655**, 660–671 (2007)
S. Basu, W.J. Chaplin, Y. Elsworth, R. New, A.M. Serenelli, Astrophys. J. **699**, 1403–1417 (2009)
S. Basu, A.-M. Broomhall, W.J. Chaplin, Y. Elsworth, Astrophys. J. **758**, 43 (2012)
P.G. Beck, J. Montalban, T. Kallinger, J. De Ridder, C. Aerts, R.A. García, S. Hekker, M.-A. Dupret, B. Mosser, P. Eggenberger, D. Stello, Y. Elsworth, S. Frandsen, F. Carrier, M. Hillen, M. Gruberbauer, J. Christensen-Dalsgaard, A. Miglio, M. Valentini, T.R. Bedding, H. Kjeldsen, F.R. Girouard, J.R. Hall, K.A. Ibrahim, Nature **481**, 55–57 (2012)
K. Belkacem, *Lecture Notes in Physics*, vol. 832 (Springer, Berlin, 2011)
K. Belkacem, R. Samadi, M.J. Goupil, M.A. Dupret, A.S. Brun, F. Baudin, Astron. Astrophys. **494**, 191–204 (2009)
C. Blancard, P. Cossé, G. Faussurier, Astrophys. J. **745**, 10 (2012)
E. Böhm-Vitense, Z. Astrophys. **46**, 108 (1958)
A. Bonaca, J.D. Tanner, S. Basu, W.J. Chaplin, T.S. Metcalfe, M.J.P.F.G. Monteiro, J. Ballot, T.R. Bedding, A. Bonanno, A.-M. Broomhall, H. Bruntt, T.L. Campante, J. Christensen-Dalsgaard, E. Corsaro, Y. Elsworth, R.A. García, S. Hekker, C. Karoff, H. Kjeldsen, S. Mathur, C. Régulo, I. Roxburgh, D. Stello, R. Trampedach, T. Barclay, C.J. Burke, D.A. Caldwell, Astrophys. J. Lett. **755**, 12 (2012)
J. Bouvier, Astron. Astrophys. **489**, 53–56 (2008)
J. Braithwaite, Mon. Not. R. Astron. Soc. **386**, 1947–1958 (2008)
J. Braithwaite, Mon. Not. R. Astron. Soc. **397**, 763–774 (2009)
J. Braithwaite, Å. Nordlund, Astron. Astrophys. **450**, 1077–1095 (2006)
J. Braithwaite, H.C. Spruit, Nature **431**, 819–821 (2004)
M.K. Browning, M.S. Miesch, A.S. Brun, J. Toomre, Astrophys. J. Lett. **648**, 157–160 (2006)
A.S. Brun, in *EAS Publications Series*, vol. 44, ed. by H. Wozniak, G. Hensler (2011), pp. 81–95
A.S. Brun, J.-P. Zahn, Astron. Astrophys. **457**, 665–674 (2006)
A.S. Brun, S. Turck-Chièze, J.P. Zahn, Astrophys. J. **525**, 1032–1041 (1999)
A.S. Brun, M.S. Miesch, J. Toomre, Astrophys. J. **742**, 79 (2011)
E. Caffau, H.-G. Ludwig, P. Bonifacio, R. Faraggiana, M. Steffen, B. Freytag, I. Kamp, T.R. Ayres, Astron. Astrophys. **514**, 92 (2010)
E. Caffau, H.-G. Ludwig, M. Steffen, B. Freytag, P. Bonifacio, Sol. Phys. **268**, 255–269 (2011)
T. Ceillier, P. Eggenberger, R.A. García, S. Mathis, Astron. Nachr. **333**, 971 (2012)
W.J. Chaplin, A.M. Serenelli, S. Basu, Y. Elsworth, R. New, G.A. Verner, Astrophys. J. **670**, 872–884 (2007)

C. Charbonnel, S. Talon, Science **309**, 2189–2191 (2005)
J. Christensen-Dalsgaard, Rev. Mod. Phys. **74**, 1073–1129 (2002)
J. Christensen-Dalsgaard, IFA print (2003). http://astro.phys.au.dk/~jcd/oscilnotes/
J. Christensen-Dalsgaard, W. Daeppen, Astron. Astrophys. Rev. **4**, 267–361 (1992)
J. Christensen-Dalsgaard, W. Dappen, S.V. Ajukov, E.R. Anderson, H.M. Antia, S. Basu, V.A. Baturin, G. Berthomieu, B. Chaboyer, S.M. Chitre, A.N. Cox, P. Demarque, J. Donatowicz, W.A. Dziembowski, M. Gabriel, D.O. Gough, D.B. Guenther, J.A. Guzik, J.W. Harvey, F. Hill, G. Houdek, C.A. Iglesias, A.G. Kosovichev, J.W. Leibacher, P. Morel, C.R. Proffitt, J. Provost, J. Reiter, E.J. Rhodes Jr., F.J. Rogers, I.W. Roxburgh, M.J. Thompson, R.K. Ulrich, Science **272**, 1286–1292 (1996)
J. Christensen-Dalsgaard, M.P. Di Mauro, G. Houdek, F. Pijpers, Astron. Astrophys. **494**, 205 (2009)
E.R. Cohen, B.N. Taylor, Rev. Mod. Phys. **59**, 1121–1148 (1987)
S. Couvidat, S. Turck-Chièze, A.G. Kosovichev, Astrophys. J. **599**, 1434–1448 (2003)
A.N. Cox, J.A. Guzik, Astrophys. J. Lett. **613**, 169–171 (2004)
W. Dappen, D.O. Gough, Progress report on helium abundance determination, in *Seismology of the Sun and the Distant Stars; Proceedings of the NATO Advanced Research Workshop*, Cambridge, England, June 17–21, 1985, (A87-20101 07-92) (Reidel, Dordrecht, 1986), pp. 275–280
W. Dappen, L. Anderson, D. Mihalas, Astrophys. J. **319**, 195–206 (1987)
T. Decressin, S. Mathis, A. Palacios, L. Siess, S. Talon, C. Charbonnel, J.-P. Zahn, Astron. Astrophys. **495**, 271–286 (2009)
S. Deheuvels, R.A. García, W.J. Chaplin, S. Basu, H.M. Antia, T. Appourchaux, O. Benomar, G.R. Davies, Y. Elsworth, L. Gizon, M.J. Goupil, D.R. Reese, C. Regulo, J. Schou, T. Stahn, L. Casagrande, J. Christensen-Dalsgaard, D. Fischer, S. Hekker, H. Kjeldsen, S. Mathur, B. Mosser, M. Pinsonneault, J. Valenti, J.L. Christiansen, K. Kinemuchi, F. Mullally, Astrophys. J. **756**, 19 (2012)
F. Delahaye, M.H. Pinsonneault, Astrophys. J. **649**, 529–540 (2006)
L. Delbouille, G. Roland, L. Neven, *Photometric Atlas of the Solar Spectrum from 3000 to 10000 A* (Univ. Liege/Obs. Roy. Belg., Liege, 1973) 380 pp.
L. Delbouille, G. Roland, J. Brault, L. Testerman, *Photometric Atlas of the Solar Spectrum from 1850 to 10000 cm*$^{-1}$ (Kitt Peak Natl. Obs./Univ. Liège, Liege, 1981) 172 pp.
B. Dintrans, A. Brandenburg, Å. Nordlund, R.F. Stein, Astron. Astrophys. **438**, 365–376 (2005)
J.-F. Donati, J.D. Landstreet, Annu. Rev. Astron. Astrophys. **47**, 333–370 (2009)
V. Duez, S. Mathis, Astron. Astrophys. **517**, 58 (2010)
V. Duez, J. Braithwaite, S. Mathis, Astrophys. J. Lett. **724**, 34–38 (2010a)
V. Duez, S. Mathis, S. Turck-Chièze, Mon. Not. R. Astron. Soc. **402**, 271–281 (2010b)
W.A. Dziembowski, A.A. Pamiatnykh, R. Sienkiewicz, Mon. Not. R. Astron. Soc. **249**, 602–605 (1991)
A. Eff-Darwich, S.G. Korzennik, Sol. Phys. **149** (2012)
P. Eggenberger, A. Maeder, G. Meynet, Astron. Astrophys. **440**, 9–12 (2005)
P. Eggenberger, J. Montalbán, A. Miglio, Astron. Astrophys. **544**, 4 (2012)
P.P. Eggleton, J. Faulkner, B.P. Flannery, Astron. Astrophys. **23**, 325 (1973)
J.R. Elliott, A.G. Kosovichev, Astrophys. J. Lett. **500**, 199 (1998)
Y. Elsworth, R. Howe, G.R. Isaak, C.P. McLeod, R. New, Nature **345**, 322–324 (1990)
D. Fabbian, E. Khomenko, F. Moreno-Insertis, Å. Nordlund, Astrophys. J. **724**, 1536–1541 (2010)
D. Fabbian, F. Moreno-Insertis, E. Khomenko, Å. Nordlund, Astron. Astrophys. **548**, 35 (2012)
C. Farmer, R. Norton, *A High-Resolution Atlas of the Infrared Spectrum of the Sun and the Earth Atmosphere from Space: A Compilation of ATMOS Spectra of the Region from 650 to 4800 cm*$^{-1}$ *(2.3 to 16 μm). Vol. I. The Sun* (Natl. Aeronautics Space Admin., Washington, 1989) 535 pp.
S.T. Fletcher, A.-M. Broomhall, D. Salabert, S. Basu, W.J. Chaplin, Y. Elsworth, R.A. Garcia, R. New, Astrophys. J. Lett. **718**, 19–22 (2010)
C. Fröhlich, J. Lean, Geophys. Res. Lett. **25**, 4377–4380 (1998)
A.H. Gabriel, G. Grec, J. Charra, J.-M. Robillot, T. Roca Cortés, S. Turck-Chièze, R. Bocchia, P. Boumier, M. Cantin, E. Cespédes, B. Cougrand, J. Crétolle, L. Damé, M. Decaudin, P. Delache, N. Denis, R. Duc, H. Dzitko, E. Fossat, J.-J. Fourmond, R.A. García, D. Gough, C. Grivel, J.M. Herreros, H. Lagardère, J.-P. Moalic, P.L. Pallé, N. Pétrou, M. Sanchez, R. Ulrich, H.B. van der Raay, Sol. Phys. **162**, 61–99 (1995)
A.H. Gabriel, F. Baudin, P. Boumier, R.A. García, S. Turck-Chièze, T. Appourchaux, L. Bertello, G. Berthomieu, J. Charra, D.O. Gough, P.L. Pallé, J. Provost, C. Renaud, J.-M. Robillot, T. Roca Cortés, S. Thiery, R.K. Ulrich, Astron. Astrophys. **390**, 1119–1131 (2002)
P. Garaud, J.-D. Garaud, Mon. Not. R. Astron. Soc. **391**, 1239–1258 (2008)
R.A. García, S. Turck-Chièze, S.J. Jiménez-Reyes, J. Ballot, P.L. Pallé, A. Eff-Darwich, S. Mathur, J. Provost, Science **316**, 1591 (2007)
R.A. García, A. Jiménez, S. Mathur, J. Ballot, A. Eff-Darwich, S.J. Jiménez-Reyes, P.L. Pallé, J. Provost, S. Turck-Chièze, Astron. Nachr. **329**, 476 (2008)

R.A. García, D. Salabert, J. Ballot, A. Eff-Darwich, R. Garrido, A. Jiménez, S. Mathis, S. Mathur, A. Moya, P.L. Pallé, C. Régulo, K. Sato, J.C. Suárez, S. Turck-Chièze, J. Phys. Conf. Ser. **271**(1), 012046 (2011a)
R.A. García, D. Salabert, J. Ballot, A. Eff-Darwich, R. Garrido, A. Jiménez, S. Mathis, S. Mathur, A. Moya, P.L. Pallé, C. Régulo, K. Sato, J.C. Suárez, S. Turck-Chièze, J. Phys. Conf. Ser. **271**, 012046 (2011b)
R.J. Garcia Lopez, H.C. Spruit, Astrophys. J. **377**, 268–277 (1991)
B. Gelly, M. Lazrek, G. Grec, A. Ayad, F.X. Schmider, C. Renaud, D. Salabert, E. Fossat, Astron. Astrophys. **394**, 285–297 (2002)
P. Goldreich, P.D. Nicholson, Astrophys. J. **342**, 1075–1078 (1989)
Z. Gong, W. Däppen, L. Zejda, Astrophys. J. **546**, 1178–1182 (2001)
D.O. Gough, M.E. McIntyre, Nature **394**, 755–757 (1998)
N. Grevesse, A. Noels, in *Origin and Evolution of the Elements*, ed. by N. Prantzos, E. Vangioni-Flam, M. Casse (Cambridge University Press, Cambridge, 1993), pp. 15–25, 561 pp.
N. Grevesse, A.J. Sauval, Space Sci. Rev. **85**, 161–174 (1998)
B. Gustafsson, B. Edvardsson, K. Eriksson, U.G. Jørgensen, Å. Nordlund, B. Plez, Astron. Astrophys. **486**, 951–970 (2008)
J.A. Guzik, K. Mussack, Astrophys. J. **713**, 1108–1119 (2010)
J.A. Guzik, F.J. Swenson, Astrophys. J. **491**, 967 (1997). doi:10.1086/304988
J.A. Guzik, L.S. Watson, A.N. Cox, Astrophys. J. **627**, 1049–1056 (2005). doi:10.1086/430438
F. Hase, L. Wallace, S.D. McLeod, J.J. Harrison, P.F. Bernath, J. Quant. Spectrosc. Radiat. Transf. **111**, 521–528 (2010)
H. Holweger, E.A. Mueller, Sol. Phys. **39**, 19–30 (1974)
D.G. Hummer, D. Mihalas, Astrophys. J. **331**, 794–814 (1988)
C.A. Iglesias, F.J. Rogers, Astrophys. J. **464**, 943 (1996)
S.D. Kawaler, Astrophys. J. **333**, 236–247 (1988)
E.-J. Kim, K.B. MacGregor, Astrophys. J. **588**, 645–654 (2003)
A.G. Kosovichev, J. Christensen-Dalsgaard, W. Daeppen, W.A. Dziembowski, D.O. Gough, M.J. Thompson, Mon. Not. R. Astron. Soc. **259**, 536–558 (1992)
C.J. Lada, *IAU Symposium*, vol. 115 (1987)
D. Lecoanet, E. Quataert, Mon. Not. R. Astron. Soc. **430**, 2363 (2013). doi:10.1093/mnras/stt055
L.H. Li, F.J. Robinson, P. Demarque, S. Sofia, D.B. Guenther, Astrophys. J. **567**, 1192–1201 (2002)
K.G. Libbrecht, M.F. Woodard, Nature **345**, 779–782 (1990)
K. Lodders, Solar system abundances of the elements, in *Principles and Perspectives in Cosmochemistry, Astrophysics and Space Science Proceedings* (Springer, Berlin, 2010), p. 379. ISBN 978-3-642-10351-3, doi:10.1007/978-3-642-10352-0_8
I. Lopes, S. Turck-Chièze, Astrophys. J. **765**(1), 14–23 (2013)
A. Maeder, *Physics, Formation and Evolution of Rotating Stars*. Astron. Astrophys. Library (Springer, Berlin, 2009). doi:10.1007/978-3-540-76949-1
A. Maeder, J.-P. Zahn, Astron. Astrophys. **334**, 1000–1006 (1998)
J.P. Marques, M.J. Goupil, Y. Lebreton, S. Talon, A. Palacios, K. Belkacem, R.-M. Ouazzani, B. Mosser, A. Moya, P. Morel, B. Pichon, S. Mathis, J.-P. Zahn, S. Turck-Chièze, P.A.P. Nghiem, Astron. Astrophys. **549**, 74 (2013)
S. Mathis, Astron. Astrophys. **506**, 811–828 (2009)
S. Mathis, in *Lecture Notes in Physics*, vol. 865, ed. by M. Goupil, K. Belkacem, C. Neiner, F. Lignières, J.J. Green (Springer, Berlin, 2013), p. 23
S. Mathis, N. de Brye, Astron. Astrophys. **526**, 65 (2011)
S. Mathis, N. de Brye, Astron. Astrophys. **540**, 37 (2012)
S. Mathis, J.-P. Zahn, Astron. Astrophys. **425**, 229–242 (2004)
S. Mathis, J.-P. Zahn, Astron. Astrophys. **440**, 653–666 (2005)
S. Mathis, S. Talon, F.-P. Pantillon, J.-P. Zahn, Sol. Phys. **251**, 101–118 (2008)
S. Matt, R.E. Pudritz, Astrophys. J. Lett. **632**, 135–138 (2005)
S.P. Matt, K.B. MacGregor, M.H. Pinsonneault, T.P. Greene, Astrophys. J. Lett. **754**, 26 (2012)
L. Mestel, in *IAU Symposium*, vol. 22, ed. by R. Lust (1965), p. 420
L. Mestel, R.J. Tayler, D.L. Moss, Mon. Not. R. Astron. Soc. **231**, 873–885 (1988)
D. Mihalas, W. Dappen, D.G. Hummer, Astrophys. J. **331**, 815–825 (1988)
D. Moss, Mon. Not. R. Astron. Soc. **257**, 593–601 (1992)
B. Mosser, M.J. Goupil, K. Belkacem, J.P. Marques, P.G. Beck, S. Bloemen, J. De Ridder, C. Barban, S. Deheuvels, Y. Elsworth, S. Hekker, T. Kallinger, R.M. Ouazzani, M. Pinsonneault, R. Samadi, D. Stello, R.A. García, T.C. Klaus, J. Li, S. Mathur, R.L. Morris, Astron. Astrophys. **548**, 10 (2012)
H. Neckel, D. Labs, Sol. Phys. **90**, 205–258 (1984)
Å. Nordlund, R.F. Stein, M. Asplund, Living Rev. Sol. Phys. **6**, 2 (2009)
G.I. Ogilvie, D.N.C. Lin, Astrophys. J. **661**, 1180–1191 (2007)

L. Piau, R. Collet, R.F. Stein, R. Trampedach, P. Morel, S. Turck-Chieze, Mon. Not. R. Astron. Soc. **437**, 164 (2014). doi:10.1093/mnras/stt1866
M.H. Pinsonneault, S.D. Kawaler, S. Sofia, P. Demarque, Astrophys. J. **338**, 424–452 (1989)
R.F. Pinto, A.S. Brun, L. Jouve, R. Grappin, Astrophys. J. **737**, 72 (2011)
F. Remus, S. Mathis, J.-P. Zahn, Astron. Astrophys. **544**, 132 (2012)
O. Richard, S. Vauclair, C. Charbonnel, W.A. Dziembowski, Astron. Astrophys. **312**, 1000–1011 (1996)
O. Richard, W.A. Dziembowski, R. Sienkiewicz, P.R. Goode, Astron. Astrophys. **338**, 756–760 (1998)
M. Rieutord, Astron. Astrophys. **451**, 1025–1036 (2006)
J. Robrade, J.H.M.M. Schmitt, F. Favata, Astron. Astrophys. **486**, 995–1002 (2008)
T.M. Rogers, G.A. Glatzmaier, Mon. Not. R. Astron. Soc. **364**, 1135–1146 (2005)
T.M. Rogers, K.B. MacGregor, Mon. Not. R. Astron. Soc. **410**, 946–962 (2011)
F.J. Rogers, A. Nayfonov, Astrophys. J. **576**, 1064–1074 (2002)
F.J. Rogers, F.J. Swenson, C.A. Iglesias, Astrophys. J. **456**, 902 (1996)
C.S. Rosenthal, J. Christensen-Dalsgaard, Å. Nordlund, R.F. Stein, R. Trampedach, Astron. Astrophys. **351**, 689–700 (1999)
G. Rudiger, L.L. Kitchatinov, Astron. Nachr. **318**, 273 (1997)
G. Rüdiger, L.L. Kitchatinov, D. Elstner, Mon. Not. R. Astron. Soc. **425**, 2267–2276 (2012)
D. Salabert, R.A. García, P.L. Pallé, S.J. Jiménez-Reyes, Astron. Astrophys. Lett. **504**, 1–4 (2009)
E. Schatzman, Astron. Astrophys. **279**, 431–446 (1993a)
E. Schatzman, Astron. Astrophys. **271**, 29 (1993b)
P.H. Scherrer, R.S. Bogart, R.I. Bush, J.T. Hoeksema, A.G. Kosovichev, J. Schou, W. Rosenberg, L. Springer, T.D. Tarbell, A. Title, C.J. Wolfson, I. Zayer, MDI Engineering Team, Sol. Phys. **162**, 129–188 (1995)
J.T. Schmelz, K. Nasraoui, J.K. Roames, L.A. Lippner, J.W. Garst, Astrophys. J. Lett. **634**, 197–200 (2005)
J. Schou, H.M. Antia, S. Basu, R.S. Bogart, R.I. Bush, S.M. Chitre, J. Christensen-Dalsgaard, M.P. di Mauro, W.A. Dziembowski, A. Eff-Darwich, D.O. Gough, D.A. Haber, J.T. Hoeksema, R. Howe, S.G. Korzennik, A.G. Kosovichev, R.M. Larsen, F.P. Pijpers, P.H. Scherrer, T. Sekii, T.D. Tarbell, A.M. Title, M.J. Thompson, J. Toomre, Astrophys. J. **505**, 390–417 (1998)
H. Schwabe, Astron. Nachr. **21**(495), 233–236 (1844)
A.M. Serenelli, S. Basu, Astrophys. J. **719**, 865–872 (2010)
R. Simoniello, K. Jain, S.C. Tripathy, S. Turck-Chièze, C. Baldner, W. Finsterle, F. Hill, M. Roth, Astrophys. J. (2013). doi:10.1088/0004-637X/765/2/100
F. Spada, A.C. Lanzafame, A.F. Lanza, Mon. Not. R. Astron. Soc., **404**, 641–660 (2010)
E.A. Spiegel, J.-P. Zahn, Astron. Astrophys. **265**, 106–114 (1992)
H.C. Spruit, Astron. Astrophys. **349**, 189–202 (1999)
H.C. Spruit, Astron. Astrophys. **381**, 923–932 (2002)
R.F. Stein, A. Nordlund, Astrophys. J. **499**, 914 (1998)
A. Strugarek, A.S. Brun, J.-P. Zahn, Astron. Astrophys. **532**, 34 (2011a)
A. Strugarek, A.S. Brun, J.-P. Zahn, Astron. Nachr. **332**, 891 (2011b)
M. Takata, H. Shibahashi, in *IAU Symposium*, vol. 185, ed. by F.-L. Deubner, J. Christensen-Dalsgaard, D. Kurtz, (1998), p. 21
S. Talon, C. Charbonnel, Astron. Astrophys. **440**, 981–994 (2005)
R.J. Tayler, Mon. Not. R. Astron. Soc. **161**, 365 (1973)
I. Thaler, Private communication, 2012
S. Turck-Chièze, S. Couvidat, Rep. Prog. Phys. **74**(8), 086901 (2011)
S. Turck-Chièze, P. Lambert, Adv. Space Res. **40**, 907–914 (2007)
S. Turck-Chieze, I. Lopes, Astrophys. J. **408**, 347–367 (1993)
S. Turck-Chièze, I. Lopes, Res. Astron. Astrophys. **12**, 1107–1138 (2012)
S. Turck-Chièze, The OPAC consortium. High Energy Density Phys. **9**, 473 (2013). doi:10.1016/j.hedp.2013.04.004
S. Turck-Chièze, W. Däppen, E. Fossat, J. Provost, E. Schatzman, D. Vignaud, Phys. Rep. **230**, 57–235 (1993)
S. Turck-Chièze, S. Couvidat, A.G. Kosovichev, A.H. Gabriel, G. Berthomieu, A.S. Brun, J. Christensen-Dalsgaard, R.A. García, D.O. Gough, J. Provost, T. Roca-Cortes, I.W. Roxburgh, R.K. Ulrich, Astrophys. J. Lett. **555**, 69–73 (2001)
S. Turck-Chièze, S. Couvidat, L. Piau, J. Ferguson, P. Lambert, J. Ballot, R.A. García, P. Nghiem, Phys. Rev. Lett. **93**(21), 211102 (2004a)
S. Turck-Chièze, R.A. García, S. Couvidat, R.K. Ulrich, L. Bertello, F. Varadi, A.G. Kosovichev, A.H. Gabriel, G. Berthomieu, A.S. Brun, I. Lopes, P. Pallé, J. Provost, J.M. Robillot, T. Roca Cortés, Astrophys. J. **604**, 455–468 (2004b)
S. Turck-Chièze, F. Delahaye, D. Gilles, G. Loisel, L. Piau, High Energy Density Phys. **5**, 132–138 (2009)
S. Turck-Chièze, A. Palacios, J.P. Marques, P.A.P. Nghiem, Astrophys. J. **715**, 1539–1555 (2010a)
S. Turck-Chièze, A. Palacios, J.P. Marques, P.A.P. Nghiem, Astrophys. J. **715**, 1539–1555 (2010b)

S. Turck-Chièze, L. Piau, S. Couvidat, Astrophys. J. Lett. **731**, 29 (2011)
S. Turck-Chièze, R.A. García, I. Lopes, J. Ballot, S. Couvidat, S. Mathur, D. Salabert, J. Silk, Astrophys. J. Lett. **746**, 12–16 (2012)
S.V. Vorontsov, V.A. Baturin, S.V. Ayukov, V.K. Gryaznov, Mon. Not. R. Astron. Soc. **430**, 1636–1652 (2013). doi:10.1093/mnras/sts701
T.S. Wood, J.O. McCaslin, P. Garaud, Astrophys. J. **738**, 47 (2011)
P.R. Young, Astron. Astrophys. **444**, 45–48 (2005)
J.P. Zahn, Ann. Astrophys. **29**, 489 (1966)
J.-P. Zahn, Astron. Astrophys. **41**, 329–344 (1975)
J.-P. Zahn, Astron. Astrophys. **252**, 179–188 (1991)
J.-P. Zahn, Astron. Astrophys. **265**, 115–132 (1992)
J.-P. Zahn, L. Bouchet, Astron. Astrophys. **223**, 112–118 (1989)
J.-P. Zahn, S. Talon, J. Matias, Astron. Astrophys. **322**, 320–328 (1997)
J.-P. Zahn, A.S. Brun, S. Mathis, Astron. Astrophys. **474**, 145–154 (2007)
C. Zanni, J. Ferreira, Astron. Astrophys. **550**, 99 (2013)
C. Zhang, L. Deng, D. Xiong, J. Christensen-Dalsgaard, Astrophys. J. Lett. **759**, 14 (2012)

Space Sci Rev (2015) 196:79–99
DOI 10.1007/s11214-015-0144-0

Solar Dynamics, Rotation, Convection and Overshoot

S. Hanasoge[1] · M.S. Miesch[2] · M. Roth[3] · J. Schou[4] · M. Schüssler[4] · M.J. Thompson[2]

Received: 18 June 2014 / Accepted: 4 March 2015 / Published online: 25 March 2015

Abstract We discuss recent observational, theoretical and modeling progress made in understanding the Sun's internal dynamics, including its rotation, meridional flow, convection and overshoot. Over the past few decades, substantial theoretical and observational effort has gone into appreciating these aspects of solar dynamics. A review of these observations, related helioseismic methodology and inference and computational results in relation to these problems is undertaken here.

Keywords Sun · Dynamics · Rotation · Convection · Overshoot

1 Introduction

Rotation and convection play an important role in solar and stellar evolution. They also play a crucial role in generating the magnetic activity exhibited by the Sun and other stars.

Rotationally induced instabilities, convection and overshoot mix chemical elements within stars, with possible consequences both for the observed chemical abundances at the surface of the star and for the availability of fuel for the nuclear energy generation in the star's core. Overshoot too can affect the chemical balance of the star, in the case of the Sun for example transporting a fraction of the fragile elements lithium and beryllium to regions where the temperature is sufficiently high for these elements to be destroyed.

Rotation and convection, and their interplay, can create dynamo action and generate magnetic field in a star. In the Sun, the overshoot layer at the base of the solar convective envelope may also play a crucial role in storing the magnetic field until it is strong enough to become

✉ M.J. Thompson
michael.thompson@sheffield.ac.uk

1 Tata Institute of Fundamental Research, Mumbai 400-005, India

2 High Altitude Observatory, National Center for Atmospheric Research, Boulder, CO 80301, USA

3 Kiepenheuer-Institut für Sonnenphysik, 79104 Freiburg, Germany

4 Max-Planck-Institut für Sonnensystemforschung, Justus-von-Liebig-Weg 3, 37077 Göttingen, Germany

buoyantly unstable and rise through the convection zone to the surface. The convection interacts with the rising magnetic field and largely determines how the magnetic field appears at the surface.

Observational data on the rotation, convection and other internal dynamics of the Sun were formerly restricted to what could be observed at the surface: granulation and supergranulation set up by the convective motions, and the surface rotation rate inferred from spectroscopic measurements and from observing the motion of tracers of the rotation such as sunspots and other surface features. This situation has been revolutionized by helioseismology, the observation and analysis of global oscillations and more localized acoustic wave propagation. At the same time as helioseismology has revolutionized the observational study of the Sun's internal dynamics, advances in numerical simulations—greatly assisted by the massive increase in computational power of modern supercomputers—have enabled great advances in the modeling and consequent understanding of the internal dynamics of the Sun and other stars. This chapter will discuss these observational and theoretical advances.

2 Flows in the Solar Interior

The global-scale flows in the Sun are its rotation and meridional circulation. Flows on smaller scales include convection, from granular scales to the putative giant-cell convection, and outflows and inflows around and beneath active regions. This section gives a brief overview of these different flows. The succeeding sections will then focus on what has been learned observationally about these flows from helioseismology, and the theoretical advances being made in modeling them.

2.1 Rotation

Like other stars, the Sun acquired its angular momentum from the interstellar gas cloud from which it formed. As the gas contracts to form a star, unless there is some extremely efficient mechanism for losing angular momentum, the proto-star spins up so as to rotate much faster than the parent cloud. Thus young stars are typically observed to be fast rotators, and that was presumably the case for the Sun also. Over the next 4–5 billion years, the Sun would have lost angular momentum from its surface layers via the solar wind (see, e.g., Bouvier 2013; Gallet and Bouvier 2013, for models). If there were no angular-momentum transport mechanisms at play in the solar interior, this would result in a fast-rotating core and a slowly rotating envelope. The evidence from helioseismology (Sect. 3) does not support such a picture, indicating that the radiative interior is rotating nearly uniformly at a rate intermediate between the polar and equatorial rate of the convection zone. This indicates that one or more mechanisms—magnetic fields, transport by rotationally induced instabilities and by gravitational waves (see review by, e.g. Mathis 2013)—have systematically extracted angular momentum from the radiative interior and redistributed the residual, suppressing rotational gradients. The solar surface has long been observed to rotate differentially, with the mid- and high-latitude regions rotating more slowly than the equatorial region. That this differential rotation largely persists with depth throughout the solar convective envelope is now established observationally by helioseismology. Angular momentum transport must maintain this differential rotation, and the mechanisms for that are addressed below in Sect. 4.1.

2.2 Meridional Circulation

There is a poleward meridional circulation in the near-surface layers of the Sun in both the north and south hemispheres. Mass conservation dictates that their must be a return, equatorward, flow somewhere in the solar interior, and this is often presumed to occur near the base of the convection zone. However, whether a single meridional cell spans the whole depth of the convection zone or whether it is shallower, with perhaps other meridional cells stacked beneath the one nearest the surface, is a matter of active discussion. There is also uncertainty about whether the poleward flow extends right up to the poles, or whether there might be a counter cell at high latitudes. There are marginal observations at best to say whether or not there is any meridional circulation in the Sun's radiative interior, though there are theoretical arguments for the existence of a slow, Eddington-Sweet circulation (e.g. Zahn 1992).

Determining the meridional flow as a function of radius and latitude is of considerable interest. This is to a large extent driven by the importance of the meridional flow in certain dynamo theories.

2.3 Convection and Convective Overshoot

Energy transport in the optically thin solar photospheric layers transitions from being effected by convection to free-streaming radiation. A spatio-temporal power spectrum of photospheric flows reveals granular and supergranular scales. Observed properties of granules, such as spatial scales, radiative intensity and spectral-line formation are highly accurately reproduced by numerical simulations (e.g., Stein and Nordlund 2000; Vögler et al. 2005; Nordlund et al. 2009). One may conjecture that the success of the simulations in spite of being in an entirely different parameter regime is due to the small scale height ($\sim$200 km, compared with radius of the Sun, $R_\odot \sim 700{,}000$ km), leading to the strong expansion of convective upflows, thereby smoothing out large fluctuations and making them almost laminar (Nordlund et al. 1997). The effects of turbulence are restricted to the downflows, where they appear to not have a big impact on the observable near-surface dynamics. Thus incorporating the ingredients of an accurate equation of state and background stratification lead to a high-fidelity reproduction of line formation and spatial scales.

Ostensibly, modeling convection in the solar interior presents a more formidable challenge, since there appear to be no dominant physical ingredients that are concurrently computationally tractable. Interior convection is likely governed by aspects more difficult to model, such as the integrity of descending plumes to diffusion and various instabilities (Rast 1998). Further, solar convection is governed by extreme parameters (Prandtl number $\sim 10^{-6}$–10^{-4}, Rayleigh number $\sim 10^{19}$–10^{24}, and Reynolds number $\sim 10^{12}$–10^{16}; Miesch 2005), which makes fully resolved three-dimensional direct numerical simulations impossible for the foreseeable future. It is likewise difficult to reproduce solar parameter regimes in laboratory experiments.

Phenomenology such as mixing-length theory (MLT) treats convective transport as being effected by parcels of fluid of specified spatial and velocity scales, coherent over a length scale termed the *mixing length*. Despite its overt simplicity (Weiss et al. 2004), it has been remarkably successful as an integral component of stellar structure models (e.g. di Mauro et al. 2010) and in describing the sub-photospheric stratification and heat transport in local solar convection simulations (Trampedach and Stein 2011). Because density and pressure scale heights increase with depth, MLT posits a corresponding increase in spatial convective scale (while velocities reduce), suggesting the existence of large convective cells, the so-called *giant cells*. Three-dimensional simulations of global convection have been performed at increasing resolutions over the past few years (Miesch et al. 2008; Ghizaru et al. 2010; Käpylä et al. 2010, 2011; Guerrero et al. 2013a; Hotta et al. 2014a, 2014b), most invoking the *anelastic approximation* (Gough 1969), a regime describing low-Mach number strongly non-Boussinesq stratified convection. These simulations predict large-scale convective turnover and velocity amplitudes comparable to those of MLT. Considerable effort has been spent in attempting surface (Hathaway et al. 2000, 2013) and interior detection (Duvall et al. 1993; Duvall 2003) of giant cells but unambiguous identification remains elusive. This difficulty in detection suggests that the amplitudes of giant cells may be significantly less than predicted by MLT and convection simulations, potentially posing serious challenges to our understanding of deep solar convection (Sect. 3.3).

At the base of the convection zone, convective motions are quickly decelerated by the steep subadiabatic stratification of the radiative interior. This thin overshoot region (Sect. 3.3) coincides with the rotational shear layer known as the solar tachocline (Sect. 3.1) as first posited by Spiegel and Zahn (1992), possibly suggesting a causal connection. The dynamics in this region are complex, involving penetrative convection and internal waves interacting with a stably-stratified, magnetized shear flow and the much longer dynamical time scales of the radiative zone (Hughes et al. 2007).

3 Helioseismic Constraints on the Sun's Internal Dynamics

One of the great successes of helioseismology has been to measure the rotation of much of the solar interior, excluding the solar core and a region around the Sun's rotation axis. Much of this has been achieved using global-mode helioseismology (e.g., Deubner and Gough 1984), which proceeds by the analysis of the observed properties of global resonant modes of oscillation of the Sun, in particular the mode frequencies. See for example the excellent review by Howe (2009). Global-mode helioseismology has been complemented by the development and application of a number of other approaches known collectively as local helioseismology. These include time-distance helioseismology (Duvall et al. 1993), acoustic holography (Lindsey and Braun 1997), and ring-diagram analysis (Hill 1988). Local helioseismology has been particularly used to probe the Sun's meridional (i.e. north-south) circulation and convective motions in the upper part of the Sun's convection zone, and motions around and under active regions (e.g., Komm et al. 2005).

3.1 Rotation

The global-mode frequencies ω_{nlm} of the Sun are labeled by the radial order n, and the degree l and azimuthal order m of the spherical harmonic that describes the horizontal structure of the mode. In the absence of rotation and other departures from perfect spherical symmetry, the frequencies would be independent of m. The Sun's rotation lifts that degeneracy. The difference in frequency between modes with like values of n and l but different values of m is called the splitting. For a star with internal rotation rate $\Omega(r,\theta)$ (with respect to spherical polar coordinates r, θ, ϕ), the rotational splitting is given by Hansen et al. (1977)

$$\delta\omega_{nlm} \equiv \omega_{nlm} - \omega_{nl0} = -i \int \rho\left(\xi^*_{nlm} \cdot \frac{\partial}{\partial\phi}\xi_{nlm}\right)\Omega\, dV \Big/ \int \rho\xi^*_{nlm} \cdot \xi_{nlm} dV \tag{1}$$

where $\xi_{n;m}$ is the radial displacement eigenvector of the mode, ρ is the density, and the integrals are over the interior volume of the star. This can be conveniently rewritten as

$$\delta\omega_{nlm} = m \int_0^R \int_0^\pi K_{nlm}(r,\theta)\Omega(r,\theta) r d\theta dr \tag{2}$$

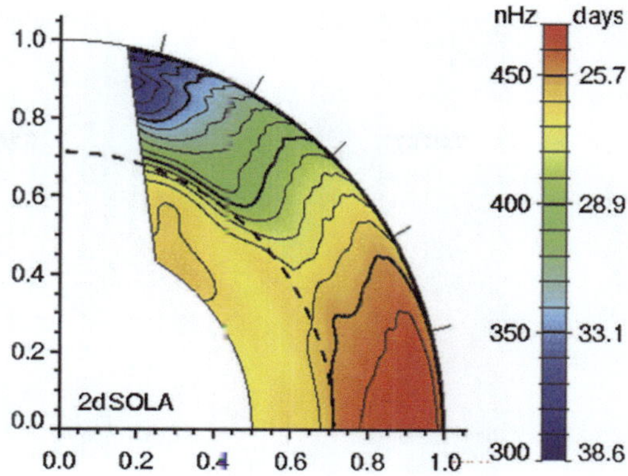

Fig. 1 Solar internal rotation as inferred from MDI observations. Contours of isorotation are shown. (Adapted from Schou et al. (1998))

where the K_{nlm} are known as rotational splitting kernels. If the static structure of the stellar interior is known, the kernels can be calculated, so that the only unknown in Eq. (2) is the rotation rate $\Omega(r, \theta)$. Measurements of the Sun's mode frequencies therefore provide a set of observational constraints of the form (2), which may be used to infer (or at least constrain) the solar internal rotation rate.

The solar internal rotation rate is thus fairly well determined from the photosphere down to beneath the convection zone, excluding a region around the poles. See e.g. Thompson et al. (1996), Schou et al. (1998), Howe et al. (2011), Eff-Darwich and Korzennik (2013) for examples of results using data from, respectively, GONG (Harvey et al. 1996), MDI (Scherrer et al. 1995) and HMI (Schou et al. 2012).

A typical rotation profile determined from helioseismology is shown in Fig. 1. The latitudinal variation of rotation observed at the surface largely persists through the convection zone, while in the radiative interior the results are consistent with solid-body rotation. But there are two rotational shear layers, a near-surface shear layer and another shear layer termed the tachocline that is located near the base of the convection zone.

Superimposed on the mean rotation profile at the surface and in the convection zone are weak but apparently coherent bands of faster and slower zonal flow that migrate in latitude over the course of the solar cycle (Fig. 2). These have been dubbed "torsional oscillations" (Howard and Labonte 1980; Schou 1999; Howe et al. 2000; Vorontsov et al. 2002). The causal connections between the flows and the magnetic manifestations of the solar cycle are unclear.

3.2 Meridional Circulation

Unlike the rotation, the meridional flow does not perturb the mode frequencies to first order. As a consequence most measurements of the meridional flow have been made using local helioseismology techniques, such as time-distance and ring-diagram analysis.

In principle, the time-distance measurements are quite straightforward. The N-S and S-N travel times are measured for a selection of latitude pairs and the results inverted to obtain the meridional flow as a function of latitude, depth and time. In reality, and as illustrated in Zhao et al. (2012), measurements of wave travel times (and seismic measurements in general) are prone to systematic errors, especially when they are taken at spatial locations away from the disk center. One manifestation of these errors is that the derived flows in the interior depend on the type of observations used, an unsatisfactory state of affairs. Making the assumption that the error is due to a center-to-limb time shift, Zhao et al. were able

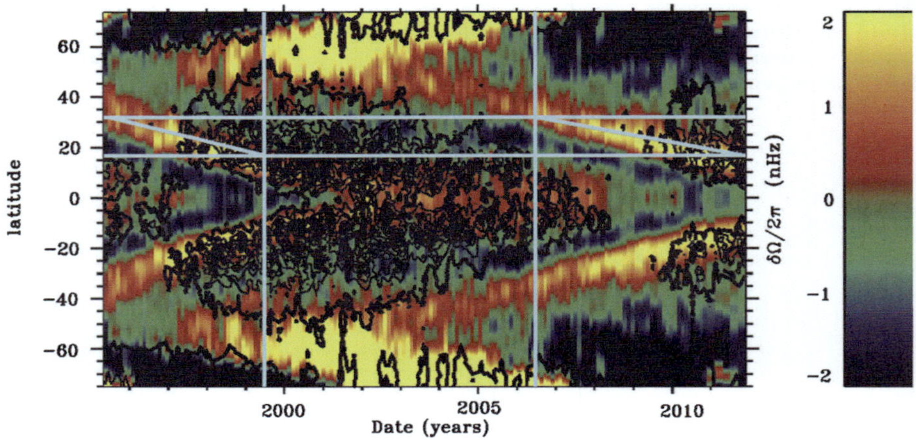

Fig. 2 Torsional oscillations in the form of faster (*red* and *yellow colors*) and slower (*green* and *blue/black colors*) zonal flows relative to the long-term average rotation, as inferred from helioseismology in the near subsurface layers of the Sun using MDI and GONG observations. Contours of photospheric magnetic field strength are superimposed to indicate the location of contemporaneous surface activity. The *left vertical blue line* marks the date 1997.3, when low-latitude flows most closely matched the inference from the 2009.2 data. The *vertical blue line* on the right signifies 2006.4, most closely resembling the analysis from the earliest data (1996.5). The *horizontal lines* mark locations of the flow bands, while the *slanted lines* indicate equatorward migration. (After Howe et al. (2009), courtesy of Rachel Howe.)

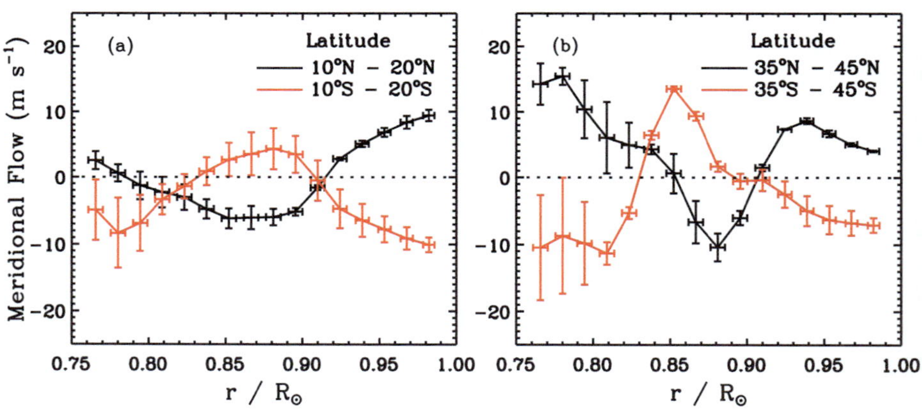

Fig. 3 Meridional flow profile as a function of radius at selected latitudes in the northern (*black*) and southern (*red*) hemispheres, obtained from time-distance helioseismology using HMI observations (Zhao et al. 2013). In the near-surface region the flows are poleward in both hemispheres, but the results indicate that there is a counter-cell beneath radius $0.90R_{\odot}$. There may be yet another cell beneath about $0.82R_{\odot}$, but the errors on the inferred flow are larger at such depths. (Figure courtesy of Junwei Zhao)

to diminish the discrepancy between observables and obtain an estimate of the meridional circulation over much of the solar convection zone. Remarkably, Zhao et al. (2013) found that the meridional circulation pattern consists of two circulation cells in the depth direction (Fig. 3), which if it were to hold up, will challenge some solar dynamo models (e.g. Jouve and Brun 2007). While suggestions have been made regarding the origin of the systematic errors (e.g., Baldner and Schou 2012), no definite conclusion has been reached so far.

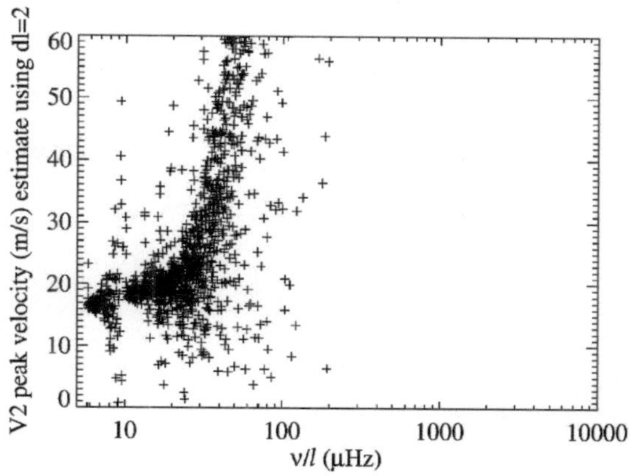

Fig. 4 Surface peak velocities of the horizontal meridional flow component of degree 2 as a function of ν/L (Woodard et al. 2013). (ν/L relates directly to the radial location r_t of the lower turning point of the mode: e.g., for $\nu/L = 10$ μHz, $r_t \simeq 0.98 R_\odot$, while for $\nu/L = 100$ μHz, $r_t \simeq 0.6 R_\odot$.) The analysis is based on fitting a model function to the cross-spectra between p-modes and is based on 500 days of HMI data

Ring-diagram analysis has also produced a number of interesting results on the meridional circulation (e.g., González Hernández et al. 1999, 2006, 2008, 2010; Haber et al. 2006). The inferred meridional circulation exhibits temporal variations, including the occasional appearance of an equatorward cell at high latitudes. Part of those variations is thought to be the result of a systematic error caused by uncertainty in the Carrington elements: the inferred equatorward meridional flow at high latitudes shows an annual variation with the B_0 angle (Zaatri et al. 2006). Nevertheless, a multi-cell structure in latitude is also seen in the Mount Wilson surface Doppler measurements (Ulrich 2010).

Lavely and Ritzwoller (1992) described a perturbation theoretical approach that included the study of general bulk flows. Following this theoretical framework, Roth and Stix (1999, 2008) studied the effect of giant cells and the meridional flow on the global mode frequencies. The effects are small, however, making it difficult for them to be measured and to be evaluated for helioseismic mappings of the solar interior. Recently, new approaches concentrate on investigating the perturbation of the eigenfunctions rather than the eigenfrequencies when studying the effect of the meridional flow on solar oscillations (Woodard 2000; Schad et al. 2011, 2012, 2013; Woodard et al. 2013). Since the meridional flow perturbs the solar model, the new oscillatory eigenstates expressed in the basis of the unperturbed eigenstates are mixtures of modes, which is often called mode coupling. Given an azimuthally symmetric meridional flow, such couplings only occur between modes of identical azimuthal order m. Depending on the complexity of the flow when expanded in terms of Legendre polynomials as a function of latitude, mode coupling may occur between modes that differ in l by the harmonic degree of the flow. As a consequence, power of one mode is expected to leak into another mode and vice versa. The cross-spectrum allows measuring this coupling of the modes. As systematic leakage is a dominant error source in the cross-spectra due to observational and instrumental constraints, detailed knowledge of the instrument and observation conditions must be taken into account.

Currently two approaches exist that estimate the meridional flow via cross-spectral analysis. The first one employs a fit of a model function to the observed cross-spectrum of modes that couple along the same ridge (Woodard et al. 2013). Figure 4 shows the resulting peak velocity of the meridional flow component with a harmonic degree of 2 as a function of the ratio $\nu/(l + 1/2)$, which is related to the inner turning point of the modes. Figure 4 would appear to indicate a rapid increase in the meridional flow below the surface. However, this may be an artifact related to the center-to-limb effect suggested by Zhao et al. (2012). The other approach evaluates ratios of the cross-spectral amplitudes (Schad et al. 2011, 2012) rather

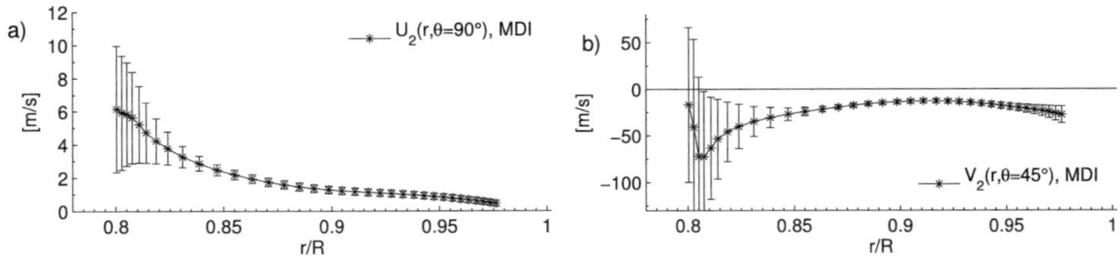

Fig. 5 The radial and horizontal amplitudes of the meridional flow component with harmonic degree 2 as a function of radius r/R at colatitude $\theta = 90°$ (equator; *left*) and $\theta = 45°$ (mid-latitude; *right*), respectively. The flow was measured by evaluating cross-spectral amplitude ratios obtained from MDI data covering the period 2004–2010. Positive values of U_2 indicate an upward radial flow, negative values of V_2 refer to a poleward flow (Schad et al. 2012)

than a fit to the cross-spectra and includes inter-ridge couplings, too. Here the radial component of the meridional flow is measured and the horizontal meridional flow is obtained via mass conservation. Through the use of ratios, this method appears robust against systematic errors present in the cross-spectra. Figure 5 displays the result for the radial meridional flow measured at the equator and the horizontal meridional flow measured at 45° latitude as a function of fractional solar radius. Again, the harmonic degree of the flow component is 2.

Evaluating mode couplings caused from flow components with a harmonic degree up to 10, Schad et al. (2013) were able to measure some of these components down to a fractional solar radius of approximately 0.5. Only the flow components with harmonic degrees 2 and 8 deviate significantly from zero, and give a first hint on a meridional flow that is confined between the tachocline region and the solar surface and that exhibits a multi-cellular structure as a function of radius and latitude. Figure 6 shows a composite of the even meridional flow components.

With both techniques the inferred solar near-surface horizontal flow is consistent with subsurface flow measurements from local helioseismology, with a poleward directed flow with a peak amplitude of order 20 m/s. However, we note that at greater depths the results shown in Fig. 3 and those derived by Hathaway (2012) are inconsistent with Fig. 6. Both Zhao et al. (2013) and Hathaway (2012) infer a return (equatorward) flow at $r \approx 0.9R_\odot$ whereas Fig. 6 suggests a flow that remains poleward down to $r = 0.8R_\odot$. Further, although qualitatively similar down to $r = 0.9R_\odot$ or so, the amplitudes of the return flow inferred by Hathaway (2012) and Zhao et al. (2013) are quantitatively different. Some of these discrepancies could be attributed to the fact that the analyzed data sets cover different time frames. Nevertheless, this is indicating that the results should be treated with some caution and that further work is needed to determine the source of the discrepancies.

3.3 Convection and Overshoot

Using techniques of time-distance helioseismology (Duvall et al. 1993; Duvall 2003; Gizon et al. 2010), Hanasoge et al. (2012) placed stringent bounds on the interior convective velocity spectrum. Two-point correlations measured from finite temporal segments of length of the observed line-of-sight photospheric Doppler velocities, taken by HMI, were used in the analysis. These correlations were spatially averaged according to a deep-focusing geometry (Hanasoge et al. 2010) in order to image the interior. Convective coherence timescales (Spruit 1974; Gough 1977; Miesch et al. 2008) were taken into account in choosing the temporal length of the data. By construction, these measurements are sensitive to the 3 components of the underlying flowfield, i.e., longitudinal, latitudinal or radial, at specific depths

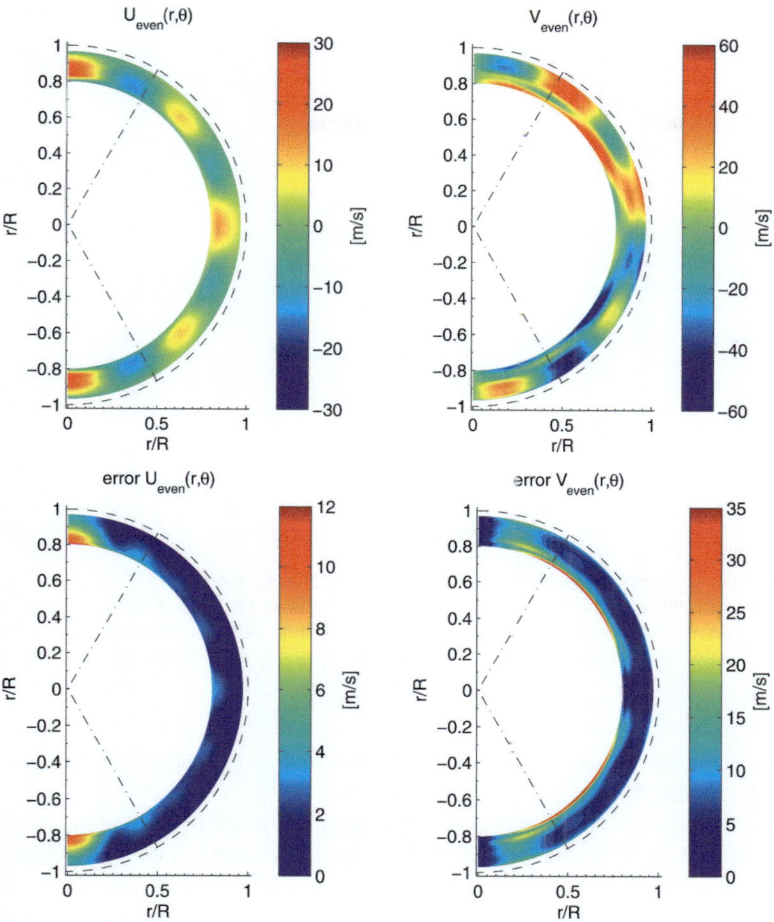

Fig. 6 Cross-sections through the meridional flow composed of the harmonic degrees 2, 4, 6, 8 (*top*) as a function of fractional radius and latitude based on evaluating cross-spectral amplitude ratios obtained from MDI data covering the period 2004–2010 (Schad et al. 2013). The *dashed-dotted lines* mark the latitudes of 60°. The radial flow is displayed on the left where positive (negative) values correspond to outward (inward) directed flows. The horizontal flow is displayed on the *right*; positive (negative) values correspond to northward (southward) directed flows. The 1σ standard error of the composite flow is given in the *lower panels*

of the solar interior ($r/R_\odot = 0.92, 0.96$). The constraints are shown in Fig. 7. The stark difference between observations and simulations suggests that the convection in the Sun may be operating in a strongly non-MLT regime. A plume-based thermal transport mechanism in this alternative regime has been the subject of speculation by, e.g., Spruit (1997) and was explored further by Rempel (2005a). The alternate mechanism put forward by Spruit (1997) would be able to account for the outward thermal transport of a solar luminosity's worth of heat flux at extremely low flow speed. However, the transport of angular momentum, i.e., the maintenance of differential rotation and meridional circulation, is not as easily explained. Based on scaling arguments, Miesch et al. (2012) estimate the minimum convective kinetic energy (and associated Reynolds' stresses) required to sustain these large-scale flow cir-

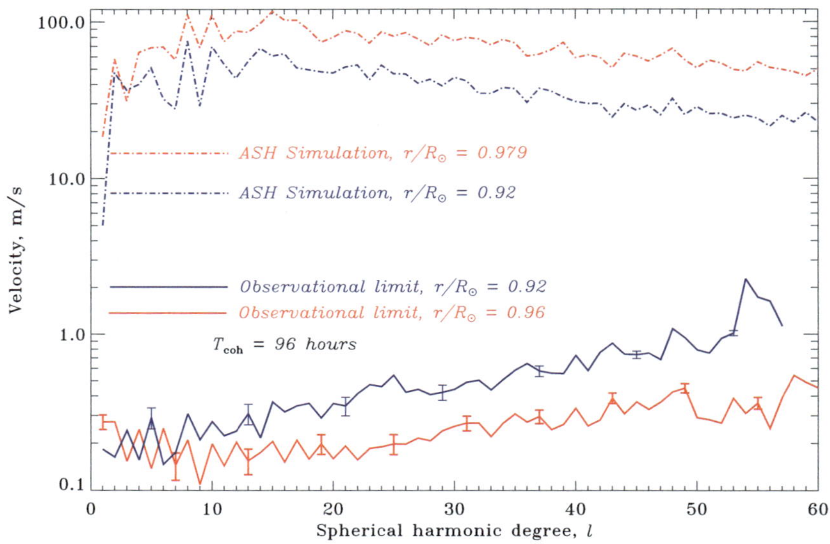

Fig. 7 Seismic constraints obtained by Hanasoge et al. (2012) using data from HMI (Schou et al. 2012). Assuming a convective coherence time of 96 hours, Hanasoge et al. (2012) obtained upper bounds on the observed convective spectrum. Shown in comparison is the Anelastic Spherical Harmonic (ASH) convective spectrum. These differences suggest that convection in the Sun may be operating in a strongly non-MLT regime

culations. Gizon and Birch (2012) provide an interesting comparison between seismology, simulation, and the phenomenology of Miesch et al. (2012).

Just as convection leads to an essentially adiabatically stratified envelope in the outer thirty per cent of the Sun, convective overshooting is expected to modify the stratification of the region in which it takes place (Skaley and Stix 1991; Deng and Xiong 2008). In the simplest picture, overshoot at the base of the convection zone leads to an extension of the adiabatically stratified region, with a more-or-less abrupt transition beneath that to the subadiabatic stratification of the radiative interior. Such a signature of a sharp transition in the sound-speed profile of the Sun has been sought with helioseismology (Basu et al. 1994; Monteiro et al. 1994; Roxburgh and Vorontsov 1994; Christensen-Dalsgaard et al. 1995). In principle, the amplitude of the abrupt transition can then be used to infer the extent of the overshoot region, and upper bounds have been quoted on the extent of the overshooting of about $0.005 R_\odot$ (Monteiro et al. 1994).

In fact, it appears that the transition between the convection zone and the radiative interior in the Sun is actually smoother even than in solar models that have no convective overshooting, which is a challenge to the simple picture above. A different model of convective overshoot has been proposed by Rempel (2004) wherein the overshoot is modeled with discrete plumes with a spectrum of strengths and hence depths of penetration into the radiative interior (also see Zahn 1991, for models of overshoot). Depending on the spectrum adopted, this can give a smoother transition, indeed the subadiabatic stratification can occur even towards the bottom of the convective region. Christensen-Dalsgaard et al. (2011) investigated the seismic signature of such models, compared with helioseismic observations, and found that some models of this class fitted the observations better than the simpler models without (the "Standard solar model") or with overshoot. They concluded that overshoot is necessary to improve the agreement between models and helioseismic constraints, that the required stratification profiles are outside the realm of classic "ballistic" overshoot models, and that the lower part of the convection zone is likely substantially subadiabatic. Overshoot has also been studied in detail using numerical simulations, with local regions (Brummell et al. 2002; Rogers and Glatzmaier 2005) and global 3-D domains (Brun et al. 2011).

4 Advances in Modeling the Sun's Internal Dynamics

Having discussed the observational findings from helioseismology about the Sun's internal dynamics, we consider now what is understood from theory and numerical simulations.

4.1 Solar Convection and Mean Flows

Recent insights into the nature of global-scale solar convection and mean flows have centered around two simple but powerful concepts. The first is thermal wind balance, which expresses the force balance in the meridional plane between the inertia of the differential rotation (Coriolis/centrifugal terms) and thermal gradients in latitude (baroclinic term):

$$\frac{\partial \Omega^2}{\partial z} = \frac{g}{r\lambda C_P} \frac{\partial \langle S \rangle}{\partial \theta}. \tag{3}$$

Here we use a mixture of spherical polar coordinates (r, θ, ϕ) and cylindrical coordinates (λ, ϕ, z), with S denoting the specific entropy. Angular brackets denote averages over longitude and time while g and C_P are the gravitational acceleration and the specific heat at constant pressure. Equation (3) holds if the meridional components of the convective Reynolds stress, the Lorentz force, and the viscous diffusion are small relative to the Coriolis and baroclinic terms. This is supported by a diverse range of theoretical and numerical modeling efforts and points to the central role of baroclinicity in accounting for the conical nature of the Ω isosurfaces inferred from helioseismic rotational inversions, as seen in Fig. 1 (Kitchatinov and Rüdiger 1995; Elliott et al. 2000; Robinson and Chan 2001; Rempel 2005b; Miesch et al. 2006; Balbus et al. 2009).

The second concept is that of gyroscopic pumping, which can be expressed by the following balance in the zonal component of the momentum equation:

$$\langle \rho \mathbf{v}_m \rangle \cdot \nabla \mathcal{L} = \mathcal{F} \tag{4}$$

where $\mathbf{v}_m$ denotes the meridional components of the velocity, and $\mathcal{L} = \lambda^2 \Omega$ is the specific angular momentum. The right-hand side of Eq. (4) is expressed as a generalized torque $\mathcal{F}$ but it can be loosely regarded as the negative divergence of the convective Reynolds stress. Lorentz forces and viscous torques may also contribute to $\mathcal{F}$ but they are unlikely to be the central mechanism for establishing the solar differential rotation. For a derivation of Eq. (4) and a thorough discussion of its implications, see Miesch and Hindman (2011).

Together these two concepts, reflected by the dynamical balances in Eqs. (3) and (4), provide a theoretical foundation for interpreting helioseismic inversions and numerical models. For example, by making the additional ansatz that Ω and $\langle S \rangle$ isosurfaces coincide, Balbus and colleagues have shown that solutions to Eq. (3) coincide remarkably well with helioseismic rotational inversions. However, there are two caveats to this result. First, it does not explain why the solar equator rotates faster than the poles; adding an arbitrary cylindrical angular velocity component $\Omega'(\lambda)$ to Ω leaves equation (3) unchanged (geostrophic degeneracy) so a given $\langle S \rangle$ profile is consistent with an infinite number of Ω profiles, some

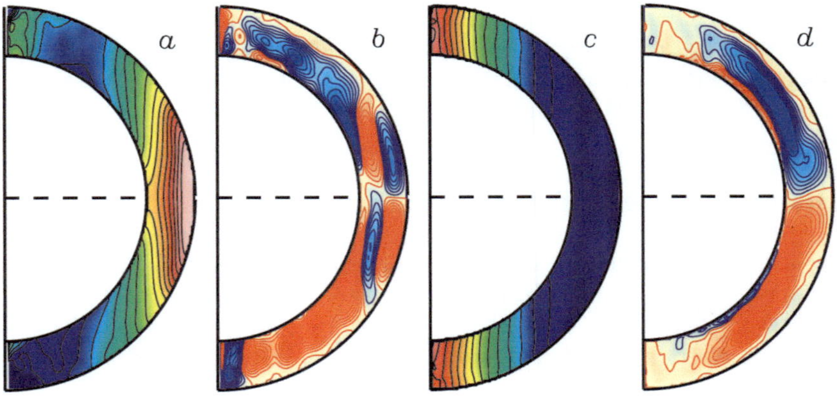

Fig. 8 Mean flow regimes in spherical convection, from Featherstone and Miesch (2014). Shown are the differential rotation (**a**, **c**) and meridional circulation profiles (**b**, **d**) in two simulations of global solar convection. *Pink/yellow* and *blue/black* tones denote faster and slower rotation in frames **a**, **c** while *red* and *blue* denote clockwise and counter-clockwise circulation in **b**, **d**. On the left (**a**, **b**) is the rapidly-rotating regime characterized by a solar-like Ω profile and multi-celled meridional circulation. On the right (**c**, **d**) is the slowly-rotating regime characterized by an anti-solar differential rotation and a single dominant circulation cell per hemisphere. The Sun may be near the transition

solar-like (fast equator, slow poles), some anti-solar (slow equator, fast poles). Second, a compelling theoretical justification of why Ω and $\langle S \rangle$ surfaces should coincide remains an outstanding challenge (though see Balbus and Schaan 2012, for one perspective).

A puzzle that has received much attention recently is that of the subsurface structure of the meridional circulation. It has been realized since the pioneering work of Gilman (1977) that spherical convection simulations exhibit two rotation regimes, delineated by the Rossby number $R_o = V_c/(2\Omega L_c)$ where V_c and L_c are characteristic velocity and length scales for the convection. As R_o is increased across values of order unity, the differential rotation undergoes a transition from being solar-like to anti-solar, as illustrated in Fig. 8. More recent modeling efforts are clarifying this transition and assessing its implications for the meridional circulation (Käpylä et al. 2011, 2014; Gastine et al. 2013, 2014; Guerrero et al. 2013b; Hotta et al. 2014a, 2014b; Featherstone and Miesch 2014). The fast-rotating regime generally exhibits multiple-cell profiles while the slow-rotating regime exhibits circulation profiles dominated by a single cell per hemisphere (Fig. 8). This transition can be understood in terms of a shift in the nature of the convective Reynolds stress $\mathcal{F}$ as discussed above in connection with Eq. (4). The Sun is likely near the transition so it is unclear which meridional flow regime it may be in (Featherstone and Miesch 2014).

Gyroscopic pumping relies on the zonal component of the convective Reynolds stress, which induces meridional circulation by means of the Coriolis force. The meridional components of the Reynolds stress and Lorentz force can also establish meridional circulation by breaking the thermal wind balance (TWB; Eq. (3)). This occurs in many mean-field models that represent meridional Reynolds stresses as a turbulent diffusion and solve for the steady flow profiles (Kitchatinov 2013; Dikpati 2014).[1] Departures from TWB occur particularly in the boundary layers, which can exert a disproportionate influence on the global meridional circulation profile. In particular, the shallow equatorward return flow inferred

[1]Though see Rempel (2005b) for an example of a time-dependent mean field model in which gyroscopic pumping is the dominant meridional flow driver.

from recent helioseismic inversions and photospheric feature tracking (Sect. 3.2) may be a boundary layer phenomenon, reflecting the penetration depth of surface-driven convective plumes (Featherstone and Miesch 2014).

Other current puzzles that are actively being investigated include the origin of the thermal gradients necessary for TWB (Eq. (3)) and the nature of the near-surface shear layer (NSSL). Proposed mechanisms for the former include the influence of rotation on convective heat transport (Kitchatinov and Rüdiger 1995; Brun and Toomre 2002; Käpylä et al. 2011), the influence of rotational shear on convective heat transport (Balbus et al. 2009; Balbus and Schaan 2012), and thermal coupling to the subadiabatic tachocline (Rempel 2005b; Miesch et al. 2006). For recent perspectives on the latter puzzle (the NSSL), see Miesch and Hindman (2011), Gastine et al. (2013), Guerrero et al. (2013b), and Hotta et al. (2014b).

4.2 Magnetoconvection and Surface Magnetism

Magnetoconvective processes, i.e., the interaction between magnetic field and convective flows, play a central role in the generation, intensification, transport, and dissipation of magnetic flux in the convection zone and in the lower atmosphere of the Sun. Significant progress in our understanding of these processes has been brought about during the last decade by the combination of high-resolution observations and sophisticated numerical simulations (for recent reviews of various aspects of solar magnetoconvection, see Miesch 2005; Stein 2012; Weiss 2012; Schüssler 2013). Prominent examples are the formation of intergranular magnetic flux concentrations (Bercik et al. 1998; Vögler et al. 2005; Stein and Nordlund 2006; Schaffenberger et al. 2006) and magnetized vortices (Vögler 2004; Shelyag et al. 2011; Moll et al. 2012; Wedemeyer-Böhm et al. 2012; Kitiashvili et al. 2012; Shelyag et al. 2013), the near-surface structure and dynamics of sunspot umbrae and penumbrae (Schüssler and Vögler 2006; Heinemann et al. 2007; Rempel et al. 2009a, 2009b; Rempel 2011a, 2011b, 2012), the emergence of active regions (Cheung et al. 2007; Martínez-Sykora et al. 2008; Cheung et al. 2010), as well as small-scale dynamo action in the deep convection zone (Brun et al. 2004) and in its near-surface layers (Vögler and Schüssler 2007; Pietarila Graham et al. 2010).

Recent developments concern magnetoconvection simulations in computational boxes of down to 50 Mm depth, so that the effect of flows at supergranular scales can be studied. These flow patterns imprint their signature on the distribution of magnetic flux at the visible surface, leading to mesogranular and supergranular network patterns. If a sufficient amount of vertical background flux is present, bigger flux concentrations resembling observations of dark pores are formed by flux expulsion and suppression of convective energy transport. An important requirement for this kind of simulation is a sufficiently big aspect ratio of the computational box (ratio of horizontal size to depth): since the largest flow structures typically have a horizontal extension of the order of the box depth, the horizontal size of the box should be at least 3–4 times its depth, so that a sufficiently large number of cells is present, thus minimizing the effect of the (typically periodic) side boundaries. Results based on simulations with aspect ratios of unity (e.g., Kitiashvili et al. 2010) should therefore be considered with some caution.

Figure 9 shows magnetic flux concentrations and pore-like structures forming in simulations in a box of 24 Mm × 24 Mm horizontal size and 6 Mm depth. The simulations differ by the amount of imposed vertical background flux: in the case with a horizontally averaged vertical field of $\langle B_z \rangle = 100$ G (upper panels), a patchy network on mesogranular scales and a few micropores formed; for $\langle B_z \rangle = 400$ G (lower panels), the network is more pronounced and a number of dark pores of the size of a few granules are present. Note that $\langle B_z \rangle = 400$ G

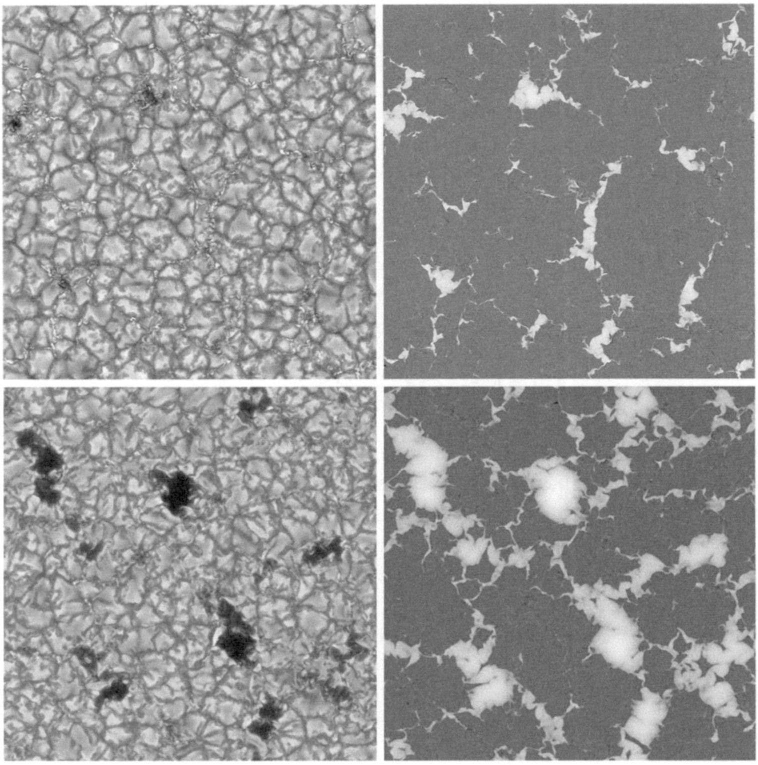

Fig. 9 Maps of bolometric brightness (*left panels*) and vertical magnetic field at the optical surface (*right panels*; *dark grey* represents very weak field) from magnetoconvection simulations with average vertical background fields of 100 G (*upper row*) and 400 G (*lower row*). The horizontal extension of the computational box was 24 Mm × 24 Mm and the depth 6 Mm. The bigger amount of magnetic flux in the case with 400 G background field leads to the formation of proper pores while only micropores are present in the case with 100 G. The latter case has a twice higher horizontal grid resolution (20.8 km), which leads to a better representation of the small-scale bright flux concentrations in the intergranular lanes

is already at the high end for solar plage areas; consequently we do expect even bigger structures (sunspots) to form spontaneously from existing background flux. This is consistent with the well-known observational fact that sunspots invariably form in the course of flux emergence and never from pre-existing flux in a mature plage region.

Figure 10 shows a comparison of flows and magnetic field patterns for various depths. The corresponding magnetoconvection simulation ($\langle B_z \rangle = 100$ G) was carried out by M.C.M. Cheung in a computational box with 49.2 Mm × 49.2 Mm horizontal size and reaching down to about 14 Mm below the optical surface. The various vertical flow patterns present at the different depths up to about supergranular scale are reflected in the multi-cellular distribution of vertical magnetic flux at the surface, owing to flux expulsion by the corresponding horizontal flows patterns. This offers the possibility to compare the simulation results with actual observations. Since the average properties of the simulated convection (e.g., depth profiles of horizontally averaged thermodynamic quantities, horizontal scales and velocities) are in good agreement with mixing-length models, this would also provide a consistency check for such models and shed light on the question whether solar convection might actually work in a completely different, essentially unmixed regime

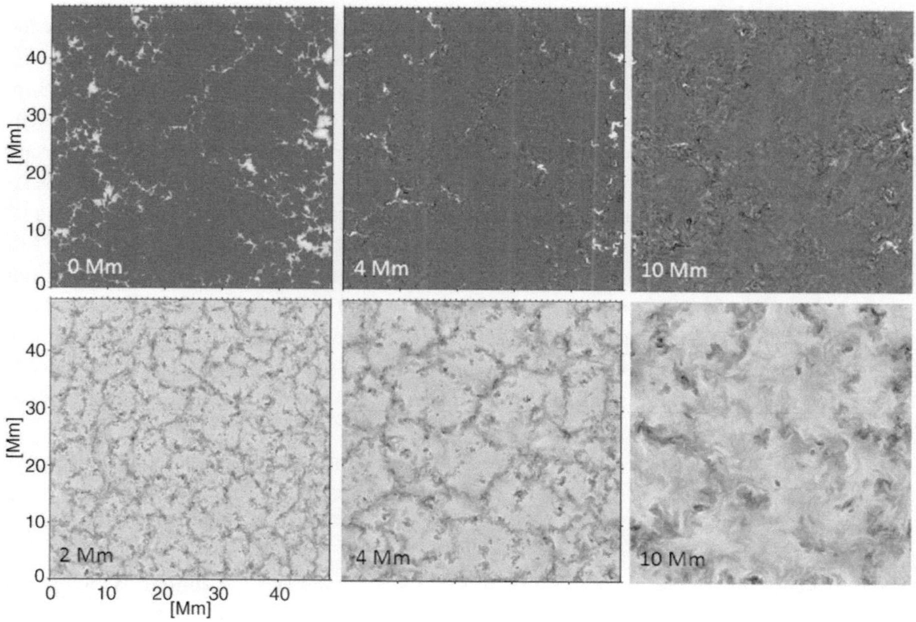

Fig. 10 Maps of the vertical magnetic field (*upper panels*; *dark grey* represents very weak field) and vertical flow velocity (*lower panels*; *darker shades* represent downflows, *brighter shades* upflows) at various depths below the optical surface from a magnetoconvection simulation with average vertical background fields of 100 G (courtesy of M.C.M. Cheung). The horizontal extension of the computational box was 49.2 Mm × 49.2 Mm and the depth 15.4 Mm. The spatial patterns of the observable surface field (*upper left panel*) reflect the horizontal scales covered by the flow patterns in the depth range of the simulation

with cool, narrow downdrafts traversing the whole convection zone and driving a very slow, broad upflow (Spruit 1997). Such a regime appears to be favored by recent indications from helioseismology (Fig. 7). Such 'slow', unmixed solar convection would have severe consequences for models of the solar dynamo and for the generation of differential rotation and meridional flow (Miesch et al. 2012, see also the discussion in Sect. 3.3 above). Existing numerical simulations invariably show well mixed convection with velocities consistent with mixing-length models. This is the case for anelastic simulations of convection in the deep convection zone (e.g., Miesch 2005) as well as for compressible simulations in the upper layers of the convection zone, which include the driving by radiative losses in the photosphere (e.g., Trampedach and Stein 2011). If solar convection actually would work in an essentially unmixed regime, then which are the critical Reynolds (and perhaps Prandtl) numbers for the transition away from the mixing-length regime? When can we expect to see numerical simulations of this transition, if it exists?

5 Future Directions

Reliable seismic inferences on the interior structure of prominent solar phenomena such as sunspots, supergranulation, convection and meridional circulation would have significant consequences not only for our understanding of the way the Sun operates but also for Sun-like stars. Understanding magneto-convection and the emergence and sustenance of large-scale field in the Sun would provide important constraints for dynamo theory. Thus the seismic science of the Sun can play a critical role in advancing astrophysics as a whole.

While global helioseismology, the study of large-scale axisymmetric structure in the Sun, has met with great success, local helioseismology, which deals with spatially localized features such as sunspots and supergranules and convection in the interior, has a long road ahead. Multi-variable inverse problems are inevitable in local helioseismology, where seismic measurements, obtained at the photosphere, must be related to a large number of parameters, e.g., the sound speed, flows and magnetic fields in the interior. In particular, the full accounting of systematic effects, finite wavelengths and inversion non-linearity, which can significantly influence and bias results, would greatly improve their trustworthiness and accuracy of local helioseismic inferences. The future of a theoretically sound seismology holds the promise of revealing important insights such as the distribution of large-scale Reynolds stresses, the drivers of differential rotation and meridional circulation, and ultimately, possibly the source of global magnetism in the Sun itself.

Meanwhile, comprehensive simulations of radiative (magneto-)convection in the near-surface layers of the Sun have achieved an impressive realism. They compare well with observational results (e.g., Nordlund et al. 2009; Stein 2012) as well as between different codes (Beeck et al. 2012).

In the deeper layers, the average properties of such simulations are similar to the results of mixing-length models and we can expect them to make contact with the anelastic simulations soon as well. Does that mean that the solar convection zone is in a well-mixed regime in the sense that the low-entropy downflows are well mixed into the high-entropy upflows? Or are simulations in the wrong regime owing to their much too small Reynolds numbers? The latter view seems to be supported by some results of local helioseismology, but these need to be confirmed before definite conclusions can be drawn. If they turn out to be correct, how can we then reconcile the fact that the near-surface simulations (which have much too low Reynolds numbers as well) are in such an excellent agreement with observations, so that even significant corrections of element abundances of fundamental astrophysical importance can be derived from them (Asplund et al. 2009).

We dedicate this paper to our colleague and friend Dr. Irene González Hernández (1969–2014), a pioneer of ring-diagram analysis and far-side imaging, who passed away on 14 February 2014.

Acknowledgements MSM is supported by NASA grants NNH09AK14I (Heliophysics SR&T) and NNX08AI57G (Heliophysics Theory Program). The National Center for Atmospheric Research is sponsored by the National Science Foundation. MR acknowledges support from the European Research Council under the European Union's Seventh Framework Program (FP/2007-2013)/ERC Grant Agreement no. 307117.

References

M. Asplund, N. Grevesse, A.J. Sauval, P. Scott, The chemical composition of the Sun. Annu. Rev. Astron. Astrophys. **47**, 481–522 (2009). doi:10.1146/annurev.astro.46.060407.145222

S.A. Balbus, E. Schaan, The stability of stratified, rotating systems and the generation of vorticity in the Sun. Mon. Not. R. Astron. Soc. **426**, 1546–1557 (2012)

S.A. Balbus, J. Bonart, H.N. Latter, N.O. Weiss, Differential rotation and convection in the Sun. Mon. Not. R. Astron. Soc. **400**, 176–182 (2009)

C.S. Baldner, J. Schou, Effects of asymmetric flows in solar convection on oscillation modes. Astrophys. J. Lett. **760**, 1 (2012). doi:10.1088/2041-8205/760/1/L1

S. Basu, H.M. Antia, D. Narasimha, Helioseismic measurement of the extent of overshoot below the solar convection zone. Mon. Not. R. Astron. Soc. **267**, 209 (1994)

B. Beeck, R. Collet, M. Steffen, M. Asplund, R.H. Cameron, B. Freytag, W. Hayek, H.-G. Ludwig, M. Schüssler, Simulations of the solar near-surface layers with the CO5BOLD, MURaM, and Stagger codes. Astron. Astrophys. **539**, 121 (2012). doi:10.1051/0004-6361/201118252

D.J. Bercik, S. Basu, D. Georgobiani, A. Nordlund, R.F. Stein, Solar magneto-convection, in *Cool Stars, Stellar Systems, and the Sun 10*, ed. by R.A. Donahue, J.A. Bookbinder. ASP Conf. Ser., vol. 154 (Astronomical Society of the Pacific, San Francisco, 1998), p. 568

J. Bouvier, Observational studies of stellar rotation, in *EAS Publications Series*, vol. 62, ed. by P. Hennebelle, C. Charbonnel (2013), pp. 143–168. doi:10.1051/eas/1362005

N.H. Brummell, T.L. Clune, J. Toomre, Penetration and overshooting in turbulent compressible convection. Astrophys. J. **570**, 825–854 (2002). doi:10.1086/339626

A.S. Brun, M.S. Miesch, J. Toomre, Global-scale turbulent convection and magnetic dynamo action in the solar envelope. Astrophys. J. **614**, 1073–1098 (2004)

A.S. Brun, M.S. Miesch, J. Toomre, Modeling the dynamical coupling of solar convection with the radiative interior. Astrophys. J. **742**, 79 (2011). doi:10.1088/0004-637X/742/2/79

A.S. Brun, J. Toomre, Turbulent convection under the influence of rotation: sustaining a strong differential rotation. Astrophys. J. **570**, 865–885 (2002)

M.C.M. Cheung, M. Schüssler, F. Moreno-Insertis, The origin of the reversed granulation in the solar photosphere. Astron. Astrophys. **461**, 1163–1171 (2007)

M.C.M. Cheung, M. Rempel, A.M. Title, M. Schüssler, Simulation of the formation of a solar active region. Astrophys. J. **720**, 233–244 (2010)

J. Christensen-Dalsgaard, M.J.P.F.G. Monteiro, M.J. Thompson, Helioseismic estimation of convective overshoot in the Sun. Mon. Not. R. Astron. Soc. **276**, 283–292 (1995)

J. Christensen-Dalsgaard, M.J.P.F.G. Monteiro, M. Rempel, M.J. Thompson, A more realistic representation of overshoot at the base of the solar convective envelope as seen by helioseismology. Mon. Not. R. Astron. Soc. **414**, 1158–1174 (2011). doi:10.1111/j.1365-2966.2011.18460.x

L. Deng, D.R. Xiong, How to define the boundaries of a convective zone, and how extended is overshooting? Mon. Not. R. Astron. Soc. **386**, 1979–1989 (2008). doi:10.1111/j.1365-2966.2008.12969.x

F.-L. Deubner, D. Gough, Helioseismology: oscillations as a diagnostic of the solar interior. Annu. Rev. Astron. Astrophys. **22**, 593–619 (1984). doi:10.1146/annurev.aa.22.090184.003113

M.P. di Mauro, D. Cardini, M. Marconi, *Synergies Between Solar and Stellar Modelling* (2010)

M. Dikpati, Generating the Suns global meridional circulation from differential rotation and turbulent Reynolds stress. Mon. Not. R. Astron. Soc. **438**, 2380–2394 (2014)

T.L. Duvall Jr., Nonaxisymmetric variations deep in the convection zone, in *GONG+ 2002. Local and Global Helioseismology: The Present and Future*, ed. by H. Sawaya-Lacoste, ESA Special Publication, vol. 517 (ESA, Noordwijk, 2003), pp. 259–262

T.L. Duvall Jr., S.M. Jefferies, J.W. Harvey, M.A. Pomerantz, Time-distance helioseismology. Nature **362**, 430–432 (1993). doi:10.1038/362430a0

A. Eff-Darwich, S.G. Korzennik, The dynamics of the solar radiative zone. Sol. Phys. **287**, 43–56 (2013). doi:10.1007/s11207-012-0048-z

J.R. Elliott, M.S. Miesch, J. Toomre, Turbulent solar convection and its coupling with rotation: the effect of Prandtl number and thermal boundary conditions on the resulting differential rotation. Astrophys. J. **533**, 546–556 (2000)

N.A. Featherstone, M.S. Miesch, Meridional circulation in solar and stellar convection zones (2014, submitted). http://arxiv.org/abs/1501.06501

F. Gallet, J. Bouvier, Improved angular momentum evolution model for solar-like stars. Astron. Astrophys. **556**, 36 (2013). doi:10.1051/0004-6361/201321302

T. Gastine, J. Wicht, J.M. Aurnou, Zonal flow regimes in rotating anelastic spherical shells: an application to giant planets. Icarus **225**, 156–172 (2013)

T. Gastine, R.K. Yadav, J. Morin, A. Reiners, J. Wicht, From solar-like to antisolar differential rotation in cool stars. Mon. Not. R. Astron. Soc. Lett. **438**, 76–80 (2014)

M. Ghizaru, P. Charbonneau, P.K. Smolarkiewicz, Magnetic cycles in global large-eddy simulations of solar convection. Astrophys. J. Lett. **715**, 133–137 (2010)

P.A. Gilman, Nonlinear dynamics of Boussinesq convection in a deep rotating spherical shell I. Geophys. Astrophys. Fluid Dyn. **8**, 93–135 (1977)

L. Gizon, A.C. Birch, Helioseismology challenges models of solar convection. Proc. Nat. Acad. Sci. **109**, 11896–11897 (2012)

L. Gizon, A.C. Birch, H.C. Spruit, Local helioseismology: three-dimensional imaging of the solar interior. Annu. Rev. Astron. Astrophys. **48**, 289–338 (2010). doi:10.1146/annurev-astro-082708-101722

I. González Hernández, J. Patrón, R.S. Bogart, T. SOI Ring Diagram Team, Meridional flows from ring diagram analysis. Astrophys. J. Lett. **510**, 153–156 (1999). doi:10.1086/311811

I. González Hernández, R. Komm, F. Hill, R. Howe, T. Corbard, D.A. Haber, Meridional circulation variability from large-aperture ring-diagram analysis of global oscillation network group and Michelson Doppler Imager data. Astrophys. J. **638**, 576–583 (2006). doi:10.1086/498642

I. González Hernández, S. Kholikov, F. Hill, R. Howe, R. Komm, Subsurface meridional circulation in the active belts. Sol. Phys. **252**, 235–245 (2008). doi:10.1007/s11207-008-9264-y

I. González Hernández, R. Howe, R. Komm, F. Hill, Meridional circulation during the extended solar minimum: another component of the torsional oscillation? Astrophys. J. Lett. **713**, 16–20 (2010). doi:10.1088/2041-8205/713/1/L16

D.O. Gough, The anelastic approximation for thermal convection. J. Atmos. Sci. **26**, 448–456 (1969). doi:10.1175/1520-0469(1969)026<0448:TAAFTC>2.0.CO;2

D.O. Gough, Mixing-length theory for pulsating stars. Astrophys. J. **214**, 196–213 (1977). doi:10.1086/155244

G. Guerrero, P.K. Smolarkiewicz, A.G. Kosovichev, N.N. Mansour, Differential rotation in solar-like stars from global simulations. Astrophys. J. **779**, 176 (2013a), 13 pp.

G. Guerrero, P.K. Smolarkiewicz, A. Kosovichev, N. Mansour, Solar differential rotation: hints to reproduce a near-surface shear layer in global simulations, in *Solar and Astrophysical Dynamos and Magnetic Activity*, ed. by A.G. Kosivichev, E.M. de Gouveia Dal Pino, Y. Yan (IAU, Paris, 2013b), p. 9

D. Haber, B. Hindman, J. Toomre, R.S. Bogart, Large-scale circulations using ring analysis, in *Proceedings of SOHO 18/GONG 2006/HELAS I, Beyond the Spherical Sun*. ESA Special Publication, vol. 624 (2006)

S.M. Hanasoge, T.L. Duvall, M.L. DeRosa, Seismic constraints on interior solar convection. Astrophys. J. **712**, 98–102 (2010)

S.M. Hanasoge, T.L. Duvall, K.R. Sreenivasan, Anomalously weak solar convection. Proc. Nat. Acad. Sci. (2012). doi:10.1073/pnas.1206570109

C.J. Hansen, J.P. Cox, H.M. van Horn, The effects of differential rotation on the splitting of nonradial modes of stellar oscillation. Astrophys. J. **217**, 151–159 (1977). doi:10.1086/155564

J.W. Harvey, F. Hill, R.P. Hubbard, J.R. Kennedy, J.W. Leibacher, J.A. Pintar, P.A. Gilman, R.W. Noyes, A.M. Title, J. Toomre, R.K. Ulrich, A. Bhatnagar, J.A. Kennewell, W. Marquette, J. Patron, O. Saa, E. Yasukawa, The global oscillation network group (GONG) project. Science **272**, 1284–1286 (1996). doi:10.1126/science.272.5266.1284

D.H. Hathaway, Supergranules as probes of the Sun's meridional circulation. Astrophys. J. **760**, 84 (2012). doi:10.1088/0004-637X/760/1/84

D.H. Hathaway, L. Upton, O. Colegrove, Giant convection cells found on the sun. Science **342**, 1217–1219 (2013). doi:10.1126/science.1244682

D.H. Hathaway, J.G. Beck, R.S. Bogart, K.T. Bachmann, G. Khatri, J.M. Petitto, S. Han, J. Raymond, The photospheric convection spectrum. Sol. Phys. **193**, 299–312 (2000). doi:10.1023/A:1005200809766

T. Heinemann, Å. Nordlund, G.B. Scharmer, H.C. Spruit, MHD simulations of penumbra fine structure. Astrophys. J. **669**, 1390–1394 (2007)

F. Hill, Rings and trumpets—three-dimensional power spectra of solar oscillations. Astrophys. J. **333**, 996–1013 (1988). doi:10.1086/166807

H. Hotta, M. Rempel, T. Yokoyama, High-resolution calculations of the solar global convection with the reduced speed of sound technique. I. The structure of the convection and the magnetic field without the rotation. Astrophys. J. **786**, 24 (2014a), 18 pp.

H. Hotta, M. Rempel, T. Yokoyama, High-resolution calculations of the solar global convection with the reduced speed of sound technique. II. Near surface shear layer with the rotation (2014b). doi:10.1088/0004-637X/786/1/24

R. Howard, B.J. Labonte, The sun is observed to be a torsional oscillator with a period of 11 years. Astrophys. J. Lett. **239**, 33–36 (1980). doi:10.1086/183286

R. Howe, Solar interior rotation and its variation. Living Rev. Sol. Phys. **6**, 1 (2009). doi:10.12942/lrsp-2009-1

R. Howe, J. Christensen-Dalsgaard, F. Hill, R.W. Komm, R.M. Larsen, J. Schou, M.J. Thompson, J. Toomre, Deeply penetrating banded zonal flows in the solar convection zone. Astrophys. J. Lett. **533**, 163–166 (2000). doi:10.1086/312623

R. Howe, J. Christensen-Dalsgaard, F. Hill, R. Komm, J. Schou, M.J. Thompson, A note on the torsional oscillation at solar minimum. Astrophys. J. Lett. **701**, 87–90 (2009). doi:10.1088/0004-637X/701/2/L87

R. Howe, T.P. Larson, J. Schou, F. Hill, R. Komm, J. Christensen-Dalsgaard, M.J. Thompson, First global rotation inversions of HMI data. J. Phys. Conf. Ser. **271**, 012061 (2011)

D.W. Hughes, R. Rosner, N.O. Weiss (eds.), *The Solar Tachocline* (Cambridge Univ. Press, Cambridge, 2007)

L. Jouve, A.S. Brun, On the role of meridional flows in flux transport dynamo models. Astron. Astrophys. **474**, 239–250 (2007). doi:10.1051/0004-6361:20077070

P.J. Käpylä, M.J. Korpi, A. Brandenburg, D. Mitra, R. Tavakol, Convective dynamos in spherical wedge geometry. Astron. Nachr. **331**, 73 (2010). doi:10.1002/asna.200911252

P.J. Käpylä, M.J. Mantere, G. Guerrero, A. Brandenburg, P. Chatterjee, Reynolds stress and heat flux in spherical shell convection. Astron. Astrophys. **531**, 162 (2011). doi:10.1051/0004-6361/201015884

P.J. Käpylä, M.J. Mantere, A. Brandenburg, Confirmation of bistable stellar differential rotation profiles. Astron. Astrophys. (2014). doi:10.1051/0004-6361/201423412

P.J. Käpylä, M.J. Mantere, G. Guerrero, A. Brandenburg, P. Chatterjee, Reynolds stress and heat flux in spherical shell convection. Astron. Astrophys. **531**, 162 (2011), 17 pp.

L.L. Kitchatinov, Theory of differential rotation and meridional circulation, in *Solar and Astrophysical Dynamos and Magnetic Activity*, ed. by A.G. Kosovichev, E.M. de Gouveia Dal Pino, Y. Yan. Proc. IAU Symposium, vol. 294 (2013), pp. 399–410. doi:10.1017/S1743921313002834. http://adsabs.harvard.edu/abs/2013IAUS..294..399K. arXiv:1210.7041. Provided by the SAO/NASA Astrophysics Data System

L.L. Kitchatinov, G. Rüdiger, Differential rotation in solar-type stars: revisiting the Taylor-number puzzle. Astron. Astrophys. **299**, 446–452 (1995)

I.N. Kitiashvili, A.G. Kosovichev, A.A. Wray, N.N. Mansour, Mechanism of spontaneous formation of stable magnetic structures on the Sun. Astrophys. J. **719**, 307–312 (2010)

I.N. Kitiashvili, A.G. Kosovichev, N.N. Mansour, A.A. Wray, Dynamics of magnetized vortex tubes in the solar chromosphere. Astrophys. J. **751**, 21 (2012)

R. Komm, R. Howe, F. Hill, I. González Hernández, C. Toner, Kinetic Helicity density in solar subsurface layers and flare activity of active regions. Astrophys. J. **630**, 1184–1193 (2005). doi:10.1086/432031

E.M. Lavely, M.H. Ritzwoller, Philos. Trans. R. Soc. Lond. A **339**, 431 (1992)

C. Lindsey, D.C. Braun, Helioseismic holography. Astrophys. J. **485**, 895–903 (1997)

J. Martínez-Sykora, V. Hansteen, M. Carlsson, Twisted flux tube emergence from the convection zone to the corona. Astrophys. J. **679**, 871–888 (2008)

S. Mathis, Transport processes in stellar interiors, in *Lecture Notes in Physics, vol. 865*, ed. by M. Goupil, K. Belkacem, C. Neiner, F. Lignières, J.J. Green (Springer, Berlin, 2013), p. 23

M.S. Miesch, Large-scale dynamics of the convection zone and tachocline. Living Rev. Sol. Phys. **2**, 1 (2005)

M.S. Miesch, A.S. Brun, M.L. De Rosa, J. Toomre, Structure and evolution of giant cells in global models of solar convection. Astrophys. J. **673**, 557–575 (2008). doi:10.1086/523838

M.S. Miesch, N.A. Featherstone, M. Rempel, R. Trampedach, On the amplitude of convective velocities in the deep solar interior. Astrophys. J. **757**, 128 (2012)

M.S. Miesch, Large-scale dynamics of the convection zone and tachocline. Living Rev. Sol. Phys. **2**, 1 (2005). http://www.livingreviews.org/lrsp-2005-1

M.S. Miesch, B.W. Hindman, Gyroscopic pumping in the solar near-surface shear layer. Astrophys. J. **743**, 79 (2011), 25 pp.

M.S. Miesch, A.S. Brun, J. Toomre, Solar differential rotation influenced by latitudinal entropy variations in the tachocline. Astrophys. J. **641**, 618–625 (2006)

M.S. Miesch, N.A. Featherstone, M. Rempel, R. Trampedach, On the amplitude of convective velocities in the deep solar interior. Astrophys. J. **757**, 128 (2012), 14 pp.

R. Moll, R.H. Cameron, M. Schüssler, Vortices, shocks, and heating in the solar photosphere: effect of a magnetic field. Astron. Astrophys. **541**, 68 (2012)

M.J.P.F.G. Monteiro, J. Christensen-Dalsgaard, M.J. Thompson, Seismic study of overshoot at the base of the solar convective envelope. Astron. Astrophys. **283**, 247–262 (1994)

Å. Nordlund, R.F. Stein, M. Asplund, Solar surface convection. Living Rev. Sol. Phys. **6**, 2 (2009)

A. Nordlund, H.C. Spruit, H.-G. Ludwig, R. Trampedach, Is stellar granulation turbulence? Astron. Astrophys. **328**, 229–234 (1997)

J. Pietarila Graham, R. Cameron, M. Schüssler, Turbulent small-scale dynamo action in solar surface simulations. Astrophys. J. **714**, 1606–1616 (2010)

M.P. Rast, Compressible plume dynamics and stability. J. Fluid Mech. **369**, 125–149 (1998)

M. Rempel, Overshoot at the base of the solar convection zone: a semianalytical approach. Astrophys. J. **607**, 1046–1064 (2004). doi:10.1086/383605

M. Rempel, Solar differential rotation and meridional flow: the role of a subadiabatic tachocline for the Taylor-Proudman balance. Astrophys. J. **622**, 1320–1332 (2005a). doi:10.1086/428282

M. Rempel, Solar differential rotation and meridional flow: the role of a subadiabatic tachocline for the Taylor-Proudman balance. Astrophys. J. **622**, 1320–1332 (2005b)

M. Rempel, Penumbral fine structure and driving mechanisms of large-scale flows in simulated sunspots. Astrophys. J. **729**, 5 (2011a)

M. Rempel, Subsurface magnetic field and flow structure of simulated sunspots. Astrophys. J. **740**, 15 (2011b)

M. Rempel, Numerical sunspot models: robustness of photospheric velocity and magnetic field structure. Astrophys. J. **750**, 62 (2012)

M. Rempel, M. Schüssler, M. Knölker, Radiative magnetohydrodynamic simulation of sunspot structure. Astrophys. J. **691**, 640–649 (2009a)
M. Rempel, M. Schüssler, R.H. Cameron, M. Knölker, Penumbral structure and outflows in simulated sunspots. Science **325**, 171 (2009b)
F.J. Robinson, K.L. Chan, A large-eddy simulation of turbulent compressible convection: differential rotation in the solar convection zone. Mon. Not. R. Astron. Soc. **321**, 723–732 (2001)
T.M. Rogers, G.A. Glatzmaier, Penetrative convection within the anelastic approximation. Astrophys. J. **620**, 432–441 (2005). doi:10.1086/423415
M. Roth, M. Stix, Coupling of solar p modes: quasi-degenerate perturbation theory. Astron. Astrophys. **351**, 1133–1138 (1999)
M. Roth, M. Stix, Meridional circulation and global solar oscillations. Sol. Phys. **251**, 77–89 (2008). doi:10.1007/s11207-008-9232-6
I.W. Roxburgh, S.V. Vorontsov, Seismology of the solar envelope—the base of the convective zone as seen in the phase shift of acoustic waves. Mon. Not. R. Astron. Soc. **268**, 880 (1994)
A. Schad, J. Timmer, M. Roth, A unified approach to the helioseismic inversion problem of the solar meridional flow from global oscillations. Astrophys. J. **734**, 97 (2011). doi:10.1088/0004-637X/734/2/97
A. Schad, J. Timmer, M. Roth, Measuring the solar meridional flow from perturbations of eigenfunctions of global oscillations. Astron. Nachr. **333**, 991 (2012). doi:10.1002/asna.201211815
A. Schad, J. Timmer, M. Roth, Global helioseismic evidence for a deeply penetrating solar meridional flow consisting of multiple flow cells. Astrophys. J. Lett. **778**, 38 (2013). doi:10.1088/2041-8205/778/2/L38
W. Schaffenberger, S. Wedemeyer-Böhm, O. Steiner, B. Freytag, Holistic mhd-simulation from the convection zone to the chromosphere, in *Solar MHD Theory and Observations: A High Spatial Resolution Perspective*, ed. by J. Leibacher, R.F. Stein, H. Uitenbroek. Astronomical Society of the Pacific Conference Series, vol. 354 (2006), p. 345
P.H. Scherrer, R.S. Bogart, R.I. Bush, J.T. Hoeksema, A.G. Kosovichev, J. Schou, W. Rosenberg, L. Springer, T.D. Tarbell, A. Title, C.J. Wolfson, I. Zayer, MDI Engineering Team, The solar oscillations investigation—Michelson Doppler Imager. Sol. Phys. **162**, 129–188 (1995). doi:10.1007/BF00733429
J. Schou, Migration of zonal flows detected using Michelson Doppler Imager F-mode frequency splittings. Astrophys. J. Lett. **523**, 181–184 (1999). doi:10.1086/312279
J. Schou, H.M. Antia, S. Basu, R.S. Bogart, R.I. Bush, S.M. Chitre, J. Christensen-Dalsgaard, M.P. Di Mauro, W.A. Dziembowski, A. Eff-Darwich, D.O. Gough, D.A. Haber, J.T. Hoeksema, R. Howe, S.G. Korzennik, A.G. Kosovichev, R.M. Larsen, F.P. Pijpers, P.H. Scherrer, T. Sekii, T.D. Tarbell, A.M. Title, M.J. Thompson, J. Toomre, Helioseismic studies of differential rotation in the solar envelope by the solar oscillations investigation using the Michelson Doppler Imager. Astrophys. J. **505**, 390–417 (1998)
J. Schou, P.H. Scherrer, R.I. Bush, R. Wachter, S. Couvidat, M.C. Rabello-Soares, R.S. Bogart, J.T. Hoeksema, Y. Liu, T.L. Duvall, D.J. Akin, B.A. Allard, J.W. Miles, R. Rairden, R.A. Shine, T.D. Tarbell, A.M. Title, C.J. Wolfson, D.F. Elmore, A.A. Norton, S. Tomczyk, Design and ground calibration of the helioseismic and magnetic imager (HMI) instrument on the solar dynamics observatory (SDO). Sol. Phys. **275**, 229–259 (2012). doi:10.1007/s11207-011-9842-2
M. Schüssler, Solar magneto-convection, in *Solar and Astrophysical Dynamos and Magnetic Activity*, ed. by A.G. Kosovichev, E.M. de Gouveia Dal Pino, Y. Yan. Proc. IAU Symp., vol. 294 (Cambridge University Press, Cambridge, 2013)
M. Schüssler, A. Vögler, Magnetoconvection in a sunspot umbra. Astrophys. J. **641**, 73–76 (2006)
S. Shelyag, P. Keys, M. Mathioudakis, F.P. Keenan, Vorticity in the solar photosphere. Astron. Astrophys. **526**, 5 (2011)
S. Shelyag, P.S. Cally, A. Reid, M. Mathioudakis, Alfvén waves in simulations of solar photospheric vortices. Astrophys. J. Lett. **776**, 4 (2013). doi:10.1088/2041-8205/776/1/L4
D. Skaley, M. Stix, The overshoot layer at the base of the solar convection zone. Astron. Astrophys. **241**, 227–232 (1991)
E.A. Spiegel, J.-P. Zahn, The solar tachocline. Astron. Astrophys. **265**, 106–114 (1992)
H. Spruit, Convection in stellar envelopes: a changing paradigm. Mem. Soc. Astron. Ital. **68**, 397 (1997)
H.C. Spruit, A model of the solar convection zone. Sol. Phys. **34**, 277–290 (1974). doi:10.1007/BF00153665
R.F. Stein, Solar surface magneto-convection. Living Rev. Sol. Phys. **9**, 4 (2012). http://www.livingreviews.org/lrsp-2012-4
R.F. Stein, Å. Nordlund, Realistic solar convection simulations. Sol. Phys. **192**, 91–108 (2000). doi:10.1023/A:1005260918443
R.F. Stein, Å. Nordlund, Solar small-scale magnetoconvection. Astrophys. J. **642**, 1246–1255 (2006)
M.J. Thompson, J. Toomre, E.R. Anderson, H.M. Antia, G. Berthomieu, D. Burtonclay, S.M. Chitre, J. Christensen-Dalsgaard, T. Corbard, M. De Rosa, C.R. Genovese, D.O. Gough, D.A. Haber, J.W. Harvey, F. Hill, R. Howe, S.G. Korzennik, A.G. Kosovichev, J.W. Leibacher, F.P. Pijpers, J. Provost, E.J.

Rhodes Jr., J. Schou, T. Sekii, P.B. Stark, P.R. Wilson, Differential rotation and dynamics of the solar interior. Science **272**, 1300–1305 (1996). doi:10.1126/science.272.5266.1300

R. Trampedach, R.F. Stein, The mass mixing length in convective stellar envelopes. Astrophys. J. **731**, 78 (2011)

R.K. Ulrich, Solar meridional circulation from Doppler shifts of the Fe I line at 5250 Å as measured by the 150-foot solar tower telescope at the Mt. Wilson Observatory. Astrophys. J. **725**, 658–669 (2010). doi:10.1088/0004-637X/725/1/658

A. Vögler, S. Shelyag, M. Schüssler, F. Cattaneo, T. Emonet, T. Linde, Simulations of magneto-convection in the solar photosphere. Equations, methods, and results of the MURaM code. Astron. Astrophys. **429**, 335–351 (2005). doi:10.1051/0004-6361:20041507

A. Vögler, Simulating radiative magneto-convection in the solar photosphere, in *Reviews in Modern Astronomy 17: The Sun and Planetary Systems—Paradigms for the Universe*, ed. by E. Schielicke (Wiley-VCH, Weinheim, 2004), pp. 69–86

A. Vögler, M. Schüssler, A solar surface dynamo. Astron. Astrophys. **465**, 43–46 (2007)

S.V. Vorontsov, J. Christensen-Dalsgaard, J. Schou, V.N. Strakhov, M.J. Thompson, Helioseismic measurement of solar torsional oscillations. Science **296**, 101–103 (2002). doi:10.1126/science.1069190

S. Wedemeyer-Böhm, E. Scullion, O. Steiner, L. Rouppe van der Voort, J. de La Cruz Rodriguez, V. Fedun, R. Erdélyi, Magnetic tornadoes as energy channels into the solar corona. Nature **486**, 505–508 (2012)

A. Weiss, W. Hillebrandt, H. Thomas, H. Ritter, *Cox and Giuli's Principles of Stellar Structure* (Princeton Publishing Associates, Cambridge, 2004)

N.O. Weiss, Reflections on magnetoconvection. Geophys. Astrophys. Fluid Dyn. **106**, 353–371 (2012)

M.F. Woodard, Theoretical signature of solar meridional flow in global seismic data. Sol. Phys. **197**, 11–20 (2000). doi:10.1023/A:1026508211960

M. Woodard, J. Schou, A.C. Birch, T.P. Larson, Global-oscillation eigenfunction measurements of solar meridional flow. Sol. Phys. **287**, 129–147 (2013). doi:10.1007/s11207-012-0075-9

A. Zaatri, R. Komm, I. González Hernández, R. Howe, T. Corbard North South asymmetry of zonal and meridional flows determined from ring diagram analysis of gong ++ data. Sol. Phys. **236**, 227–244 (2006). doi:10.1007/s11207-006-0106-5

J.-P. Zahn, Convective penetration in stellar interiors. Astron. Astrophys. **252**, 179–188 (1991)

J.-P. Zahn, Circulation and turbulence in rotating stars. Astron. Astrophys. **265**, 115–132 (1992)

J. Zhao, K. Nagashima, R.S. Bogart, A.G. Kosovichev, T.L. Duvall Jr., Systematic center-to-limb variation in measured helioseismic travel times and its effect on inferences of solar interior meridional flows. Astrophys. J. Lett. **749**, 5 (2012). doi:10.1088/2041-8205/749/1/L5

J. Zhao, R.S. Bogart, A.G. Kosovichev, T.L. Duvall Jr., T. Hartlep, Detection of equatorward meridional flow and evidence of double-cell meridional circulation inside the Sun. Astrophys. J. Lett. **774**, 29 (2013). doi:10.1088/2041-8205/774/2/L29

Space Sci Rev (2015) 196:101–136
DOI 10.1007/s11214-013-0028-0

Recent Advances on Solar Global Magnetism and Variability

A.S. Brun · M.K. Browning · M. Dikpati · H. Hotta · A. Strugarek

Received: 17 July 2013 / Accepted: 30 October 2013 / Published online: 27 November 2013

Abstract We discuss recent observational, theoretical and numerical progress made in understanding the solar global magnetism and its short and long term variability. We discuss the physical process thought to be at the origin of the solar magnetic field and its 22-yr cycle, namely dynamo action, and the nonlinear interplay between convection, rotation, radiation and magnetic field, yielding modulations of the solar constant or of the large scale flows such as the torsional oscillations. We also discuss the role of the field parity and dynamo families in explaining the complex multipolar structure of the solar global magnetic field. We then present some key MHD processes acting in the deep radiative interior and discuss the probable topology of a primordial field there. Finally we summarize how helioseismology has contributed to these recent advances and how it could contribute to resolving current unsolved problems in solar global dynamics and magnetism.

A.S. Brun (✉) · A. Strugarek
AIM, CEA/CNRS/University of Paris 7, Bat 709, CEA-Saclay, 91191 Gif-sur-Yvette, France
e-mail: sacha.brun@cea.fr

A. Strugarek
e-mail: antoine.strugarek@cea.fr

M.K. Browning
Physics and Astronomy, University of Exeter, Stocker Road, EX4 4QL Exeter, UK
e-mail: browning@astro.ex.ac.uk

M. Dikpati
HAO/NCAR, Boulder, CO, USA
e-mail: dikpati@hao.ucar.edu

H. Hotta
Department of Earth and Planetary Science, University of Tokyo, 7-3-1 Hongo, Bunkyo-ku, Tokyo 113-0033, Japan
e-mail: hotta.h@eps.s.u-tokyo.ac.jp

Present address:
A. Strugarek
Astronomy Dept., University of Montreal, Montreal, Canada

Keywords Sun · Magnetism · Dynamo · Solar constant · Torsional Oscillations · Field parity · MHD instabilities

1 Latest Observations of the Sun's Global Magnetism and Variability

The Sun is the archetype of main sequence magnetic stars possessing a large spectrum of magnetic phenomena such as active regions, magnetic prominences, hot corona and an activity cycle. This 22-yr magnetic activity cycle is composed of two consecutive sunspot cycles of about 11 years each. The proximity of the Sun has allowed us to observe it continuously with steadily increasing spatial and temporal resolutions for the last four centuries or so. Many features of its observed variability are linked to the presence of intense and constantly evolving magnetic fields. Surely the presence of a turbulent surface convection zone coupled to intense magnetism contributes to the observed multi-scale dynamics and variability of the large scale flows and solar irradiance. Yet explaining how magnetism, convection and radiation interact, and determining what sets the period of solar magnetic activity to 11 years, has remained a challenging task over the last 50 years. Currently there is a general understanding that a dynamo process is occurring inside and at the base of the convection zone, the details of which including its impact on the variations of the solar luminosity and large scale flows, are still under discussion in the scientific community. It is also important to realize that observations of surface magnetic field do not directly inform us about the possible existence of more deeply anchored magnetic fields. The Sun may possess a primordial magnetic field deep in its radiative interior, which may interact with the magnetic field generated by dynamo action nearer the surface. Hence developing a global vision of inner solar magnetism from deep inside up to the photosphere is challenging but necessary. Helioseismology, with its ability to probe the inner layers of the Sun is, without doubt, one, if not the only, technique that permits direct access to the dynamics of the solar interior, hence allowing us to make a clearer connection with surface phenomena. We will thus discuss the latest progress made in describing and understanding the global magnetic Sun, emphasizing what helioseismology has helped resolve, and what is currently still missing in our description of our star.

1.1 Helioseismic Constraints on Global Solar Magnetism and 11-yr Cycle

Helioseismic measurements have now quite well defined the average differential rotation of the Sun as a function of depth and latitude in and below the convection zone (Thompson et al. 2003; Howe 2009). From these measurements both the tachocline and near surface shear layers are well measured. Both play a central role in most solar dynamo models through the so-called Ω-effect (Moffatt 1978).

Beyond differential rotation, helioseismology has given us some measures of meridional circulation within the convection zone, mostly in the uppermost layers, but results from different data sets and different analysis methods show considerable variation (Hathaway 1996; Giles et al. 1997; Haber et al. 2002; Howe et al. 2006; Howe 2009; Basu and Antia 2010; Rightmire-Upton et al. 2012; Komm et al. 2012). There does not yet appear to be a consensus from such measurements about the details of meridional flow much below the solar surface. However over the past recent years more evidence seem to suggest that the Sun may have a multi-cellular meridonal circulation profile per hemisphere (Haber et al. 2002; Mitra-Kraev and Thompson 2007; Hathaway 2012; Zhao et al. 2013) with a return flow much closer to the surface as it was actually anticipated by 3-D numerical simulations (Brun and Toomre 2002; Miesch et al. 2006; Brun and Rempel 2009). Solar dynamo models would benefit greatly from improvements in measurement of meridional circulation, including its variations with time as it has severe consequences for the resulting cycle period, phasing of field components, dominant mode symmetry and butterfly diagram (Jouve and Brun 2007; DeRosa et al. 2012; Belucz and Dikpati 2013, see Sect. 2.1.1).

Very little is yet known about other global motions at depth in the convection zone, such as global waves in the tachocline, or 'giant cells' in the convection zone. Knowledge of these would be particularly helpful for dynamo models that go beyond 2D (latitude-radius) to incorporate global 3D effects.

Direct inferences from helioseismic data of, for example, toroidal fields thought to be produced by the solar dynamo near the base of the convection zone remain beyond reach. However, some information is available on fields, or their thermal effects, in and near sunspots and active regions provide their complex horizontal and vertical structure is well taken into account (Ilonidis et al. 2011; Gizon and Birch 2005; Gizon et al. 2010; Schunker et al. 2013).

1.2 11-yr Cycle and Magnetic Parity

Many observational characteristics of the solar cycle are well established (Clette and Lefèvre 2012), and provide rather detailed constraints for solar dynamo models (see Fig. 1). Hale's polarity law has been known since the early twentieth century. The butterfly diagram, which shows the pattern of sunspot occurrence with latitude and time, was first established in the late nineteenth century and is continually updated (Maunder 1904; Hale and Nicholson 1925, Fig. 2). For the past few cycles, this diagram has also been available from averaging magnetograms (Fig. 2, Hathaway 2010), which shows both active region and global poloidal fields at all latitudes.

It is clear from these fundamental measures of the solar magnetic field that there is a phase relation between sunspot and surface poloidal fields, in which the predominantly unipolar fields near the solar poles (which have opposite sign in North and South hemispheres), reverse sign near maximum in the sunspot cycle (e.g., a quadrature of phase between the poloidal and toroidal component of the solar magnetic field). This phase relation is a particularly important feature that dynamo models must account for (see Sect. 2).

It is also well established that surface poloidal magnetic flux migrates from latitudes of solar activity toward both the poles and the equator, presumably under the influence of meridional flow and turbulent diffusion (Wang and Sheeley 1991; Schrijver and Title 2001; Hathaway 2010, see Fig. 5), among other effects. Because of the 'tilt' of active regions with respect to latitude circles, such that leading polarity spots are at lower latitudes than those of follower polarity in the same active region, more follower flux migrates toward the poles; this migration is the agent for reversing the polar field that was brought there during the previous cycle (Babcock 1961; Benevolenskaya 2004). Measurement of and statistical description of the distribution of tilt angles is an active area of research (Kosovichev and Stenflo 2008; Dasi-Espuig et al. 2010; Li and Ulrich 2012; McClintock and Norton 2013). Hemispherical rules for various quantities such as magnetic helicity, twist, rotation of active regions during flux emergence are also studied in detail in order to find systematic trends and possibly clues for pre-eruptive configurations (Pevtsov et al. 2001; Démoulin and Pariat 2009; Hao and Zhang 2011).

Because the polar fields of the Sun are usually of opposite polarity on North and South hemispheres, the Sun is described as having primarily an antisymmetric (dipolar-like symmetry) magnetic field about the equator. This is a somewhat oversimplified picture, since polar fields in the two hemispheres do not change sign at the same time, just within 1–2 years

Fig. 1 AIA/SDO images of the Sun. *Top Left*: continuum at 4600 Å; 6000 K photosphere, *Top Right*: continuum at 1700 Å; 10,000 K chromosphere, *Bottom Left*: He II at 304 Å; 100,000 K transition region, *Bottom Right*: Fe IX at 171 Å; 1,000,000 K corona. Use helioviewer.org on Jan 11th 2013

of each other (Shiota et al. 2012), and because solar activity, as measured by peak annual average sunspot number, is not necessarily of the same amplitude in the two hemispheres. Analysis of observational data indicates there is a fairly strong correlation between North and South peaks (Dikpati et al. 2007), but the occurrence of these peaks is often separated by up to two years. These differences between hemispheres suggest that the Sun's surface field most likely has quadrupolar and higher components too. DeRosa et al. (2012) have shown that the quadrupolar component (as well as the $\ell = 4$ and 6 modes) is dominant during reversal and that the symmetric (secondary family) accounts for up to 25 % of magnetic field energy during most of the solar cycle. The role of parity and dynamo families (symmetric vs antisymmetric) is key in understanding the nonlinear solar dynamo (see Sect. 2.1.2).

Simulating and understanding the origins of the differences between hemispheres of solar magnetic fields is currently a major challenge for solar dynamo models, which is just now beginning to be addressed. All measures of differences in differential rotation, meridional circulation and magnetic patterns are of particular interest. It is thus important to go beyond classical global seismology inversions that tend to symmetrize the flow with respect to the equator.

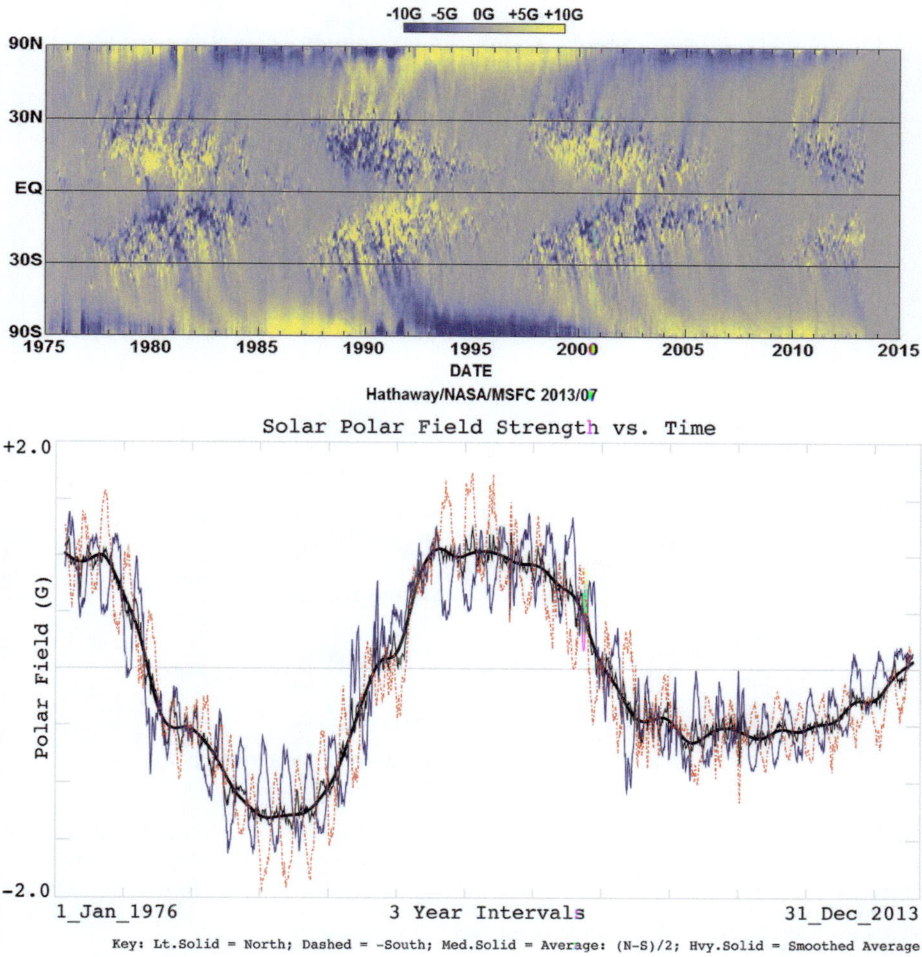

Fig. 2 *Top*: Solar butterfly diagram showing the longitudinally averaged line of sight component of the magnetic field. We clearly see the equatorial and polar branches of solar activity during several cycles, and the reversal of polarities from one sunspot cycle to another (courtesy of D. Hathaway). *Bottom*: Polar field integrated flux, showing the North–South asymmetry and time lag between reversals, as well as the average 11-yr solar cycle. Note that we the polar cap reversal is near completion for cycle 24 (courtesy of T. Hoeksema)

1.3 The Solar Constant

The solar luminosity varies on a secular time scale due to its slow evolution on the main sequence. For instance the Sun was dimmer by 30 % when it arrived on the ZAMS about 4.6 Gyr ago. In 1-Gyr it will be about 10 % more luminous than now. On the human time scale, the solar luminosity can thus be considered constant and is found to be about 1366 W/m^2 at Earth's orbit. Note however that this value has recently be reduced to 1361 W/m^2 by recent studies (Kopp and Lean 2011; Schmutz et al. 2012). The Total Solar Irradiance (TSI), defined as the spectrally integrated solar radiation at 1 AU, is important for characterizing Earth's radiation budget (Domingo et al. 2009).

Still, over the course of the 11-yr cycle the TSI is found to vary by 0.08 % from max to min, being maximal near the peak of the activity cycle, when the Sun is covered by

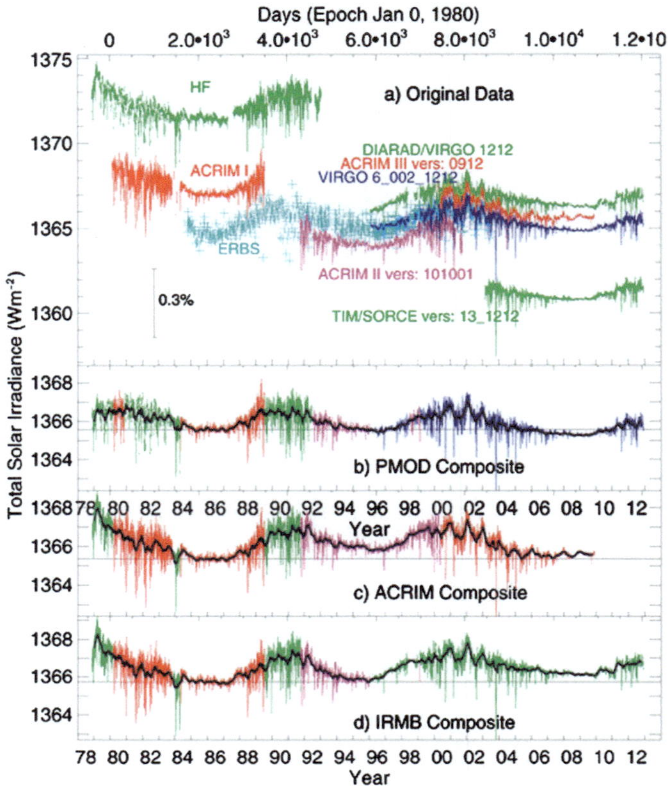

Fig. 3 *Top panel*: Daily averaged values of the Sun's total irradiance TSI from radiometers on different space platforms since November 1978. The data are plotted as published by the corresponding instrument teams. *Bottom Panels*: The PMOD, ACRIM and IRMB composite TSI as daily values plotted in different colors to indicate where the data are coming from. Fröhlich and Lean (1998), Fröhlich (2000, 2012); http://www.pmodwrc.ch/pmod.php?topic=tsi/composite/SolarConstant

sunspots. Indeed, contrary to what might be naively expected, the presence of dark sunspots on the solar surface does not lead to a dimmer Sun because the deficit of radiation due to the dark sunspot is overcompensated by the bright faculae and small scale field surrounding them (Spruit 2000). This is clearly evident when comparing in Fig. 1 the two images of the top row. One notes that the faculae cover a larger surface area than the sunspots. In order to get an absolute measure of the TSI, one must observe from space, given the filtering effect of the Earth's atmosphere on the incoming solar radiation on the ground. Over the last 30 years this has been possible and several satellites have monitored the TSI. We show in Fig. 3 a synthesis of the collected data and three reconstructions of the TSI depending on the calibration used Fröhlich (2000, 2012).

To assess the influence on the Earth of the solar irradiance and of its associated 11-yr modulation, one must evaluate the variability of the solar spectrum through all wavelengths. This detailed version of the TSI is called the Solar Spectral Irradiance (SSI) (Pap 2003; Pap and Fox 2003). Ermolli et al. (2012) have recently made an analysis showing that in the UV range the variations can be as large as a factor 2 and in the X-ray up to a factor of 100 (see also: Lean 1997; Krivova et al. 2007; Pevtsov et al. 2003). UV-light is known to have a direct impact on the ozone layer (Palmer et al. 2004) and a large variation of the UV

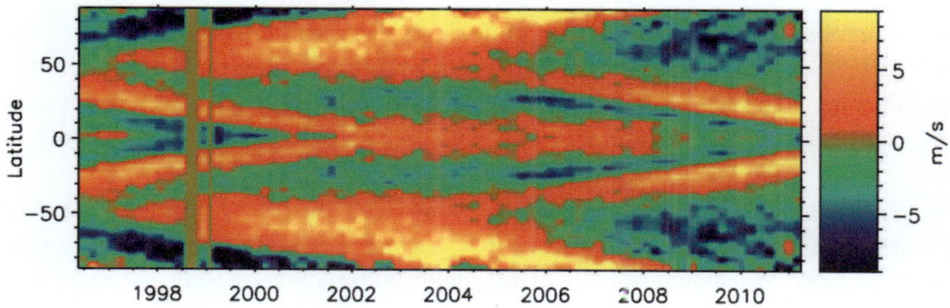

Fig. 4 Torsional oscillations: *color contours* showing the departure in m/s from the temporally averaged solar differential rotation over the last 16 years inverted using MDI data. (Courtesy of T. Corbard)

radiation will thus influence the atmosphere's chemistry and radiative forcing (Haigh et al. 2010). Still, it is currently believed that the Sun's variability, even with its high amplitude in the high energy range, cannot account for more than 10 to 20 % (or about 0.2°) of the current rise (of up to 1°) of the Earth's mean temperature (Foukal et al. 2006; Lockwood 2012). Theoretical studies to understand the physical mechanism that creates the relationship between TSI and the 11-year cycle are still being carried out (Domingo et al. 2009; Spruit 2012). It is worth noting that the influence of short temporal scale trends of the SSI on Earth's atmosphere (such as the one due to solar rotation) and of intermediate temporal scale trends (such as the one linked to the 11-yr cycle in particular in the 150 to 400 nm UV range) are nowadays reasonably well understood (Bolduc et al. 2012; Ball et al. 2012; Thuillier et al. 2012). Longer trends, on centennial or millennium time scales, are, however, still hard to reproduce. Also, the variability of the atmospheric response to various solar forcing models seems to be higher than the intrinsic variations observed, hence pointing to an overestimation of the various effects or feedback loops. Hence more work is needed on both model reconstruction of the SSI and calibration of the observational data. We defer to Sect. 2.1.4 for a brief account of the current physical understanding behind such variabilities and the corresponding models.

1.4 Observational Evidence of Torsional Oscillations

Helioseismic measurements also have documented certain time dependent features of the differential rotation, in particular the so-called torsional oscillations (Howard and Labonte 1980; Labonte and Howard 1982; Vorontsov et al. 2002; Howe et al. 2009). Their role and importance for the dynamo is somewhat less clear, but they are likely to be measures of the feedback of induced magnetic fields on the flows that maintain them, as well as on related thermal patterns (Howe et al. 2009; González Hernández et al. 2010). These time variations occur within a sunspot cycle; we are just beginning to see how torsional oscillations and even the average differential rotation vary from one cycle to the next. The torsional oscillations at the beginning of cycle 24 are found to be weaker than in the previous cycle, possibly the consequence of the observed weaker polar field (see Figs. 2 and 4 and Howe et al. 2013). We delay to Sect. 2.1.4 a discussion of their physical origin.

1.5 Fossil Field and MHD Instabilities

The Sun may have a 'fossil' magnetic field buried deep in its interior, below the convection zone, present in some form since it was condensed out of the primordial interstellar medium.

There are no observations of surface magnetic fields that clearly demonstrate the existence of such a field. Some signature of it would have to be transmitted through the very turbulent convection zone, to be seen at the surface (Strugarek et al. 2011a). At best, it would have to be extracted from the data that describes the solar cycle fields, which are clearly not fossil fields.

On the other hand, surface observations do provide possible evidence of at least two MHD instabilities occurring within and/or just below the convection zone. Active regions themselves could be a manifestation of magnetic buoyancy instability (Moreno-Insertis 1983) of toroidal flux generated by the dynamo at or near the base of the convection zone (Parker 1993). Tubes there become buoyant by virtue of their lower density relative to their surroundings, and rise to the surface. Coriolis forces act on the rising (E-W oriented) tubes to create the tilts seen in active regions. Convection may also contribute to the rise of such tubes to the surface (Jouve and Brun 2009; Weber et al. 2011; Nelson et al. 2013b; Weber et al. 2013; Jouve et al. 2013) their relative contribution being still under study and likely depends on the actual inner field amplitude of the rising structure (Petrovay and Christensen 2010).

The distribution of active regions in latitude also show certain distribution in longitude through a solar cycle, which contain repeated emergence events at those selected longitudes, called 'active longitudes'. The systematic nature of active longitudes may indicate the evidence of another instability, namely a global MHD instability of the toroidal field and differential rotation in the tachocline (Dikpati and Gilman 2005). This instability can create upward bulges in the top of the tachocline, which, if they contain toroidal field, could become favored sites for magnetic buoyancy and/or convection to bring the magnetic flux in the bulge to the surface. More generally, global HD and MHD instabilities of differential rotation in the tachocline, which should create low longitudinal wave number patterns of flow there, could imprint similar global velocity patterns on the bulk of the convection zone. Various attempts have been made to find such velocity patterns at the surface (Komm et al. 2011), but the existence may be very difficult to distinguish from so-called 'giant cell' convection, for which detection efforts have also been on-going (Beck et al. 1998; Komm et al. 2011).

2 A Magnetic and Variable Sun

We now turn to briefly present current theories and models to describe and explain the origin and influence of solar magnetism.

2.1 Global Models of Solar Magnetism and Surface Variability

2.1.1 2D Mean Field Dynamo Models

To date the solar dynamo models that have been most extensively invoked in simulating and explaining principal features of the solar cycle have been 2D mean-field kinematic models. They are an axisymmetric version of mean-field electrodynamic systems, first derived in the late 1960's (Steenbeck et al. 1966). In the form most commonly used, the equations for this 2D system are given by

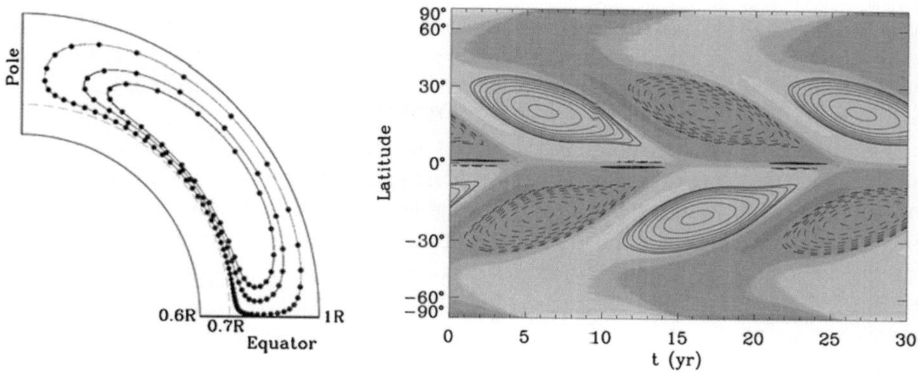

Fig. 5 *Left*: Large meridional circulation used in most of the Babcock-Leighton flux transport dynamo models (each dot represent an interval of 1 yr). *Right*: Butterfly diagram obtained in standard mean field dynamo models (Dikpati et al. 2004)

$$\frac{\partial B}{\partial t} = -\frac{1}{r}\left[\frac{\partial}{\partial r}(r u_r B) + \frac{\partial}{\partial \theta}(u_\theta B)\right] + r\sin\theta(\mathbf{B}_p.\nabla)\Omega$$

$$- \hat{\mathbf{e}}_\phi.[\nabla\eta \times \nabla \times B\hat{\mathbf{e}}_\phi] + \eta\left(\nabla^2 - \frac{1}{r^2\sin^2\theta}\right)B, \tag{1}$$

$$\frac{\partial A}{\partial t} = -\frac{1}{r\sin\theta}(\mathbf{u}.\nabla)(r\sin\theta A) + \eta\left(\nabla^2 - \frac{1}{r^2\sin^2\theta}\right)A$$

$$+ \frac{S(r,\theta)B}{1+(B/B_0)^2} + \frac{\alpha B}{1+(B/B_0)^2} \tag{2}$$

in which the notations have their usual meaning: A and B are respectively the poloidal field vector potential and toroidal field strength, u_r and u_θ are r and θ components of meridional flow, Ω the differential rotation, η the diffusivity, S the surface poloidal source (works as a nonlocal α-effect on toroidal flux tubes risen to the surface from the tachocline) and α an α-effect that can be located with the tachocline or the convective envelope. B_0 is the quenching field strength, which may or may not be the same for the nonlocal and local poloidal field sources; however, in most calculations the value of B_0 is the same for both α-effects (Steenbeck et al. 1966; Moffatt 1978; Ossendrijver 2003; Charbonneau 2005; Tobias 2009).

Here the velocities include differential rotation and meridional circulation (Fig. 5 left panel). Also included is the turbulent magnetic diffusivity and an 'alpha-effect' that captures the turbulent production of poloidal field from toroidal field, sometimes also represented as a surface Babcock-Leighton poloidal source term (Babcock 1961; Leighton 1969) that captures the effect of active region tilts combined with surface diffusion of flux. Other effects are sometimes also included, such as 'quenching' of the alpha-effect and/or the magnetic diffusivity due to strong induced fields. In recent years so-called 'turbulent pumping' has been shown to efficiently advect the field toward the base of the convection zone and could be an alternative to the large scale meridional flow (Tobias et al. 2001; Mason et al. 2008; Guerrero and de Gouveia Dal Pino 2008; Do Cao and Brun 2011).

Among 2D mean field dynamo models, the most successful ones for reproducing a typical solar cycle have been the flux-transport dynamos (Wang et al. 1991), in which a specified meridional circulation plays a prominent role. These models simultaneously simulate the equatorward migration of the peak toroidal field near the bottom of the convection zone, responsible for spots in the butterfly diagram, and the poleward surface migration of weak poloidal fields. They correctly simulate the reversal of polar fields at or near cycle maximum, with the correct change in sign of fields (Dikpati and Charbonneau 1999 cf. Fig. 5 right panel).

Flux-transport dynamo models also show that the shape of the butterfly diagram depends sensitively on the depth of penetration of the meridional flow and the induced fields into the tachocline (controlled in these models by the turbulent magnetic diffusivity profile there). The best fit to observations is found when this penetration is not much below 0.7 R, in which case the latitudinal gradient of differential rotation there is more important than the radial gradient.

In flux-transport type 2D mean-field dynamo models, the period of the magnetic cycle is determined largely by the strength and latitudinal extent of the primary meridional circulation cell included (cf. Fig. 5). Eleven year periods are obtained using meridional circulation whose peak amplitude at the surface is consistent with current observations of this poleward flow. Recently, the fact that cycle 23 was close to 12 years while the previous three cycles had periods around 10 years was explained in these models by the fact that only during cycle 23 did the poleward meridional flow reach all the way to the poles, while during the previous three cycles it stopped near 60° latitude (Ulrich 2010; Dikpati et al. 2010). Alternatively a variation of meridional flow amplitude can also explain the delay of cycle 23-yr and the overlap of active region between consecutive cycles (Dikpati et al. 2010; Nandy et al. 2011). However such changes in the meridional flow intensity should closely follow observations (Hathaway and Rightmire 2010) and not just be adjusted as a free parameter to fit the solar cycle (Karak and Choudhuri 2012). Of course, because these flux transport models are so highly dependent on the meridional flow profile and amplitude, it is not surprising that variations in that flow induce large variations in the cycle length and butterfly diagram (Jouve and Brun 2007). There are several alternatives to the problem of having to take into account a multi-cellular meridional circulation in flux-transport dynamo models. One could invoke for instance either a large radial magnetic diffusion to shortcut the meridional advection path (Yeates et al. 2008; Hazra et al. 2013) or intense magnetic field pumping by turbulent convection to transport the field at the base of the convection zone and in the tachocline further below via penetrative convection (Tobias et al. 2001; Guerrero and de Gouveia Dal Pino 2008; Mason et al. 2008; Do Cao and Brun 2011). Another possibility would be that the solar dynamo is only operating in a shallow layer near the surface as proposed by Brandenburg (2005). As of today none of these models have yet replaced as the "standard model of the solar dynamo" the more classical meridional circulation based mean field dynamo models but clearly they question the current paradigm.

Another limitation of flux transport dynamo models is that observers have often noted that the phenomenology of the solar cycle is hard to reconcile with the notions of turbulent mean field models. The emergence and development of the active regions, the key source of information about the cycle, takes place largely independent of any observed convective flows, for example. Information such as this is discarded in the dynamo models discussed above. An alternate view more consistent with the observations is that the cycle operates on buoyant instability of a strong ordered magnetic field at depth, rather than convective flows. Convective turbulence plays a role only indirectly, by maintaining the differential rotation that is the energy source of the ordered magnetic field. This view, dating back to Cowling (1953) has been formulated particularly clearly by Leighton (1969) and Piddington (1971). It has received substantial support from emerging flux tube calculations such as D'Silva and Choudhuri (1993) and Schussler et al. (1994), see review in Schüssler (2002).

In this view there is a direct connection between the source of the field at the base of the convection zone and the surface phenomenology. In other words, a description in terms of local turbulent dynamo equations misses the essence of the processes responsible for the cycle (see discussion in Spruit 2011, 2012). Attempts to reconcile parts of this picture with a local mean field view, for example by adding 'Babcock-Leighton source terms' to a dynamo equation, ignore the essence of the clues provided by the observations.

Turning back to the classical mean field flux transport solar dynamo model and their likely shortcoming given the complex problem they try to explain, there now exist in the community several flux-transport dynamo models that use different numerical methods. These have all been benchmarked together and shown to give similar results for specified cases (Jouve et al. 2008).

The relative success of these mean field dynamo models in simulating many important features of an average or typical solar cycle has motivated a growing effort to use these models to make prediction of solar 11-yr magnetic activity (Petrovay 2010) in order to go beyond classical geomagnetic precursor based for instance on *aa* or *F10.7* indices (Hathaway 2009). The idea is to implement modern data-assimilation, a technique that has been previously applied in atmospheric and oceanic prediction models, into solar mean field dynamo models. In the solar case, it started with very simple 'data nudging' (Dikpati and Gilman 2006) for simulating the sequence of past cycles and has now moved to increasingly sophisticated variational and sequential methods, developed previously for atmospheric and oceanic prediction systems (Brun 2007; Jouve et al. 2011). Implementation of this method in dynamo models requires solar magnetic and velocity data. Developing data-assimilation techniques for solar dynamo models while in parallel improving the models may eventually lead to the ability to predict future cycles. Currently this method is being used to reconstruct the past meridional flow profile and its time variation, which could explain the observed sequence of solar cycle peaks, durations, and shapes. Following this approach several groups have tried to predict the next solar 11-yr cycle. As of today they have not been successful at all at predicting a weak solar cycle 24 (Dikpati et al. 2006), and the models that did predict a weak cycle required a total reset of the poloidal field at minimum to do so (Choudhuri et al. 2007). Nevertheless it is worthwhile pursuing this approach as it took many decades for the weather community to improve their prediction skills.

How far ahead prediction could be possible depends on the 'memory length' of the model chosen, which in turn depends on the strength of the meridional flow and both the amplitude and the profile with radius of the turbulent magnetic diffusivity (Charbonneau and Dikpati 2000). The larger the diffusivity, the shorter the memory (Yeates et al. 2008). Current estimates of the memory length range from about one to two sunspot cycles. Another important time scale for prediction models that use data assimilation is the response time of the model to time-varying data inputs. Recent results (Dikpati and Anderson 2012) suggest that this time scale is in the neighborhood of six months, depending on details of the model. The important point is that it is short enough to have the model respond to time varying inputs of observational data on time scales quite short compared to a solar cycle, so that in principle some details of the ascent to maximum and descent to the following minimum can be simulated.

Efforts are beginning to be aimed at understanding the origin of differences in amplitude and phase between North and South hemispheres in a particular cycle, by studying the role of equator asymmetry, dynamo family mode coupling or coupling to velocity field with a different time scale (Tobias 2002; DeRosa et al. 2012). Mean field models can be extended to take into account the effect of the large-scale Lorentz force, with the so-called Malkus-Proctor effect (Malkus and Proctor 1975). This approach has the advantage that the nonlinear saturation is a natural outcome of the model via the quenching of the flow by Lorentz force (Brandenburg et al. 1992; Moss and Brooke 2000; Bushby 2006). A direct consequence of the feedback of the magnetic field on the flow is to obtain torsional-like oscillations (Covas et al. 2005; Rempel 2006). Still, Spruit (2000) and Rempel (2006) advocate that thermal effects due to either small scale or large scale magnetic field could also explain such torsional oscillations. In particular, the equatorward branch could be difficult to explain with only a direct feedback of the Lorentz force on the flow (see Sect. 2.14 for more details). Flux-transport models can be further generalized to include global features that are dependent on longitude, such as the observed 'active longitudes' (Berdyugina et al. 2006). Finally, mean field models should have a better treatment of the surface boundary conditions, by going beyond the usual potential field extrapolation. Coupling to a dynamic corona or to a solar wind are in progress (Yeates et al. 2010; Pinto et al. 2011; Warnecke et al. 2012).

2.1.2 Parity of Solar Global Magnetic Field with the Mean Field Flux Transport Dynamo

The study of parity (symmetry properties with respect to the equator) of solar global magnetic field is essential to understand the Sun, its dynamo and the coupling between dynamo mode families (McFadden et al. 1991; Gubbins and Zhang 1993; Pétrélis et al. 2009). Observationally, it is confirmed that the solar global magnetic field is roughly anti-symmetric about the equator. The polarity of the polar field is almost always opposite and Hale's law (Hale and Nicholson 1925) says that the sunspot pairs have opposite polarity rule between hemispheres. Note that polarity reversal in one hemisphere can be shifted in time by up to 2 years, hence hinting at the existence of modes of higher degree than the dipole. Most recently, DeRosa et al. (2012) have conducted a detailed observational study using both Wilcox and MDI data for the spherical harmonic component of the solar global magnetic field. It is here worth recalling that simple parity rules of spherical harmonics imply that when the sum $\ell + m$ is respectively even (odd; with ℓ the spherical harmonic degree and m the azimuthal wavenumber), the mode is symmetric (anti-symmetric) with respect to the equator (see Fig. 6). Hence the axisymmetric (polar) dipole is anti-symmetric and the axisymmetric quadrupole is symmetric. So in axisymmetric mean field dynamo models (since $m = 0$) the parity of ℓ is enough to tell in which family a mode belongs. Note however that the equatorial dipole is symmetric since $\ell + m = 2$, and hence is part of the secondary/quadrupolar-like dynamo family, not of the primary/dipolar-like family as its name would have implied! Looking carefully at the spherical harmonics decomposition over the last 3 cycles, DeRosa et al. (2012) conclude that the symmetric components have of the order of 25 % of the anti-symmetric component amplitude, most of the cycle, with the axisymmetric quadrupole becoming dominant (see Fig. 6) during reversal as for the Earth's magnetic field (Leonhardt and Fabian 2007). In addition, there are irregularities about the parity. During the Maunder minimum, there was a significant asymmetry of the sunspot number between hemispheres, i.e., the southern hemisphere had a larger number of sunspots (Sokoloff and Nesme-Ribes 1994). Such hemispherical dynamo/activity can be easily explained if symmetric and anti-symmetric modes (such as the $m = 0$ dipole and quadrupole modes) are of the same amplitude (Gallet and Pétrélis 2009). Asymmetric meridional circulation and/or poloidal source terms can also lead to different behavior in the northern vs the southern hemisphere and explain time delay between the reversal of the polar caps (DeRosa et al. 2012; Belucz and Dikpati 2013). Recent Hinode observations confirm old observations (Babcock 1959; Shiota et al. 2012) that the reversal of one hemisphere's polar magnetic field (in this case the northern hemisphere's) can be faster than that of the opposite pole. These observations raise

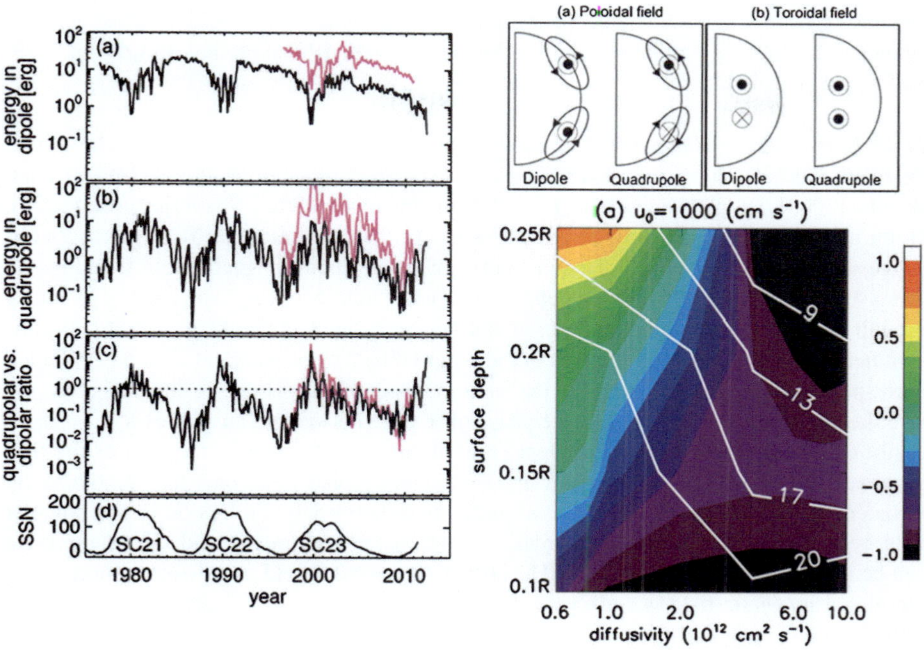

Fig. 6 *Left*: Solar dipole/quadrupole evolution and ratio over the last 3 solar cycles (Wilcox observatory (*black*); MDI data (*magenta*); DeRosa et al. 2012). *Right*, *top*: Cartoon representation of field parity/symmetry; *Right*, *bottom*: Symmetry parameter (between −1 (antisymmetric) and 1) for mean field dynamo solutions as a function of magnetic diffusivity and surface depth for a given amplitude of the meridional flow. Superimposed white contours indicate the cycle period of the dynamo (Hotta and Yokoyama 2010b; Gubbins and Zhang 1993)

several questions, such as 1. How does the Sun maintain a dipolar dominant global magnetic field? 2. How is the irregularity in parity caused? 3. What is the coupling between dynamo mode families?

Although there are many three-dimensional studies about the solar magnetic field including the effect of the thermal convection that can address many of these aspects (e.g., Brun et al. 2004a; Browning et al. 2006; Miesch and Brown 2012; Strugarek et al. 2013), the mean-field approach to this problem is useful and complementary, since the time evolution of parity can occur on long time scales with respect to the convective overturning time (up to centuries). Aspects of this problem can thus be addressed with flux transport dynamo models, which reproduce a number of features of the solar magnetic cycle as was discussed in the previous section (see also, Choudhuri et al. 1995; Dikpati and Charbonneau 1999; Dikpati and Gilman 2001). One can for the sake of clarity divide parity studies into two categories: One category has a long time scale (∼1000–10000 years), and the other has a short time scale (∼10 years). The former is caused by the turbulent diffusion between hemispheres and the latter is caused by the irregularity of the meridional flow, the Babcock-Leighton α-effect, asymmetries with respect to the equator or non linearities (Tobias 1997). Indeed it is worth recalling that in simple kinematic mean field dynamos the symmetry properties of the main physical ingredients (α, Ω, S) favors (or not) the coupling of dynamo family modes. In the classical $\alpha - \Omega$ dynamo, if α is chosen to be anti-symmetric and Ω symmetric, then both dynamo families are a solution of the problem but are disjoint (e.g. independent). Only if α is symmetric are the primary and secondary dynamo families coupled as observed in the Sun. Of course when for instance asymmetries or nonlinearities are considered this coupling genuinely happens (McFadden et al. 1991; Gubbins and Zhang 1993; DeRosa et al. 2012).

Long Time Scale (~1000–10000 years) Investigations of the parity of the solar global magnetic field on a long time scale using a flux-transport dynamo were done by Moss and Brooke (2000), Dikpati and Gilman (2001), Bonanno et al. (2002), Chatterjee et al. (2004), Hotta and Yokoyama (2010b), and DeRosa et al. (2012). Hotta and Yokoyama (2010b) suggest a simple explanation of maintenance of solar global magnetic field (see Fig. 6). They suggest that the diffusion of poloidal (toroidal) field between hemispheres promotes growth of anti-symmetric (symmetric) global magnetic field. When the global field is anti-symmetric about the equator, the sign of the longitudinal vector potential A (see (2)), which corresponds to the poloidal field, is the same between hemispheres. In this situation, the turbulent diffusion can smooth out the phase difference between hemispheres which can be easily generated in the highly turbulent convection zone. In contrast the sign of the longitudinal vector potential is opposite in a symmetric global field, thus the smoothing by turbulent diffusion cannot work. This suggests that the turbulent diffusion of the poloidal field between hemispheres tends to make an anti-symmetric global field. The same logic can be adopted for the toroidal field. The sign of the toroidal field is opposite (same), when the global magnetic field is anti-symmetric (symmetric). The efficiencies of the turbulent diffusion on the poloidal and toroidal magnetic field determine the parity of the solar global magnetic field on long time scale. The transition time scale from one parity to the other is roughly the diffusion time scale from a pole to the other pole, which is 1500 years when the value of turbulent diffusivity is $10^{12}\ \mathrm{cm^2\,s^{-1}}$. Hotta and Yokoyama (2010b) conclude that a strong turbulent diffusivity only near the surface promotes the anti-symmetric global field with efficient diffusion of poloidal field, since the Babcock-Leighton α-effect generates the poloidal magnetic field only near the surface. A fast meridional flow without the α-effect in the convection zone suppresses the growth of anti-symmetric global field, since the meridional flow transports poloidal field poleward at the surface and toroidal field is transported equatorward around the base of the convection zone. This promotes (suppresses) the diffusion of toroidal (poloidal) field. Dikpati and Gilman (2001), Bonanno et al. (2002), and Chatterjee et al. (2004) suggest that the α-effect in the convection zone caused by magnetic buoyancy instability or by the unstable shallow-water modes significantly promotes the anti-symmetric global field. This can be explained by noting that when the poloidal field is generated in the convection zone the meridional flow transports it equatorward and this promotes the diffusion of poloidal field between hemispheres. DeRosa et al. (2012) find additional factor which determines the global magnetic field. In their parameters, the anti-symmetric component of the global field dominates the symmetric component by about 5 orders of magnitude. When they introduce a small asymmetry (10^3 times smaller than the main component) in the meridional flow or the α-effect, the amplitude of the symmetric component becomes larger which can explain the real solar global magnetic field.

Short Time Scale (~10 years) The investigations of the parity of global magnetic field on a short time scale is studied for instance in Goel and Choudhuri (2009), Choudhuri and Karak (2009), and Karak (2010). Irregularities of Babcock-Leighton α-effect and the meridional flow can introduce this short time scale fluctuation of the parity. There is possibility that this type of short time scale parity issue has important role in the context of the grand minima. Choudhuri and Karak (2009) suggest that a large sunspot which does not obey Joy's law causes a weak poloidal magnetic field in global scale and the grand minima. Karak (2010) suggests that a decrease in the meridional flow causes a grand minima in diffusion-dominated regime (see also Yeates et al. 2008; Hotta and Yokoyama 2010a). A slow meridional flow generates long cycle period which is consistent with the observation of the carbon-14 by Miyahara et al. (2004) which suggests that the cycle period is relatively long (∼14 years) in the Maunder minima. However other studies have found shorter or less well defined cycle period lengths (Beer et al. 1998; Solanki et al. 2006). Time changes in surface meridional flow within a solar cycle that are different in North and South hemispheres are observed (Ulrich 2010), and, while there are no direct observations of time changes in the alpha-effect, from the differences in active region patterns between North and South within a cycle, we can infer that such time changes are likely to be present. Therefore the possibility that these changes occur simultaneously in both hemispheres is extremely small. When this type of anomaly occurs only in one hemisphere, the global parity is significantly affected. So the detailed observational study of solar magnetic field parity (e.g., Sokoloff and Nesme-Ribes 1994; Shiota et al. 2012), and the asymmetry of the solar mean field and flow (e.g., Rightmire-Upton et al. 2012) can further help assessing the triggering of or the exit from grand minima (Tobias 1997; Abreu et al. 2008).

2.1.3 3-D Global Dynamo Models

Computational resources do not permit a complete 3–D simulation of the global solar dynamo. All the nondimensional parameters traditionally used in fluid mechanics to characterize various properties of the flow relative to dissipation—e.g., the Reynolds number and Rayleigh number, measuring inertia and buoyancy driving relative to dissipation, or the magnetic Reynolds number, measuring magnetic induction relative to Ohmic dissipation—are many orders of magnitude larger than in any tractable numerical simulation. Equivalently, the Sun possesses flows and magnetic fields that span a far greater range of spatial scales than can be captured in present-day simulations. Thus any numerical simulation of solar dynamo processes must choose which scales to resolve explicitly, and which to parameterize or ignore. Typically these simulations have fallen into two broad groups: one seeks to model localized domains with great fidelity, capturing for instance flux emergence (e.g., Rempel et al. 2009; Cheung et al. 2010) or small-scale convection (e.g., Stein and Nordlund 1998; Cattaneo 1999; Vögler et al. 2005). Another seeks to model the largest scales of flows and fields, which feel the overall geometry and the effects of rotation, while employing subgrid-scale (SGS) parameterizations of the smallest scales which remain unresolved. Recent progress in improving the SGS treatment of global spherical codes such as ASH or Eulag-MHD have lead to interesting results with magnetic field being more fibril in nature with extended wings in their pdf's, characteristic of intermittent and intense magnetic flux bundle (Nelson et al. 2013b; Charbonneau 2013).

Whether these two modeling approaches truly capture all the interactions governing the evolution of the scales they model can be questioned (e.g., Tobias and Cattaneo 2008). This can be probed by studying the detailed spectral transfers of energy (Bullard and Gellman 1954; Biskamp 1993; Lesieur 2008; Mininni et al. 2005; Strugarek et al. 2013) to characterize, e.g., which scales are responsible for the growth and saturation of the largest magnetic scales, or what sets the spectral slopes at the smallest scales. Various research groups in recent years have expended special effort towards developing such studies. All these works highlight the importance of considering the interactions of the largest magnetic and velocity scales (i.e., at the scale of the Sun), together with a sufficiently broad range of somewhat smaller (often referred to as *intermediate*) scales, in order to accurately describe the largest scales' evolution (Livermore et al. 2010; Strugarek et al. 2013;

Hughes and Proctor 2013). Nevertheless, the studies of Livermore et al. (2010) and Strugarek et al. (2013) both showed that in some cases, particular scales dominated the energy transfer towards and from the largest scales, providing some hope that it may be possible to parameterize the basic mechanism leading to large scale magnetic field development in non-linear turbulent dynamos. Interestingly, Strugarek et al. (2013) showed that the large-scale axisymmetric dipole was saturated through its interaction with the magnetic scale at the peak of the magnetic energy spectrum. This scale is highly sensitive to the parameters of the numerical simulation, and in particular to the magnetic Prandtl number $Pm = \nu/\eta$. Hence, this study clearly pointed out the reasons for the apparently different results obtained in the literature on the large scale magnetic field dynamo solution when varying the magnetic Prandtl number. Detailed studies of spectral transfers in dynamo cases in super-equipartition (Brown et al. 2010) and which possess magnetic cycles (Racine et al. 2011; Nelson et al. 2013b) may reveal the important pieces of large scale magnetic field generation and sustainment in the near future.

Here, we describe briefly some principal results and recent developments in global 3-D simulations of solar dynamo processes, leaving discussion of more localized calculations to other contributions within this volume. In most conceptual models of the large-scale solar dynamo, a central role is played by global-scale flows: differential rotation, which can stretch poloidal fields into toroidal fields, and perhaps meridional circulations, which can advect magnetic fields and in some models figure prominently in setting the strength and period of solar cycles as described in Sect. 2.1.1. Many of the most fascinating observational features of solar magnetism—the equatorward propagation of sunspots, the global reversal of the dipole field every 11 years—are likewise global-scale phenomena. Because of this, substantial effort over the past several decades has been dedicated to developing 3–D simulations that encompass these largest scales of flow and magnetism.

Early global simulations that sought to model the solar convection zone and its magnetism were conducted by Gilman and collaborators (e.g., Gilman and Miller 1981; Gilman 1983; Glatzmaier 1985). The earliest of these models were linear, Boussinesq calculations; later simulations adopted the anelastic approximation (which captures the overall density stratification but filters out sound waves) and considered highly nonlinear (though still quite laminar) flows. Some of these simulations generated global-scale magnetic fields that showed latitudinal propagation and cyclical polarity reversals. These cyclical dynamos were distinguished from non-cyclical simulations in part by the presence of reasonably strong differential rotation (which comprised the majority of the total kinetic energy): simulations in which the differential rotation was too weak, either because of dissipative effects or strong Lorentz force feedbacks, tended to exhibit more chaotic magnetism. Even in the cyclical cases, field propagation in these simulations was towards the poles—in accordance with kinematic dynamo predictions for this combination of differential rotation and kinetic helicity, but opposite to that observed in the Sun. The cycle periods were also either irregular (Gilman and Miller 1981) or, when orderly, much shorter than the observed solar cycle (e.g., Gilman 1983).

With increasing computational power, later global-scale simulations of the convection zone are able to model increasingly turbulent flows extending over multiple density scale heights first in a purely hydrodynamical setting and then by taking into account magnetism. 3-D global-scale convection models have for instance identified the key role played by baroclinic effects to explain the peculiar (conical) profile of the solar differential rotation and the importance of the thermal, dynamical and magnetic coupling of the surface convective envelope with the subadiabatic layer below (Miesch et al. 2006; Brun et al. 2011). These models have further predicted that the meridional circulation should possess many cells per hemisphere and not just one as many mean field 2-D models are finding. Of course many issues remain to be understood, such as the near surface shear layer, the existence or dynamical properties of various scales of convection such as giant cells, supergranules or mesogranules and the exact profile in latitude and depth of the meridional circulation. We refer the interested reader to Chap. 4 of this book for a recent account of the latest findings and limitations of these global hydrodynamical models, focusing here on describing the global solar dynamo.

Until recently, many global-scale models exhibited predominantly small-scale magnetism that reversed in polarity chaotically, after relatively brief (about 500 hundred day) intervals (Brun et al. 2004b; as shown in Fig. 7a). The mean axisymmetric fields in these simulations reach of the order of 2–3 % of the total magnetic energy which represents about 10 % of the total kinetic energy. In these purely convective dynamo simulations the magnetic boundary condition (open or closed) influenced the organization of the field and in some cases yielded substantially more ordered fields; e.g., Brown (2009). Increasingly turbulent simulations in more localized domains likewise highlighted some of the difficulty of building a strong, spatially and temporally organized field by turbulent convection alone (e.g., Cattaneo and Hughes 2009; Hughes et al. 2011).

Over much of the past two decades, attention turned to the possibility that the global solar dynamo might be seated not throughout the convection zone, but primarily at its base (e.g., Parker 1993; Spiegel and Weiss 1980). That region was revealed by helioseismology to be a site of strong shear (e.g., Thompson et al. 2003), so might plausibly allow strong toroidal fields to be generated from seed poloidal fields. It was also realized that the stable stratification below the convection zone might allow fields to be greatly amplified before becoming susceptible to magnetic buoyancy instabilities, which might otherwise lead to loss of fields on timescales short compared to the 11-year solar cycle (or to the time thought necessary to build strong toroidal fields).

Global dynamo simulations in turn have attempted to capture elements of this "interface dynamo," albeit in a highly simplified fashion. The simulations of Browning et al. (2006) and Browning et al. (2007), for example, encompassed both a convective envelope and a stably stratified region below it; a "tachocline" of shear was essentially imposed, by driving the stably stratified region towards solid-body rotation while allowing the convection zone to build differential rotation. The basic picture that emerged from these simulations was of intense small-scale magnetism in the convection zone, accompanied by somewhat more organized and long-lived field structures below its base (see Fig. 7b). The toroidal fields generated in the calculations were antisymmetric about the equator, as inferred in the Sun, independent of the symmetry of the initial imposed magnetic fields. But in sharp contrast to observations of the solar magnetic cycle (which reverses in polarity roughly every 11 years) and earlier simulations of the convection zone alone (which typically reversed over much shorter intervals), these simulations showed no clear polarity reversals. This may have been due partly to the weak differential rotation in these simulations, which is known to affect the presence or absence of oscillating solutions (Gilman 1983). Later simulations by Charbonneau and collaborators (Ghizaru et al. 2010; Racine et al. 2011; Beaudoin et al. 2013), using a different numerical method to model a similar geometry, *have* yielded orderly magnetic cycles: the simulations show large-scale fields that reverse in polarity over timescales of about 36 years (cf. Fig. 7c). The fields imprint through both the convection zone and portions of the underlying radiative region, and are strongest just below the core-envelope interface. This is arguably the closest contact any 3–D simulation has yet made with the global solar cycle. But several major discrepancies with observations persist: the toroidal field peaks at higher latitudes than in the Sun (Beaudoin et al. 2013),

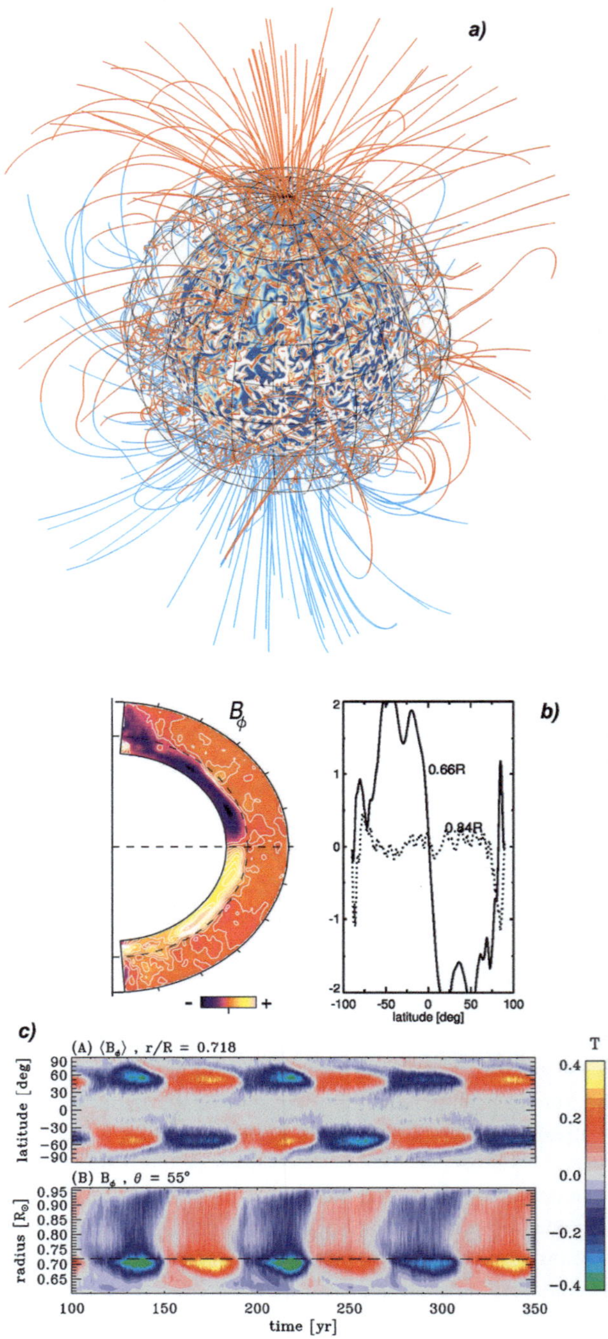

Fig. 7 (**a**) 3-D reconstruction of magnetic field lines in a global turbulent convective dynamo (Brun et al. 2004). (**b**) *Color contours* showing the azimuthal and temporal averaged of the toroidal magnetic field in a global dynamo simulation with a tachocline along with two latitudinal cuts in the convection zone ($r = 0.84$ R) and in the tachocline ($r = 0.66$ R) (Browning et al. 2006). (**c**) Butterfly diagram realized in a 3-D dynamo simulation with a tachocline possessing a cycle period of about 40 years (Racine et al. 2011)

and no equatorward migration of the field is realized. The poloidal and toroidal fields in these calculations also appear to oscillate in phase, whereas in the Sun they appear to be phase-lagged. More fundamentally, it is not yet clear what delineates the cyclical dynamo solutions from steady or chaotic ones (an issue we return to below).

These simulations provide some support for the view that a layer akin to the solar tachocline may be *helpful* for building organized, cyclical magnetism—though even this basic statement remains controversial. But is a tachocline *essential* for building highly ordered fields in the Sun and other stars? Some recent simulations have essentially gone "full circle," focusing on the properties of magnetic fields generated by turbulent convection alone (without a tachocline). Global simulations are now suggesting that organized fields can be built even amidst reasonably turbulent convection, provided other conditions are met. In the anelastic simulations of Brown et al. (2010, 2011), for instance, strong "wreaths" of toroidal field were generated by convection in a spherical shell rotating fairly rapidly (of the order of 3 times solar rotation rate); in some parameter regimes these were steady, and in others cyclic. In another recent development, Käpylä et al. (2012) found equatorward propagation of magnetic fields, and cyclic polarity reversals on roughly decadal timescales, in "wedge" simulations that modeled portions of a spherical convective envelope. The mechanism for equatorward propagation in these simulations is unclear: they do not possess a near-surface shear layer, but do exhibit equatorward meridional circulations at the surface (in contrast to the observed solar surface flows). (These circulations may be too weak to advect the fields equatorward, though.) Most recently, Käpylä et al. (2013) have argued that the role of density stratification is crucial in setting the direction of field propagation, though from other simulations (e.g., Gastine et al. 2012) it is clear that strong stratification alone is not enough to get equatorward propagation. Further the radial shear in the tachocline may not be able to sustain the forces required to operate the solar cycle as discussed in Spruit (2011, 2012).

At first glance, many of these results (e.g., Brown et al. 2010, 2011; Käpylä et al. 2010, 2012) are reminiscent of what was found by Gilman and collaborators long ago: cycles are realized, particularly when rotation is relatively rapid, but neither the timescale of reversals nor the sense of latitudinal propagation have generally matched what is observed in the Sun. But in other ways these models are less similar: the flows being probed in the latest simulations are significantly more turbulent, and admit stronger density stratifications, than in the models of yesteryear. That even such complex flows can build organized fields—even if those fields are still not altogether solar-like—is striking. The loss of magnetic fields through buoyancy instabilities has been less severe than anticipated, even with strong (~40 kG) toroidal fields generated within the convection zone. Part of this may be a numerical effect, for fibril structures (like a rising flux tube) may be rapidly diffused in most feasible numerical simulations. But various physical effects, like the magnetic pumping of fields downward (e.g., Tobias et al. 2001), the turbulent pressure of the surrounding medium, and the fundamentally non-axisymmetric nature of the dynamo, also appear to play roles. The most recent global-scale calculations (e.g., Nelson et al. 2013b), probing increasingly turbulent parameter regimes, have demonstrated polarity reversals of organized magnetic structures at more modest rotation rates than some earlier work (Brown et al. 2011). In these simulations, quasi-cyclical reversals are delineated from steady solutions partly by the diffusion: lowering the diffusivity (i.e., more turbulent flows) helps lead to reversals. In the most turbulent cases, resolved Ohmic and viscous diffusion are at long last demonstrably playing less of a role than the resolved convective motions.

In even more recent work (Nelson et al. 2013a), Ω-loops have been self-consistently generated. They formed out of the most intense large scale magnetic wreaths. This is a direct consequence of the more intermittent and less diffusive nature of the wreaths yielding concentrated bundles of field that become buoyant. These simulations, coupled with systematic parameter studies of flux tube emergence in global 3-D simulations (Jouve and Brun 2009; Weber et al. 2011, 2013; Jouve et al. 2013; Pinto and Brun 2013), confirm that magnetic layers (wreaths) straddling the base of the convection zone and the tachocline can be the origin of the observed active region complex seen at the surface (Petrovay and Christensen 2010). Of course many issues remain to be sorted out in detail, such as the maximum degree of twist and intensity of the field allowed, the latitude of 'emergence' of the structures and their tilt angle, but real progress has been realized in recent years. This fact is cause for cautious optimism, since it suggests that the simulations *may* be approaching a regime in which the details of the numerical implementation, and where exactly in parameter space the simulations lie, may matter less and less. This in turn would allow for more quantitative statements about how cycles and field propagation arise in some simulations—and perhaps ultimately how they arise in the Sun, too. To achieve this goal significant efforts should be devoted to obtaining the first solar *spot-dynamo*, i.e. a nonlinear global solar dynamo that has self-consistently emerging magnetic structures.

2.1.4 Theoretical Understanding of Solar Modulations: Torsional Oscillations & the Solar Constant

In order to model the variabilities associated to the TSI or SSI and to the torsional oscillations several approaches have been developed. The basic idea is to understand how the solar plasma responds to a variation and intensification of the magnetic field. What is for instance the nature of Lorentz force feedback on the flow or the influence of magnetic fields on thermal energy transfer and radiative output? To explain most of the observed modulation of the TSI, variations in surface emissivity, i.e. the reduced cooling in sunspots and enhanced emission by small scale magnetic fields, are the most effective mechanisms (Spruit 2003). In order to model the TSI, the simplest approach consists of developing proxy models. Such models aim at developing a description of the variation of the radiative output in a given frequency range by taking into account observed correlations between various indicators (F10.7, aa index, etc...; Fox 2004; Usoskin 2013). More physically based models have also been developed, which aim at computing the solar spectrum in details using radiative transfer calculations (Kurucz 1993; Fontenla et al. 2011). Improvements in the models include treating the gas in NLTE, considering better atmosphere's equilibrium and dynamics, taking into account more lines, multi-level atoms and the degree of ionization, computing better opacity, etc... (Unruh et al. 2000). Inspired by the increasing accuracy of magnetograms and photospheric and chromospheric images, they use various filling factors for faculae and spots coverage. Some models even predict the center to limb variations while others just provide a full disk integration. Another approach is to develop local radiation-convection models of the solar surface. Indeed such simulations of surface convection and magnetism coupled to radiation have now reached a high level of realism (Asplund et al. 2004; Vögler et al. 2005; Stein and Nordlund 2006). They show that in plages or faculae there is an enhanced radiation. They thus confirm that the presence of magnetic field is the cause of irradiance variations on short time scales (shorter than the 11-yr cycle).

The easiest way to explain torsional oscillations is to consider the feedback of the Lorentz force on the flow. Several studies have looked at this process and it is in qualitative agreement with observations (Moss and Brooke 2000; Covas et al. 2005; Bushby 2006; Beaudoin et al. 2013). Maxwell stresses tend to oppose Reynolds stresses influencing the maintenance of the differential rotation in rotating convective envelope. With the band of magnetic activity propagating both equatorward and poleward as the 11-yr cycle progress, the strongest feedback from the Lorentz force moves likewise, yielding two branches. Another explanation of the physical origin of torsional oscillations was proposed by Spruit (2003). In this work he proposes that the deviations in the solar rotation rate are due to a geostrophic flow driven by temperature change near the surface due to enhanced emission of radiation by the small-scale magnetic field and not by the direct action of the Lorentz force. Rempel (2006) has shown that the equatorward and poleward branches differ in both strength and depth. In his nonlinear dynamo model the Lorentz force can explain the poleward branch but not the equatorward one. Only when including thermal forcing in the activity belt can he recover the latter. The required temperature variations are found to be of order 0.1 K.

However in flux transport dynamo model of the Babcock-Leighton type as used in the study of Rempel (2006) there is a tendency for the magnetic flux to be concentrated near the poles if a low diffusivity of order $10^{11}\ \text{cm}^2/\text{s}$ is used. This tends to concentrate the Lorentz force near the surface to high latitude. If one instead enhances the diffusivity near the surface to values close to the observed decay of active regions $\sim 2 \times 10^{12}\ \text{cm}^2/\text{s}$, the ratio between the amplitude of the poloidal field at the poles and near the equator is much closer to observations (Yeates et al. 2008). This has the consequence that the equatorward branch in nonlinear dynamo models (i.e with a feedback coming from the large scale Lorentz force, e.g., the so-called Malkus-Proctor effect) is of comparative strength to the poleward torsional oscillation branch (Brun and Do Cao 2013 private communication). Thus the direct action of the Lorentz force may still be a plausible solution for both branches of the torsional oscillations even though the thermal origin of the equatorward branch is currently the preferred solution. Recent papers by Jiang et al. (2010) and Cameron and Schüssler (2012) have further considered how active regions can locally influence the meridional circulation (Jouve and Brun 2009) and impact the solar dynamo by perturbing the advection of the field to the polar caps. Finally we mention that a third explanation of the origin of torsional oscillations could involve the microquenching model of the differential rotation by active regions (Petrovay and Forgács-dajka 2002).

2.2 Deep Interior Magnetism

Intense magnetism manifests itself in a wide spectrum of spatial and temporal scales at the solar surface as we have just discussed. However one can wonder if a more deeply buried magnetic field is not present and if some of the surface magnetic properties do not have a deeper root than the convective envelope and the tachocline at its base. Indeed, as the young Sun formed, it likely did so in a weakly magnetized interstellar medium (Shu et al. 1987; Nakano et al. 2002; Padoan et al. 2011; Vázquez-Semadeni et al. 2011; Crutcher et al. 2009; Joos et al. 2012). Through the pre main sequence (PMS) phase (e.g., Palla and Stahler 1993; Moss 2003), such inner magnetic field has most likely been influenced by the presence of deep turbulent convective motions by being either destroyed by enhanced magnetic diffusion, expelled by convective motions or amplified by dynamo action. As the deep (full) convective envelope recessed toward the surface, a magnetic field has most certainly been left behind. Given the very slow Ohmic diffusion time in the radiative interior of the Sun ($\tau_\eta \gtrsim 10^9$ Gyr), such a fossil field is certainly still present within the Sun. We can thus ask what is its amplitude, topology and dynamical influence on the Sun (Gough and McIntyre 1998).

2.2.1 On Tachocline Confinement

Deep solar interior magnetism is still unconstrained by observations today. The rotation profile observations (e.g., Schou et al. 1998; Thompson et al. 2003) are the only observational contraints that may guide our intuition on which kind of magnetic field could exist in the radiation zone, if there is any. For instance, the solid body rotation of the radiation zone and the extreme thinness of the tachocline are still not satisfactorily understood today. In particular, Spiegel and Zahn (1992) showed that the solar tachocline (as we know it presently) should slowly diffuse into the radiation zone on a hyperdiffusive Eddington-Sweet time scale. If the tachocline had the same thickness today as when the sun entered the Main Sequence, it should have spread over 30 % of the solar radius by now (see also Wood and Brummell 2012, for a detailed parametric study of this so-called *radiative spreading*). Hence, one has to come up with a way to confine the solar tachocline and a way of maintaining an almost solid-body rotation in the radiative interior. One way can be to rely on horizontal turbulence or on internal gravity waves (Zahn et al. 1997; Charbonnel and Talon 2005). Another is to rely on the existence of an inner magnetic field. Indeed a large-scale axisymmetric magnetic field in the radiation zone could in principle explain why it rotates roughly like a solid body (Rüdiger and Kitchatinov 1997). This can be achieved through the homogenization of the rotational profile along the magnetic field lines (the so-called Ferraro's law of iso-rotation, Ferraro 1937). Hence, many studies are aimed at exploring the possible existence of such a hypothetical magnetic field and its stability and dynamical evolution, to assess if it could explain the observed solar inner rotation profile.

Despite the appeal of the magnetic scenario to explain the solar internal rotation profile, it has to be noted that an organized large-scale magnetic field does not remain naturally confined and tends to diffuse out on ohmic diffusion time-scales. The ohmic diffusion time $\tau_\eta \sim R^2/\eta$ (where R is the characteristic length over which the field diffuses and η is the ohmic diffusion coefficient) is very long in the solar radiation zone ($\sim 10^9$ yr). Therefore, at the age of the Sun a significant part of such magnetic field would have diffused out and encountered the tachocline (and eventually the convection zone above), which could severely affect both the topology of the magnetic field and its effect on the rotation profile. Gough and McIntyre (1998) (noted GM98 hereafter) proposed a scenario in which the interior magnetic field could be confined by a penetrating meridional flow coming from the convection zone. A significant amount of work has been carried out since to assess the reality of this scenario non-linearly, with the help of numerical simulations. The first numerical attempt to reproduce the GM98 scenario was carried out by Garaud (2002) in 2D. Taking into account the radiation zone only, she showed that the magnetic field lines should open into the tachocline and transport angular momentum throughout the solar interior. Similar results were obtained by Brun and Zahn (2006) in 3D. They also observed the development of magnetic instabilities (see Sect. 2.2.2), confirming that a purely axisymmetric dipolar field was unstable in 3D and could not be a likely candidate for the tachocline confinement scenario. In order to better test the GM98 scenario, *ad-hoc* meridional circulations (supposedly originating from the unresolved convection zone) were added to the numerical models (Sule et al. 2005; Rüdiger and Kitchatinov 2007; Garaud and Garaud 2008). They confirmed that in order for the GM98 scenario to be valid the amplitude and shape of the meridional circulation must be chosen with care otherwise the tachocline spreads. The meridional flow penetration at the base of the convection zone is also an important ingredient of that model but it is currently poorly constrained by observations. In order to test the GM98 scenario more self-consistently, two paths have been recently followed.

First, a better understanding of the meridional circulation penetration was developed in reduced models by Garaud and Brummell (2008), Garaud and Acevedo Arreguin (2009) and Wood and Brummell (2012). These studies shed light on the importance of the parameter $\sigma = \frac{N}{2\Omega}\sqrt{\mathrm{Pr}}$ (where Ω is the stellar rotation rate, N the Brunt-Väisälä frequency and $\mathrm{Pr} = \nu/\kappa$ the Prandtl number). They identified various regimes of penetration of meridional flows depending on the value of σ at the top of the tachocline. In the Sun, this parameter is much lower than 1, which is a difficult regime (very low Prandtl number) to achieve in numerical simulations.

Global models coupling a convection zone to a radiation zone were recently developed by Rogers (2011), Wood and Garaud (2011), Acevedo-Arreguin et al. (2013) in 2D and by Brun et al. (2011) and Strugarek et al. (2011a) in 3D. A self-consistent tachocline radiatively spreading in time naturally develops in those models. Studying the GM98 scenario in a Cartesian geometry where they force large scale flows (mimicking the convection without taking into account the convective motions), Wood and Garaud (2011) showed they could obtain the magnetic confinement of the tachocline for values of σ in a certain limited range. Acevedo-Arreguin et al. (2013) performed more recently 2D spherical axisymmetric numerical simulations and explored a large (σ) parameter range. They obtained various steady-state solutions in which the internal rotation profile varies significantly. However, none of their solutions fully recover the close-to solid body rotation for the radiation zone constrained by helioseismology observations. Other studies, taking into account the dynamical convective flows, reached similar conclusions. Both Rogers (2011) and Strugarek et al. (2011a) report the lack of magnetic field confinement inside the radiative zone when the convection zone flows (both large-scale means flows and small-scale turbulent flows) are modeled self-consistently. It is worth noting that Rogers (2011) reports a tachocline confinement without magnetic field, though the tachocline may still be slowly thickening due to viscous spreading in her simulations (Wood and Brummell 2012). The axisymmetric magnetic field considered by Strugarek et al. (2011a) was shown to pervade through the tachocline into the convection zone and to easily transport angular momentum between the two zones (see Fig. 8 bottom right panel). In order to explore whether other magnetic configurations could couple the two zones so easily, Strugarek et al. (2011b) considered the case of tilted magnetic dipoles, up to the singular case of the perpendicular dipole (Strugarek, private communications). In all cases, they established that the magnetic field was able to pervade the tachocline and that both Maxwell torques and stresses transported angular momentum in such a way that the final rotation profile could not be reconciled with observations. An alternative magnetic scenario could involve the 11-yr cyclic variations of the solar magnetic field and the associated feedback of the Lorentz force on the differential rotation. By quenching the latitudinal shear at the base of the convection zone, the result would be to get rid of the problem of radiative spreading of the tachocline altogether (Forgács-Dajka and Petrovay 2002; Forgács-Dajka 2004; Karak and Petrovay 2012).

2.2.2 MHD Instabilities

Below the tachocline, by definition there is very little or no differential rotation, so there should not be any instabilities that have their origin in shearing flow such as differential rotation. Because this domain has very subadiabatic stratification, governed by radiative energy balance, any instability that does occur on a dynamical timescale will tend to have flow confined to spherical surfaces, with no radial component. But on much longer timescales, governed by diffusive processes, negative buoyancy can be neutralized at least partly by radiative diffusion. There must still be an energy source, which would most likely be from the primordial magnetic fields.

However it was long ago recognized by Tayler and collaborators that the field topology is key to determine if a magnetic field is stable or unstable in a purely radiative interior. Purely poloidal field are known to undergo a high-m wavenumber instability (Markey and Tayler 1973, Fig. 8 top left panel). Likewise purely axisymmetric toroidal fields undergo a $m = 1$ instability if their amplitudes go beyond a certain threshold (Tayler 1973; Acheson 1978;

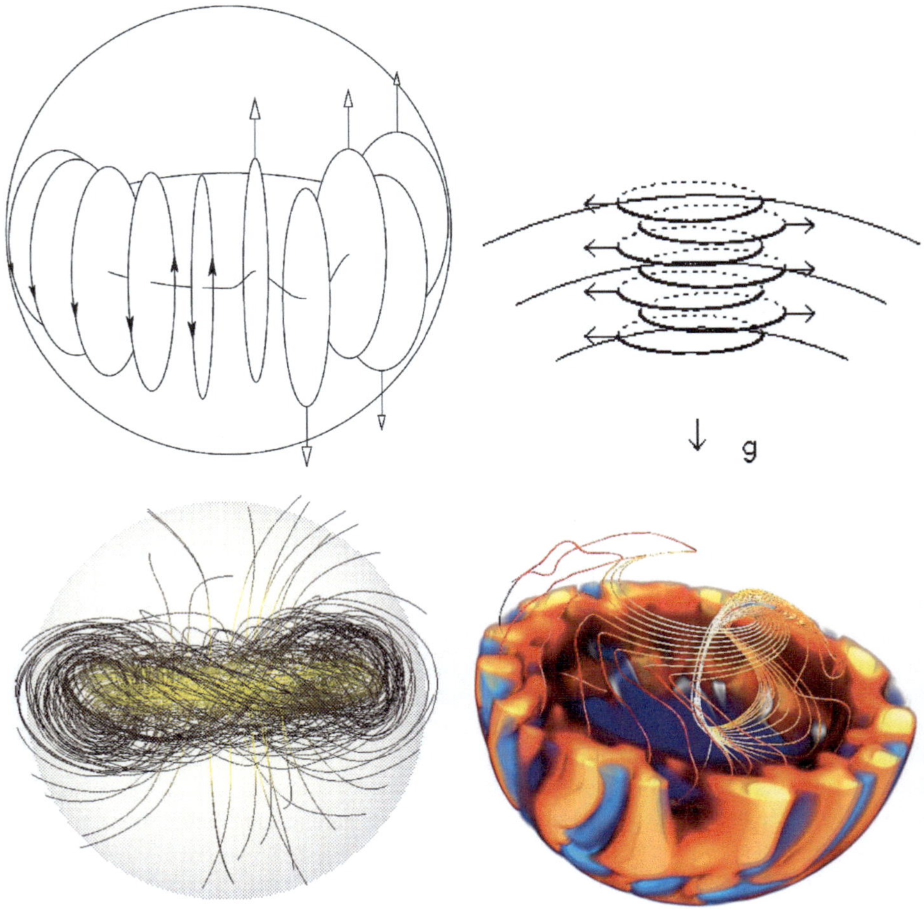

Fig. 8 *Top*; *Left*: Cartoon of the high m instability developing for a purely poloidal magnetic field in a stellar radiative zone (Markey and Tayler 1973; Braithwaite 2009). *Right*: Cartoon of Tayler's $m = 1$ toroidal instability (Tayler 1973; Spruit 1999). *Bottom*; *Left*: Stable magnetic field configuration in a stellar radiative zone (Braithwaite and Spruit 2004; Braithwaite 2009). *Right*: Interaction between a fossil field and a convective envelope in seeking to explain the solar tachocline confinement (Gough and McIntyre 1998; Strugarek et al. 2011a, 2011b)

Spruit 2002; Zahn et al. 2007, Fig. 8 top right panel). To be stable the magnetic configuration in a stellar radiative interior must be of mixed poloidal/toroidal topology. For the poloidal field instability it has been shown that an amplitude of the toroidal field of order at least 0.24 of the poloidal field strength is required (Wright 1973; Markey and Tayler 1973). A more systematic study of the Bpol/Btor ratio has been computed numerically by Braithwaite (2009) for both instabilities. This work confirms that the ratio of the poloidal magnetic energy to the total magnetic energy must be comprised between 1 to 5 % and 80 %. While these ratios are important to assess, it is unlikely that the magnetic field trapped by the forming star or reprocessed by dynamo action during the PMS phase will have at the onset a simple topology. Braithwaite and Spruit (2004) and Braithwaite and Nordlund (2006) have searched for numerical stable solutions while starting from random seed field. They found in the non rotating case that the random field evolved into a stable twisted torus (see Fig. 8

bottom left panel). Duez et al. (2010) have further derived analytically a minimum energy field topology configuration and they too find a twisted torus like configuration. However the presence of bulk rotation reduces the growth rate with $\sigma \sim \omega_A$ changing to $\sigma \sim \omega_A^2/\Omega_0$ and increases the critical wavenumber of the instability (Pitts and Tayler 1985; Brun 2007; Braithwaite 2008; Bonanno and Urpin 2013). Duez (2011) has shown that in the presence of rotation the axis of the stable twisted torus is likely not aligned with the rotation axis with tilt angle as large as 90°.

The interaction with a large scale differential rotation may also influence the field configuration (Spruit 2002). In between the radiative deep interior and the overshoot part of the tachocline, there is a radiative tachocline with some differential rotation, within which both HD and MHD global instabilities with dynamical timescales are possible, both in the Sun (Gilman and Fox 1997) and in other F, G and K stars (Dikpati et al. 2009; Dikpati and Cally 2011 and earlier references therein). These can occur for both solar and anti-solar type differential rotation, and for both axisymmetric and longitude-dependent modes. Unlike instabilities in the deep interior, these radiative tachocline instabilities could be a part of the solar dynamo, contributing departures from axisymmetry as well as production of poloidal fields from toroidal fields in the tachocline.

Finally, Spruit (2002) has proposed the existence of a dynamo in radiative interiors/envelopes, subject to a large scale differential rotation. The large scale differential rotation generates a toroidal magnetic field from an existing poloidal field that goes unstable via the $m = 1$ Tayler instability, hence possibly regenerating the original poloidal field. Zahn et al. (2007) have argued that while a dynamo based on large scale differential rotation and Tayler's $m = 1$ instability can exist, the dynamo loop must involve a mean electromotive force (*emf*) between non-axisymmetric velocity and magnetic fields triggered by Tayler's instability, the amplitude of which is unclear at present. Braithwaite and Nordlund (2006) did find, using a cylindrical setting with an imposed large scale shear and magnetic flux, a small parameter range where magnetic energy is sustained over about a diffusive time scale. However, Gellert et al. (2008), found, in a similar setting, that dynamo action is observed only when a magnetic flux through the domain is maintained as was done in Braithwaite's simulations. In spherical geometry, Zahn et al. (2007) found that some *emf* does develop due to Tayler's $m = 1$ instability but not to a level high enough to overcome Ohmic dissipation. Similar studies by Rüdiger et al. (2012) indicate that α^2 dynamos in radiative zones are unlikely if based on toroidal magnetic field instability but that in some specific conditions an $\alpha - \Omega$ dynamo confined to the polar cap may be possible. All these attempts, however are hampered by the fact that the small length scales on which the process is predicted to take place cannot be resolved for realistic conditions in stellar interiors. The currently published simulations are still outside the parameter range where the dynamo effect is predicted to be present. Since the magnetic Reynolds number, the amplitude of the differential rotation and the domain geometry play a key role here, further investigations are required to settle this issue.

3 Future Directions

As our discussion here has suggested, a complete theory of how the Sun builds its cyclical magnetism remains elusive. There is still considerable disagreement about which dynamical processes are most important in generating and shaping the magnetism, about the relative roles of each of these, and about what this implies for predictability of solar magnetic phenomena. Even among the authors of this contribution it has occasionally proved difficult to find statements about the solar dynamo that are uncontroversial. At the same time, though, great progress has been made in examining (both observationally and theoretically) many different mechanisms that likely contribute to building the observed fields. It is also clear that helioseismology holds the promise of resolving many current uncertainties in the modeling.

We therefore close here by first noting some aspects of solar magnetism on which we think there is broad (though by no means universal) *agreement*, then by listing some issues on which there is still considerable *disagreement* or uncertainty. Finally, we note some helioseismic measurements that would be particularly useful for constraining theories of the solar dynamo.

3.1 Agreements and Disagreements About the Solar Dynamo

The solar modeling community—or at least the subset of it represented here—is in broad *agreement* about a number of specific issues in our current theoretical picture of the solar dynamo. We list some of these below:

- Mechanisms for flux emergence are very important, but not yet well understood or included in dynamo models. Progress has been made in that direction recently, however Nelson et al. (2013b).
- While 2D mean-field kinematic and dynamical models have successfully modeled many important solar cycle features, including torsional oscillation (Rempel 2006), 3-D models are needed for longitude-dependent features.
- Meridional circulation is likely critical to the dynamo; therefore good measures of it are necessary.
- Surface flux is carried poleward by a combination of diffusion and advection, largely determining the amplitude and evolution of polar fields.
- The relative amplitude and phase of the parity with respect to the equator of the solar magnetic field plays an important role in the reversal process and is a key feature of nonlinear dynamos. The solar magnetic field is predominantly anti-symmetric (dipolar-like) but symmetric (quadrupolar like) components play an important role in explaining delays between the south and north poles or hemispherical dynamo state such as during the Maunder minimum.

However there are still a number of issues on which we do not think there is a clear consensus in the community. The most profound of these is probably the role played by the tachocline: in some models and simulations the radial shear in the tachocline is critical, whereas in many mean-field models the radial shear is unimportant (and in fact can disorganize the large-scale fields). There are also conceptual difficulties associated with tapping into the radial shear across the convective-radiative boundary (as discussed in Sect. 2.1.3 above). Among other critical uncertainties are the following:

- Are robust predictions possible with current solar dynamo models?
- What are the relative roles of diffusion and advection? In the language of mean-field theory, is the solar dynamo advection or diffusion dominated?
- Is the memory length of the solar dynamo less than or greater than one solar cycle?
- What are the origins of cyclic activity in simulations that show cycles? What sets the direction of field propagation in these simulations?
- Is the return flow of meridional circulation near the bottom of the convection zone, or much higher up? How many cells (in radius and latitude) are there?

- What are the relative roles of thermal effects and direct Lorentz force feedback in explaining the amplitude and depth of the torsional oscillation poleward and equatorward branches?
- Is there a (primordial) large scale magnetic field in the solar radiative zone? If so, what is its topology and amplitude? Can it prevent the spread of the tachocline or not?
- Can a dynamo in a stably stratified radiative zone subject to a large scale shear exist in stars (Spruit 2002)? If so, what physical processes close the dynamo loop from the toroidal back to the poloidal field (Zahn et al. 2007)?

3.2 What Do We Need from Helioseismology in Order to Progress?

Given these uncertainties, it is clear that much remains to be done! We close with a "wish list" of measurements from helioseismology that would be particularly useful for the modeling community:

- Knowledge of the depth of penetration of meridional circulation below the base of the convection zone
- Although the mean differential rotation is known well, improvements in precision measurements of torsional oscillations will help with evaluating MHD dynamo models.
- Depth at which the latitudinal component of meridional circulation reverses to equatorward
- Estimates of turbulent viscosity and magnetic diffusivity with depth, especially near the bottom of the convection zone
- Radial extent of latitudinal inflows into active region latitudes
- Measures of global but longitude-dependent flows at any depth, but especially near the bottom of the convection zone.
- Departures from spherical shape of the tachocline, in both velocity and thermal structure, including any changes with time.
- g-mode detections to constrain radiative interior dynamics (rotation, mixing) and the existence and amplitude of a fossil field.
- Further constraints on the amplitude of large-scale convective motions at depth, including the presence or absence of "giant cells."
- Robust, early detection below the surface of upcoming magnetic flux emergence.

Acknowledgements We acknowledge critical comments on solar global magnetism and dynamo by Prof. H.C. Spruit. We thank two anonymous referees for detailed and constructive comments. One of us (A.S.B) thanks the University of Kyoto, RIMS and Professors M. Yamada, K. Shibata, H. Isobe and S. Takehiro for their kind invitation in Fall 2013 during which this paper was finished. This work is partially supported by the NASA LWS grant NNX08AQ34G, by the ERC STARS2 project by the French national program on Sun-Earth relationships (INSU/PNST) and by CNES Sun-Heliosphere program, and by a Consolidated Grant by the UK STFC (ST/J001627/1STFC). We acknowledge the support from ISSI Bern, for our participation in the workshop. The National Center for Atmospheric Research is sponsored by the National Science Foundation.

References

J.A. Abreu, J. Beer, F. Steinhilber, S.M. Tobias, N.O. Weiss, For how long will the current grand maximum of solar activity persist? Geophys. Res. Lett. **35**, 20109 (2008). doi:10.1029/2008GL035442

L.A. Acevedo-Arreguin, P. Garaud, T.S. Wood, Dynamics of the solar tachocline—III. Numerical solutions of the Gough and McIntyre model. Mon. Not. R. Astron. Soc. **434**, 720–741 (2013). doi:10.1093/mnras/stt1065

D. Acheson, Instability by magnetic buoyancy. Sol. Phys. **62**, 23–235050 (1978)

M. Asplund, N. Grevesse, A.J. Sauval, C. Allende Prieto, D. Kiselman, Line formation in solar granulation. IV. [O I], O I and OH lines and the photospheric O abundance. Astron. Astrophys. **417**, 751–768 (2004). doi:10.1051/0004-6361:20034328

H.D. Babcock, The Sun's polar magnetic field. Astrophys. J. **130**, 364 (1959). doi:10.1086/146726

H.W. Babcock, The topology of the Sun's magnetic field and the 22-YEAR cycle. Astrophys. J. **133**, 572 (1961). doi:10.1086/147060

W.T. Ball, Y.C. Unruh, N.A. Krivova, S. Solanki, T. Wenzler, D.J. Mortlock, A.H. Jaffe, Reconstruction of total solar irradiance 1974–2009. Astron. Astrophys. **541**, 27 (2012). doi:10.1051/0004-6361/201118702

S. Basu, H.M. Antia, Characteristics of solar meridional flows during solar cycle 23. Astrophys. J. **717**, 488–495 (2010). doi:10.1088/0004-637X/717/1/488

P. Beaudoin, P. Charbonneau, E. Racine, P.K. Smolarkiewicz, Torsional oscillations in a global solar dynamo. Sol. Phys. **282**, 335–360 (2013). doi:10.1007/s11207-012-0150-2

J.G. Beck, T.L. Duvall, P.H. Scherrer, Long-lived giant cells detected at the surface of the Sun. Nature **394**, 653–655 (1998). doi:10.1038/29245

J. Beer, S. Tobias, N. Weiss, An active Sun throughout the maunder minimum. Sol. Phys. **181**, 237–249 (1998)

B. Belucz, M. Dikpati, Role of asymmetric meridional circulation in producing North–South asymmetry in a solar cycle dynamo model. Astrophys. J. **777**, 1–9 (2013). doi:10.1088/0004-637X/777/1

E.E. Benevolenskaya, Polar magnetic flux on the Sun in 1996–2003 from SOHO/MDI data. Astron. Astrophys. **428**, 5–8 (2004). doi:10.1051/0004-6361:200400092

S.V. Berdyugina, D. Moss, D. Sokoloff, I.G. Usoskin, Active longitudes, nonaxisymmetric dynamos and phase mixing. Astron. Astrophys. **445**, 703–714 (2006). doi:10.1051/0004-6361:20053454

D. Biskamp, Nonlinear magnetohydrodynamics, in *Cambridge Monographs on Plasma Physics*, Cambridge [England] (Cambridge University Press, New York, 1993)

C. Bolduc, P. Charbonneau, V. Dumoulin, M.S. Bourqui, A.D. Crouch, A fast model for the reconstruction of spectral solar irradiance in the near- and mid-ultraviolet. Sol. Phys. **279**, 383–409 (2012). doi:10.1007/s11207-012-0019-4

A. Bonanno, V. Urpin, Rotation and stability of the toroidal magnetic field in stellar radiation zones. Astrophys. J. **766**, 52 (2013). doi:10.1088/0004-637X/766/1/52

A. Bonanno, D. Elstner, G. Rüdiger, G. Belvedere, Parity properties of an advection-dominated solar $\alpha^2\Omega$-dynamo. Astron. Astrophys. **390**, 673–680 (2002). doi:10.1051/0004-6361:20020590

J. Braithwaite, On non-axisymmetric magnetic equilibria in stars. Mon. Not. R. Astron. Soc. **386**, 1947–1958 (2008). doi:10.1111/j.1365-2966.2008.13218.x

J. Braithwaite, Axisymmetric magnetic fields in stars: relative strengths of poloidal and toroidal components. Mon. Not. R. Astron. Soc. **397**, 763–774 (2009). doi:10.1111/j.1365-2966.2008.14034.x

J. Braithwaite, Å. Nordlund, Stable magnetic fields in stellar interiors. Astron. Astrophys. **450**(3), 1077–1095 (2006)

J. Braithwaite, H.C. Spruit, A fossil origin for the magnetic field in a stars and white dwarfs. Nature **431**, 819–821 (2004). doi:10.1038/nature02934

A. Brandenburg, The case for a distributed solar dynamo shaped by near-surface shear. Astrophys. J. **625**, 539–547 (2005). doi:10.1086/429584

A. Brandenburg, D. Moss, I. Tuominen, Stratification and thermodynamics in mean-field dynamos. Astron. Astrophys. **265**, 328–344 (1992)

B.P. Brown, Convection and Dynamo Action in Rapidly Rotating Suns, Ph.D. thesis, University of Colorado at Boulder (2009)

B.P. Brown, M.K. Browning, A.S. Brun, M.S. Miesch, J. Toomre, Persistent magnetic wreaths in a rapidly rotating Sun. Astrophys. J. **711**, 424–438 (2010). doi:10.1088/0004-637X/711/1/424

B.P. Brown, M.S. Miesch, M.K. Browning, A.S. Brun, J. Toomre, Magnetic cycles in a convective dynamo simulation of a Young solar-type star. Astrophys. J. **731**, 69 (2011). doi:10.1088/0004-637X/731/1/69

B.P. Brown, M.K. Browning, A.S. Brun, M.S. Miesch, J. Toomre, Persistent magnetic wreaths in a rapidly rotating Sun. Astrophys. J. **711**(1), 424–438 (2010)

M.K. Browning, M.S. Miesch, A.S. Brun, J. Toomre, Dynamo action in the solar convection zone and tachocline: pumping and organization of toroidal fields. Astrophys. J. **648**, 157–160 (2006). doi:10.1086/507869

M.K. Browning, A.S. Brun, M.S. Miesch, J. Toomre, Dynamo action in simulations of penetrative solar convection with an imposed tachocline. Astron. Nachr. **328**, 1100 (2007). doi:10.1002/asna.200710849

M.K. Browning, M.S. Miesch, A.S. Brun, J. Toomre, Dynamo action in the solar convection zone and tachocline: pumping and organization of toroidal fields. Astrophys. J. **648**(2), 157–160 (2006)

A.S. Brun, Towards using modern data assimilation and weather forecasting methods in solar physics. Astron. Nachr. **328**, 329 (2007). doi:10.1002/asna.200610739

A.S. Brun, O. Do Cao, Private communication (2013)

A.S. Brun, M. Rempel, Large scale flows in the solar convection zone. Space Sci. Rev. **144**, 151–173 (2009). doi:10.1007/s11214-008-9454-9
A.S. Brun, M.S. Miesch, J. Toomre, Global-scale turbulent convection and magnetic dynamo action in the solar envelope. Astrophys. J. **614**, 1073–1098 (2004a). doi:10.1086/423835
A.S. Brun, M.S. Miesch, J. Toomre, Global-scale turbulent convection and magnetic dynamo action in the solar envelope. Astrophys. J. **614**, 1073–1098 (2004b). doi:10.1086/423835
A.S. Brun, Nonlinear simulations of magnetic instabilities in stellar radiation zones: the role of rotation and shear. Astron. Nachr. **328**(10), 1137–1141 (2007)
A.S. Brun, J. Toomre, Turbulent convection under the influence of rotation: sustaining a strong differential rotation. Astrophys. J. **570**, 865 (2002)
A.S. Brun, J.-P. Zahn, Magnetic confinement of the solar tachocline. Astron. Astrophys. **457**, 665 (2006)
A.S. Brun, M.S. Miesch, J. Toomre, Global-scale turbulent convection and magnetic dynamo action in the solar envelope. Astrophys. J. **614**(2), 1073–1098 (2004)
A.S. Brun, M.S. Miesch, J. Toomre, Modeling the dynamical coupling of solar convection with the radiative interior. Astrophys. J. **742**(2), 79 (2011)
E. Bullard, H. Gellman, Homogeneous dynamos and terrestrial magnetism. Philos. Trans. R. Soc. Lond. Ser. A, Math. Phys. Sci. **247**(9), 213–278 (1954)
P.J. Bushby, Zonal flows and grand minima in a solar dynamo model. Mon. Not. R. Astron. Soc. **371**, 772–780 (2006). doi:10.1111/j.1365-2966.2006.10706.x
R.H. Cameron, M. Schüssler, Are the strengths of solar cycles determined by converging flows towards the activity belts? Astron. Astrophys. **548**, 57 (2012). doi:10.1051/0004-6361/201219914
F. Cattaneo, On the origin of magnetic fields in the quiet photosphere. Astrophys. J. Lett. **515**, 39–42 (1999). doi:10.1086/311962
F. Cattaneo, D.W. Hughes, Problems with kinematic mean field electrodynamics at high magnetic Reynolds numbers. Mon. Not. R. Astron. Soc. **395**, 48–51 (2009). doi:10.1111/j.1745-3933.2009.00639.x
P. Charbonneau, Dynamo models of the solar cycle. Living Rev. Sol. Phys. **2**, 2 (2005)
P. Charbonneau, Where is the solar dynamo? J. Phys. Conf. Ser. **440**(1), 012014 (2013). doi:10.1088/1742-6596/440/1/012014
P. Charbonneau, M. Dikpati, Stochastic fluctuations in a Babcock–Leighton model of the solar cycle. Astrophys. J. **543**, 1027–1043 (2000). doi:10.1086/317142
C. Charbonnel, S. Talon, Influence of gravity waves on the internal rotation and Li abundance of solar-type stars. Science **309**, 2189–2191 (2005). doi:10.1126/science.1116849
P. Chatterjee, D. Nandy, A.R. Choudhuri, Full-sphere simulations of a circulation-dominated solar dynamo: exploring the parity issue. Astron. Astrophys. **427**, 1019–1030 (2004). doi:10.1051/0004-6361:20041199
M.C.M. Cheung, M. Rempel, A.M. Title, M. Schüssler, Simulation of the formation of a solar active region. Astrophys. J. **720**, 233–244 (2010). doi:10.1088/0004-637X/720/1/233
A.R. Choudhuri, B.B. Karak, A possible explanation of the Maunder minimum from a flux transport dynamo model. Res. Astron. Astrophys. **9**, 953–958 (2009). doi:10.1088/1674-4527/9/9/001
A.R. Choudhuri, P. Chatterjee, J. Jiang, Predicting solar cycle 24 with a solar dynamo model. Phys. Rev. Lett. **98**(13), 131103 (2007). doi:10.1103/PhysRevLett.98.131103
A.R. Choudhuri, M. Schüssler, M. Dikpati, The solar dynamo with meridional circulation. Astron. Astrophys. **303**, 29 (1995)
F. Clette, L. Lefèvre, Are the sunspots really vanishing? Anomalies in solar cycle 23 and implications for long-term models and proxies. J. Space Weather Space Clim. **2**(26), 260000 (2012). doi:10.1051/swsc/2012007
E. Covas, D. Moss, R. Tavakol, Dynamo models and differential rotation in late-type rapidly rotating stars. Astron. Astrophys. **429**, 657–665 (2005). doi:10.1051/0004-6361:20041741
T.G. Cowling, in *Solar Electrodynamics*, ed. by G.P. Kuiper (1953) p. 532
R.M. Crutcher, N. Hakobian, T.H. Troland, Testing magnetic star formation theory. Astrophys. J. **692**, 844–855 (2009). doi:10.1088/0004-637X/692/1/844
M. Dasi-Espuig, S.K. Solanki, N.A. Krivova, R. Cameron, T. Peñuela, Sunspot group tilt angles and the strength of the solar cycle. Astron. Astrophys. **518**, 7 (2010). doi:10.1051/0004-6361/201014301
P. Démoulin, E. Pariat, Modelling and observations of photospheric magnetic helicity. Adv. Space Res. **43**, 1013–1031 (2009). doi:10.1016/j.asr.2008.12.004
M.L. DeRosa, A.S. Brun, J.T. Hoeksema, Solar magnetic field reversals and the role of dynamo families. Astrophys. J. **757**, 96 (2012). doi:10.1088/0004-637X/757/1/96
M. Dikpati, J.L. Anderson, Evaluating potential for data assimilation in a flux-transport dynamo model by assessing sensitivity and response to meridional flow variation. Astrophys. J. **756**, 20 (2012). doi:10.1088/0004-637X/756/1/20

M. Dikpati, P.S. Cally, Analyses of three-dimensional magnetohydrodynamic instability of antisolar latitudinal differential rotation in F, G, and K stars. Astrophys. J. **739**, 4 (2011). doi:10.1088/0004-637X/739/1/4

M. Dikpati, P. Charbonneau, A Babcock–Leighton flux transport dynamo with solar-like differential rotation. Astrophys. J. **518**, 508–520 (1999). doi:10.1086/307269

M. Dikpati, P.A. Gilman, Flux-transport dynamos with α-effect from global instability of tachocline differential rotation: a solution for magnetic parity selection in the Sun. Astrophys. J. **559**, 428–442 (2001). doi:10.1086/322410

M. Dikpati, P.A. Gilman, A Shallow–Water theory for the Sun's active longitudes. Astrophys. J. Lett. **635**, 193–196 (2005). doi:10.1086/499626

M. Dikpati, P.A. Gilman, Simulating and predicting solar cycles using a flux-transport dynamo. Astrophys. J. **649**, 498–514 (2006). doi:10.1086/506314

M. Dikpati, G. de Toma, P.A. Gilman, Predicting the strength of solar cycle 24 using a flux-transport dynamo-based tool. Geophys. Res. Lett. **33**, 5102 (2006). doi:10.1029/2005GL025221

M. Dikpati, G. de Toma, P.A. Gilman, C.N. Arge, O.R. White, Diagnostics of polar field reversal in solar cycle 23 using a flux transport dynamo model. Astrophys. J. **601**, 1136–1151 (2004). doi:10.1086/380508

M. Dikpati, P.A. Gilman, G. de Toma, S.S. Ghosh, Simulating solar cycles in northern and southern hemispheres by assimilating magnetic data into a calibrated flux-transport dynamo. Sol. Phys. **245**, 1–17 (2007). doi:10.1007/s11207-007-9016-4

M. Dikpati, P.A. Gilman, G. de Toma, R.K. Ulrich, Impact of changes in the Sun's conveyor-belt on recent solar cycles. Geophys. Res. Lett. **37**, 14107 (2010). doi:10.1029/2010GL044143

M. Dikpati, P.A. Gilman, R.K. Ulrich, Physical origin of differences among various measures of solar meridional circulation. Astrophys. J. **722**(1), 774–778 (2010)

M. Dikpati, P.A. Gilman, P.S. Cally, M.S. Miesch, Axisymmetric MHD instabilities in Solar/Stellar tachoclines. Astrophys. J. **692**(2), 1421–1431 (2009)

O. Do Cao, A.S. Brun, Effects of turbulent pumping on stellar activity cycles. Astron. Nachr. **332**, 907 (2011). doi:10.1002/asna.201111623

V. Domingo, I. Ermolli, P. Fox, C. Fröhlich, M. Haberreiter, N. Krivova, G. Kopp, W. Schmutz, S.K. Solanki, H.C. Spruit, Y. Unruh, A. Vögler, Solar surface magnetism and irradiance on time scales from days to the 11-year cycle. Space Sci. Rev. **145**, 337–380 (2009). doi:10.1007/s11214-009-9562-1

S. D'Silva, A.R. Choudhuri, A theoretical model for tilts of bipolar magnetic regions. Astron. Astrophys. **272**, 621 (1993)

V. Duez, Numerical simulations of magnetic relaxation in rotating stellar radiation zones. Astron. Nachr. **332**(9), 983 (2011)

V. Duez, J. Braithwaite, S. Mathis, On the stability of non-force-free magnetic equilibria in stars. Astrophys. J. Lett. **724**(1), 34–38 (2010)

I. Ermolli, K. Matthes, T. Dudok de Wit, N.A. Krivova, K. Tourpali, M. Weber, Y.C. Unruh, L. Gray, U. Langematz, P. Pilewskie, E. Rozanov, W. Schmutz, A. Shapiro, S.K. Solanki, G. Thuillier, T.N. Woods, Recent variability of the solar spectral irradiance and its impact on climate modelling. Atmos. Chem. Phys. Discuss. **12**, 24557–24642 (2012). doi:10.5194/acpd-12-24557-2012

V.C.A. Ferraro, The non-uniform rotation of the Sun and its magnetic field. Mon. Not. R. Astron. Soc. **97**, 458 (1937)

J.M. Fontenla, J. Harder, W. Livingston, M. Snow, T. Woods, High-resolution solar spectral irradiance from extreme ultraviolet to far infrared. J. Geophys. Res., Atmos. **116**, 20108 (2011). doi:10.1029/2011JD016032

E. Forgács-Dajka, Dynamics of the fast solar tachocline. II. Migrating field. Astron. Astrophys. **413**, 1143 (2004)

E. Forgács-Dajka, K. Petrovay, Dynamics of the fast solar tachocline. I. Dipolar field. Astron. Astrophys. **389**, 629–640 (2002). doi:10.1051/0004-6361:20020586

P. Foukal, C. Fröhlich, H. Spruit, T.M.L. Wigley, Variations in solar luminosity and their effect on the Earth's climate. Nature **443**, 161–166 (2006). doi:10.1038/nature05072

P. Fox, Solar activity and irradiance variations, in *Solar Variability and Its Effects on Climate. Geophysical Monograph 141*, ed. by J.M. Pap, P. Fox, C. Frohlich, H.S. Hudson, J. Kuhn, J. McCormack, G. North, W. Sprigg, S.T. Wu Washington DC American Geophysical Union Geophysical Monograph Series, vol. 141, 2004, p. 141

C. Fröhlich, Observations of irradiance variations. Space Sci. Rev. **94**, 15–24 (2000)

C. Fröhlich, Total solar irradiance observations. Surv. Geophys. **33**, 453–473 (2012). doi:10.1007/s10712-011-9168-5

C. Fröhlich, J. Lean, The Sun's total irradiance: cycles, trends and related climate change uncertainties since 1976. Geophys. Res. Lett. **25**, 4377–4380 (1998). doi:10.1029/1998GL900157

B. Gallet, F. Pétrélis, From reversing to hemispherical dynamos. Phys. Rev. E **80**(3), 035302 (2009)

P. Garaud, Dynamics of the solar tachocline—I. An incompressible study. Mon. Not. R. Astron. Soc. **329**, 1 (2002)

P. Garaud, L. Acevedo Arreguin, On the penetration of meridional circulation below the solar convection zone. II. Models with convection zone, the Taylor–Proudman constraint, and applications to other stars. Astrophys. J. **704**, 1 (2009)

P. Garaud, N.H. Brummell, On the penetration of meridional circulation below the solar convection zone. Astrophys. J. **674**(1), 498–510 (2008)

P. Garaud, J.-D. Garaud, Dynamics of the solar tachocline—II. The stratified case. Mon. Not. R. Astron. Soc. **391**, 1239 (2008)

T. Gastine, L. Duarte, J. Wicht, Dipolar versus multipolar dynamos: the influence of the background density stratification. Astron. Astrophys. **546**, A19 (2012). doi:10.1051/0004-6361/201219799

M. Gellert, G. Rüdiger, D. Elstner, Helicity generation and α-effect by Tayler instability with z-dependent differential rotation. Astron. Astrophys. **479**, 33–36 (2008). doi:10.1051/0004-6361:20077781

M. Ghizaru, P. Charbonneau, P. Smolarkiewicz, Magnetic cycles in global large-eddy simulations of solar convection. Astrophys. J. Lett. **715**(2), 133–137 (2010)

P.M. Giles, T.L. Duvall, P.H. Scherrer, R.S. Bogart, A subsurface flow of material from the Sun's equator to its poles. Nature **390**, 52–54 (1997). doi:10.1038/36294

P.A. Gilman, J. Miller, Dynamically consistent nonlinear dynamos driven by convection in a rotating spherical shell. Astrophys. J. Suppl. Ser. **46**, 211–238 (1981). doi:10.1086/190743

P.A. Gilman, Dynamically consistent nonlinear dynamos driven by convection in a rotating spherical shell. II—Dynamos with cycles and strong feedbacks. Astrophys. J. Suppl. Ser. **53**, 243 (1983)

P.A. Gilman, P.A. Fox, Joint instability of latitudinal differential rotation and toroidal magnetic fields below the solar convection zone. Astrophys. J. **484**, 439 (1997)

L. Gizon, A.C. Birch, H.C. Spruit, Local helioseismology: three-dimensional imaging of the solar interior. Annu. Rev. Astron. Astrophys. **48**, 289–338 (2010). doi:10.1146/annurev-astro-082708-101722

L. Gizon, A.C. Birch, Local helioseismology. Living Rev. Sol. Phys. **2**, 6 (2005)

G.A. Glatzmaier, Numerical simulations of stellar convective dynamos. II—Field propagation in the convection zone. Astrophys. J. **291**, 300 (1985)

A. Goel, A.R. Choudhuri, The hemispheric asymmetry of solar activity during the last century and the solar dynamo. Res. Astron. Astrophys. **9**, 115–126 (2009). doi:10.1088/1674-4527/9/1/010

I. González Hernández, R. Howe, R. Komm, F. Hill, Meridional circulation during the extended solar minimum: another component of the torsional oscillation? Astrophys. J. Lett. **713**, 16–20 (2010). doi:10.1088/2041-8205/713/1/L16

D.O. Gough, M.E. McIntyre, Inevitability of a magnetic field in the Sun's radiative interior. Nature **394**, 755 (1998)

D. Gubbins, K. Zhang, Symmetry properties of the dynamo equations for palaeomagnetism and geomagnetism. Phys. Earth Planet. Inter. **75**(4), 225–241 (1993)

G. Guerrero, E.M. de Gouveia Dal Pino, Turbulent magnetic pumping in a Babcock–Leighton solar dynamo model. Astron. Astrophys. **485**, 267–273 (2008). doi:10.1051/0004-6361:200809351

D.A. Haber, B.W. Hindman, J. Toomre, R.S. Bogart, R.M. Larsen, F. Hill, Evolving submerged meridional circulation cells within the upper convection zone revealed by ring-diagram analysis. Astrophys. J. **570**, 855–864 (2002). doi:10.1086/339631

J.D. Haigh, A.R. Winning, R. Toumi, J.W. Harder, An influence of solar spectral variations on radiative forcing of climate. Nature **467**, 696–699 (2010). doi:10.1038/nature09426

G.E. Hale, S.B. Nicholson, The law of Sun-spot polarity. Astrophys. J. **62**, 270 (1925). doi:10.1086/142933

J. Hao, M. Zhang, Hemispheric helicity trend for solar cycle 24. Astrophys. J. Lett. **733**, 27 (2011). doi:10.1088/2041-8205/733/2/L27

D.H. Hathaway, Doppler measurements of the Sun's meridional flow. Astrophys. J. **460**, 1027 (1996). doi:10.1086/177029

D.H. Hathaway, Solar cycle forecasting. Space Sci. Rev. **144**, 401–412 (2009). doi:10.1007/s11214-008-9430-4

D.H. Hathaway, The solar cycle. Living Rev. Sol. Phys. **7**, 1 (2010)

D.H. Hathaway, Supergranules as probes of the Sun's meridional circulation. Astrophys. J. **760**, 84 (2012). doi:10.1088/0004-637X/760/1/84

D.H. Hathaway, L. Rightmire, Variations in the Sun s meridional flow over a solar cycle. Science **327**, 1350 (2010). doi:10.1126/science.1181990

G. Hazra, B.B. Karak, A.R. Choudhuri, Is a deep one-cell meridional circulation essential for the flux transport Solar Dynamo? ArXiv e-prints (2013)

H. Hotta, T. Yokoyama, Importance of surface turbulent diffusivity in the solar flux-transport dynamo. Astrophys. J. **709**, 1009–1017 (2010a). doi:10.1088/0004-637X/709/2/1009

H. Hotta, T. Yokoyama, Solar parity issue with flux-transport dynamo. Astrophys. J. **714**, 308–312 (2010b). doi:10.1088/2041-8205/714/2/L308

R. Howard, B.J. Labonte, The Sun is observed to be a torsional oscillator with a period of 11 years. Astrophys. J. Lett. **239**, 33–36 (1980). doi:10.1086/183286

R. Howe, Solar interior rotation and its variation. Living Rev. Sol. Phys. **6**, 1 (2009)

R. Howe, R. Komm, F. Hill, R. Ulrich, D.A. Haber, B.W. Hindman, J. Schou, M.J. Thompson, Large-scale zonal flows near the solar surface. Sol. Phys. **235**, 1–15 (2006). doi:10.1007/s11207-006-0117-2

R. Howe, J. Christensen-Dalsgaard, F. Hill, R. Komm, J. Schou, M.J. Thompson, A note on the torsional oscillation at solar minimum. Astrophys. J. Lett. **701**, 87–90 (2009). doi:10.1088/0004-637X/701/2/L87

R. Howe, J. Christensen-Dalsgaard, F. Hill, R. Komm, T.P. Larson, M. Rempel, J. Schou, M.J. Thompson, The high-latitude branch of the solar torsional oscillation in the rising phase of cycle 24. Astrophys. J. Lett. **767**, 20 (2013). doi:10.1088/2041-8205/767/1/L20

D.W. Hughes, M.R.E. Proctor, The effect of velocity shear on dynamo action due to rotating convection. J. Fluid Mech. **717**, 395–416 (2013). doi:10.1017/jfm.2012.584

D.W. Hughes, M.R.E. Proctor, F. Cattaneo, The α-effect in rotating convection: a comparison of numerical simulations. Mon. Not. R. Astron. Soc. **414**, 45–49 (2011). doi:10.1111/j.1745-3933.2011.01053.x

S. Ilonidis, J. Zhao, A. Kosovichev, Detection of emerging sunspot regions in the solar interior. Science **333**, 993 (2011). doi:10.1126/science.1206253

J. Jiang, E. Işik, R.H. Cameron, D. Schmitt, M. Schüssler, The effect of activity-related meridional flow modulation on the strength of the solar polar magnetic field. Astrophys. J. **717**, 597–602 (2010). doi:10.1088/0004-637X/717/1/597

M. Joos, P. Hennebelle, A. Ciardi, Protostellar disk formation and transport of angular momentum during magnetized core collapse. Astron. Astrophys. **543**, 128 (2012). doi:10.1051/0004-6361/201118730

L. Jouve, A.S. Brun, G. Aulanier, Global dynamics of subsurface solar active regions. Astrophys. J. **762**, 4 (2013). doi:10.1088/0004-637X/762/1/4

L. Jouve, A.S. Brun, R. Arlt, A. Brandenburg, M. Dikpati, A. Bonanno, P.J. Käpylä, D. Moss, M. Rempel, P. Gilman, M.J. Korpi, A.G. Kosovichev, A solar mean field dynamo benchmark. Astron. Astrophys. **483**, 949–960 (2008). doi:10.1051/0004-6361:20078351

L. Jouve, A.S. Brun, On the role of meridional flows in flux transport dynamo models. Astron. Astrophys. **474**(1), 239–250 (2007)

L. Jouve, A.S. Brun, Three-dimensional nonlinear evolution of a magnetic flux tube in a spherical shell: influence of turbulent convection and associated mean flows. Astrophys. J. **701**(2), 1300–1322 (2009)

L. Jouve, A.S. Brun, O. Talagrand, Assimilating data into an $\alpha\Omega$ dynamo model of the Sun: a variational approach. Astrophys. J. **735**(1), 31 (2011)

P.J. Käpylä, M.J. Mantere, A. Brandenburg, Cyclic magnetic activity due to turbulent convection in spherical wedge geometry. Astrophys. J. Lett. **755**, 22 (2012). doi:10.1088/2041-8205/755/1/L22

P.J. Käpylä, M.J. Korpi, A. Brandenburg, D. Mitra, R. Tavakol, Convective dynamos in spherical wedge geometry. Astron. Nachr. **331**, 73 (2010). doi:10.1002/asna.200911252

P.J. Käpylä, M.J. Mantere, A. Brandenburg, Cyclic magnetic activity due to turbulent convection in spherical wedge geometry. Astrophys. J. **755**(1), 22 (2012)

P.J. Käpylä, M.J. Mantere, E. Cole, J. Warnecke, A. Brandenburg, Effects of strong stratification on equatorward dynamo wave propagation. Astrophys. J. **778**, 41 (2013). doi:10.1088/0004-637X/778/1/41

B.B. Karak, Importance of meridional circulation in flux transport dynamo: the possibility of a maunder-like grand minimum. Astrophys. J. **724**, 1021–1029 (2010). doi:10.1088/0004-637X/724/2/1021

B.B. Karak, A.R. Choudhuri, Quenching of meridional circulation in flux transport dynamo models. Sol. Phys. **278**, 137–148 (2012). doi:10.1007/s11207-012-0142-2

B.B. Karak, K. Petrovay, On the Compatibility of a Flux Transport Dynamo with a Fast Tachocline Scenario. Sol. Phys. (2012)

R. Komm, R. Howe, F. Hill, A search for coherent structures in subsurface flows. J. Phys. Conf. Ser. **271**(1), 012065 (2011). doi:10.1088/1742-6596/271/1/012065

R. Komm, I. González Hernández, F. Hill, R. Bogart, M.C. Rabello-Soares, D. Haber, Subsurface meridional flow from HMI using the ring-diagram pipeline. Sol. Phys. (2012). doi:10.1007/s11207-012-0073-y

G. Kopp, J.L. Lean, A new, lower value of total solar irradiance: evidence and climate significance. Geophys. Res. Lett. **38**, 1706 (2011). doi:10.1029/2010GL045777

A.G. Kosovichev, J.O. Stenflo, Tilt of emerging bipolar magnetic regions on the Sun. Astrophys. J. Lett. **688**, 115–118 (2008). doi:10.1086/595619

N.A. Krivova, L. Balmaceda, S.K. Solanki, Reconstruction of solar total irradiance since 1700 from the surface magnetic flux. Astron. Astrophys. **467**, 335–346 (2007). doi:10.1051/0004-6361:20066725

R.L. Kurucz, Atomic data for interpreting stellar spectra: isotopic and hyperfine data. Phys. Scr. T **47**, 110–117 (1993). doi:10.1088/0031-8949/1993/T47/017

B.J. Labonte, R. Howard, Torsional waves on the Sun and the activity cycle. Sol. Phys. **75**, 161–178 (1982). doi:10.1007/BF00153469

J. Lean, The Sun's variable radiation and its relevance for Earth. Annu. Rev. Astron. Astrophys. **35**, 33–67 (1997). doi:10.1146/annurev.astro.35.1.33

R.B. Leighton, A Magneto–Kinematic model of the solar cycle. Astrophys. J. **156**, 1 (1969). doi:10.1086/149943

R. Leonhardt, K. Fabian, Paleomagnetic reconstruction of the global geomagnetic field evolution during the Matuyama/Brunhes transition: iterative Bayesian inversion and independent verification. Earth Planet. Sci. Lett. **253**, 172–195 (2007). doi:10.1016/j.epsl.2006.10.025

M. Lesieur, *Turbulence in Fluids*, Series: Fluid Mechanics and Its Applications. Berlin: Springer, 2008

J. Li, R.K. Ulrich, Long-term measurements of sunspot magnetic tilt angles. Astrophys. J. **758**, 115 (2012). doi:10.1088/0004-637X/758/2/115

P.W. Livermore, D.W. Hughes, S.M. Tobias, Nonlinear generation of large-scale magnetic fields in forced spherical shell dynamos. PoF **22**(3), 7101 (2010)

M. Lockwood, Solar influence on global and regional climates. Surv. Geophys. **33**, 503–534 (2012). doi:10.1007/s10712-012-9181-3

W.V.R. Malkus, M.R.E. Proctor, The macrodynamics of alpha-effect dynamos in rotating fluids. J. Fluid Mech. **67**, 417–443 (1975). doi:10.1017/S0022112075000390

P. Markey, R.J. Tayler, The adiabatic stability of stars containing magneticfields-II. Poloidal fields. Mon. Not. R. Astron. Soc. **163**, 77 (1973)

J. Mason, D.W. Hughes, S.M. Tobias, The effects of flux transport on interface dynamos. Mon. Not. R. Astron. Soc. **391**(1), 467–480 (2008)

E.W. Maunder, Note on the distribution of Sun-spots in heliographic latitude, 1874–1902. Mon. Not. R. Astron. Soc. **64**, 747–761 (1904)

B.H. McClintock, A.A. Norton, Recovering Joy's law as a function of solar cycle, hemisphere, and longitude. Sol. Phys. **287**, 215–227 (2013). doi:10.1007/s11207-013-0338-0

P.L. McFadden, R.T. Merrill, M.W. McElhinny, S. Lee, Reversals of the Earth's magnetic field and temporal variations of the dynamo families. J. Geophys. Res. **96**, 3923–3933 (1991). doi:10.1029/90JB02275

M.S. Miesch, B.P. Brown, Convective Babcock–Leighton dynamo models. Astrophys. J. **746**, 26 (2012). doi:10.1088/2041-8205/746/2/L26

M.S. Miesch, A.S. Brun, J. Toomre, Solar differential rotation influenced by latitudinal entropy variations in the tachocline. Astrophys. J. **641**, 618–625 (2006). doi:10.1086/499621

P.D. Mininni, A. Alexakis, A. Pouquet, Shell-to-shell energy transfer in magnetohydrodynamics. II. Kinematic dynamo. Phys. Rev. E **72**(4), 46302 (2005)

U. Mitra-Kraev, M.J. Thompson, Meridional flow profile measurements with SOHO/MDI. Astron. Nachr. **328**, 1009–1012 (2007). doi:10.1002/asna.200710873

H. Miyahara, K. Masuda, Y. Muraki, H. Furuzawa, H. Menjo, T. Nakamura, Cyclicity of solar activity during the maunder minimum deduced from radiocarbon content. Sol. Phys. **224**, 317–322 (2004). doi:10.1007/s11207-005-6501-5

H.K. Moffatt, *Magnetic Field Generation in Electrically Conducting Fluids* (Cambridge University Press, Cambridge, 1978)

F. Moreno-Insertis, Rise times of horizontal magnetic flux tubes in the convection zone of the Sun. Astron. Astrophys. **122**, 241–250 (1983)

D. Moss, The survival of fossil magnetic fields during pre-main sequence evolution. Astron. Astrophys. **403**, 693–697 (2003). doi:10.1051/0004-6361:20030431

D. Moss, J. Brooke, Towards a model for the solar dynamo. Mon. Not. R. Astron. Soc. **315**, 521–533 (2000). doi:10.1046/j.1365-8711.2000.03452.x

T. Nakano, R. Nishi, T. Umebayashi, Mechanism of magnetic flux loss in molecular clouds. Astrophys. J. **573**, 199–214 (2002). doi:10.1086/340587

D. Nandy, A. Muñoz-Jaramillo, P.C.H. Martens, The unusual minimum of sunspot cycle 23 caused by meridional plasma flow variations. Nature **471**, 80–82 (2011). doi:10.1038/nature09786

N.J. Nelson, B.P. Brown, A. Sacha Brun, M.S. Miesch, J. Toomre, Buoyant magnetic loops generated by global convective dynamo action. Sol. Phys. (2013a). doi:10.1007/s11207-012-0221-4

N.J. Nelson, B.P. Brown, A.S. Brun, M.S. Miesch, J. Toomre, Magnetic wreaths and cycles in convective dynamos. Astrophys. J. **762**, 73 (2013b)

M. Ossendrijver, The solar dynamo. Astron. Astrophys. Rev. **11**, 287–367 (2003). doi:10.1007/s00159-003-0019-3

P. Padoan, T. Lunttila, M. Juvela, Å. Nordlund, D. Collins, A. Kritsuk, M. Normal, S. Ustyugov, Magnetic fields in molecular clouds, in *IAU Symposium*, ed. by N.H. Brummell, A.S. Brun, M.S. Miesch, Y. Ponty IAU Symposium, vol. 271, 2011, pp. 187–196. doi:10.1017/S1743921311017601

F. Palla, S.W. Stahler, The pre-main-sequence evolution of intermediate-mass stars. Astrophys. J. **418**, 414 (1993). doi:10.1086/173402

M.A. Palmer, L.J. Gray, M.R. Allen, W.A. Norton, Solar forcing of climate: model results. Adv. Space Res. **34**, 343–348 (2004). doi:10.1016/j.asr.2003.02.039

J.M. Pap, Total solar and spectral irradiance variations from near-uv to infrared, in *The Sun's Surface and Subsurface: Investigating Shape*, ed. by J.-P. Rozelot Lecture Notes in Physics, Berlin Springer Verlag, vol. 599, 2003, pp. 129–155

J.M. Pap, P.A. Fox, Solar variability and its effects on climate. Washington DC American Geophysical Union Geophysical Monograph Series **141** (2003)

E.N. Parker, A solar dynamo surface wave at the interface between convection and nonuniform rotation. Astrophys. J. **408**, 707–719 (1993). doi:10.1086/172631

F. Pétrélis, S. Fauve, E. Dormy, J.-P. Valet, Simple mechanism for reversals of Earth's magnetic field. Phys. Rev. Lett. **102**(1), 144503 (2009)

K. Petrovay, Solar cycle prediction. Living Rev. Sol. Phys. **7**, 6 (2010). doi:10.12942/lrsp-2010-6

K. Petrovay, U.R. Christensen, The magnetic Sun: reversals and long-term variations. Space Sci. Rev. **155**, 371–385 (2010). doi:10.1007/s11214-010-9657-8

K. Petrovay, E. Forgács-dajka, The role of active regions in the generation of torsional oscillations. Sol. Phys. **205**, 39–52 (2002). doi:10.1023/A:1013833709489

A.A. Pevtsov, R.C. Canfield, S.M. Latushko, Hemispheric helicity trend for solar cycle 23. Astrophys. J. Lett. **549**, 261–263 (2001). doi:10.1086/319179

A.A. Pevtsov, G.H. Fisher, L.W. Acton, D.W. Longcope, C.M. Johns-Krull, C.C. Kankelborg, T.R. Metcalf, The relationship between X-ray radiance and magnetic flux. Astrophys. J. **598**, 1387–1391 (2003). doi:10.1086/378944

J.H. Piddington, Theory of the solar 22-year cycle. Proc. Astron. Soc. Aust. **2**, 7 (1971)

R.F. Pinto, A.S. Brun, Flux emergence in a magnetized convection zone. Astrophys. J. **772**, 55 (2013). doi:10.1088/0004-637X/772/1/55

R.F. Pinto, A.S. Brun, L. Jouve, R. Grappin, Coupling the solar dynamo and the corona: wind properties, mass, and momentum losses during an activity cycle. Astrophys. J. **737**(2), 72 (2011)

E. Pitts, R.J. Tayler, The adiabatic stability of stars containing magnetic fields. IV—The influence of rotation. Mon. Not. R. Astron. Soc. **216**, 139–154 (1985)

É. Racine, P. Charbonneau, M. Ghizaru, A. Bouchat, P.K. Smolarkiewicz, On the mode of dynamo action in a global large-eddy simulation of solar convection. Astrophys. J. **735**, 46 (2011). doi:10.1088/0004-637X/735/1/46

M. Rempel, Flux-transport dynamos with Lorentz force feedback on differential rotation and meridional flow: saturation mechanism and torsional oscillations. Astrophys. J. **647**, 662–675 (2006). doi:10.1086/505170

M. Rempel, M. Schüssler, M. Knölker, Radiative magnetohydrodynamic simulation of sunspot structure. Astrophys. J. **691**, 640–649 (2009). doi:10.1088/0004-637X/691/1/640

L. Rightmire-Upton, D.H. Hathaway, K. Kosak, Measurements of the Sun's high-latitude meridional circulation. Astrophys. J. **761**, 14 (2012). doi:10.1088/2041-8205/761/1/L14

T.M. Rogers, On limiting the thickness of the solar tachocline. Astrophys. J. **733**(1), 12 (2011)

G. Rüdiger, L.L. Kitchatinov, D. Elstner, Helicity and dynamo action in magnetized stellar radiation zones. Mon. Not. R. Astron. Soc. **425**, 2267–2276 (2012). doi:10.1111/j.1365-2966.2012.21569.x

G. Rüdiger, L.L. Kitchatinov, The slender solar tachocline: a magnetic model. Astron. Nachr. **318**, 273 (1997)

G. Rüdiger, L.L. Kitchatinov, Structure and stability of the magnetic solar tachocline. New J. Phys. **9**, 302 (2007)

W. Schmutz, A. Fehlmann, W. Finsterle, the PREMOS team, First Light of PREMOS/PICARD, Annual Report 2011, Physikalisch-Meteorologisches Observatorium, World Radiation Center, Davos, Switzerland 44 pp., 24574, 24575, 24576, 24608 (2012). http://www.pmodwrc.ch/annualreport/annualreport2011.pdf,

J. Schou, H.M. Antia, S. Basu, R.S. Bogart, R.I. Bush, S.M. Chitre, J. Christensen-Dalsgaard, M.P. di Mauro, W.A. Dziembowski, A. Eff-Darwich, D.O. Gough, D.A. Haber, J.T. Hoeksema, R. Howe, S.G. Korzennik, A.G. Kosovichev, R.M. Larsen, F.P. Pijpers, P.H. Scherrer, T. Sekii, T.D. Tarbell, A.M. Title, M.J. Thompson, J. Toomre, Helioseismic studies of differential rotation in the solar envelope by the solar oscillations investigation using the Michelson Doppler imager. Astrophys. J. **505**, 390 (1998)

C.J. Schrijver, A.M. Title, On the formation of polar spots in Sun-like stars. Astrophys. J. **551**, 1099–1106 (2001). doi:10.1086/320237

H. Schunker, L. Gizon, R.H. Cameron, A.C. Birch, Helioseismology of sunspots: how sensitive are travel times to the Wilson depression and to the subsurface magnetic field? Astron. Astrophys. (2013)

M. Schüssler, The formation of sunspots and starspots. Astron. Nachr. **323**, 377–382 (2002). doi:10.1002/1521-3994(200208)323:3/4<377::AID-ASNA377>3.0.CO;2-A

M. Schussler, P. Caligari, A. Ferriz-Mas, F. Moreno-Insertis, Instability and eruption of magnetic flux tubes in the solar convection zone. Astron. Astrophys. **281**, 69–72 (1994)

D. Shiota, S. Tsuneta, M. Shimojo, N. Sako, D. Orozco Suárez, R. Ishikawa, Polar field reversal observations with hinode. Astrophys. J. **753**, 157 (2012). doi:10.1088/0004-637X/753/2/157

F.H. Shu, F.C. Adams, S. Lizano, Star formation in molecular clouds—observation and theory. Annu. Rev. Astron. Astrophys. **25**, 23–81 (1987). doi:10.1146/annurev.aa.25.090187.000323

D. Sokoloff, E. Nesme-Ribes, The Maunder minimum: a mixed-parity dynamo mode? Astron. Astrophys. **288**, 293–298 (1994)

S.K. Solanki, B. Inhester, M. Schüssler, The solar magnetic field. Rep. Prog. Phys. **69**, 563–668 (2006). doi:10.1088/0034-4885/69/3/R02

E.A. Spiegel, N.O. Weiss, Magnetic activity and variations in solar luminosity. Nature **287**, 616 (1980)

E.A. Spiegel, J.-P. Zahn, The solar tachocline. Astron. Astrophys. **265**, 106 (1992)

H.C. Spruit, Differential rotation and magnetic fields in stellar interiors. Astron. Astrophys. **349**, 189–202 (1999)

H. Spruit, Theory of solar irradiance variations. Space Sci. Rev. **94**, 113–126 (2000)

H. Spruit, Theories of the solar cycle and its effect on climate. Prog. Theor. Phys. Suppl. **195**, 185–200 (2012)

H.C. Spruit, Dynamo action by differential rotation in a stably stratified stellar interior. Astron. Astrophys. **381**, 923–932 (2002). doi:10.1051/0004-6361:20011465

H.C. Spruit, Origin of the torsional oscillation pattern of solar rotation. Sol. Phys. **213**, 1–21 (2003)

H.C. Spruit, in *Theories of the Solar Cycle: a Critical View*, ed. by M.P. Miralles, 2011, p. 39. J. Sánchez Almeida

M. Steenbeck, F. Krause, K.-H. Rädler, Berechnung der mittleren LORENTZ-feldstärke ???? X ???? Für ein elektrisch leitendes medium in turbulenter, durch CORIOLIS-kräfte beeinflußter bewegung. Z. Naturforsch. Teil A, Phys. Phys. Chem. Kosmophys. **21**, 369 (1966)

R.F. Stein, A. Nordlund, Simulations of solar granulation. I. General properties. Astrophys. J. **499**, 914 (1998). doi:10.1086/305678

R.F. Stein, Å. Nordlund, Solar small-scale magnetoconvection. Astrophys. J. **642**, 1246–1255 (2006). doi:10.1086/501445

A. Strugarek, A.S. Brun, J.-P. Zahn, Magnetic confinement of the solar tachocline: II. Coupling to a convection zone. Astron. Astrophys. **532**, 34 (2011a)

A. Strugarek, A.S. Brun, J.-P. Zahn, Magnetic confinement of the solar tachocline: the oblique dipole. Astron. Nachr. **332**(9), 891 (2011b)

A. Strugarek, A.S. Brun, S. Mathis, Y. Sarazin, Magnetic energy cascade in spherical geometry. I. The stellar convective dynamo case. Astrophys. J. **764**, 189 (2013). doi:10.1088/0004-637X/764/2/189

A. Sule, G. Rüdiger, R. Arlt, A numerical MHD model for the solar tachocline with meridional flow. Astron. Astrophys. **437**(3), 1061–1067 (2005)

R.J. Tayler, The adiabatic stability of stars containing magnetic fields-I. Toroidal fields. Mon. Not. R. Astron. Soc. **161**, 365 (1973)

M.J. Thompson, J. Christensen-Dalsgaard, M.S. Miesch, J. Toomre, The internal rotation of the Sun. Annu. Rev. Astron. Astrophys. **41**, 599–643 (2003). doi:10.1146/annurev.astro.41.011802.094848

M.J. Thompson, J. Christensen-Dalsgaard, M.S. Miesch, J. Toomre, The internal rotation of the Sun. Annu. Rev. Astron. Astrophys. **41**, 599 (2003)

G. Thuillier, M. Deland, A. Shapiro, W. Schmutz, D. Bolsée, S.M.L. Melo, The solar spectral irradiance as a function of the Mg ii index for atmosphere and climate modelling. Sol. Phys. **277**, 245–266 (2012). doi:10.1007/s11207-011-9912-5

S.M. Tobias, The solar cycle: parity interactions and amplitude modulation. Astron. Astrophys. **322**, 1007–1017 (1997)

S.M. Tobias, Modulation of solar and stellar dynamos. Astron. Nachr. **323**, 417–423 (2002). doi:10.1002/1521-3994(200208)323:3/4<417::AID-ASNA417>3.0.CO;2-U

S.M. Tobias, The solar dynamo: the role of penetration, rotation and shear on convective dynamos. Space Sci. Rev. **144**, 77–86 (2009). doi:10.1007/s11214-008-9442-0

S.M. Tobias, N.H. Brummell, T.L. Clune, J. Toomre, Transport and storage of magnetic field by overshooting turbulent compressible convection. Astrophys. J. **549**, 1183–1203 (2001). doi:10.1086/319448

S.M. Tobias, F. Cattaneo, Limited role of spectra in dynamo theory: coherent versus random dynamos. Phys. Rev. Lett. **101**(12), 125003 (2008)

R.K. Ulrich, Solar meridional circulation from Doppler shifts of the Fe I line at 5250 Å as measured by the 150-foot solar tower telescope at the Mt. Wilson observatory. Astrophys. J. **725**(1), 658–669 (2010)

Y.C. Unruh, S.K. Solanki, M. Fligge, Modelling solar irradiance variations: comparison with observations, including line-ratio variations. Space Sci. Rev. **94**, 145–152 (2000)

I.G. Usoskin, A history of solar activity over millennia. Living Rev. Sol. Phys. **10**, 1 (2013). doi:10.12942/lrsp-2013-1

E. Vázquez-Semadeni, R. Banerjee, G.C. Gómez, P. Hennebelle, D. Duffin, R.S. Klessen, Molecular cloud evolution—IV. Magnetic fields, ambipolar diffusion and the star formation efficiency. Mon. Not. R. Astron. Soc. **414**, 2511–2527 (2011)

A. Vögler, S. Shelyag, M. Schüssler, F. Cattaneo, T. Emonet, T. Linde, Simulations of magneto-convection in the solar photosphere. Equations, methods, and results of the MURaM code. Astron. Astrophys. **429**, 335–351 (2005). doi:10.1051/0004-6361:20041507

S.V. Vorontsov, J. Christensen-Dalsgaard, J. Schou, V.N. Strakhov, M.J. Thompson, Helioseismic measurement of solar torsional oscillations. Science **296**, 101–103 (2002). doi:10.1126/science.1069190

Y. Wang, N.R. Sheeley Jr., Magnetic flux transport and the Sun's dipole moment—new twists to the Babcock–Leighton model. Astrophys. J. **375**, 761–770 (1991). doi:10.1086/170240

Y.-M. Wang, N.R. Sheeley Jr., A.G. Nash, A new solar cycle model including meridional circulation. Astrophys. J. **383**, 431–442 (1991). doi:10.1086/170800

J. Warnecke, P.J. Käpylä, M.J. Mantere, A. Brandenburg, Ejections of magnetic structures above a spherical wedge driven by a convective dynamo with differential rotation. Sol. Phys. **280**, 299–319 (2012). doi:10.1007/s11207-012-0108-4

M.A. Weber, Y. Fan, M.S. Miesch, The rise of active region flux tubes in the turbulent solar convective envelope. Astrophys. J. **741**, 11 (2011)

M.A. Weber, Y. Fan, M.S. Miesch, Comparing simulations of rising flux tubes through the solar convection zone with observations of solar active regions: constraining the dynamo field strength. Sol. Phys. **287**, 239–263 (2013). doi:10.1007/s11207-012-0093-7

T.S. Wood, N.H. Brummell, Transport by meridional circulations in solar-type stars. Astrophys. J. **755**(2), 99 (2012)

T.S. Wood, P. Garaud, The Sun's meridional circulation and interior magnetic field. Astrophys. J. **738**(1), 47 (2011)

G.A.E. Wright, Pinch instabilities in magnetic stars. Mon. Not. R. Astron. Soc. **162**, 339 (1973)

A.R. Yeates, D. Nandy, D.H. Mackay, Exploring the physical basis of solar cycle predictions: flux transport dynamics and persistence of memory in advection- versus diffusion-dominated solar convection zones. Astrophys. J. **673**, 544–556 (2008). doi:10.1086/524352

A.R. Yeates, D.H. Mackay, A.A. van Ballegooijen, J.A. Constable, A nonpotential model for the Sun's open magnetic flux. J. Geophys. Res. **115**, 9112 (2010). doi:10.1029/2010JA015611

J.-P. Zahn, S. Talon, J. Matias, Angular momentum transport by internal waves in the solar interior. Astron. Astrophys. **322**, 320–328 (1997)

J.-P. Zahn, A.S. Brun, S. Mathis, On magnetic instabilities and dynamo action in stellar radiation zones. Astron. Astrophys. **474**, 145 (2007)

J. Zhao, R.S. Bogart, A.G. Kosovichev, T.L. Duvall Jr., T. Hartlep, Detection of equatorward meridional flow and evidence of double-cell meridional circulation inside the Sun. Astrophys. J. Lett. **774**, 29 (2013). doi:10.1088/2041-8205/774/2/L29

Space Sci Rev (2015) 196:137–166
DOI 10.1007/s11214-015-0212-5

The Importance of Long-Term Synoptic Observations and Data Sets for Solar Physics and Helioseismology

Yvonne Elsworth[1] · Anne-Marie Broomhall[2] · Sanjay Gosain[3] · Markus Roth[4] · Stuart M. Jefferies[5] · Frank Hill[3]

Received: 6 June 2015 / Accepted: 14 October 2015 / Published online: 24 November 2015

Abstract A casual single glance at the Sun would not lead an observer to conclude that it varies. The discovery of the 11-year sunspot cycle was only made possible through systematic daily observations of the Sun over 150 years and even today historic sunspot drawings are used to study the behavior of past solar cycles. The origin of solar activity is still poorly understood as shown by the number of different models that give widely different predictions for the strength and timing of future cycles. Our understanding of the rapid transient phenomena related to solar activity, such as flares and coronal mass ejections (CMEs) is also insufficient and making reliable predictions of these events, which can adversely impact technology, remains elusive. There is thus still much to learn about the Sun and its activity that requires observations over many solar cycles. In particular, modern helioseismic observations of the solar interior currently span only 1.5 cycles, which is far too short

✉ F. Hill
fhill@nso.edu

Y. Elsworth
y.p.elsworth@bham.ac.uk

A.-M. Broomhall
A-M.Broomhall@warwick.ac.uk

S. Gosain
sgosain@nso.edu

M. Roth
mroth@kis.uni-freiburg.de

S.M. Jefferies
stuartj@ifa.hawaii.edu

1 School of Physics and Astronomy, U. Birmingham, Edgbaston, Birmingham B15 2TT, UK

2 Institute of Advanced Study, U. Warwick, Millburn House, Millburn Hill Road, University of Warwick Science Park, Coventry CV4 7HS, UK

3 National Solar Observatory, 950 N. Cherry Ave., Tucson, AZ 85719, USA

4 Kiepenheuer-Institut für Sonnenphysik, Schöneck str. 6, 79104 Freiburg, Germany

5 Institute for Astronomy, U. Hawaii, 34 'Ōhi'a Kū St., Pukalani, HI 96768-8288, USA

to adequately sample the characteristics of the plasma flows that govern the dynamo mechanism underlying solar activity. In this paper, we review some of the long-term solar and helioseismic observations and outline some future directions.

Keywords Sun · Observations · Helioseismology

1 The Need for Synoptic observations

Although we may consider the Sun to be a constant star, there is considerable evidence that the level and character of its output changes with time. It is of particular interest to know if these changes are purely superficial or whether they are due to changes in the solar interior. There are only two ways that we currently know to probe the interior: the detection of solar neutrinos or helioseismology, the study of the acoustic waves that permeate the Sun. Of these, helioseismology is by far the more flexible and feasible. The solar output varies on both short (seconds to days) and long (months to millennia) time scales, with an emphasis on the approximately 11-year length of the sunspot cycle. There is thus a need for observations that are continuous over many cycles. Once cross calibration is established, there is also the possibility of using proxies of solar variation which can be used to follow behavior over millennia. The ^{10}Be and ^{14}C concentrations in the Earth's crust are examples of these proxies. We will come back to these in the section on activity but here we just point out that these proxies are by their nature only approximate and lacking in detail.

The solar physics community is currently developing the next generation of solar telescopes with large aperture sizes ($\sim$4 meters) such as the Daniel K. Inouye Solar Telescope (DKIST, originally known as the Advanced Technology Solar Telescope or ATST) in the US and the European Solar Telescope (EST) in the European Union. The large aperture of these telescopes will provide solar observations at an unprecedented angular resolution and will allow precision measurements of the physical parameters of the solar plasma at high spatial and temporal resolution. These detailed measurements will provide important clues about physical processes and structures at the finest spatial scales resolvable, such as the evolution of the quiet Sun magnetic flux, its emergence and concentration in the intergranular flux tubes, their role in chromospheric heating, fine structure of Evershed flow, sunspot umbral and penumbral flux tubes, solar filaments and prominences etc.

Having detailed observations at hand allows us to confront numerical solar models with reality and therefore allows us to refine the models. Highly refined models could provide a better understanding about the solar variability in parameters such as the total solar irradiance. However, a bigger challenge is to understand the solar dynamo, which operates inside the Sun and makes the number of sunspots wax and wane on a 10- to 13-year time scale. Our current understanding of this operating mechanism behind the solar cycle is rather poor and various models (empirical or physical) gave a wide range of predictions for the strength and timing of cycle 24 (Pesnell 2008).

In order to understand the solar activity cycle, i.e., the generation and decay of the magnetic field, we need high-resolution observations of the Sun together with full-disk synoptic (continuous and comprehensive) observations of the Sun in different wavelengths and polarizations. While the high-resolution observations shed light on the details of small scale magnetic fields, such as local dynamo processes, flux emergence, cancellation and surface diffusion, the synoptic observations of solar oscillations allow us to study the subsurface dynamics of the Sun, such as the profile of solar differential rotation, meridional and zonal flows, and their variability. These phenomena play important roles in the solar dynamo operation.

In addition, the field of space weather is increasingly important for our technology-dependent world, and the useful forecasting of space weather impacts from flares and CMEs requires a reliable source of continuous real-time measurements of the magnetic field vector in solar active regions. It is therefore desirable to complement the quest for high-resolution observations of the Sun with a parallel effort that seeks to modernize and extend our existing sources (e.g. BiSON, GONG and HMI) of synoptic observations that provide essential information about the solar interior through helioseismology.

1.1 Helioseismology

The physical foundation of helioseismology rests on the natural resonances of the Sun arising from the propagation of acoustic-gravity waves in the interior. While the amplitude of these standing waves peaks at a frequency ν of around 3 mHz (periods of around 5 minutes), their periods are observed to range from 200 to more than 1000 seconds. The sampling theorem requires that, if a harmonic signal is to be faithfully recorded, it must be sampled at least twice per period of the wave. In order to increase the signal-to-noise ratio (SNR), data are usually collected at intervals of no longer than one minute. Further, the resolution in ν is $1/T$, where T is the total length of the observations. There is thus a need to make T as large as possible.

Global helioseismic oscillations are evident in both "Sun-as-a-star" (unresolved) and resolved observations: Sun-as-a-star data are only sensitive to modes with the largest horizontal length scales (modes with low values of spherical harmonic degree $\ell < 5$). However, these are the modes that sample the solar core. Resolved observations broaden the spatial spectrum of modes that can be observed, with degrees ℓ up to and even greater than 1000 being detected.

The sound waves are not purely sinusoidal signals but are stochastically excited and intrinsically damped, which produces a very highly structured power spectrum as discussed in Anderson et al. (1990). The spectrum of the oscillations contains a wealth of features arising from both the Sun and the observing method. This makes helioseismology a powerful diagnostic tool but it carries with it penalties. In order to clearly resolve and track solar properties over multiple sunspot cycles, the data need to be taken at a relatively rapid cadence as just indicated, over very long periods of time and, for reasons that will be explained, the data sets should be nearly continuous. These constraints can be satisfied but they require long-term commitments to funding and an understanding of the nature of synoptic observations.

1.2 The Sunspot Cycle

Astronomy is an observationally-based science, and it is not possible to absolutely perfectly repeat a measurement. When we analyze solar data we typically look for long term cyclic behavior. However, there will also be trends that may repeat on timescales that are long by comparison with human lives but short compared with the thermal timescale of the Sun. The most obvious time scale to be considered here is the period of the solar sunspot cycle. The earliest known records of sunspot observations come from China around 364 B.C. and sporadic records of naked-eye observations were made in both Western and Eastern civilizations over the subsequent 1900 years. In 1610 through 1613, the invention of the telescope resulted in a flurry of nearly-simultaneous European observations of and publications about sunspots from Thomas Harriot, David and Johannes Fabricius, Christoph Scheiner and, most

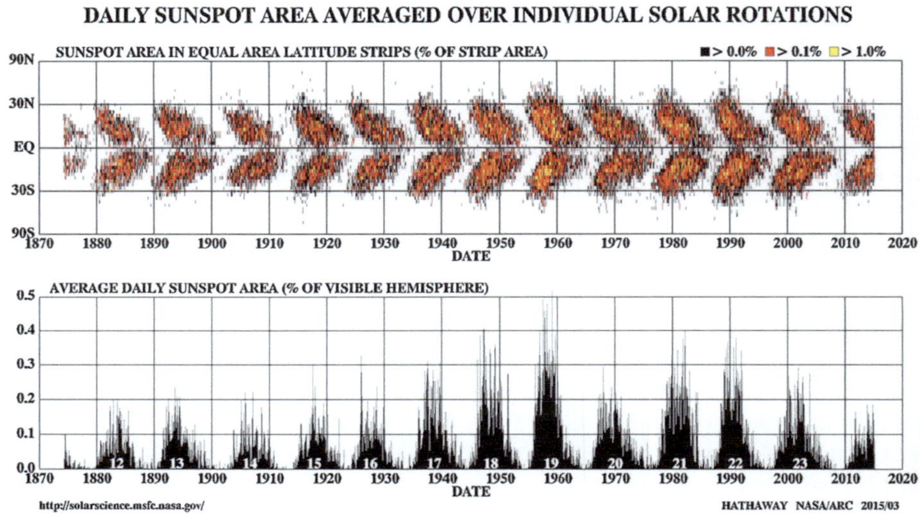

Fig. 1 Sunspot cycles from 1875 through March 2015. Top: The butterfly diagram, showing the locations where sunspots emerge as a function of solar latitude and time. Lower: The time history of the number of sunspots. Image courtesy of David Hathaway, NASA/ARC

famously, Galileo Galilei. The realization that the spots were actually on the Sun and were not clouds or other planets led to the first measurements of the solar rotation and the discovery that the Sun did not rotate rigidly but that its rotation rate is a function of solar latitude.

Since then the appearances and solar heliographic locations of sunspots have been recorded, forming the fundamental set of the first synoptic observations of the Sun. As a result, the daily number of sunspots is now a 400-year long time series displaying the variable solar magnetic activity. In the first half of the nineteenth century, Heinrich Schwabe and Rudolf Wolf detected the sunspot cycle, the discovery of which had been delayed by the Maunder Minimum, the period of 1645 to 1715 during which the Sun was nearly spotless. This long record of observations clearly shows the approximately 11-year period of the number of sunspots on the solar disk (known as the Schwabe cycle), but it was not until Hale was able to measure the solar magnetic field in the early twentieth century that it was realized that the activity cycle is actually around 22 years since the polarity of the Sun's magnetic field reverses during each Schwabe cycle. A longer period cycle of 87 years, known as the Gleissberg cycle, can be seen in the amplitude envelope of the Schwab cycle, and periods of pronounced sunspot maxima and extended grand minima can be identified. Figure 1 shows the last 140 years of the sunspot number record along with a label identifying each cycle. We are in cycle 24 as of the writing of this paper.

Sunspots are locations of activity in the form of flares and CMEs. These events are responsible for creating the extreme type of space weather, which otherwise is a mostly steady solar wind. Flares and CMEs produce high levels of radiation primarily in the form of charged particles and X-rays that are two or more orders of magnitude higher than those typically present in the background solar wind. These blasts can strike the Earth and its magnetic field, creating geomagnetic storms, generating currents in the Earth's crust, and disrupting the ionosphere. Space weather can adversely impact many common technologies including GPS, telecommunications, space activities, military operations, navigation, power grids, financial markets, emergency services, surveying, agriculture, and air travel. A fundamental goal of synoptic observations is understanding the physical processes inside the Sun with sufficient detail so that accurately predicting the level of solar activity becomes possible.

2 Signatures of Solar Activity

The number of sunspots visible on the disk is the most obvious signature of solar activity, but it is by no means the only variable that is observed. In addition, it is surprisingly hard to actually count the number of sunspots on a given day. Variations in the Earth's atmosphere can hide small spots, human observers can make different choices as to how many individual spots are in a group, and telescope optics can degrade with time (Svalgaard 2012). The effects of these issues are mostly avoided in the other signatures of activity discussed here.

2.1 Sunspot Number

We are interested in the variations in the magnetic field inside the Sun but, since this is very difficult to measure directly, we resort to using proxies for the magnetic field. The obvious and original proxy is the variation in the number of sunspots that are observed on the visible disk of the Sun. Since it is not so simple to just count the number of spots, this method is open to considerable personal interpretation and there has recently been a revision in the historical record to remove some of the discrepancies. Svalgaard (2012) has shown that only two adjustments are necessary to reconcile the Group Sunspot Number with the Zürich Sunspot Number. The first of these adjustments involves increasing the values after the 1940s by 20 % due to weighting of the sunspot counts according to spot size. Svalgaard also shows that the Group Sunspot Number before about 1885 is too low by about 50 %. The application of these adjustments results in a single sunspot number series and has the consequence that there is no longer a distinct Modern Grand Maximum (Clette et al. 2015).

At the beginning of each activity cycle, the location where sunspots apparently emerge from the solar interior starts at a solar latitude of about 40°. This emergence latitude moves towards the solar equator as the cycle progress. Thus, plots that record the location of sunspot emergence as a function of solar latitude and time display a pattern known as the butterfly diagram. This is shown, along with the sunspot number, in Fig. 1.

2.2 Sunspot Field Strength, Polar Fields

Sunspots are the most visible and familiar aspect of the solar cycle, but it is the magnetic field that is actually the fundamental controlling factor of solar activity. Sunspots are very large regions of very strong magnetic field that alters the thermal structure of the solar atmosphere and redirects energy into a number of complex paths, including explosive reconnection events in the form of flares. The maximum magnetic field strength observed in sunspots is typically around 3500 G, although there may have been a spot with 6000 G strength observed in the 1940s (Livingston et al. 2006). There has been considerable interest in the possibility that the central magnetic field intensity and relative darkness of sunspot umbras has been decreasing over the last two solar cycles, implying that sunspots would disappear in the next cycle (Livingston et al. 2012) and lead to a Grand Minimum perhaps akin to the Maunder Minimum. This conclusion was contradicted by the work of Kiess et al. (2014), and has since been retracted (Watson et al. 2014).

While the sunspot field strength is now thought to be constant across solar cycles, it has been established that the strength of the fields at the Sun's north and south poles is not. Indeed, the strength of the polar field at solar minimum is highly correlated with the peak amplitude of the next cycle (e.g. Svalgaard et al. 2005). The reversal of the polar fields, driven by the meridional flow (north-south) of the solar plasma on the surface, is now taken to be one of the major indications that solar maximum has been reached (Petrie et al. 2014;

Upton and Hathaway 2014). However, the timing of the reversal can be complicated as the northern and southern solar hemispheres typically have asymmetric behavior, as seen in the current cycle 24 (Svalgaard and Kamide 2013).

2.3 Irradiance Measurements

The amount of solar radiative energy reaching the Earth, integrated over all wavelengths of light, is called the total solar irradiance (TSI), with an observed value of 1361 Watts/m^2 above the Earth's atmosphere and at a Sun-Earth distance of 1 AU. The TSI varies on time scales of days due to the presence of sunspots as the solar rotation carries them across the disk, however the most significant variation is an increase of about 0.1 % during solar maximum (Fröhlich 2013). Many models of the detailed variation in TSI have been constructed based on the surface magnetic field of the Sun, some of which can be found in Velasco Herrera et al. (2015), Yeo et al. (2014a,b), Dasi-Espuig et al. (2014), Fontenla et al. (2014), Chapman et al. (2013), Foukal (2012), Ulrich et al. (2010), Pap et al. (2010), Jones et al. (2008), Wang et al. (2005) and Lean et al. (1998).

Measurements of the TSI have been made by a number of satellites, starting with electrical substitution radiometers (ESRs) on the Nimbus 7 satellite in November 1978 (Hickey et al. 1980; Kyle et al. 1994). These have been continued with the Active Cavity Radiometer Irradiance Measurement (ACRIM) instrument on SMM (Willson et al. 1981), the Earth Radiation Budget Experiment (ERBE, Barkstrom and Smith 1986), two more ACRIM experiments on UARS and ACRIMSAT, the VIRGO instrument on SOHO (Fröhlich et al. 1995), and SORCE (Woods et al. 2000). There are considerable differences between these measurements even when they overlap in time and reconciling the differences has been challenging (e.g. Fröhlich 2006).

While the TSI is integrated over all wavelengths of light, there are also variations of solar output that depend on the spectral region being studied. The Spectral Solar Irradiance (SSI) has been measured since 2003 with the SORCE space mission (Woods et al. 2000), and these measurements have shown that variations in the SSI can be either in or out of phase with the TSI changes depending on the spectral band (e.g. Harder et al. 2010). This complex behavior is thought to be linked to the detailed effect of magnetic fields on the radiative transfer of the light as it passes through the photosphere (Haberreiter et al. 2010; Fontenla et al. 2011).

Long term variability in both TSI and SSI can potentially lead to climatic variations via changes in Earth's upper atmosphere, global sea temperatures etc. Thus, climatologists try to understand the role of solar variability among other causes which lead to climate changes on the Earth. Therefore, in order to make progress it is imperative that we keep consistent records of solar irradiance measurements systematically, and this has been challenging as outlined above. All of the models of TSI and SSI depend on synoptic measurements of the solar magnetic field.

2.4 Chromospheric Indices

Chromospheric indices measure the intensity integrated over specific and narrow wavelength bands in spectral lines that are formed mainly in the chromosphere, and are thus special cases of the SSI discussed above. The most popular spectral line that is used for these measurements is the Ca II K line at 3933.7 Å. This line is very broad, and displays features in its core that vary in strength over the solar cycle. The most prevalent chromospheric index is the 1-Å Ca K emission index (Keil and Worden 1984; Scargle et al. 2013), which is essentially the excess intensity of the central 1 Å of the spectral line as compared to some reference line profile. This index varies in phase with the solar cycle, and is also easy to observe in more distant stars, which makes it useful for comparisons of solar and stellar activity cycles.

Another widely used measure of the level of the solar activity that is emitted from the chromosphere is the radio flux measured at a wavelength of 10.7 cm (known as F10.7). This has been systematically recorded since 1947 and clearly shows a correlation with the level of solar activity. Unlike the sunspot number, the F10.7 signal does not drop to zero at solar minimum and the relationship between these two proxies for the solar activity is not linear. Livingston et al. (2012) discuss this relationship and show that it has changed since about 1990 in a manner that is consistent with their measurement that the intense field in the center of dark sunspots has been declining. They use this to argue for the approach of a Grand Minimum.

2.5 Further Activity Proxies

There are other ways of monitoring the solar activity level which are either spectroscopic (for example the core to wing ratio in Mg II at 280 nm or the equivalent width of the He II line at 1083 nm) or involve direct measurement of the strong magnetic field. In addition, the long-term record of solar activity covering thousands of years has been inferred by measuring the relative concentration of certain radioactive isotopes, particularly ^{14}C and ^{10}Be, in ice cores. The production of these isotopes is modulated by the flux of cosmic rays hitting the Earth's atmosphere, and this flux is anticorrelated with the solar activity. Studies of these records suggest that the level of solar activity since the middle of the twentieth century is amongst the highest of the past 10,000 years, and that Maunder minimum-like epochs of suppressed activity of varying durations have occurred repeatedly over that time span (Solanki et al. 2004). It also shows that, during the Maunder minimum, the cyclical change of the solar magnetic field persisted even without apparent sunspots.

The temporal correlation between several activity indices displays hysteresis, as shown by Bachmann and White (1994). This suggests that there are delays between the onset and decline of activity reflected in variables that are sensitive to conditions in various solar layers, such as the corona, chromosphere, photosphere, and interior.

2.6 The Response of the Oscillations to Solar Activity

The normal modes of oscillation of the Sun fundamentally depend on the speed of sound in the solar interior. While this is primarily a function of the local temperature and composition, the magnetic field can also be important. This is particularly so in the outer regions of the Sun where the magnetic pressure and the gas pressure are of comparable magnitude. It is therefore to be expected that the properties of solar oscillations change in phase with the levels of solar activity. A key point here is that the examination of the changes in the observed p-mode properties throughout the solar cycle provides information about solar-cycle-related processes that occur beneath the photosphere.

2.6.1 Frequency Shifts

Soon after the first detection and identification of the oscillations, people looked for signs of the frequencies being sensitive to the level of solar activity. The earliest such evidence was found in the mid 1980s when it became clear that the frequencies of oscillations in intensity as observed from space using the ACRIM instrument on the Solar Maximum Mission

Fig. 2 Typical "Sun-as-a-star" BiSON spectrum of low-degree p-mode solar oscillations

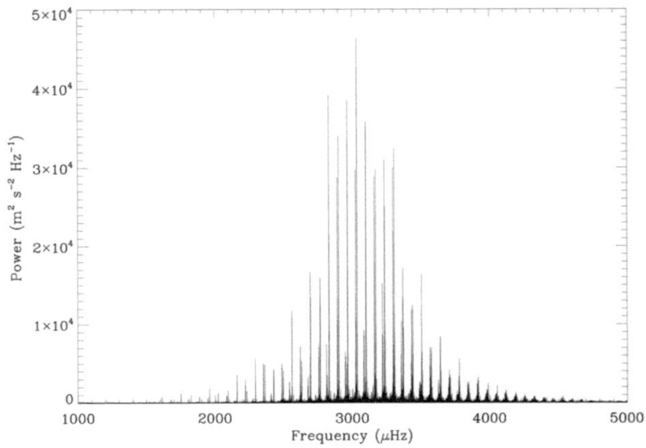

(SMM) satellite varied throughout the solar cycle with the frequencies being largest when the solar activity is at maximum (Woodard and Noyes 1985). The measurement was based on a comparison of the signal near solar maximum in 1980 and near minimum in 1984. This was an important step forward but what was really needed were continuous measurements throughout the cycle to show reversals in the frequency shift when the activity reverses. This was provided about five years later using low-degree data collected with the early phase of the Birmingham Solar Oscillations Network (BiSON). The features of BiSON are discussed later in Sect. 4.3.1. These data followed the activity through the minimum of 1984/5 towards the maximum in about 1990 (Pallé et al. 1989; Elsworth et al. 1990). Further years of data, much improved fill and better analysis techniques emphasized the clear relationship between solar activity and the frequencies of the low-degree p-mode oscillations with spherical harmonic degree $\ell < 5$ (Elsworth et al. 1994; Chaplin et al. 2001; Jiménez-Reyes et al. 2003, 2004, 2007; Verner et al. 2004, 2006). For a historical perspective and a discussion about the impact of the treatment of line asymmetry on the determination of the individual frequency shifts see Howe et al. (2015).

Solar-activity related frequency shifts at higher values of ℓ were originally observed at Big Bear Observatory (Libbrecht and Woodard 1990) and have been studied extensively ever since (Howe et al. 1999; Rhodes et al. 2002, 2011; Rabello-Soares et al. 2008; Rabello-Soares 2011; Tripathy et al. 2007, 2010, 2013, 2015; Jain et al. 2000, 2009, 2011, 2012, 2013).

2.6.2 Extracting the Frequency Shifts from Sun-as-a-Star Data

BiSON is in a unique position to study the changes in oscillation frequencies that accompany the solar cycle as it has now been collecting data for well over 30 years. The best data start from when the full network was operational in the early 1990s but the older data have unique historical importance.

In order to obtain the shifts in the p-mode frequencies with time the BiSON data must be split into short subsets. However, the longer the time series the more precisely the mode frequencies can be obtained. Therefore, the duration of the subsets must represent a balance between being long enough to accurately and precisely extract mode frequencies and short enough to resolve solar cycle variations.

Commonly chosen lengths of data are 108 d, 182.5 d or 365 d. A typical BiSON spectrum is shown in Fig. 2. Each of the peaks represents a different mode of oscillation. Around

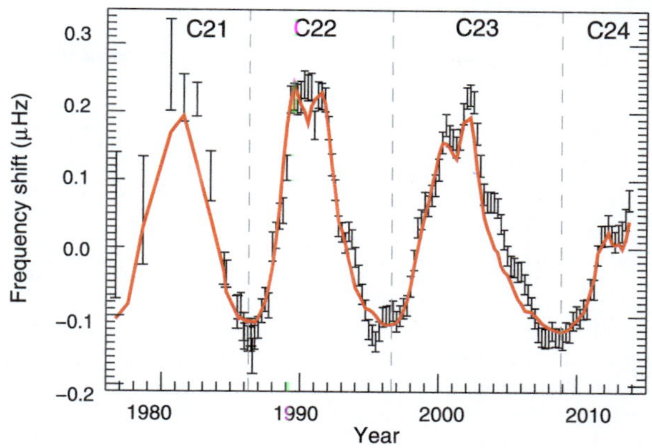

Fig. 3 Average frequency shifts of "Sun-as-a-star" modes with frequencies between 2.4 and 3.5 mHz. The results were obtained from 365 d time series that overlapped by 273.75 d. Also plotted in *red* is a scaled version of the 10.7 cm flux, a well known proxy for solar activity

3000 μHz the SNR of the oscillations is large and the individual peaks are well resolved. At lower frequencies, the SNR decreases and sets a lower limit on the range of frequencies one can consider. At higher frequencies the modes become more rapidly damped, meaning they have broad profiles in a frequency spectrum (see right panel of Fig. 4). At high frequencies, therefore, it is difficult to resolve individual peaks and obtain precise mode frequencies. Thus there is also an upper limit on the range of frequencies that one can consider. Estimates of the mode frequencies are extracted from each subset by fitting a modified series of Lorentzian profile models to the data using a standard maximum likelihood estimation (MLE) method, which was applied in the manner described in Fletcher et al. (2009). One could also use Bayesian methods, which give more robust estimates of the parameters. However, it has been shown that the frequencies from MLE are reliable and are much easier to obtain (Howe et al. 2015).

For each investigation we first define a frequency range of interest. For this range, a reference frequency set is chosen by averaging the frequencies in subsets covering a predetermined time period. While the choice of subsets used to make the reference frequency set determines the absolute values of the frequency shifts, the main results described here are insensitive to the exact choice: A different choice will transpose the frequency shifts to higher or lower values but the variation with time will remain the same. Frequency shifts are then defined as the differences between frequencies given in the reference set and the frequencies of the corresponding modes observed at different epochs (Broomhall et al. 2009). This gives one value of the frequency shift for each epoch and frequency range. Note that it is, in general, not possible to get an instantaneous reading of the frequency shift. They are, by definition, averages.

Figure 3 shows mean frequency shifts of the p modes observed by BiSON (also see Broomhall et al. 2009; Salabert et al. 2009; Fletcher et al. 2010). Here the reference frequency set was created by averaging over all subsets. A scaled version of the 10.7-cm flux is also plotted in Fig. 3. The flux has been scaled by fitting a linear relationship between the 10.7-cm flux and the frequency shifts. The 11-year cycle is clearly visible.

For a low-degree mode at about 3000 μHz, the change in frequency between solar maximum and minimum is about 0.4 μHz. At higher frequencies, the frequency changes are larger but they become more difficult to detect as the line-widths of the modes themselves also increase. For frequencies below 3000 μHz, the response to the activity is reduced. Although the mode line width is also decreasing, at some point no cycle-related changes in the mode frequencies are detectable. The data for this are shown in Fig. 4.

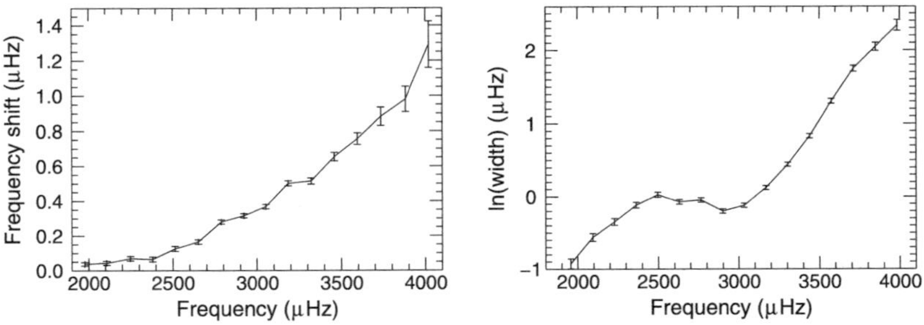

Fig. 4 Left: Maximum-minimum frequency shift as a function of frequency obtained from BiSON data. The average frequencies observed during the maximum of cycle 23 were compared to the frequencies observed during the minimum between cycles 23 and 24. Right: Natural logarithm of typical widths of Lorentzian profiles fitted to mode peaks in a BiSON spectrum, plotted as a function of frequency

2.6.3 Physical Causes of the Frequency Shifts

The dependencies of the solar cycle frequency shifts, $\delta\nu_{n,\ell}$, on both angular degree, ℓ, and frequency, $\nu_{n,\ell}$, are well known (see e.g. Libbrecht and Woodard 1990; Elsworth et al. 1994; Chaplin et al. 1998, 2001; Howe et al. 1999; Jiménez-Reyes et al. 2001). These dependencies indicate that the observed 11-year signal must be the result of changes in acoustic properties in the few hundred kilometers just beneath the photosphere. This is a region where higher-frequency modes are much more sensitive to conditions than their lower-frequency counterparts because of differences in the location of the upper boundaries of the cavities in which the modes are trapped (Libbrecht and Woodard 1990; Christensen-Dalsgaard and Berthomieu 1991). The detail of this is further discussed in Elsworth et al. (2012) and we briefly summarize it here.

Solar p modes are trapped in cavities in the solar interior. The upper turning point is relatively independent of ℓ (for low and intermediate ℓ) since, near the photosphere, all waves travel in an approximately radial direction. However, the position of the upper turning point depends on frequency. At fixed frequency the lower-ℓ modes penetrate more deeply into the solar interior than do the higher-ℓ modes. Hence, higher-frequency modes are more sensitive to surface perturbations. Libbrecht and Woodard (1990) discuss the origin of the perturbation. If the perturbations were to extend over a significant fraction of the radius, asymptotic theory implies that the fractional mode frequency shift would depend mainly on $\nu_{n,\ell}/\ell$. However, the observed $\delta\nu_{n,\ell}$ is not well described by a function of $\nu_{n,\ell}/\ell$. This implies that the relevant structural changes occur mainly in a thin layer. This thin layer must be near the surface because if it were buried an oscillatory frequency dependence in $\delta\nu_{n,\ell}$ would be observed (Thompson 1988). Libbrecht and Woodard (1990), therefore, conclude that the oscillations are responding to changes in the strength of the solar magnetic activity near the Sun's surface. As modes below approximately 1800 μHz experience almost no solar cycle frequency shift it is reasonable to conclude that the origin of the perturbation is concentrated in a region above the upper turning points of these modes.

Measurements of the strong magnetic field on the surface of the Sun can be made in a manner that allows one to see the distribution of the magnetic field with latitude on the Sun. This is particularly interesting as the low-degree modes also show a latitudinal distribution. Chaplin et al. (2004) explore the variations which were observed in the low-degree solar p-mode frequencies during solar cycles 22 and 23, and their relation to global and spatially decomposed magnetic-field proxies of the surface activity. They regressed the BiSON data sets against the longitudinal magnetic field measurement made by Kitt Peak Observatory. The magnetic data were projected on to different spherical harmonic components in order to search for enhanced correlation between the mode-frequency shifts and the measured fields when the latitudinal dependence of both components was considered. They found some evidence for this. For the higher-ℓ modes this agreement is much more pronounced (see Howe et al. 2002).

None of the activity indices discussed earlier perfectly follow the variation in the frequency of the low-degree, solar p-modes but they all show rough agreement. It is found that in some cycles one particular index is best, but in another cycle the best choice is a different activity index (see Chaplin et al. 2004). In addition, frequency changes over the solar cycle also display hysteresis (Tripathy et al. 2001).

In summary, we have the situation where a variety of surface and solar-atmosphere proxies of solar activity agree with the variations that are seen in the p-mode frequencies. The ultimate goal is to link these phenomena together with some fundamental physics. To take this further, we now look at the periodicities that are observed in the frequency changes and discuss ways that the seat of the variation can be localized.

2.6.4 *Two-Year and Eleven-Year Signals in Frequency Variations*

Over the past twenty years it has become apparent that significant quasi-periodic variability in activity is also seen on timescales shorter than 11 years, particularly between 1 and 4 years (e.g. Benevolenskaya 1995; Mursula et al. 2003; Valdés-Galicia and Velasco 2008). This same phenomenon is visible in the p-mode frequencies (Broomhall et al. 2009, 2010; Fletcher et al. 2010; Simoniello et al. 2012, 2013) and is seen very clearly in Fig. 3. The period of this structure is approximately two years and so it is known as the quasi-biennial signal. Fletcher et al. (2010) showed that the signal was visible in both BiSON and GOLF data and Broomhall et al. (2010) demonstrated that the quasi-biennial signal was also evident in Variability of solar IRradiance and Gravity Oscillations (VIRGO; Fröhlich et al. 1995) data. VIRGO consists of three sun photometers (SPMs), that observe at different wavelengths, namely the blue channel (402 nm), the green channel (500 nm), and the red channel (862 nm). The results are similar for each individual channel. Simoniello et al. (2012) demonstrated that the signal is also present in the resolved GONG data. We can conclude from this that the signal is not a function of making Sun-as-a-star observations from the ground or in velocity. It is likely to be a solar phenomenon.

In order to extract these shorter periodicities, we subtracted a smooth trend from the average total shifts by applying a boxcar filter of width 3 years. This removed the dominant 11-year signal of the solar cycle. Note that, although the width of this boxcar is only slightly larger than the periodicity we are examining here, wider filters produce similar results. The resulting residuals, shown in Fig. 5, show a periodicity on a timescale of about two years. Comparison with the 10.7-cm flux shows that the signal is consistent with surface and atmospheric measures of solar activity.

Using unresolved data observed during cycle 23 Broomhall et al. (2012) found that the QBO shows some frequency dependence but that this dependence is not as strong as that observed in the 11-year cycle. Simoniello et al. (2013) found similar results with resolved helioseismic data, also observed during cycle 23. Salabert et al. (2015) recently observed a periodicity, longer than the QBO but shorter than the 11-year cycle, in the frequency variations of low-frequency (1.8 to 2.4 mHz) modes during the rising phase of cycle 24 (originating in 2006). This periodicity was not present in the frequency variations observed in higher-frequency modes. However, the amplitude of the QBO in cycle 24 cannot truly be

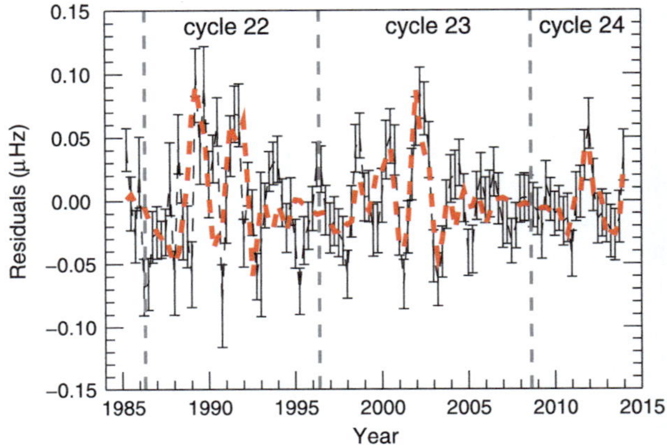

Fig. 5 Rapidly varying component of the oscillation frequency shifts obtained from 182.5-d BiSON data sets that overlap by 91.25 d. The 11-yr component was removed by subtracting a smoothed component (obtained by applying a 3-yr box car smoothing). Plotted in *red*, for comparison, is a scaled version of the 10.7-cm flux

assessed until the next minimum, since the amplitude of the signal itself varies with time and is largest around solar activity maximum.

The envelope of the two-year signal is modulated by the 11-year signal. However there must be some additive component because it is still present when the 11-year signal is at minimum. Furthermore, because it is more visible at lower frequencies than is the 11-year signal, the two-year signal must have its origin in significantly deeper layers (Broomhall et al. 2012), which may be positioned below the upper turning point of the lowest frequency modes examined (since the depth of a mode's upper turning point increases with decreasing frequency). The upper turning point of modes with frequencies of 1.8 mHz occurs at a depth of approximately 1000 km, whereas the influence of the 11-year cycle is concentrated in the upper few 100 km of the solar interior. Put together, this all points to a phenomenon that is separate from, but nevertheless susceptible to, the influence of the 11-year cycle. This is discussed in more detail in Elsworth et al. (2012) where the results from structure inversions are used to illuminate the discussion.

The activity minimum between cycle 23 and cycle 24 was very unusual in both its duration and its depth. It was also, at the time, unexpected. A closer analysis of the BiSON data shows that there was evidence that could have predicted that something strange was happening. Basu et al. (2012) applied boxcar smoothing to the frequency shifts to remove the influence of the two-year signal and then compared the behavior of the F10.7 activity proxy with the smoothed shifts in three frequency bands. They showed that, although in the two higher frequency bands, the frequency shifts followed the activity proxy, for the low-frequency band this was not the case. It was noted that, for this third band, the frequency shift responded less well to the variations in the activity and the frequencies varied less than expected. This effect started during the declining phase of cycle 23. They interpreted this as evidence for a thinning in the extent of the magnetic layer which influences the mode frequencies. It will be interesting to see if this effect persists into cycle 24.

2.6.5 Variations in Mode Amplitude, Line Width, and Energy Supply Rate

In addition to conditions in the interior, the solar acoustic power spectrum also provides information on the physics of the modes themselves over the solar cycle. The key measured parameters are the amplitudes A of the modes, their line widths in the solar spectrum, Γ, which measures the lifetimes of the modes, and the energy supply rate $\dot{E}$, which is proportional to $A\Gamma^2$. At low degrees, the change in $\dot{E}$ is small and consistent with a zero change, as found by (e.g. Chaplin et al. 2000; Jiménez-Reyes et al. 2003). Studies of the variations of these parameters for modes with $\ell >\sim 20$ over the cycle show that, as activity increases, A decreases, Γ increases (indicating shorter mode lifetimes), and $\dot{E}$ decreases (Komm et al. 2000a,b; Rhodes et al. 2011). These results suggest that the energy that excites the intermediate and high-degree modes is slightly reduced at high activity. Since convective downdrafts are thought to be the main driver of the oscillations, this further suggests that granulation is suppressed as activity increases. This is supported by the study of Komm et al. (2002) where the source of the width and amplitude variations is found to be concentrated in the locations of activity.

3 Solar Interior and Helioseismology

Since the magnetic field and activity cycle must originate from inside the Sun, knowledge of the changes in the dynamics and structure of the solar interior on decadal time scales is vital if we are to understand the physical mechanism of the solar dynamo. This information can only be obtained from long-term helioseismic observations. In this section we briefly review the status of some of these topics.

3.1 Internal Dynamics

Information about the changes in the solar rotation rate is obtained from frequency splittings, which are the differences in ν between global modes with the same values of spherical harmonic degree ℓ and radial order n but with values of azimuthal degree m differing by 1. The splittings contain information about solar internal dynamics and departures of the solar shape from a perfect sphere. The dominant source of the splittings is the differential rotation of the Sun as a function of depth and latitude. Solar cycle variations of the splittings are primarily the result of the zonal flow known as the torsional oscillation, which is discussed in Brun et al. (2013). Another source may be a periodic 1.3-year fluctuation in the rotation rate at the base of the convection zone (Howe et al. 2000, 2011), but this is unconfirmed (Antia and Basu 2000). However, there is evidence that the position as well as the transition of the rotation rate across the tachocline, where the internal solar rotation rate changes from differential to rigid, show some temporal variations (Antia and Basu 2011).

A major research area of solar physics concentrates on local helioseismology, where the oscillations are analyzed in a variety of techniques that do not consider the Sun as a global entity but instead focuses on specific localized areas (Gizon and Birch 2005). In terms of long-term solar cycle variations, these techniques have the great advantage that they are not restricted to results that are symmetric across the solar equator. They can thus reveal north-south hemispheric asymmetries in both the zonal and meridional flows (González Hernández et al. 2010; Komm et al. 2014, 2015) that can be correlated with the hemispheric asymmetries in the magnetic field discussed earlier.

3.2 Internal Structure

Helioseismic studies of the internal structure focus primarily on inversions for the wave speed, which depends on the temperature, composition, and magnetic field of the internal solar plasma. Unfortunately, it is not possible to reliably separate the effects of these three physical quantities since, unlike for velocity, they do not depend on the propagation direction of the wave. Nonetheless, it is possible to detect changes in the subsurface wave speed over the cycle (Basu and Mandel 2004; Basu et al. 2007, 2013; Baldner and Basu 2008; Baldner et al. 2010). These changes are usually attributed to the surface magnetic field, but they may well have an additional thermal component. In addition, deeply situated magnetic or thermal variations at the base of the convection zone could explain some of the observed variations (Baldner and Basu 2008).

4 Observing Strategies

4.1 Getting Continuous Data

In making observations of solar oscillations we must consider the detailed structure of the Sun's acoustic spectrum. The fundamental repeat period of the different radial orders is about 137 μHz and the typical splitting between the $\ell = 0$ and $\ell = 2$ modes is around 10 μHz. Great care must be taken to minimize any artifacts in the spectrum that could overlap with the solar features. It is therefore unfortunate that 24-hour diurnal interruptions from the Earth's rotation produce features at multiples of 11.57 μHz. Early measurements from single geographical sites were very vulnerable to this problem. The solutions are three fold. One can make the measurements from a suitable orbit in space which has a constant view of the Sun (Scherrer et al. 1995), observe from Antarctica during the Austral summer (Jefferies et al. 1988), or invest in a ground-based network. This latter solution was the one adopted by BiSON and by GONG.

4.2 Advantages and Disadvantages for Working from Space and from the Ground

Space-based observations have the advantage of higher spatial resolution, image stability and access to high-energy ultra-violet and X-ray wavelength regions. There are numerous space-based solar programs that are currently in operation such as the *Hinode* Solar Optical telescope (SOT), Solar Dynamics Observatory (SDO), Interface Region Imaging Spectrograph (IRIS), Solar TErrestrial RElations Observatory (STEREO), SOlar Heliospheric Observatory (SOHO), and Reuven Ramaty High Energy Solar Spectroscope Imager (RHESSI). More are planned in the near future including Solar Orbiter, Solar Probe Plus, Aditya, and Solar-C. While satellites provide high-quality observations, they always have a risk of failures that cannot be easily fixed. Further, they tend to have associated high costs of development and operation, no possibility to upgrade, and limited lifetimes. For long-term variability studies of the Sun this can be a disadvantage. One such example is the TSI measurements by various satellites since 1978. Although all data sets show solar cycle variations, they disagree due to differences in sensitivity and calibrations. Therefore, only relative scaling can be applied (assuming one of the TSI series to be correct) to derive a continuous TSI temporal profile. This non-uniformity in data sets makes it difficult to compare, for example, the TSI value during the minima of three solar cycles.

Ground-based observations have potentially unlimited lifetimes, and thorough calibrations can be carried out when instruments are upgraded to keep track of long-term variability of the parameters. Ground-based instruments can use the latest in technology while space-based technologies have to be proven to be space qualified. High rates of data telemetry are easy to achieve on the ground, and ground-based instrumentation can be quite large in size, such as the Daniel K. Inyoue Solar Telescope (DKIST) and European Solar Telescope (EST) projects of 4-m apertures that are in planning now. The biggest challenge for

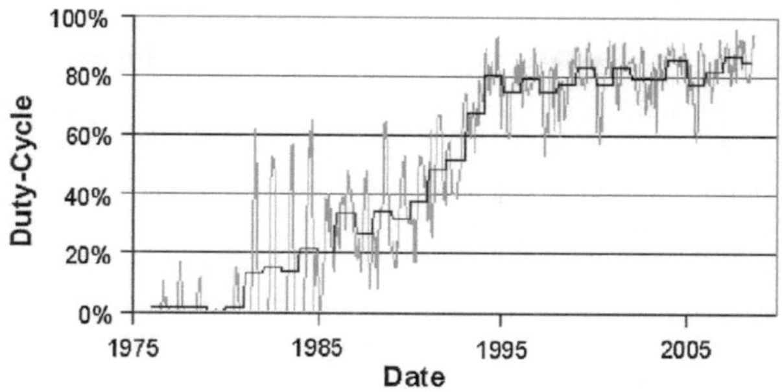

Fig. 6 Evolution of the BiSON duty cycle 1976–2009 in 4-week blocks. The *staircase black lines* are yearly averages. Adapted from Allison et al. (2009)

ground-based observations is the atmospheric seeing caused by turbulence. New techniques are being developed to address these challenges, such as adaptive optics (including multi-conjugate adaptive optics for extended field-of-view), active temperature control of the telescope optics and mechanics to control instrument seeing, detectors with low readout noise and high frame rates such as pn-CCD detectors (being tested by IMPRS for fast polarimetry to reduce seeing variations), large aperture etalon development (e.g. the Visible Tunable Filter being developed for the DKIST by the Kiepenheuer-Institut für Sonnenphysik (KIS)), post processing of images to enhance spatial resolution (speckle interferometry) and so on.

4.3 Networks

4.3.1 BiSON

The Birmingham Solar-Oscillations Network (BiSON) (Chaplin et al. 1996; Davies et al. 2014) makes Sun-as-a-star (unresolved) Doppler velocity observations, which are sensitive to the p modes with the largest horizontal scales (or the lowest spherical harmonic degrees, ℓ). Consequently, the frequencies measured by BiSON are of the truly global modes of the Sun. These modes travel to the Sun's core, however, their dwell time at the surface is much longer than at the solar core because the sound speed inside the Sun increases with depth. Therefore, the low-ℓ modes are most sensitive to variations in regions of the interior that are close to the surface and so are able to give a picture of the influence of near-surface activity.

BiSON is a network of autonomous ground-based observatories that are strategically positioned at various longitudes and latitudes in order to provide nearly continuous coverage of the Sun. The network began with just one station and gradually expanded over the course of 17 years, finally becoming a six-site network in 1992. There are four sites in the southern hemisphere and two in the northern hemisphere. The current set of sites is Izaña, Tenerife; Sutherland, South Africa; Carnarvon, Western Australia; Narrabri, New South Wales; Mount Wilson, California and Las Campanas, Chile.

The quality of the early data is poor compared to more recent data because of limited time coverage but is invaluable because it represents an epoch where other observations are sparse. Calculations have shown that 6 sites is the minimum required to give data fills which are above 80 % over periods of months and which lack very regular breaks that give rise to strongly periodic features in the spectra (Hill and Newkirk 1985). Mid-latitude sites are chosen because they represent a reasonable compromise between having the Sun high in the sky and typically high insolation levels. The evolution of the BiSON duty cycle is shown in Fig. 6 from 1976 to 2009.

4.3.2 GONG

The Global Oscillation Network Group (GONG; Harvey et al. 1996) obtains full-disk solar data with a nominal resolution of 1024 × 1024 pixels at a cadence of one image per minute. GONG is a six-site network located at Big Bear Solar Observatory, California; Mauna Loa Solar Observatory, Hawaii; Learmonth Solar Observatory, Australia; Udiapur Solar Observatory, India; Observatorio del Teide, Spain, and the Cerro Tololo Interamerican Observatory, Chile. These sites were selected after a six-year site survey that measured the transparency of the sky at 15 locations (Hill et al. 1994a,b). GONG was deployed in 1995 with a small-format detector that was upgraded to the current camera in 2001 (Harvey et al. 1998). The median daily duty cycle for GONG over its lifetime to date is 91 %.

4.3.3 Other Networks

Three other networks have been deployed for helioseismology, but are no longer operational. The International Research on the Interior of the Sun (IRIS) network operated from 1990 to 2001 with a single-pixel integrated-light resonance cell like that of BiSON but operating with a sodium vapor (Fossat 1988, 2013). IRIS had six stations at Kumbel, Uzbekistan; La Silla, Chile; Stanford, California; Oukaimeden, Morocco; Izaña, Tenerife; and Narrabri, Australia and had a typical duty cycle of 80 %. The Taiwan Oscillation Network (TON) used a 1024 × 1024 camera to obtain full-disk solar intensity images in the Ca K spectral line from 1994 to 2002 with three sites at Teide Observatory (Tenerife), Huairou Solar Observing Station (near Beijing), and Big Bear Solar Observatory (California) (Chou et al. 1995). Finally, the Experiment for Coordinated Helioseismic Observations (ECHO) network was a more developed version of the LOWL instrument (Tomczyk et al. 1993, 1995), which consisted of a magneto-optical filter using sodium. ECHO was composed of two sites, at Teide Observatory and Mauna Loa, and operated briefly in the mid 1990s.

4.4 Space Observations

Purposely designed continuous helioseismic observations from space have been obtained since 1995. In that year, the Solar and Heliospheric Observatory (SOHO) was launched carrying, among other instruments, three devices with the primary goal of probing the solar interior (Domingo et al. 1995). The Global Oscillations at Low Frequency (GOLF) instrument (Gabriel et al. 1995) is an integrated-light sodium vapor resonance cell in a longitudinal magnetic field that could sample the two wings of the sodium absorption line before one of the modulators failed. The Variability of solar IRradiance and Gravity Oscillations (VIRGO) experiment is a three-channel photometer with a 12-pixel low-resolution imager (Fröhlich et al. 1995). The final SOHO instrument is the Michelson Doppler Imager (MDI), a full-disk imager of 1024 × 1024 pixels to measure the surface Doppler shift and magnetic field (Scherrer et al. 1995). However, the orbit of SOHO at the L_1 Lagrangian point restricted the telemetry bandwidth such that MDI was only able to return full-disk measurements for two or three months per year. During the remaining time, the MDI images were binned down on board SOHO to decrease the spatial resolution and fit within the allowed telemetry rate.

SOHO was supplemented by the Solar Dynamics Observatory (SDO) mission, which launched in 2010. SDO carries the Helioseismic and Magnetic Imager (HMI), which obtains full disk observations with a 4096 × 4096 pixel detector that records an image every 45 s. Since SDO is in a geostationary orbit, it can return all of the data obtained by HMI and the other instrumentation. SDO is currently the primary source of space-based helioseismology data. SOHO continues to return data from a number of instruments, including GOLF and VIRGO.

4.5 Instrumentation

Helioseismology requires that the entire solar disk be sampled simultaneously in order to remove the phase differences in the waves that would result from scanning parts of the Sun at different times. This essentially eliminates spectrograph-based instrumentation as options for helioseismology observations. Devices that can sample the disk as a unit are filters, Fabry-Perot etalons, Michelson interferometers, magneto-optical filters, and single-pixel integrated light instruments such as resonance filters.

The BiSON and IRIS instruments use single-pixel integrated light resonance filters, which provide an accurate and stable frequency reference for Doppler shift velocity measurements. In these devices, the incoming solar light is scattered from a vapor composed of an element with a resonant atomic line. Two lines are generally used in helioseismology—the 769.9-nm line of potassium, used by BiSON, and the 589.6-nm sodium line used by IRIS. The potassium line is relatively free from interference from telluric lines and solar blends, and also is at a wavelength at which photodetectors are efficient. Sodium vaporizes at a lower temperature than potassium making it easier to work with. At the heart of these instruments is a small glass cell, containing potassium or sodium vapor permeated by a longitudinal magnetic field. The magnetic field causes the laboratory line to be split into two Zeeman components thus allowing for measurements to be made of the steepest part of the solar absorption line of potassium or sodium. See Brookes et al. (1976, 1978) for a detailed discussion of the BiSON device. The details of the instrumentation have evolved over the years to take account of improvements in electronics—particularly with the advent of microcontrollers. The IRIS instrument is very similar to the BiSON device except for replacing the potassium vapor with that of sodium (Grec et al. 1991) The GOLF instrument on SOHO is also a sodium resonance filter.

Note that there are other important consequences to the choice of spectral line for helioseismology. In particular, spectral lines are formed at different heights in the solar atmosphere, and they have different sensitivities to magnetic fields. Thus using different spectral lines can reveal how waves propagate as a function of height in the solar atmosphere, and how they interact with the magnetic fields they encounter.

Atomic resonance cells are also at the heart of magneto-optical filters, but here the transmitted rather than the scattered light is observed by placing the detector along instead of perpendicular to the magnetic field (Cimino et al. 1968; Agnelli et al. 1975; Cacciani and Fofi 1978; Lin and Kuhn 1989). In this case an image can be formed and spatially resolved, as in ECHO (Tomczyk et al. 1993) and the Magneto-Optical filters at Two Heights (MOTH) instrument (Cacciani et al. 2003; Finsterle et al. 2004a,b), which exploits the separation of the height of formation of the potassium and sodium lines to study wave propagation in the solar atmosphere.

The GONG instrument (Harvey et al. 1996) is based on the Fourier Tachometer concept (Beckers and Brown 2013) and uses a polarizing Michelson interferometer and a rotating half-wave plate to sweep and interference fringe across the Ni I 676.8-nm spectral line. The MDI and HMI instruments also use Michelson interferometers, but in combination to essentially create a tunable narrow-band filter. MDI used the same Ni I 676.8-nm line as GONG, while HMI uses the Fe I 617.3-nm line. In the future, Fabry-Perot etalons may be the instrument of choice for multi-wavelength multi-height observations, as described in the section on Future Directions.

5 Future Directions

There have been a number of recent articles about future directions for helioseismology (Hill 2009; Hill et al. 2013; Toomre and Thompson 2015), focusing mainly on multi-height and multi-viewpoint observations. The main path towards multi-viewpoint observations is the Solar Orbiter mission, covered in the article by Löptien et al. (2014). Here we cover some other possible future directions.

5.1 New Observational Techniques—Multi-Wavelength Observations

Multi-wavelength observations can be thought of as coming in two flavors: those with a small separation in wavelength between observations (different points across a single spectral line) and those with a large separation (different spectral lines). As a spectral line is formed over a range of heights in the solar atmosphere and different lines have different ranges, multi-wavelength observations are generally equivalent to multi-height observations.

Observing the Doppler velocity and intensity variations simultaneously at different heights in the solar atmosphere offers some important advantages over traditional single height observations and has been touted as a future direction for helioseismology (Hill 2009; Roth 2011). The first advantage is that multi-height observations provide a means to improve the accuracy and precision of our seismic mapping of the solar interior, in particular, in the vicinity of magnetic regions (Hill 2009). The reduction in systematic errors (i.e. improved accuracy) comes through two paths. The first is a better understanding of the physics of the oscillation modes such as the effects of magnetic field on the p modes, or the origin of the correlated component of stochastic excitation, etc. (e.g. Mitra-Kraev et al. 2008; Kosovichev et al. 2009; Cally 2009; Baldner and Schou 2012; Moradi and Cally 2013; Hansen and Cally 2014). The other path is to enforce, during the data analysis, the prior knowledge that the morphology and dynamics of the interior should not depend on the spectral line used for the observation. Whereas the former has received some attention (Jain et al. 2006a,b, 2008), the latter has yet to be developed. The gain in precision comes by leveraging, during the data analysis, the fact that the coherence of the solar background signal decreases with height in the solar atmosphere (Lefebvre et al. 2008; Espagnet et al. 1995) to improve the spectral SNR (García et al. 2004). We note this gain only applies to oscillation modes with lifetimes shorter than the observing time (Appourchaux et al. 2007). Improving the SNR for modes with lifetimes that are longer than the observing time is significantly more difficult as the challenge lies in separating the oscillation signal from the convective background signal without loss of spectral resolution (which is needed for detection due to the small line widths and amplitudes of the modes): this separation requires the measurement of several independent realizations of the solar noise. Although different solar noise realizations can be obtained during the observing period by looking at different spatial areas on the Sun (García et al. 2009), the resulting degradation in the spatial window function causes any oscillation signal to be dispersed across a number of sidebands, thus most likely negating any gain in SNR from the reduction of the solar noise.

A second advantage of multi-height observations is that they allow studies of the properties of the magneto-acoustic gravity (MAG) waves that are omnipresent in the solar atmosphere. This opens the door for seismic mapping of the atmosphere (Finsterle et al. 2004a,b; Nagashima et al. 2009) and also allows us to ascertain the role the MAG waves play in the transport of convective energy through the atmosphere (Jefferies et al. 2006; Straus et al. 2008). Both studies require knowledge of the effective heights for each of the observations.

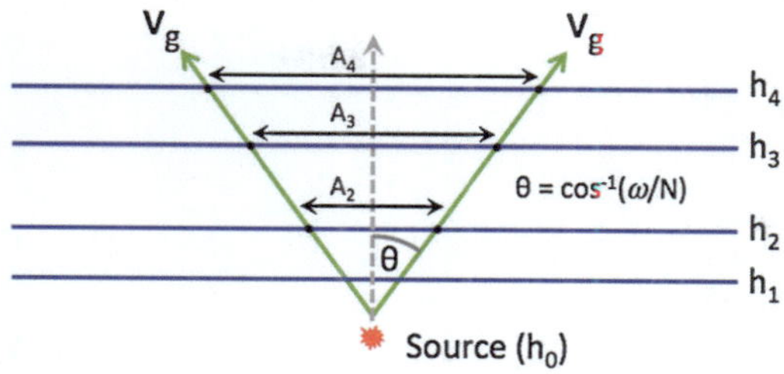

Fig. 7 This cartoon shows a cross-correlation analysis method for tracking gravity waves. The first step is to fix the radius of the annulus A_i in the data for height h_i. The next step is to frequency filter the data to admit waves in a narrow range of frequencies around ω. For gravity waves this then sets the propagation angle θ and the source height h_0 if the waves are to pass through the annulus (as gravity waves obey the relation $\omega = N\cos\theta$ where N is the Brunt-Vasaila frequency). Selecting the frequency therefore selects the radius of the second annulus for the cross-correlation, A_{i+1}. The cross-correlation between the two annuli A_i and A_{i+1} are then computed as the frequency (and thus θ and the radius of A_{i+1}) is varied. The detection of a gravity wave passing through A_i is shown as a maximum in the cross-correlation function versus frequency data. The whole analysis can be repeated for a range of different radii of A_i to build up signal-to-noise. We can gain additional confidence in the wave characterization by incorporating data from additional heights into the analysis. A plot of frequency versus $\cos\theta$ will show a linear trend if we are observing gravity waves. We note that the cross-correlation functions corresponding to gravity wave frequencies will contain information on the phase and group travel times of the waves between the observing heights (cf. Finsterle et al. 2004b)

This is readily obtained from the data via the gradient of the phase difference signal above 5 mHz in the velocity-velocity cross spectrum (Fleck and Deubner 1989). There is one small caveat, however, the heights obtained in this manner are with respect to some fiducial height (i.e. spectral line) and are not absolute. One area where multi-height observations offer significant potential to improve our understanding of the dynamics of the solar atmosphere is the study of internal (atmospheric) gravity waves (AGWs). These waves, which can transport energy and momentum over large distances and interact with large-scale flows (Mihalas and Toomre 1981), have received little attention up to now due to the difficulties associated with observing them. However, multi-height observations acquired with high-spatial resolution could enable in-depth studies of AGWs using correlation methods similar to those used in time-distance seismology of the solar interior (e.g., Fig. 7).

On a related matter, we note that recent multi-height observations acquired with SDO/HMI and MOTH (Finsterle et al. 2004b) indicate the presence of traveling waves that have the signature of internal gravity waves (opposite group and phase velocities) at low-spatial frequencies below and between the p-mode ridges. The origin and nature of these "anomalous" waves, so-called as they are unexpected by theory, is a mystery (see Fig. 8). However, it is clear that there is a likely interaction with the p modes that needs to be understood.

Currently, the majority of multi-height helioseismic studies are performed using a mixture of velocity and intensity data from disparate instruments. However, there is a price to pay for mixing observables, it complicates the interpretation of any results. Ideally, at this stage in the evolution of multi-wavelength seismology, measurements should be in Doppler velocity: the velocity spectrum has a lower level of solar noise than the spectrum of intensity measurements (Harvey 1985), which is important for the detection of the low-frequency p (and g) modes that probe the deep interior, and is generally easier to interpret. Moreover, the measurements should be made with the same (or similar) instrument(s) to minimize systematic errors.

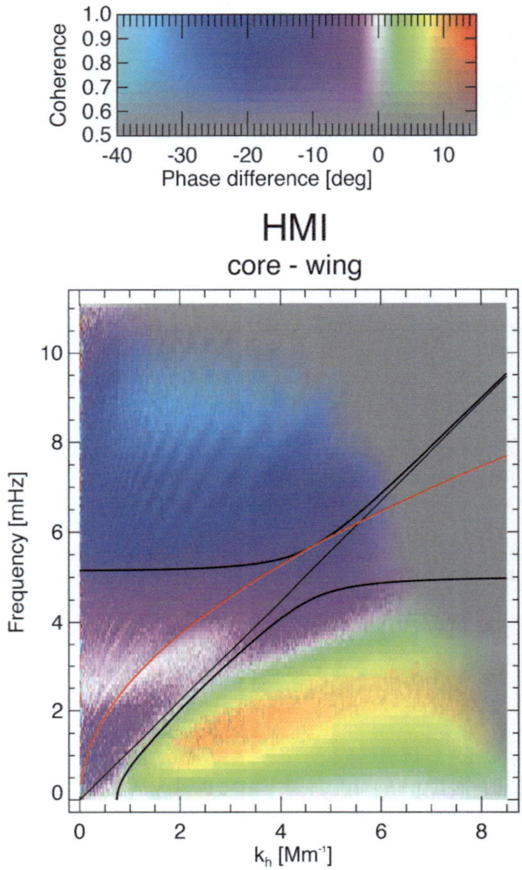

Fig. 8 The k_h—ω velocity-velocity phase difference diagrams for the HMI core-wing data (courtesy of Bernhard Fleck). The *solid black lines* indicate the propagation boundary curves for acoustic waves (top) and gravity waves (bottom). The former is a lower boundary the latter is an upper boundary. The *diagonal black line* represents the Lamb frequency (middle). The *red line* indicates the fundamental mode. The region of negative phase difference above the Lamb line, below the acoustic cut-off frequency and between the p-mode ridges indicates the presence of "anomalous" waves

Some first steps in this direction have already been taken. The next generation MOTH instruments use magneto-optical filters, which inherently have high zero-point wavelength stability, to spatially resolve the full-solar disk with 4-arcsec resolution simultaneously in velocity, intensity and line-of-sight magnetic field using the Na (589 nm), K (770 nm), Ca (422 nm) and He (1083 nm) solar spectral lines. The HELioseismic Large Region Interferometric DEvice (HELLRIDE; Staiger 2011) makes use of a Fabry-Perot based 2D spectrograph combined with a fast pre-filter exchange matrix that allows scanning through a chosen set of spectral lines in relatively short time (approximately 2 s for up to 50 scan steps in wavelength). In contrast, spatially unresolved multi-wavelength velocity measurements have been implemented using both the multiple points across a spectral line and the multiple spectral lines methods. The former approach is used by the prototype next generation GOLF instrument, which observes at multiple wavelengths across the sodium D lines. The Stellar Oscillations Network Group recently demonstrated the latter approach using an Echelle spectrograph to observe global solar oscillations with extremely low noise by monitoring the scattered daylight signal (Pallé et al. 2013). An interesting possibility for the future is a high-stability, fiber-fed, cross-dispersed Echelle spectrometer that resolves the full solar disk with high-spectral but low-spatial resolution: the wavelength stability coming through the use of a molecular magneto-optical filter (e.g., iodine).

5.2 New Observational Techniques—High-Resolution, Ground-Based, Observations

For the last twenty years helioseismology has been supported by experiments onboard a number of space missions (e.g., SOHO, SDO, *Hinode*). The likelihood of another long-term space-based helioseismology experiment in the near future, however, is small. The immediate future for new instrument concepts lies in ground-based experiments.

Ground-based observations of the Sun are always blurred due to the transmission of the solar signal through the Earth's turbulent atmosphere. The amount of blurring depends on the strength of the turbulence that in turn depends on the size of the instrument aperture (D) and the coherence length of the atmosphere (r_0) during the observations: at a good solar site $r_0 \approx 10$ cm for an observing wavelength λ of 500 nm. This is commensurate with the typical aperture size of current helioseismology instruments and allows for long integration times (10 s of seconds) without significant loss of resolution: $r_0 \approx D$ and the seeing-limited resolution ($1.22\lambda/r_0$) is close to the diffraction limited resolution ($1.22\lambda/D$). However, this will not be the case for future large aperture solar telescopes, such as the DKIST, observing at visible wavelengths. Here the turbulence strength is large ($D/r_0 \gg 1$) and the high-order adaptive optics systems that will be used with such telescopes will only partially mitigate the image blur and the resulting spatial resolution will be severely compromised. Additional methods are needed if the full resolution of the telescope is to be preserved so that we can study oscillations on the finest spatial scales.

Recent research into methods for high-resolution imaging through strong atmospheric turbulence has shown that high-quality imagery can be obtained by improving the synergy between the image acquisition and image restoration processes (Jefferies et al. 2013b) by arranging to observe the object of interest in a way that facilitates accurate estimation of both the atmospheric wave-front phase and the low-spatial frequencies of the object. One way to achieve this is to use high-cadence imaging and dual channel aperture diversity. Here the input light is split into two channels, each of which divides the full aperture into a number of sub-apertures. In one channel the aperture is divided into small rectangular patches using a micro-lens array, while in the other it is divided into concentric annuli. The signal from each sub-aperture is imaged separately onto a camera: the micro-lens data at a cadence that is shorter than the coherence time for the atmosphere, the annular data at the same cadence or less. Only the images from the largest annular sub-aperture contain spatial frequencies out to the diffraction limit of the telescope. All the other sub-aperture configurations yield imagery with lower spatial frequency content. The imagery from the micro-lens array is simply the data that are typically acquired using a correlating Shack-Hartmann wave-front sensor set up for solar adaptive optics. By taking advantage of the temporal behavior of the atmospheric wave front over short time scales (tens of milliseconds; Jefferies and Hart 2011) the micro-lens imagery can be post-processed to provide accurate estimation of the wave front phases at spatial scales smaller than the separation between micro-lenses (Jefferies et al. 2013a). The annular imagery and the micro-lens imagery both contribute to high-accuracy estimation of the low-spatial frequency of the target object during the restoration process. This, in turn, enables high-quality estimation of the high-spatial frequencies of the object (Jefferies et al. 2013b).

5.3 A New Network—SPRING

The proposal to design and build the next generation of synoptic observing instruments is currently underway (Hill et al. 2013). The Solar Physics Research Integrated Network

Group (SPRING) project, funded as a Joint Research Activity within the High-Resolution Solar Physics Network (Solarnet) by the European Union for a design study, is being currently led by KIS Freiburg. The scientific goal of SPRING is three fold. (i) To provide high quality context images of the full Sun in multiple wavelengths as a support for future generation of high resolution solar telescopes such as the DKIST and the EST. These large-aperture telescopes will only observe a small portion of the Sun (a few arcmin^2, or less than 5 % of the solar disk area) with high angular resolution, and therefore will require context images of the full Sun at decent resolution (1 arcsec) or better, at a high cadence. Here many spectral bands will be chosen to have an overlap with existing synoptic full disk images of the Sun such as Ca K, H-α imaging etc. (ii) To replace aging ground-based helioseismology instruments such as GONG with a next generation instrument capable of multi-wavelength (multi-height) velocity measurements, and operating for another twenty years or more. (iii) Finally, to enhance space weather prediction capability by making the first ever ground-based network of full-disk vector magnetographs, capable of measuring photospheric and chromospheric magnetic field vector at a cadence of few tens of minutes.

The requirement for the SPRING instrument to obtain spectral and spectropolarimetric imaging of the full disk in multiple lines puts constraints on its design. Instruments based on Michelson interferometers are optimized for a single spectral line while the design of MOTH has limited wavelength bands that can be observed. The measurement of velocity maps as well as magnetic maps of the full disk of the Sun at a cadence of one minute or better eliminates slit scanning instruments such as SOLIS, which takes about 20 minutes to complete a scan of the full solar disk in one spectral line. Even with multiple fibres and multi-slit approaches the measurement could be too cumbersome for the desired spatial resolution of about 1 arc-sec (0.5 arc-sec pixel). This leaves us with a filter-based spectro-polarimeter approach. The choice of Fabry-Perot (FP) etalons has proved to be successful for imaging spectropolarimetry of the Sun. Many designs exist such as the tandem telecentric design TESOS, the tandem collimated designs TIP, IBIS, SVM etc. The stability of the FP is an issue and for that good calibration procedures need to be devised. However, the versatility of the FP systems, rapid tunability and large wavelength range of operation, are definite advantages.

An instrument designed to obtain both vector magnetic field and high-cadence Doppler velocity measurements must necessarily compromise some aspects of the observations—either a lower signal-to-noise in the vector magnetograms, or slower cadence of the Dopplergrams. Thus one design goal of SPRING is to provide a platform that can accommodate multiple instruments that can be tailored for different scientific goals. This would allow different sets of instruments to be installed at different sites, as well as provide flexibility in funding and scheduling profiles.

5.4 Challenges of Making Observations of Long-Term Changes

We have shown that there is evidence that the near-surface but invisible layers of the Sun are changing over several timescales. To draw these conclusions required data that were available over several decades. In spite of this, there are still unanswered questions and the need for even longer data sets. We know that the underlying Hale solar activity cycle actually covers 22 years rather than the 11 years on which we see the sunspot number vary. We have not yet had consistent helioseismic measurements over a single Hale cycle. The effects that we need to measure are not large and require consistency in the data over very long periods of time. We must be confident that observed variations are genuinely due to the Sun and not a consequence of a change in the observing methods. A key attribute of any robust detection is that there are multiple, independent observations of it. We have already seen this in the case of the quasi-biennial signal. On the other hand, it is important that instrumentation changes to take advantage of new improvements in technology. These changes should be arranged so that there are always periods of overlap for inter-calibration studies. We absolutely must have long-term consistent solar observations if we are ever going to understand the solar activity that underlies the potentially harmful effects of space weather.

Acknowledgements This work utilizes data obtained by the Global Oscillation Network Group (GONG) program, managed by the National Solar Observatory, which is operated by AURA, Inc. under a cooperative agreement with the National Science Foundation. The data were acquired by instruments operated by the Big Bear Solar Observatory, High Altitude Observatory, Learmonth Solar Observatory, Udaipur Solar Observatory, Instituto de Astrofísica de Canarias, and Cerro Tololo Interamerican Observatory. MR acknowledges support from the ORIGIN project, which has received funding from the European Research Council under the European Union's Seventh Framework Program (FP/2007–2013)/ERC Grant Agreement No. 307117. YE, AMB, MR, and FH acknowledge support from the SpaceInn project, which has received funding from the European Community's Seventh Framework Programme ([FP7/2007–2013]) under grant agreement No. 312844. YE, SG, MR, and FH acknowledge support from the Solarnet project, which has received funding by the European Commission's FP7 Capacities Programme for the period April 2013–March 2017 under the Grant Agreement number 312495. YE acknowledges grant support from Science & Technology Facilities Council (UK). SJ acknowledges support from the National Science Foundation under award PLR-1341755. We acknowledge the support from ISSI Bern, for our participation in the workshop.

References

G. Agnelli, A. Cacciani, M. Fofi, The magneto-optical filter. I—preliminary observations in Na D lines. Sol. Phys. **44**, 509–518 (1975). doi:10.1007/BF00153229

J. Allison, I. Barnes, A.-M. Broomhall, W. Chaplin, G. Davies, Y. Elsworth, S. Hale, B. Jackson, B. Miller, R. New, S. Fletcher, BiSON update, in *Solar-Stellar Dynamos as Revealed by Helio- and Asteroseismology: GONG 2008/SOHO 21*, ed. by M. Dikpati, T. Arentoft, I. González Hernández, C. Lindsey, F. Hill. Astronomical Society of the Pacific Conference Series, vol. 416 (2009), pp. 227–229

E.R. Anderson, T.L. Duvall Jr., S.M. Jefferies, Modeling of solar oscillation power spectra. Astrophys. J. **364**, 699–705 (1990). doi:10.1086/169452

H.M. Antia, S. Basu, Temporal variations of the rotation rate in the solar interior. Astrophys. J. **541**, 442–448 (2000). doi:10.1086/309421

H.M. Antia, S. Basu, Revisiting the solar tachocline: average properties and temporal variations. Astrophys. J. Lett. **735**, 45 (2011). doi:10.1088/2041-8205/735/2/L45

T. Appourchaux, J. Leibacher, P. Boumier, On cross-spectrum capabilities for detecting stellar oscillation modes. Astron. Astrophys. **463**, 1211–1214 (2007). doi:10.1051/0004-6361:20065271

K.T. Bachmann, O.R. White, Observations of hysteresis in solar cycle variations among seven solar activity indicators. Sol. Phys. **150**, 347–357 (1994). doi:10.1007/BF00712896

C.S. Baldner, S. Basu, Solar cycle related changes at the base of the convection zone. Astrophys. J. **686**, 1349–1361 (2008). doi:10.1086/591514

C.S. Baldner, J. Schou, Effects of asymmetric flows in solar convection on oscillation modes. Astrophys. J. Lett. **760**, 1 (2012). doi:10.1088/2041-8205/760/1/L1

C.S. Baldner, H.M. Antia, S. Basu, T.P. Larson, Internal magnetic fields inferred from helioseismic data. Astron. Nachr. **331**, 879–882 (2010). doi:10.1002/asna.201011418

B.R. Barkstrom, G.L. Smith, The Earth radiation budget experiment—science and implementation. Rev. Geophys. **24**, 379–390 (1986). doi:10.1029/RG024i002p00379

S. Basu, A. Mandel, Does solar structure vary with solar magnetic activity? Astrophys. J. Lett. **617**, 155–158 (2004). doi:10.1086/427435

S. Basu, H.M. Antia, R.S. Bogart, Structure of the near-surface layers of the Sun: asphericity and time variation. Astrophys. J. **654**, 1146–1165 (2007). doi:10.1086/509251

S. Basu, A.-M. Broomhall, W.J. Chaplin, Y. Elsworth, Thinning of the Sun's magnetic layer: the peculiar solar minimum could have been predicted. Astrophys. J. **758**, 43 (2012). doi:10.1088/0004-637X/758/1/43

S. Basu, A.-M. Broomhall, W.J. Chaplin, Y. Elsworth, G.R. Davies, J. Schou, T.P. Larson, Comparing the internal structure of the Sun during the cycle 23 and cycle 24 minima, in *Fifty Years of Seismology of the Sun and Stars*, ed. by K. Jain, S.C. Tripathy, F. Hill, J.W. Leibacher, A.A. Pevtsov. Astronomical Society of the Pacific Conference Series, vol. 478 (2013), p. 161

J.M. Beckers, T.M. Brown, The history of the Fourier tachometer, in *Fifty Years of Seismology of the Sun and Stars*, ed. by K. Jain, S.C. Tripathy, F. Hill, J.W. Leibacher, A.A. Pevtsov. Astronomical Society of the Pacific Conference Series, vol. 478 (2013), p. 93

E.E. Benevolenskaya, Double magnetic cycle of solar activity. Sol. Phys. **161**, 1–8 (1995). doi:10.1007/BF00732080

J.R. Brookes, G.R. Isaak, H.B. van der Raay, Observation of free oscillations of the Sun. Nature **259**, 92–95 (1976). doi:10.1038/259092a0

J.R. Brookes, G.R. Isaak, H.B. van der Raay, A resonant-scattering solar spectrometer. Mon. Not. R. Astron. Soc. **185**, 1–18 (1978)

A. Broomhall, W.J. Chaplin, Y. Elsworth, S.T. Fletcher, R. New, Is the current lack of solar activity only skin deep? Astrophys. J. Lett. **700**, 162–165 (2009). doi:10.1088/0004-637X/700/2/L162

A. Broomhall, S.T. Fletcher, D. Salabert, S. Basu, W.J. Chaplin, Y. Elsworth, R.A. Garcia, A. Jimenez, R. New, Are short-term variations in solar oscillation frequencies the signature of a second solar dynamo? ArXiv e-prints (2010)

A.-M. Broomhall, W.J. Chaplin, Y. Elsworth, R. Simoniello, Quasi-biennial variations in helioseismic frequencies: can the source of the variation be localized? Mon. Not. R. Astron. Soc. **420**, 1405–1414 (2012). doi:10.1111/j.1365-2966.2011.20123.x

A.S. Brun, M.K. Browning, M. Dikpati, H. Hotta, A. Strugarek, Recent advances on solar global magnetism and variability. Space Sci. Rev. (2013). doi:10.1007/s11214-013-0028-0

A. Cacciani, M. Fofi, The magneto-optical filter. II—velocity field measurements. Sol. Phys. **59**, 179–189 (1978). doi:10.1007/BF00154941

A. Cacciani, S.M. Jefferies, W. Finsterle, P. Rapex, A. Knox, C. Giebink, V. di Martino, A new instrument for sounding the solar atmosphere, in *GONG+ 2002. Local and Global Helioseismology: The Present and Future*, ed. by H. Sawaya-Lacoste. ESA Special Publication, vol. 517 (2003), pp. 243–245

P.S. Cally, Magnetic and thermal phase shifts in the local helioseismology of sunspots. Mon. Not. R. Astron. Soc. **395**, 1309–1318 (2009). doi:10.1111/j.1365-2966.2009.14708.x

W.J. Chaplin, Y. Elsworth, R. Howe, G.R. Isaak, C.P. McLeod, B.A. Miller, H.B. van der Raay, S.J. Wheeler, R. New, BiSON performance. Sol. Phys. **168**, 1–18 (1996). doi:10.1007/BF00145821

W.J. Chaplin, Y. Elsworth, G.R. Isaak, R. Lines, C.P. McLeod, B.A. Miller, R. New, An analysis of solar p-mode frequencies extracted from BiSON data: 1991–1996. Mon. Not. R. Astron. Soc. **300**, 1077–1090 (1998)

W.J. Chaplin, Y. Elsworth, G.R. Isaak, B.A. Miller, R. New, Variations in the excitation and damping of low-l solar p modes over the solar activity cycle. Mon. Not. R. Astron. Soc. **313**, 32–42 (2000). doi:10.1046/j.1365-8711.2000.03176.x

W.J. Chaplin, T. Appourchaux, Y. Elsworth, G.R. Isaak, R. New, The phenomenology of solar-cycle-induced acoustic eigenfrequency variations: a comparative and complementary analysis of GONG, BiSON and VIRGO/LOI data. Mon. Not. R. Astron. Soc. **324**, 910–916 (2001). doi:10.1046/j.1365-8711.2001.04357.x

W.J. Chaplin, Y. Elsworth, G.R. Isaak, B.A. Miller, R. New, The solar cycle as seen by low-l p-mode frequencies: comparison with global and decomposed activity proxies. Mon. Not. R. Astron. Soc. **352**, 1102–1108 (2004). doi:10.1111/j.1365-2966.2004.07998.x

G.A. Chapman, A.M. Cookson, D.G. Preminger, Modeling total solar irradiance with San Fernando observatory ground-based photometry: comparison with ACRIM, PMOD, and RMIB composites. Sol. Phys. **283**, 295–305 (2013). doi:10.1007/s11207-013-0233-8

D.-Y. Chou, M.-T. Sun, T.-Y. Huang, S.-P. Lai, P.-J. Chi, K.-T. Ou, C.-C. Wang, J.-Y. Lu, A.-L. Chu, C.-S. Niu, T.-M. Mu, K.-R. Chen, Y.-P. Chou, A. Jimenez, M.C. Rabello-Soares, H. Chao, G. Ai, G.-P. Wang, H. Zirin, W. Marquette, J. Nenow, Taiwan oscillation network. Sol. Phys. **160**, 237–243 (1995). doi:10.1007/BF00732806

J. Christensen-Dalsgaard, G. Berthomieu, Theory of Solar Oscillations (Solar Interior and Atmosphere, 1991), pp. 401–478

M. Cimino, A. Cacciani, N. Sopranzi, An instrument to measure solar magnetic fields by an atomic-beam method. Sol. Phys. **3**, 618–622 (1968). doi:10.1007/BF00151943

F. Clette, L. Svalgaard, E.W. Cliver, J.M. Vaquero, L. Lefèvre, The new sunspot and group numbers: a full recalibration, in *IAU General Assembly*, vol. 22 (2015), p. 49591

M. Dasi-Espuig, J. Jiang, N.A. Krivova, S.K. Solanki, Modelling total solar irradiance since 1878 from simulated magnetograms. Astron. Astrophys. **570**, 23 (2014). doi:10.1051/0004-6361/201424290

G.R. Davies, W.J. Chaplin, Y. Elsworth, S.J. Hale, BiSON data preparation: a correction for differential extinction and the weighted averaging of contemporaneous data. Mon. Not. R. Astron. Soc. **441**, 3009–3017 (2014). doi:10.1093/mnras/stu803

V. Domingo, B. Fleck, A.I. Poland, The SOHO mission: an overview. Sol. Phys. **162**, 1–37 (1995). doi:10.1007/BF00733425

Y. Elsworth, A.-M. Broomhall, W. Chaplin, Helioseismology—a clear view of the interior, in *IAU Symposium*, ed. by C.H. Mandrini, D.F. Webb. IAU Symposium, vol. 286 (2012), pp. 77–87. doi:10.1017/S174392131200467X

Y. Elsworth, R. Howe, G.R. Isaak, C.P. McLeod, R. New, Variation of low-order acoustic solar oscillations over the solar cycle. Nature **345**, 322–324 (1990). doi:10.1038/345322a0

Y. Elsworth, R. Howe, G.R. Isaak, C.P. McLeod, B.A. Miller, R. New, C.C. Speake, S.J. Wheeler, Solar p-mode frequencies and their dependence on solar activity recent results from the BISON network. Astrophys. J. **434**, 801–806 (1994). doi:10.1086/174783

O. Espagnet, R. Muller, T. Roudier, N. Mein, P. Mein, Penetration of the solar granulation into the photosphere: height dependence of intensity and velocity fluctuations. Astron. Astrophys. Suppl. Ser. **109**, 79–108 (1995)

W. Finsterle, S.M. Jefferies, A. Cacciani, P. Rapex, S.W. McIntosh, Helioseismic mapping of the magnetic canopy in the solar chromosphere. Astrophys. J. Lett. **613**, 185–188 (2004a). doi:10.1086/424996

W. Finsterle, S.M. Jefferies, A. Cacciani, P. Rapex, C. Giebink, A. Knox, V. Dimartino, Seismology of the solar atmosphere. Sol. Phys. **220**, 317–331 (2004b). doi:10.1023/B:SOLA.0000031397.73790.7b

B. Fleck, F.-L. Deubner, Dynamics of the solar atmosphere. II—standing waves in the solar chromosphere. Astron. Astrophys. **224**, 245–252 (1989)

S.T. Fletcher, W.J. Chaplin, Y. Elsworth, R. New, Efficient pseudo-global fitting for helioseismic data. Astrophys. J. **694**, 144–150 (2009). doi:10.1088/0004-637X/694/1/144

S.T. Fletcher, A. Broomhall, D. Salabert, S. Basu, W.J. Chaplin, Y. Elsworth, R.A. García, R. New, A seismic signature of a second dynamo? Astrophys. J. Lett. **718**, 19–22 (2010). doi:10.1088/2041-8205/718/1/L19

J.M. Fontenla, J. Harder, W. Livingston, M. Snow, T. Woods, High-resolution solar spectral irradiance from extreme ultraviolet to far infrared. J. Geophys. Res., Atmos. **116**(D15), 20108 (2011). doi:10.1029/2011JD016032

J.M. Fontenla, E. Landi, M. Snow, T. Woods, Far- and extreme-UV solar spectral irradiance and radiance from simplified atmospheric physical models. Sol. Phys. **289**, 515–544 (2014). doi:10.1007/s11207-013-0431-4

E. Fossat, I.R.I.S.: a network for full disk helioseismology, in *Seismology of the Sun and Sun-Like Stars*, ed. by E.J. Rolfe. ESA Special Publication, vol. 286 (1988), pp. 161–162

E. Fossat, Early South Pole observations and the IRIS network: the quarter century history of a small sodium cell, in *Fifty Years of Seismology of the Sun and Stars*, ed. by K. Jain, S.C. Tripathy, F. Hill, J.W. Leibacher, A.A. Pevtsov. Astronomical Society of the Pacific Conference Series, vol. 478 (2013), p. 73

P. Foukal, A new look at solar irradiance variation. Sol. Phys. **279**, 365–381 (2012). doi:10.1007/s11207-012-0017-6

C. Fröhlich, Solar irradiance variability since 1978. Revision of the PMOD composite during solar cycle 21. Space Sci. Rev. **125**, 53–65 (2006). doi:10.1007/s11214-006-9046-5

C. Fröhlich, Total solar irradiance: what have we learned from the last three cycles and the recent minimum? Space Sci. Rev. **176**, 237–252 (2013). doi:10.1007/s11214-011-9780-1

C. Fröhlich, J. Romero, H. Roth, C. Wehrli, B.N. Andersen, T. Appourchaux, V. Domingo, U. Telljohann, G. Berthomieu, P. Delache, J. Provost, T. Toutain, D.A. Crommelynck, A. Chevalier, A. Fichot, W. Däppen, D. Gough, T. Hoeksema, A. Jiménez, M.F. Gómez, J.M. Herreros, T.R. Cortés, A.R. Jones, J.M. Pap, R.C. Willson, VIRGO: experiment for helioseismology and solar irradiance monitoring. Sol. Phys. **162**, 101–128 (1995). doi:10.1007/BF00733428

A.H. Gabriel, G. Grec, J. Charra, J. Robillot, T. Roca Cortés, S. Turck-Chièze, R. Bocchia, P. Boumier, M. Cantin, E. Cespédes, B. Cougrand, J. Crétolle, L. Damé, M. Decaudin, P. Delache, N. Denis, R. Duc, H. Dzitko, E. Fossat, J. Fourmond, R.A. García, D. Gough, C. Grivel, J.M. Herreros, H. Lagardère, J. Moalic, P.L. Pallé, N. Pétrou, M. Sanchez, R. Ulrich, H.B. van der Raay, Global oscillations at low frequency from the SOHO mission (GOLF). Sol. Phys. **162**, 61–99 (1995). doi:10.1007/BF00733427

R.A. García, S.J. Jiménez-Reyes, S. Turck-Chièze, S. Mathur, Helioseismology from the blue and red wings of the NA profile as seen by GOLF, in *SOHO 14 Helio- and Asteroseismology: Towards a Golden Future*, ed. by D. Danesy. ESA Special Publication, vol. 559 (2004), p. 432

R.A. García, S. Mathur, I. González Hernández, A. Jiménez, Spatial cross spectrum: reducing incoherent convective background of resolved helioseismic instruments, in *Solar-Stellar Dynamos as Revealed by Helio- and Asteroseismology: GONG 2008/SOHO 21*, ed. by M. Dikpati, T. Arentoft, I. González Hernández, C. Lindsey, F. Hill. Astronomical Society of the Pacific Conference Series, vol. 416 (2009), p. 297

L. Gizon, A.C. Birch, Local helioseismology. Living Rev. Sol. Phys. **2**, 6 (2005). doi:10.12942/lrsp-2005-6

I. González Hernández, R. Howe, R. Komm, F. Hill, Meridional circulation during the extended solar minimum: another component of the torsional oscillation? Astrophys. J. Lett. **713**, 16–20 (2010). doi:10.1088/2041-8205/713/1/L16
G. Grec, E. Fossat, B. Gelly, F.X. Schmider, The IRIS sodium cell instrument. Sol. Phys. **133**, 13–30 (1991). doi:10.1007/BF00149819
M. Haberreiter, S. Wedemeyer-Boehm, M. Rast, NLTE spectral synthesis based on 3D MHD convection simulations—understanding the role of the magnetic field in intensity variations, in *38th COSPAR Scientific Assembly*. COSPAR Meeting, vol. 38 (2010), p. 132
S.C. Hansen, P.S. Cally, Time-distance seismology and the solar transition region. Sol. Phys. **289**, 4425–4432 (2014). doi:10.1007/s11207-014-0600-0
J. Harder, J. Fontenla, M. Rast, P. Pilewskie, T. Woods, Measured and modeled trends in solar spectral irradiance variability in the visible and infrared, in *38th COSPAR Scientific Assembly*. COSPAR Meeting, vol. 38 (2010), p. 16
J. Harvey, R. Tucker, L. Britanik, High resolution upgrade of the GONG instruments, in *Structure and Dynamics of the Interior of the Sun and Sun-Like Stars*, ed. by S. Korzennik. ESA Special Publication, vol. 418 (1998), p. 209
J.W. Harvey, Solar/stellar seismology instruments. Trans. Int. Astron. Union **19**, 51–56 (1985)
J.W. Harvey, F. Hill, R.P. Hubbard, J.R. Kennedy, J.W. Leibacher, J.A. Pintar, P.A. Gilman, R.W. Noyes, A.M. Title, J. Toomre, R.K. Ulrich, A. Bhatnagar, J.A. Kennewell, W. Marquette, J. Patron, O. Saa, E. Yasukawa, The Global Oscillation Network Group (GONG) project. Science **272**, 1284–1286 (1996). doi:10.1126/science.272.5266.1284
J.R. Hickey, L.L. Stowe, H. Jacobowitz, P. Pellegrino, R.H. Maschhoff, F. House, T.H. Vonder Haar, Initial solar irradiance determinations from Nimbus 7 cavity radiometer measurements. Science **208**, 281–283 (1980). doi:10.1126/science.208.4441.281
F. Hill, The future of helioseismology, in *Solar-Stellar Dynamos as Revealed by Helio- and Asteroseismology: GONG 2008/SOHO 21*, ed. by M. Dikpati, T. Arentoft, I. González Hernández, C. Lindsey, F. Hill. Astronomical Society of the Pacific Conference Series, vol. 416 (2009), p. 557
F. Hill, G. Newkirk Jr., On the expected performance of a solar oscillation network. Sol. Phys. **95**, 201–219 (1985). doi:10.1007/BF00152398
F. Hill, M.J. Thompson, M. Roth, Workshop report: a new synoptic solar observing network. Space Weather **11**, 392–393 (2013). doi:10.1002/swe.20068
F. Hill, G. Fischer, J. Grier, J.W. Leibacher, H.B. Jones, P.P. Jones, R. Kupke, R.T. Stebbins, The global oscillation network group site survey. 1: data collection and analysis methods. Sol. Phys. **152**, 321–349 (1994a). doi:10.1007/BF00680443
F. Hill, G. Fischer, S. Forgach, J. Grier, J.W. Leibacher, H.P. Jones, P.B. Jones, R. Kupke, R.T. Stebbins, D.W. Clay, The global oscillation network group site survey, 2: results. Sol. Phys. **152**, 351–379 (1994b). doi:10.1007/BF00680444
F. Hill, C.S. Baldner, R.A. García, M. Roth, H. Schunker, Where to go from here: the future of helio- and astero-seismology, in *Fifty Years of Seismology of the Sun and Stars*, ed. by K. Jain, S.C. Tripathy, F. Hill, J.W. Leibacher, A.A. Pevtsov. Astronomical Society of the Pacific Conference Series, vol. 478 (2013), pp. 401–408
R. Howe, R. Komm, F. Hill, Solar cycle changes in GONG P-mode frequencies, 1995–1998. Astrophys. J. **524**, 1084–1095 (1999). doi:10.1086/307851
R. Howe, R.W. Komm, F. Hill, Localizing the solar cycle frequency shifts in global p-modes. Astrophys. J. **580**, 1172–1187 (2002). doi:10.1086/343892
R. Howe, J. Christensen-Dalsgaard, F. Hill, R.W. Komm, R.M. Larsen, J. Schou, M.J. Thompson, J. Toomre, Dynamic variations at the base of the solar convection zone. Science **287**, 2456–2460 (2000). doi:10.1126/science.287.5462.2456
R. Howe, R. Komm, F. Hill, J. Christensen-Dalsgaard, T.P. Larson, J. Schou, M.J. Thompson, J. Toomre, Rotation-rate variations at the tachocline: an update. J. Phys. Conf. Ser. **271**(1), 012075 (2011). doi:10.1088/1742-6596/271/1/012075
R. Howe, G.R. Davies, W.J. Chaplin, Y.P. Elsworth, S.J. Hale, Validation of solar-cycle changes in low-degree helioseismic parameters from the Birmingham Solar-Oscillations Network. Mon. Not. R. Astron. Soc. **454**, 4120–4141 (2015). doi:10.1093/mnras/stv2210
K. Jain, S.C. Tripathy, A. Bhatnagar, Solar cycle-induced variations in GONG P-mode frequencies and splittings. Astrophys. J. **542**, 521–527 (2000). doi:10.1086/309508
K. Jain, S.C. Tripathy, F. Hill, Solar activity phases and intermediate-degree mode frequencies. Astrophys. J. **695**, 1567–1576 (2009). doi:10.1088/0004-637X/695/2/1567
K. Jain, S.C. Tripathy, F. Hill, How peculiar was the recent extended minimum: a hint toward double minima. Astrophys. J. **739**, 6 (2011). doi:10.1088/0004-637X/739/1/6

K. Jain, F. Hill, S.C. Tripathy, H.M. Antia, J.D. Armstrong, S.M. Jefferies, E.J. Rhodes Jr., P.J. Rose, A comparison of acoustic mode parameters using multi-spectral data, in *Proceedings of SOHO 18/GONG 2006/HELAS I, Beyond the Spherical Sun*. ESA Special Publication, vol. 624 (2006a), p. 103

K. Jain, F. Hill, I. González Hernández, C.G. Toner, S.C. Tripathy, J.D. Armstrong, S.M. Jefferies, Does the inference of solar subsurface flow change with choice of the spectral line?, in *Proceedings of SOHO 18/GONG 2006/HELAS I, Beyond the Spherical Sun*. ESA Special Publication, vol. 624 (2006b), p. 127

K. Jain, F. Hill, S.C. Tripathy, I. González-Hernández, J.D. Armstrong, S.M. Jefferies, E.J. Rhodes Jr., P. Rose, Multi-spectral analysis of acoustic mode characteristics in active regions, in *Subsurface and Atmospheric Influences on Solar Activity*, ed. by R. Howe, R.W. Komm, K.S. Balasubramaniam, G.J.D. Petrie. Astronomical Society of the Pacific Conference Series, vol. 383 (2008), p. 389

K. Jain, S. Tripathy, F. Hill, T. Larson, Solar oscillations in cycle 24 ascending. J. Phys. Conf. Ser. **440**(1), 012023 (2013). doi:10.1088/1742-6596/440/1/012023

R. Jain, S.C. Tripathy, F.T. Watson, L. Fletcher, K. Jain, F. Hill, Variation of solar oscillation frequencies in solar cycle 23 and their relation to sunspot area and number. Astron. Astrophys. **545**, 73 (2012). doi:10.1051/0004-6361/201219876

S.M. Jefferies, M. Hart, Deconvolution from wave front sensing using the frozen flow hypothesis. Opt. Express **19**, 1975 (2011). doi:10.1364/OE.19.001975

S.M. Jefferies, M.A. Pomerantz, T.L. Duvall Jr., J.W. Harvey, D.B. Jaksha, Helioseismology from the South Pole: comparison of 1987 and 1981 results, in *Seismology of the Sun and Sun-Like Stars*, ed. by E.J. Rolfe. ESA Special Publication, vol. 286 (1988), pp. 279–284

S.M. Jefferies, S.W. McIntosh, J.D. Armstrong, T.J. Bogdan, A. Cacciani, B. Fleck, Magnetoacoustic portals and the basal heating of the solar chromosphere. Astrophys. J. Lett. **648**, 151–155 (2006). doi:10.1086/508165

S.M. Jefferies, D.A. Hope, M. Hart, J.G. Nagy, High-resolution imaging through strong atmospheric turbulence, in *Society of Photo-Optical Instrumentation Engineers (SPIE) Conference Series*. Society of Photo-Optical Instrumentation Engineers (SPIE) Conference Series, vol. 8890 (2013a), p. 1. doi:10.1117/12.2028193

S.M. Jefferies, D. Hope, M. Hart, J. Nagy, High-resolution imaging through strong atmospheric turbulence and over wide fields of view, in *Advanced Maui Optical and Space Surveillance Technologies Conference* (2013b), p. 52

S.J. Jiménez-Reyes, T. Corbard, P.L. Pallé, T. Roca Cortés, S. Tomczyk, Analysis of the solar cycle and core rotation using 15 years of Mark-I observations: 1984–1999 . I. The solar cycle. Astron. Astrophys. **379**, 622–633 (2001). doi:10.1051/0004-6361:20011374

S.J. Jiménez-Reyes, R.A. García, A. Jiménez, W.J. Chaplin, Excitation and damping of low-degree solar p-modes during activity cycle 23: analysis of GOLF and VIRGO Sun photometer data. Astrophys. J. **595**, 446–457 (2003). doi:10.1086/377304

S.J. Jiménez-Reyes, W.J. Chaplin, Y. Elsworth, R.A. García, Tracing the "acoustic" solar cycle: a direct comparison of BiSON and GOLF low-l p-mode variations. Astrophys. J. **604**, 969–976 (2004). doi:10.1086/381936

S.J. Jiménez-Reyes, W.J. Chaplin, Y. Elsworth, R.A. García, R. Howe, H. Socas-Navarro, T. Toutain, On the variation of the peak asymmetry of low-l solar p modes. Astrophys. J. **654**, 1135–1145 (2007). doi:10.1086/509700

H.P. Jones, G.A. Chapman, K.L. Harvey, J.M. Pap, D.G. Preminger, M.J. Turmon, S.R. Walton, A comparison of feature classification methods for modeling solar irradiance variation. Sol. Phys. **248**, 323–337 (2008). doi:10.1007/s11207-007-9069-4

S.L. Keil, S.P. Worden, Variations in the solar calcium K line 1976–1982. Astrophys. J. **276**, 766–781 (1984). doi:10.1086/161663

C. Kiess, R. Rezaei, W. Schmidt, Properties of sunspot umbrae observed in cycle 24. Astron. Astrophys. **565**, 52 (2014). doi:10.1051/0004-6361/201321119

R. Komm, R. Howe, F. Hill, Localizing width and energy of solar global p-modes. Astrophys. J. **572**, 663–673 (2002). doi:10.1086/340196

R.W. Komm, R. Howe, F. Hill, Solar-cycle changes in GONG P-mode widths and amplitudes 1995–1998. Astrophys. J. **531**, 1094–1108 (2000a). doi:10.1086/308518

R.W. Komm, R. Howe, F. Hill, Width and energy of solar p-modes observed by global oscillation network group. Astrophys. J. **543**, 472–485 (2000b). doi:10.1086/317101

R. Komm, R. Howe, I. González Hernández, F. Hill, Solar-cycle variation of subsurface zonal flow. Sol. Phys. **289**, 3435–3455 (2014). doi:10.1007/s11207-014-0490-1

R. Komm, I. González Hernández, R. Howe, F. Hill, Subsurface zonal and meridional flow derived from GONG and SDO/HMI: a comparison of systematics. Sol. Phys. **290**, 1081–1104 (2015). doi:10.1007/s11207-015-0663-6

A. Kosovichev, J. Zhao, T. Sekii, K. Nagashima, U. Mitra-Kraev, New results of high-resolution helioseismology from Hinode, in *Solar-Stellar Dynamos as Revealed by Helio- and Asteroseismology: GONG 2008/SOHO 21*, ed. by M. Dikpati, T. Arentoft, I. González Hernández, C. Lindsey, F. Hill. Astronomical Society of the Pacific Conference Series, vol. 416 (2009), p. 41

H.L. Kyle, D.V. Hoyt, J.R. Hickey, A review of the Nimbus-7 ERB solar dataset. Sol. Phys. **152**, 9–12 (1994). doi:10.1007/BF01473176

J.L. Lean, J. Cook, W. Marquette, A. Johannesson, Magnetic sources of the solar irradiance cycle. Astrophys. J. **492**, 390–401 (1998). doi:10.1086/305015

S. Lefebvre, R.A. García, S.J. Jiménez-Reyes, S. Turck-Chièze, S. Mathur, Variations of the solar granulation motions with height using the GOLF/SoHO experiment. Astron. Astrophys. **490**, 1143–1149 (2008). doi:10.1051/0004-6361:200810344

K.G. Libbrecht, M.F. Woodard, Solar-cycle effects on solar oscillation frequencies. Nature **345**, 779–782 (1990). doi:10.1038/345779a0

H. Lin, J.R. Kuhn, An imaging, tunable magneto-optical filter. Sol. Phys. **122**, 365–380 (1989). doi:10.1007/BF00913002

W. Livingston, M.J. Penn, L. Svalgaard, Decreasing sunspot magnetic fields explain unique 10.7 cm radio flux. Astrophys. J. Lett. **757**, 8 (2012). doi:10.1088/2041-8205/757/1/L8

W. Livingston, J. Harvey, O. Malanushenko, L. Webster, Sunspot magnetic fields measured up to 6000 Gauss, in *IAU Joint Discussion*. IAU Joint Discussion, vol. 3 (2006), p. 54

B. Löptien, A.C. Birch, L. Gizon, J. Schou, T. Appourchaux, J. Blanco Rodríguez, P.S. Cally, C. Dominguez-Tagle, A. Gandorfer, F. Hill, J. Hirzberger, P.H. Scherrer, S.K. Solanki, Helioseismology with Solar Orbiter. Space Sci. Rev. (2014). doi:10.1007/s11214-014-0065-3

B.W. Mihalas, J. Toomre, Internal gravity waves in the solar atmosphere. I—adiabatic waves in the chromosphere. Astrophys. J. **249**, 349–371 (1981). doi:10.1086/159293

U. Mitra-Kraev, A.G. Kosovichev, T. Sekii, Properties of high-degree oscillation modes of the Sun observed with Hinode/SOT. Astron. Astrophys. **481**, 1–4 (2008). doi:10.1051/0004-6361:20079042

H. Moradi, P.S. Cally, Helioseismic implications of mode conversion, in *Fifty Years of Seismology of the Sun and Stars*, ed. by K. Jain, S.C. Tripathy, F. Hill, J.W. Leibacher, A.A. Pevtsov. Astronomical Society of the Pacific Conference Series, vol. 478 (2013), p. 263

K. Mursula, B. Zieger, J.H. Vilppola, Mid-term quasi-periodicities in geomagnetic activity during the last 15 solar cycles: connection to solar dynamo strength to the memory of Karolen I. Paularena (1957–2001). Sol. Phys. **212**, 201–207 (2003). doi:10.1023/A:1022980029618

K. Nagashima, T. Sekii, A.G. Kosovichev, J. Zhao, T.D. Tarbell, Helioseismic signature of chromospheric downflows in acoustic travel-time measurements from Hinode. Astrophys. J. Lett. **694**, 115–119 (2009). doi:10.1088/0004-637X/694/2/L115

P.L. Pallé, C. Régulo, T. Roca Cortés, Solar cycle induced variations of the low L solar acoustic spectrum. Astron. Astrophys. **224**, 253–258 (1989)

P.L. Pallé, F. Grundahl, A. Triviño Hage, J. Christensen-Dalsgaard, S. Frandsen, R.A. García, K. Uytterhoeven, M.F. Andersen, P.K. Rasmussen, A.N. Sørensen, H. Kjeldsen, P. Spano, H. Nilsson, H. Hartman, U.G. Jørgensen, J. Skottfelt, K. Harpsøe, M.I. Andersen, Observations of the radial velocity of the Sun as measured with the novel SONG spectrograph: results from a 1-week campaign. J. Phys. Conf. Ser. **440**(1), 012051 (2013). doi:10.1088/1742-6596/440/1/012051

J.M. Pap, H. Jones, D. Parker, G. Chapman, L. Floyd, Solar irradiance variations related to intensity and magnetic flux of solar features, in *38th COSPAR Scientific Assembly*. COSPAR Meeting, vol. 38 (2010), p. 1783

W.D. Pesnell, Predictions of solar cycle 24. Sol. Phys. **252**, 209–220 (2008). doi:10.1007/s11207-008-9252-2

G.J.D. Petrie, K. Petrovay, K. Schatten, Solar polar fields and the 22-year activity cycle: observations and models. Space Sci. Rev. **186**, 325–357 (2014). doi:10.1007/s11214-014-0064-4

M.C. Rabello-Soares, Characterization of solar-cycle induced frequency shift of medium- and high-degree acoustic modes. J. Phys. Conf. Ser. **271**(1), 012026 (2011). doi:10.1088/1742-6596/271/1/012026

M.C. Rabello-Soares, S.G. Korzennik, J. Schou, Variations of the solar acoustic high-degree mode frequencies over solar cycle 23. Adv. Space Res. **41**, 861–867 (2008). doi:10.1016/j.asr.2007.03.014

E.J. Rhodes Jr., J. Reiter, J. Schou, Solar cycle variability of high-frequency and high-degree p-mode oscillation frequencies, in *From Solar Min to Max: Half a Solar Cycle with SOHO*, ed. by A. Wilson. ESA Special Publication, vol. 508 (2002), pp. 37–40

E.J. Rhodes, J. Reiter, J. Schou, T. Larson, P. Scherrer, J. Brooks, P. McFaddin, B. Miller, J. Rodriguez, J. Yoo, Temporal changes in the frequencies of the solar p-mode oscillations during solar cycle 23, in *IAU Symposium*, ed. by D. Prasad Choudhary, K.G. Strassmeier. IAU Symposium, vol. 273 (2011), pp. 389–393. doi:10.1017/S1743921311015614

M. Roth, The future of helioseismology. Contrib. Astron. Obs. Skaln. Pleso **41**, 156–166 (2011)

D. Salabert, R.A. García, S. Turck-Chièze, Seismic sensitivity to sub-surface solar activity from 18 yr of GOLF/SoHO observations. Astron. Astrophys. **578**, 137 (2015). doi:10.1051/0004-6361/201425236

D. Salabert, R.A. García, P.L. Pallé, S.J. Jiménez-Reyes, The onset of solar cycle 24. What global acoustic modes are telling us. Astron. Astrophys. **504**, 1–4 (2009). doi:10.1051/0004-6361/200912736

J.D. Scargle, S.L. Keil, S.P. Worden, Solar cycle variability and surface differential rotation from Ca II K-line time series data. Astrophys. J. **771**, 33 (2013). doi:10.1088/0004-637X/771/1/33

P.H. Scherrer, R.S. Bogart, R.I. Bush, J.T. Hoeksema, A.G. Kosovichev, J. Schou, W. Rosenberg, L. Springer, T.D. Tarbell, A. Title, C.J. Wolfson, I. Zayer, MDI Engineering Team, The solar oscillations investigation—Michelson Doppler Imager. Sol. Phys. **162**, 129–188 (1995). doi:10.1007/BF00733429

R. Simoniello, W. Finsterle, D. Salabert, R.A. García, S. Turck-Chièze, A. Jiménez, M. Roth, The quasi-biennial periodicity (QBP) in velocity and intensity helioseismic observations. The seismic QBP over solar cycle 23. Astron. Astrophys. **539**, 135 (2012). doi:10.1051/0004-6361/201118057

R. Simoniello, K. Jain, S.C. Tripathy, S. Turck-Chièze, C. Baldner, W. Finsterle, F. Hill, M. Roth, The Quasi-biennial Periodicity as a window on the solar magnetic dynamo configuration. Astrophys. J. **765**, 100 (2013). doi:10.1088/0004-637X/765/2/100

S.K. Solanki, I.G. Usoskin, B. Kromer, M. Schüssler, J. Beer, Unusual activity of the Sun during recent decades compared to the previous 11,000 years. Nature **431**, 1084–1087 (2004). doi:10.1038/nature02995

J. Staiger, A multiline spectrometer for the analysis of solar atmospheric oscillations and flows at the VTT, Tenerife. Astron. Astrophys. **535**, 83 (2011). doi:10.1051/0004-6361/201117500

T. Straus, B. Fleck, S.M. Jefferies, G. Cauzzi, S.W. McIntosh, K. Reardon, G. Severino, M. Steffen, The energy flux of internal gravity waves in the lower solar atmosphere. Astrophys. J. Lett. **681**, 125–128 (2008). doi:10.1086/590495

L. Svalgaard, How well do we know the sunspot number?, in *IAU Symposium*, ed. by C.H. Mandrini, D.F. Webb. IAU Symposium, vol. 286 (2012), pp. 27–33. doi:10.1017/S1743921312004590

L. Svalgaard, Y. Kamide, Asymmetric solar polar field reversals. Astrophys. J. **763**, 23 (2013). doi:10.1088/0004-637X/763/1/23

L. Svalgaard, E.W. Cliver, Y. Kamide, Sunspot cycle 24: smallest cycle in 100 years? Geophys. Res. Lett. **32**, 1104 (2005). doi:10.1029/2004GL021664

M.J. Thompson, Evidence for a thin perturbative layer near the base of the solar convection zone, in *Seismology of the Sun and Sun-Like Stars*, ed. by E.J. Rolfe. ESA Special Publication, vol. 286 (1988), pp. 321–324

S. Tomczyk, A. Cacciani, S.A. Veitzer, LOWL—an instrument to observe low-degree solar oscillations, in *GONG 1992. Seismic Investigation of the Sun and Stars*, ed. by T.M. Brown. Astronomical Society of the Pacific Conference Series, vol. 42 (1993), p. 469

S. Tomczyk, K. Streander, G. Card, D. Elmore, H. Hull, A. Cacciani, An instrument to observe low-degree solar oscillations. Sol. Phys. **159**, 1–21 (1995). doi:10.1007/BF00733027

J. Toomre, M.J. Thompson, Prospects and challenges for helioseismology. Space Sci. Rev. (2015). doi:10.1007/s11214-015-0147-x

S.C. Tripathy, K. Jain, F. Hill, Acoustic mode frequencies of the Sun during the minimum phase between solar cycles 23 and 24. Sol. Phys. **282**, 1–13 (2013). doi:10.1007/s11207-012-0130-6

S.C. Tripathy, K. Jain, F. Hill, Variations in high degree acoustic mode frequencies of the Sun during solar cycle 23 and 24. Astrophys. J. **812** (2015), article id. 20. doi:10.1088/0004-637X/812/1/20

S.C. Tripathy, B. Kumar, K. Jain, A. Bhatnagar, Analysis of hysteresis effect in p-mode frequency shifts and solar activity indices. Sol. Phys. **200**, 3–10 (2001). doi:10.1023/A:1010318428454

S.C. Tripathy, F. Hill, K. Jain, J.W. Leibacher, Solar oscillation frequency changes on time scales of nine days. Sol. Phys. **243**, 105–120 (2007). doi:10.1007/s11207-007-9000-z

S.C. Tripathy, K. Jain, F. Hill, J.W. Leibacher, Unusual trends in solar P-mode frequencies during the current extended minimum. Astrophys. J. Lett. **711**, 84–88 (2010). doi:10.1088/2041-8205/711/2/L84

R.K. Ulrich, D. Parker, L. Bertello, J. Boyden, Modeling total solar irradiance variations using automated classification software on Mount Wilson data. Sol. Phys. **261**, 11–34 (2010). doi:10.1007/s11207-009-9460-4

L. Upton, D.H. Hathaway, Predicting the Sun's polar magnetic fields with a surface flux transport model. Astrophys. J. **780**, 5 (2014). doi:10.1088/0004-637X/780/1/5

J.F. Valdés-Galicia, V.M. Velasco, Variations of mid-term periodicities in solar activity physical phenomena. Adv. Space Res. **41**, 297–305 (2008). doi:10.1016/j.asr.2007.02.012

V.M. Velasco Herrera, B. Mendoza, G. Velasco Herrera, Reconstruction and prediction of the total solar irradiance: from the medieval warm period to the 21st century. New Astron. **34**, 221–233 (2015). doi:10.1016/j.newast.2014.07.009

G.A. Verner, W.J. Chaplin, Y. Elsworth, Search for solar cycle changes in the signature of rapid variation in BiSON data. Mon. Not. R. Astron. Soc. **351**, 311–316 (2004). doi:10.1111/j.1365-2966.2004.07788.x

G.A. Verner, W.J. Chaplin, Y. Elsworth, BiSON data show change in solar structure with magnetic activity. Astrophys. J. Lett. **640**, 95–98 (2006). doi:10.1086/503101

Y.-M. Wang, J.L. Lean, N.R. Sheeley Jr., Modeling the Sun's magnetic field and irradiance since 1713. Astrophys. J. **625**, 522–538 (2005). doi:10.1086/429689

F.T. Watson, M.J. Penn, W. Livingston, A multi-instrument analysis of sunspot umbrae. Astrophys. J. **787**, 22 (2014). doi:10.1088/0004-637X/787/1/22

R.C. Willson, S. Gulkis, M. Janssen, H.S. Hudson, G.A. Chapman, Observations of solar irradiance variability. Science **211**, 700–702 (1981). doi:10.1126/science.211.4483.700

M.F. Woodard, R.W. Noyes, Change of solar oscillation eigenfrequencies with the solar cycle. Nature **318**, 449 (1985)

T.N. Woods, G.J. Rottman, J.W. Harder, G.M. Lawrence, W.E. McClintock, G.A. Kopp, C. Pankratz, Overview of the EOS SORCE mission, in *Earth Observing Systems V*, ed. by W.L. Barnes. Society of Photo-Optical Instrumentation Engineers (SPIE) Conference Series, vol. 4135 (2000), pp. 192–203. doi:10.1117/12.494229

K.L. Yeo, N.A. Krivova, S.K. Solanki, Solar cycle variation in solar irradiance. Space Sci. Rev. **186**, 137–167 (2014a). doi:10.1007/s11214-014-0061-7

K.L. Yeo, N.A. Krivova, S.K. Solanki, K.H. Glassmeier, Reconstruction of total and spectral solar irradiance from 1974 to 2013 based on KPVT, SoHO/MDI, and SDO/HMI observations. Astron. Astrophys. **570**, 85 (2014b). doi:10.1051/0004-6361/201423628

Space Sci Rev (2015) 196:167–199
DOI 10.1007/s11214-013-0023-5

Sub-photosphere to Solar Atmosphere Connection

Rudolf Komm · Ineke De Moortel · Yuhong Fan · Stathis Ilonidis · Oskar Steiner

Received: 24 May 2013 / Accepted: 15 September 2013 / Published online: 8 October 2013

Abstract Magnetic fields extend from the solar interior through the atmosphere. The formation and evolution of active regions can be studied by measuring subsurface flows with local helioseismology. The emergence of magnetic flux from the solar convection zone is associated with acoustic perturbation signatures. In near-surface layers, the average dynamics can be determined for emerging regions. MHD simulations of the emergence of a twisted flux tube show how magnetic twist and free energy are transported from the interior into the corona and the dynamic signatures associated with such transport in the photospheric and sub-photospheric layers. The subsurface twisted flux tube does not emerge into the corona as a whole in emerging active regions. Shear flows at the polarity inversion line and coherent vortical motions in the subsurface flux tubes are the major means by which twist is transported into the corona, leading to the formation of sigmoid-shaped coronal magnetic fields capable of driving solar eruptions. The transport of twist can be followed from the interior by using the kinetic helicity of subsurface flows as a proxy of magnetic helicity; this quantity holds great promise for improving the understanding of eruptive phenomena.

R. Komm (✉)
National Solar Observatory, Tucson, AZ, USA
e-mail: komm@noao.edu

I. De Moortel
School of Mathematics & Statistics, University of St. Andrews, North Haugh, St. Andrews, KY16 9SS, UK
e-mail: ineke@mcs.st-and.ac.uk

Y. Fan
High Altitude Observatory, Boulder, CO, USA
e-mail: yfan@hao.ucar.edu

S. Ilonidis
W.W. Hansen Experimental Physics Laboratory, Stanford University, Stanford, CA 94305-4085, USA
e-mail: ilonidis@stanford.edu

O. Steiner
Kiepenheuer-Institut für Sonnenphysik, Freiburg, Germany
e-mail: steiner@kis.uni-freiburg.de

Waves are not only vital for studying the link between the solar interior and the surface but for linking the photosphere with the corona as well. Acoustic waves that propagate from the surface into the magnetically structured, dynamic atmosphere undergo mode conversion and refraction. These effects enable atmospheric seismology to determine the topography of magnetic canopies in the solar atmosphere. Inclined magnetic fields lower the cut-off frequency so that low frequency waves can leak into the outer atmosphere. Recent high resolution, high cadence observations of waves and oscillations in the solar atmosphere, have lead to a renewed interest in the potential role of waves as a heating mechanism. In light of their potential contribution to the heating of the solar atmosphere, some of the recent observations of waves and oscillations and ongoing modelling efforts are reviewed.

Keywords Helioseismology · Solar magnetic fields · Solar interior · Solar activity · Magnetohydrodynamics in astrophysics

1 Introduction

The solar magnetic fields provide a connection between the sub-photospheric layers and the solar atmosphere. Twisted magnetic flux tubes emerge from the solar interior into the solar atmosphere and corona, building up twisted/sheared coronal magnetic fields with stored free magnetic energy to drive solar eruptions. Continuous buffeting by convection at the footpoints of the atmospheric magnetic field lines anchored in the turbulent solar convection zone generates waves and oscillations, which are sources of the mechanical energy for heating the corona and accelerating the solar wind. Local helioseismology provides an important means to infer subsurface conditions of emerging active region flux tubes and to determine the subsurface dynamics that drives the coronal magnetic field evolution. At the same time, the propagation of acoustic waves that are used as diagnostics by helioseismology are affected by the presence of strong magnetic fields.

We discuss the dynamics of subsurface layers determined by local helioseismology and their role in solar magnetic activity, such as eruptive events and flux emergence. We then discuss examples of MHD simulations of magnetic flux emergence from the top layers of the convection zone into the solar atmosphere and corona. We review recent observations of waves and oscillations in the solar atmosphere, generated by the buffeting of the magnetic field by subsurface flows and motions, and their potential contribution to heating of the solar atmosphere. We finally discuss how magnetic fields at the solar surface affect acoustic waves that travel from the convection zone outward into the solar atmosphere.

2 Subsurface Flows and Dynamics and Their Role in Solar Magnetic Activity

Local helioseismology has made it possible to measure flows below the solar surface at individual locations of latitude and longitude on time scales of one day or less and thus to study the dynamics of subsurface flows associated with active regions (Gizon and Birch 2005, review). Subsurface flows are typically determined on spatial scales comparable to the size of active regions and time scales of one day by ring-diagram analysis (Hill 1988) and on spatial scales comparable to sunspots and time scales of eight hours by time-distance analysis (Duvall et al. 1993) and by helioseismic holography (Lindsey and Braun 1990). Main data sources are space-based SOHO/MDI (Scherrer et al. 1995), which has been superseded by SDO/HMI (Scherrer et al. 2012) and ground-based GONG (Harvey et al. 1996). The most

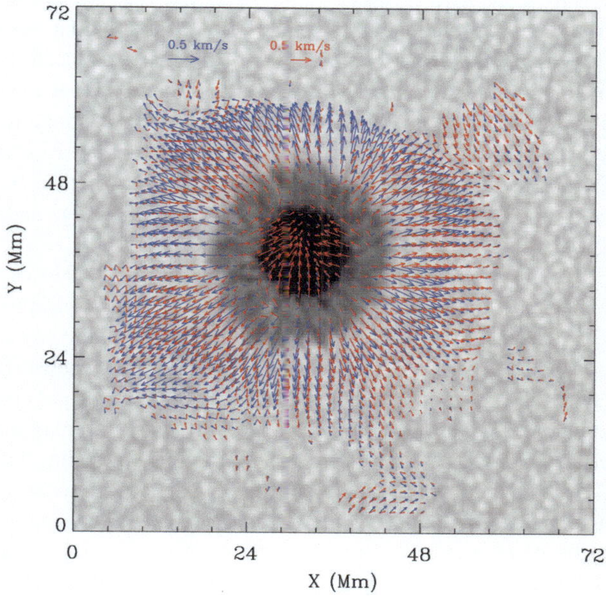

Fig. 1 Horizontal flows in AR 11084 in the photosphere (*blue arrows*) and in the −0.5 Mm layer (*red arrows*) overplotted on a continuum intensity image taken at 2 July 2010 04:00 UT. The photospheric flow is an average of eight velocity maps at a cadence of one hour from 00:00 UT to 08:00 UT derived by DAVE4VM using HMI vector magnetograms. The flow in the −0.5 Mm layer is derived by a time-distance helioseismology method applied to eight-hour HMI Dopplergrams observed from 2 July 2010 00:00 UT to 08:00 UT. Only velocities at locations where the horizontal velocity in the photosphere is greater than 0.07 km s^{-1} are plotted. Adapted from Liu et al. (2012)

detailed study of the subsurface dynamics and structure of a sunspot has been performed by Gizon et al. (2009) and Moradi et al. (2010) and the helioseismic studies of sunspots were reviewed by Kosovichev (2012). Here, we focus on the interaction of subsurface flows with magnetic activity and on the relationship between acoustic perturbation signatures and flux emergence.

2.1 Subsurface Flows of Active Regions

The flow estimates from local helioseismology are overall robust. As Fig. 1 shows, horizontal flow patterns at the surface are similar to those slightly below (Liu et al. 2012). The ring-diagram and time-distance techniques lead to similar flows when their differences, such as spatial resolution, are taken into account (Hindman et al. 2004). As a note of caution, we point out that strong magnetic fields affect acoustic modes (Sect. 5.1) and could bias the results, especially those with high spatial resolution. The subsurface flows associated with active regions show in- and outflows depending on spatial scales and depth. On scales comparable to sunspots, the most prominent feature is an outflow, which relates to the moat flow seen at the solar surface (Gizon et al. 2000; Braun and Lindsey 2003; Zhao et al. 2001, 2010; Hindman et al. 2009). On scales comparable to active regions, there are convergent horizontal flows close to the surface and divergent flows at depths greater than about 10 Mm (Gizon et al. 2000; Haber et al. 2004; Braun et al. 2004; Zhao and Kosovichev 2004; Komm et al. 2004). These inflows have been interpreted as downflows, while outflows have been interpreted as upflows.

Surface measurements have shown that the solar rotation rate derived from active regions is, in general, greater than that of quiet regions (Howard 1996, review). Helioseismic observations could help to determine whether this reflects the rotation rate at some "anchoring" depth in the solar convection zone or whether it is due to an interaction between magnetic buoyancy and drag on flux tubes coupled with the Coriolis force acting on rising flux tubes (for reviews, see, Fan 2009a; Gizon and Birch 2005). It has been established that the residual rotation rate of active regions is faster than that of quiet regions in the near-surface layers of the convection zone. Zhao et al. (2004) found with a time-distance analysis of MDI data that the residual rotational velocity of magnetic elements depends nearly linearly on their magnetic-field strength. A faster residual rotation of active regions has also been found in shallow layers within about 3 Mm of the surface by applying helioseismic-holography analysis to MDI data (Braun et al. 2004) and within about 16 Mm with ring-diagram analysis of GONG data (Komm et al. 2009b). The question whether this is different for different magnetic tracers and whether this depends on evolution, age and size of the magnetic features is largely unexplored by local helioseismology.

There are other cases where the properties of subsurface flows can be important for the behavior of magnetic activity observed in the solar atmosphere. We mention four examples. (1) There are some active regions that do not follow Joy's law. It is conceivable that the non-Joy behavior is due to a particular twist of the subsurface flows that would cause the center of gravity of the leading polarity in bipolar sunspot groups to lie farther away from the equator than that of the following polarity. A preliminary analysis by González Hernández et al. (2013) shows that the vertical vorticity of a region that does not follow Joy's law has the same sign as expected from the hemispheric rule (Sect. 2.2). Whether this region has an excessive amount of vorticity compared to other regions of its size and strength needs to be determined. (2) Some sunspots are observed to rotate around their umbral centers (Brown et al. 2003; Gopasyuk and Kosovichev 2011). The associated subsurface motions show the same sense of rotation (Zhao and Kosovichev 2003) and change significantly with time compared to regions with non-rotating sunspots (Jain et al. 2012). (3) The interaction between flows and magnetic fields plays an important role in the formation and maintenance of filaments. Hindman et al. (2006) found that the magnetic neutral line of a filament tends to lie along the centers of large convection cells and that there is a significant shear flow along the neutral line but no systematic convergent flow. (4) The horizontal subsurface flows associated with two long-lived coronal holes are weakly divergent (upflows) with small cyclonic vorticity (Komm et al. 2013b). These flows are thus similar to subsurface flows of quiet regions with regard to the vertical flows and similar to flows of active regions with regard to vorticity.

2.2 Kinetic Helicity of Subsurface Flows

Helicity plays a role in a broad range of solar phenomena from the dynamo to the eruption of flares and coronal mass ejections (CME) (e.g. Rüdiger et al. 2001; Kazachenko et al. 2012). Magnetic helicity arises from twisting and kinking of an individual flux tube and by linking and knotting of flux tubes (Berger and Field 1984). Two commonly used magnetic helicity proxies are the vertical component of the current helicity density, h_c, defined as the scalar product of the vertical current density, J_z, and the vertical magnetic field, B_z, (Bao and Zhang 1998; Hagino and Sakurai 2005) and the mean twist parameter, α_z, defined as the ratio of J_z and B_z (Pevtsov et al. 1994, 1995). The kinetic helicity density, h_k, is a scalar that corresponds to changing orientation in space of fluid particles and is thus associated with mixing and turbulence (Moffatt and Tsinober 1992). It is defined as the scalar product of the velocity and vorticity vector with the vorticity defined as the curl of the velocity vector. The vertical contribution to kinetic helicity, h_z, defined as the product of the vertical velocity and the curl of the horizontal velocity components, is then the equivalent of the vertical component of the current helicity density, h_c.

There are two effects that relate subsurface helicity to magnetic helicity. The α-effect applies to mean fields and creates helicity in the mean field with right-handed eddies leading to left-handed fields (Krause and Raedler 1980), while the Σ-effect applies to flux tubes and creates twist with right-handed eddies leading to right-handed twist (Longcope et al. 1998). The current helicity, h_c, of active regions follows, on average, a hemispheric rule with positive values in the southern and negative values in the northern hemisphere (Pevtsov et al. 1995, 2001; Hagino and Sakurai 2005; Zhang 2006). It turns out that the vertical contribution to the kinetic helicity density, h_z, follows the same rule (Zhao and Kosovichev 2004; Komm et al. 2007), as expected from the Σ effect. However, this is a statistical relationship and it does not imply that this is the case for every active region (Maurya et al. 2011; Komm et al. 2013a). A positive correlation has been found between the temporal evolution of subsurface kinetic helicity and photospheric current helicity in active regions (Gao et al. 2012). It thus seems possible to use the kinetic helicity as a proxy for the magnetic helicity.

Komm et al. (2013a) searched for a signature of helicity flux from the solar interior to the photosphere and chromosphere. Active regions that show a regular pattern of super-penumbral whirls in chromospheric and coronal images are promising candidates for such studies since the sense of helicity is clearly visible. Preliminary results show that the current helicity of the regions with whirls follows the hemispheric rule, while the corresponding subsurface kinetic helicity shows the opposite behavior, which makes them "different" from other regions.

Highly non-potential magnetic fields are very probably responsible for strong eruptive phenomena such as flares and CMEs. The strength and complexity of magnetic fields are of great importance for determining the flare activity of active regions. Many parameters based on surface magnetic-field measurements have been considered as flare-forecasting criteria but show only modest improvements over climatological or uniform forecasts (for a quantitative comparison, see Barnes and Leka 2008). The properties of subsurface flows might provide additional insights into these eruptive events. Large values of subsurface kinetic helicity and vorticity occur mainly in flare-productive active regions (Mason et al. 2006; Komm and Hill 2009). The vorticity and helicity of subsurface flows have the ability to distinguish between flaring and non-flaring active regions, as demonstrated in a statistical analysis using linear discriminant approach (Komm et al. 2011a). Reinard et al. (2010) developed a parameter that captures the variation of kinetic helicity with depth and time, the so-called Normalized Helicity Gradient Variance (NHGV). This parameter increases 2–3 days before a flare occurs and the NHGV values for flaring and non-flaring active regions

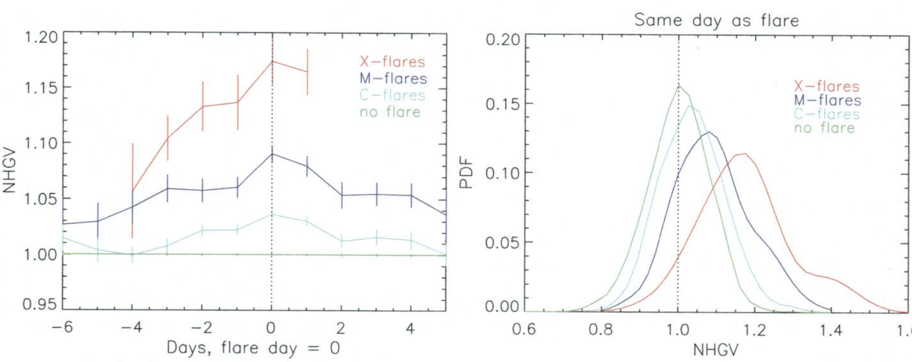

Fig. 2 *Left*: Superposed epoch analysis for the NHGV parameter of active regions associated with X-class (*red*), M-class (*blue*), and C-class flares (*cyan*). Shown in *green* is an average value for active regions that do not flare. *Right*: Probability density functions of NHGV values for active regions that produce an X-class, M-class, C-class or no flare at the same day as the flare occurrence. From Reinard et al. (2010) reproduced by permission of the AAS

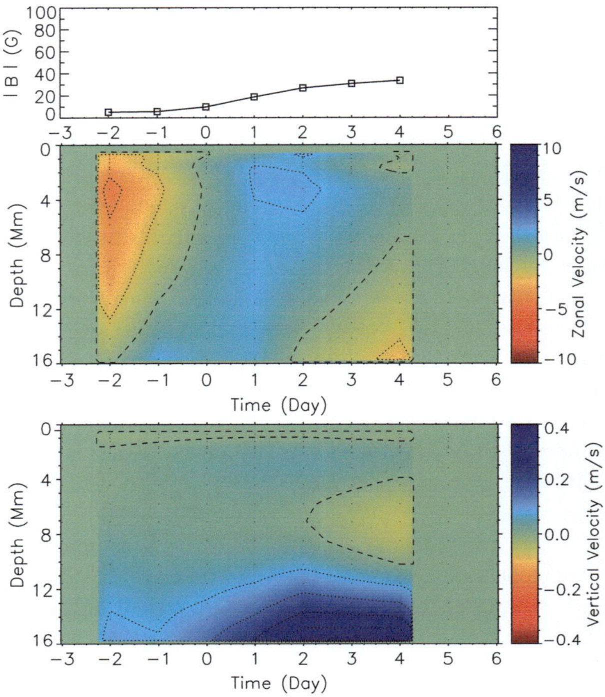

Fig. 3 Emerging active regions and their subsurface flows derived from GONG data using ring-diagram analysis. *Top*: The unsigned magnetic flux averaged over 83 regions, selected as the half of the emerging-flux subset with minimum flux values less than the median value. *Middle*: The residual zonal velocity shows faster zonal velocities (positive values) during emergence. The mean has been subtracted at every depth. *Bottom*: The residual vertical velocity shows enhanced upflows (positive values) that change to downflows (negative values) within two days in intermediate layers. Adapted from Komm et al. (2011b)

represent clearly separate populations (Fig. 2). If this can be improved, for example, by reducing the overlap of the populations that exists even at the highest NHGV values, this parameter holds great promise for quantifying the buildup of subsurface kinetic helicity and thus for an improved understanding of flare production.

2.3 Emerging and Decaying Active Regions

There have been several attempts to detect pre-emergence signatures of solar active regions. These studies can be categorized into two groups: statistical and case studies. Statistical studies with large samples of emerging-flux and quiet regions have been carried out by Komm et al. (2009a, 2011b, 2012), Leka et al. (2013), and Birch et al. (2013). Such studies are useful to determine the average subsurface properties of emerging-flux regions (Fig. 3). It was shown using ring-diagram analysis that emerging-flux regions are characterized by upflows (Komm et al. 2009a), faster rotation (Komm et al. 2011b), and increased vorticity of subsurface flows (Komm et al. 2012). Birch et al. (2013) used helioseismic holography and reported evidence of statistically significant subsurface wave-speed perturbations and plasma flows at least a day before emergence. The characteristic downflow pattern of mature active regions is established within two days after flux emergence (Fig. 3). The average behavior is thus consistent with numerical simulations of emerging flux tubes (Fan 2009a; Schüssler and Rempel 2005) where upflows indicate the beginning of flux emergence, and surface cooling leads to downflows along the emerged loops. There are hints that the increase in vorticity happens about a day later at depths below about 8 Mm compared to shallower layers (Komm et al. 2012), which might imply that this is a near-surface effect with a downward traveling response. However, the time step of the ring-diagram analysis used is comparable to the timing difference and thus too long to adequately resolve the difference in onset with depth.

There are also several case studies with individual emerging-flux and quiet regions (Braun 1995; Chang et al. 1999; Kosovichev et al. 2000; Jensen et al. 2001; Komm et al. 2008; Kosovichev and Duvall 2008; Zharkov and Thompson 2008). These studies search mainly for subsurface wave-speed perturbations and plasma flows before or during the emergence of a large active region. Although there are indications in these studies that support the detection of upward-moving magnetic structures, the signatures of these structures could not be unambiguously detected before the active region emergence in the photosphere. The fast emergence speed and the low signal-to-noise (S/N) ratio of the helioseismic measurements at large depths may have posed difficulties in the detection of pre-emergence signatures.

In addition to local helioseismic techniques, other methods have been also employed in the search of pre-emergence signatures of magnetic flux. Hartlep et al. (2011) reported an acoustic power reduction in the frequency band of 3–4 mHz before the emergence of AR 10488. Toriumi et al. (2012) detected a horizontal divergent flow about 100 minutes before the emergence of AR 11081. In another study, Toriumi et al. (2013) measured the acoustic power with use of phase-speed and ridge filters and managed to detect the rising motion of magnetic flux in the uppermost 10–15 Mm.

Recently, Ilonidis et al. (2011) used deep-focus time-distance helioseismology and detected strong perturbation signatures associated with emerging magnetic structures at depths between 42 and 75 Mm for several emerging active regions. The perturbation signal was a phase travel-time shift of the order of 12–16 s which was detected up to 6 hours (using only preceded data) before the start of emergence in the photosphere and 1–2 days before the detected magnetic structures emerge in the photosphere. The detection was made possible with extensive averaging of the oscillation signals over multiple arc configurations, a phase-speed filter with high S/N ratio, and averaging of the cross-covariance functions over a wide range of depths (Ilonidis et al. 2012b). Braun (2012b) reported that an independent analysis with helioseismic holography could not detect significant subsurface travel-time perturbations associated with flux emergence. Although the reason of this discrepancy between the two helioseismic methods is not known, there are indications that the differences in the measurement procedures may have played a role (Ilonidis et al. 2012a). More recently, Kholikov (2013) confirmed the detection of subsurface perturbation signatures before the emergence of the magnetic field at the surface using time-distance technique and a measurement procedure similar to that employed by Ilonidis et al. (2011).

Figure 4 shows the detection of AR 10488 which emerged on the solar disc on 26 October 2003 around 09:30 UT. Mean travel-time perturbation maps at a depth of 42–75 Mm show a strong and large travel-time anomaly before the start of emergence at the surface. The anomaly first appeared 10 hours before the active region emergence and developed gradually within 4–5 hours reaching a maximum size of about 50 Mm and maximum amplitude of 16.3 s. In the next 3–4 hours, the anomaly weakened and shrank until the signal fell below the detection level. During this time and until the start of photospheric emergence, there is no significant magnetic activity at the location of emergence as well as in the surrounding area which is used for the computation of cross-covariances. The noise level of the measurement procedure was estimated at 3.3 s by analyzing nine quiet regions. Therefore, the detection of a signal of 16.3 s is statistically a very significant event with an S/N ratio of 4.9 (Ilonidis et al. 2011). Travel-time maps computed at the same location every one hour for a total period of more than 2 days found no other strong phase travel-time shift. More recently, Ilonidis et al. (2013) reported that the detected phase travel-time shifts may not be caused by subsurface wave-speed perturbations or flows, but are associated with acoustic power variations and frequency shifts of the cross-covariance function measured in the emerging-flux region. Therefore, the large amplitude of the detected travel-time shifts does not contradict

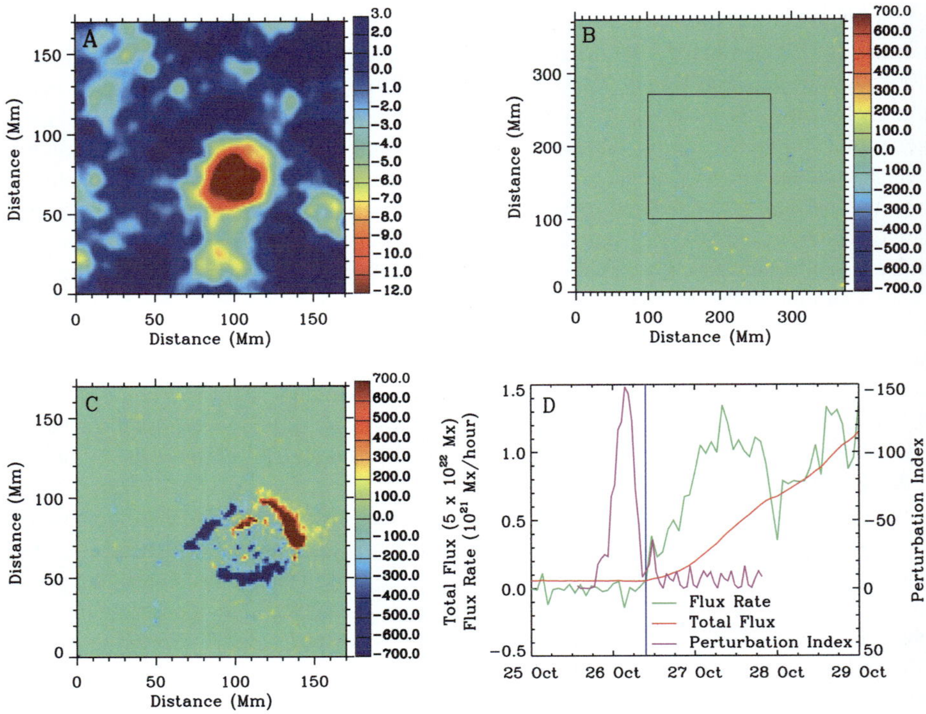

Fig. 4 (**A**) Mean travel-time perturbation map (in seconds) of AR 10488 at a depth of 42 to 75 Mm, obtained from an 8-hour data set centered at 03:30 UT, 26 October 2003. (**B**) Photospheric magnetic field (in Gauss) at the same time as (**A**). The whole map corresponds to the region where the computations were carried out, whereas the squared area at the center corresponds to the region shown in (**A**). (**C**) Photospheric magnetic field (in Gauss) at the same location as (**A**) but 24 hours later. (**D**) Total unsigned magnetic flux (*red line*) and magnetic flux rate (*green line*) of AR 10488. The *vertical blue line* marks the start of emergence. The *pink line* shows the temporal evolution of the perturbation index (in units of 125 $\mathrm{s\,Mm^2}$), which is defined as the sum of travel-time perturbations with values lower than -5.4 s, within the signature of (**A**). From Ilonidis et al. (2011)

theoretical estimates of these shifts based on the effects of wave-speed perturbations and flows.

Figure 5 shows that it is possible to track the rising motion of the magnetic flux from a depth of 75 Mm to 42 Mm and study the temporal evolution of the perturbation signatures at each depth. These signatures first appear at the deepest layer of 64–75 Mm, shift to the middle layer of 53–64 Mm a few hours later, and gradually propagate to the top layer 42–53 Mm approximately 5 hours after the initial detection (Ilonidis et al. 2011). The coexistence of moderate to strong perturbations at 03:30 UT at all three layers explains why the maximum signal is detected around this time when the cross-covariances from all depths are combined as in Fig. 4. The emergence of the helioseismic signatures from deeper to shallower regions is strong evidence that these signatures are associated with the rise of magnetic flux.

Numerical simulations have also provided important physical insight into the process of magnetic flux emergence and helped helioseismic studies in the search of pre-emergence signatures. Three-dimensional anelastic MHD simulations of rising magnetic flux tubes from the base of the convection zone up to a depth of 16 Mm have been carried out by Fan

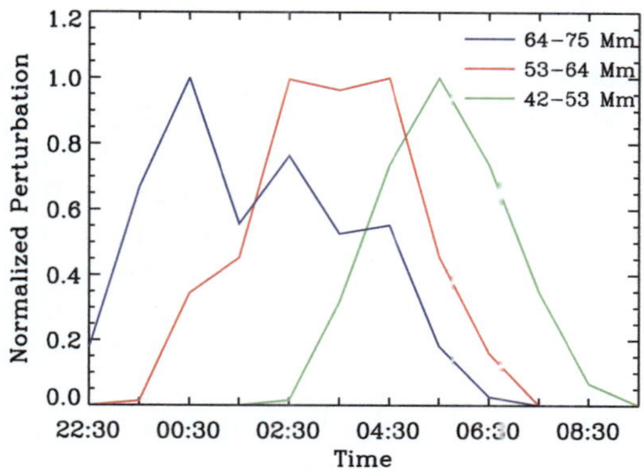

Fig. 5 Temporal evolution of the normalized perturbation at depths of 42–53 Mm (*green line*), 53–64 Mm (*red line*), and 64–75 Mm (*blue line*). The normalized perturbation is defined as the sum of travel-time perturbations with values lower than -3σ, within the detected travel-time anomalies, normalized to the maximum value at each depth. The time starts at 22:30 UT, 25 October 2003. The travel-time anomalies emerge ~20 Mm in about 5 hours with an average speed of the order of ~1 km/s. From Ilonidis et al. (2013)

(2008). Birch et al. (2010) used this simulation model to estimate the detectability of rising flux tubes by helioseismology. More recently, Braun (2012a) and Ilonidis et al. (2013) have used numerical simulation data of subsurface static sound-speed perturbations (Hartlep et al. 2011) to test the ability of helioseismic holography and time-distance technique respectively to detect signatures of these perturbations. However, it was also shown that the analysis of simulation data could not reproduce all the observed properties of the cross-covariance functions in emerging-flux regions (Ilonidis et al. 2013). Effects such as absorption and scattering as well as the flux density, the emergence speed, and the topology of the magnetic field, which are not included in the hydrodynamic numerical model with static sound-speed perturbations, may also play some role.

Active regions evolve during their life time from emergence to decay and the subsurface flows change as well. When active regions decay, the corresponding flow patterns disappear too; the zonal flow slows down to values comparable to that of quiet regions, the upflows become weaker at deeper layers, and the magnitude of vorticity decreases as well (Komm et al. 2011b, 2012). There are hints that the vorticity begins to decrease about a day earlier at layers deeper than about 8 Mm compared to shallower ones, while there is no comparable timing difference for the flows themselves. The variations in the deeper layers might imply that the outflow from the disconnected parent flux tube has stopped (Schüssler and Rempel 2005) or that the parent flux tube has submerged (van Ballegooijen and Mackay 2007). The variations at shallow layers are consistent with active regions changing from a coherent Ω loop to a passively diffusing, rapidly deformable field configuration, such as plages (Schrijver and Title 1999). A comprehensive study of the evolution of subsurface flows of active regions has yet to be performed.

3 Emergence of Magnetic Flux Tubes and the Formation of the Coronal Magnetic Field

There is observational evidence to suggest the emergence of twisted magnetic flux tubes from the solar interior into the solar atmosphere and corona. Observations of the photosphere by TRACE, Hinode, and recently SDO have shown many cases of rotating sunspots in emerging active regions, with angular speed typically of the order of 1°/hour, reaching up to as high as 8°/hour (e.g. Brown et al. 2003; Min and Chae 2009; Jiang et al. 2012; Yan et al. 2012). In some of these cases, the chromospheric fribrils observed in Hα and the EUV loops in the coronal fan also show whirls or spiral patterns that are consistent with the sense of rotation of the underlying sunspots (e.g. Brown et al. 2003). Observations have also shown that rotating sunspots are often associated with eruptive flares in the corona (see review by Schrijver 2009). In many examples, sigmoid shaped X-ray loops as well as sigmoid shaped filaments develop in rotating sunspot regions, and the region undergoes repeated eruptive flares (e.g. Gibson et al. 2002; Tian and Alexander 2006; Zhang et al. 2008; Min and Chae 2009; Jiang et al. 2012).

Other observational evidence that suggest the emergence of a twisted flux tube is implied in a certain emergence pattern of the magnetic flux on the photosphere. The development of the elongated polarities, called "magnetic tongues" or "Ying-Yang" configurations, in the vertical magnetic flux can be explained by the emergence of the apex section of an arched twisted flux tube (e.g. Luoni et al. 2011). Due to the azimuthal flux in the tube cross-section, the polarity orientation of the emerging region will appear to be north-south oriented at the beginning, and then a shear of the two polarities develops. Based on the shear pattern and the shape of the polarity inversion line, one can distinguish the sense of the twist of the emerging tube. Such emergence patterns have been observed (e.g. Canou et al. 2009; Luoni et al. 2011).

We will discuss below a few examples of MHD simulations of magnetic flux emergence from the top layers of the solar convection zone into the solar atmosphere and the corona, to illustrate how magnetic flux, twist and free energy are transported from the plasma dominated, convectively unstable solar convection zone into the stably stratified, rarefied solar atmosphere and corona. We will group the discussion into two parts. First we will discuss the results from a set of earlier "laminar" models which ignore explicit convective motions, focusing on the role of the magnetic buoyancy instability in transporting magnetic flux into the stably stratified solar atmosphere, and the critical role of the Lorentz force-driven shear and torsional motions in transporting twist and free energy into the corona. Then we will discuss results from recent more sophisticated radiation MHD simulations of magneto-convection and flux emergence to show how they have modified our views of active region flux emergence. Although these latter simulations with more realistic physics have in many ways superseded the earlier laminar models, they tend to be more complicated. To disentangle the various physical effects such as the fundamental roles of the magnetic buoyancy, shear and torsional Alfvén waves from the effects of turbulent convection, we still first highlight the basic results of the earlier laminar models.

3.1 Ideal MHD Simulations Without Convection

Pioneered by the earlier work of Shibata and collaborators (Shibata et al. 1989), 3D MHD simulations of the emergence of a twisted magnetic flux tube into the solar atmosphere (e.g. Magara and Longcope 2001; Fan 2001, 2009b; Magara 2004, 2006; Manchester et al. 2004; Archontis et al. 2004, 2009; Murray et al. 2006; Archontis and Török 2008) have been carried out, which have provided important insight into the dynamic process of flux emergence into the atmosphere. These simulations span from the top few Mm of the solar interior up to about 20 Mm into the corona. Convective motions are not modeled in the interior layer by assuming an adiabatically stratified polytrope, and the thermodynamics treatment is highly simplified by assuming adiabatic evolution of an ideal gas.

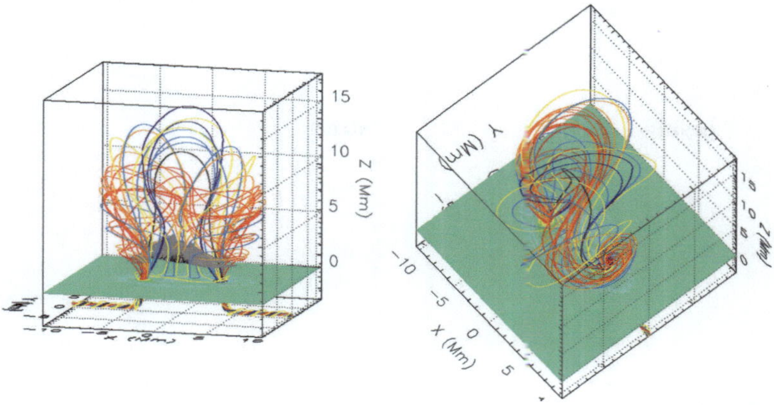

Fig. 6 The 3D coronal magnetic field resulting from the dynamic emergence of a twisted subsurface flux tube, based on an ideal MHD simulation similar to that described in Fan (2009b), but with an initial subsurface twisted flux tube having a shorter length λ for the buoyant segment. A coronal flux rope structure containing sigmoid-shaped dipped fields and with a sigmoid shaped current sheet (the gray isosurface) that extend from the lower atmosphere into the corona has formed

It has been shown that magnetic flux reaching the photosphere can undergo a dynamic expansion into the atmosphere as a result of the non-linear growth of the magnetic buoyancy instability. The condition required of the emerging magnetic field for the onset of such dynamic expansion into the atmosphere has been examined by Archontis et al. (2004) and Murray et al. (2006). It is found that the emerging field needs to build up sufficient magnetic pressure and reach a sufficiently low plasma β in the photospheric layer for the condition to be met. As a result, if either the field strength or the twist of the subsurface flux tube is too low, the magnetic pressure build-up at the photosphere may not achieve the critical condition and flux emergence into the atmosphere may fail to take place.

Figure 6 shows the resulting coronal magnetic field from an example MHD simulation of the dynamic emergence of a twisted, Ω-shaped flux tube from the interior into the solar atmosphere and corona.

It is found that the subsurface twisted flux tube does not emerge bodily into the corona as a whole due to the heavy plasma that is trapped at the bottom concave (or U-shaped) portions of the winding field lines (e.g. Fan 2001; Manchester et al. 2004; Archontis et al. 2004). While the upper parts of the helical field lines of the twisted flux tube expand into the atmosphere and the corona, the U-shaped parts of the winding field lines remain largely trapped at and below the photosphere layer. Despite this, a twisted coronal flux rope containing sigmoid-shaped, concave upturning field lines threading under a central axis that rises into coronal heights (see Fig. 6) is found to develop in the end in many simulations (e.g. Magara and Longcope 2001; Magara 2004, 2006; Manchester et al. 2004; Archontis and Török 2008; Fan 2009b). Shear flows at the polarity inversion line, and rotational motions in each of the polarity concentrations (reminiscent of sunspot rotations) are found to develop (see Fig. 7), driven by the Lorentz force, and are the major means through which twist and free magnetic energy is transported from the interior into the corona (e.g. Manchester et al. 2004; Fan 2009b).

The emergence pattern of the vertical flux in the photospheric layer also shows the "Ying-Yang" pattern (see left panel of Fig. 7) as seen in many observed flare productive emerging regions. The shear and rotational flows twist up the coronal magnetic field, causing the fields above the polarity inversion line to rotate into sigmoid shaped field lines containing dips,

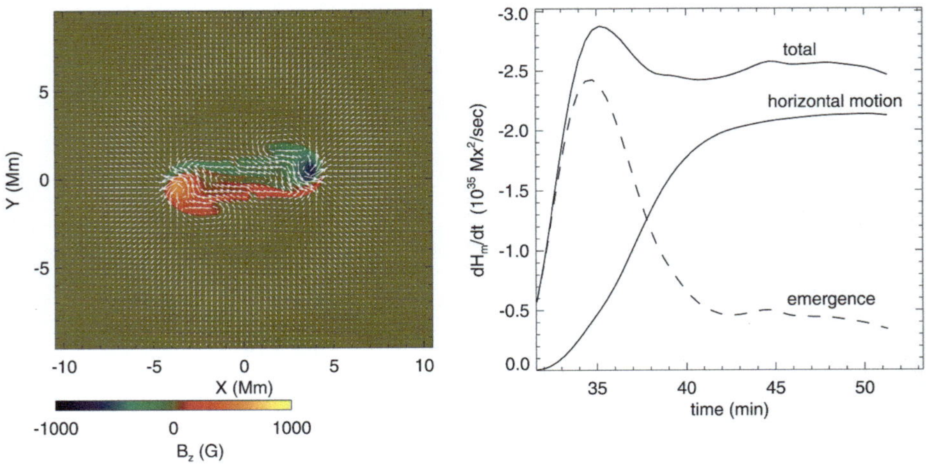

Fig. 7 The *left panel* shows the vertical magnetic field B_z and the horizontal velocity field (*arrows*) in the photosphere layer resulting from the emergence of a twisted flux tube, based on the same simulation shown in Fig. 6. The *longest arrow* length corresponds to a speed of about 2 km/s. The *right panel* shows the total magnetic helicity flux (*solid curve*) through the photosphere into the corona, and its contributions due to direct vertical flux emergence (*dashed curve*) and horizontal shear and vortical motions (*dash-dotted curve*)

and a sigmoid shaped current sheet forms that extends from the lower atmosphere into the corona (see Fig. 6). These results are consistent with the development of X-ray sigmoids and sigmoid shaped filaments (supported by the dipped field lines) in rotating sunspot regions.

3.2 MHD Simulations of Magneto-Convection and Flux Emergence

In the last couple of years, rapid progress has been made in realistic radiation MHD simulations of magneto-convection in the top layers of the solar convection zone, and flux emergence in the photospheric layer and into the solar atmosphere and corona (Abbett 2007; Cheung et al. 2010; Stein and Nordlund 2012; Fang et al. 2010, 2012a, 2012b). Specifically the recent work by Fang et al. (2010, 2012a, 2012b) provides the first set of simulations which model active region flux emergence from the top 20 Mm layer of the solar convection zone, through the photosphere, all the way into the solar corona (to about 20 Mm above the photosphere). These simulations incorporated non-adiabatic energy source terms in the energy equation, which included an empirical coronal heating that maintains the hot corona, and optically thin radiative cooling of the solar atmosphere which was artificially extended into the photosphere density and temperature to produce surface cooling to drive convection in the convection layer. The effect of partial ionization is included in the equation of state used for the convection zone. An initially twisted flux tube is placed in the granular convection in the convective layer and allowed to rise buoyantly towards the photosphere. It is found that the flux emergence pattern at the photosphere is strongly affected by convection. Initially the pattern emerges as small scale fluxes of mixed polarity, and then the small scale fluxes coalesce into larger polarity concentrations and downdrafts due to the presence of coherent subsurface magnetic roots. The roots of the sunspots in the convection zone exhibit a coherent rotational motion (see Fig. 8), transporting twist and helicity into the coronal field, similar to what was found in the laminar simulations, although the two polarities exhibit significant asymmetries in the more realistic convecting case.

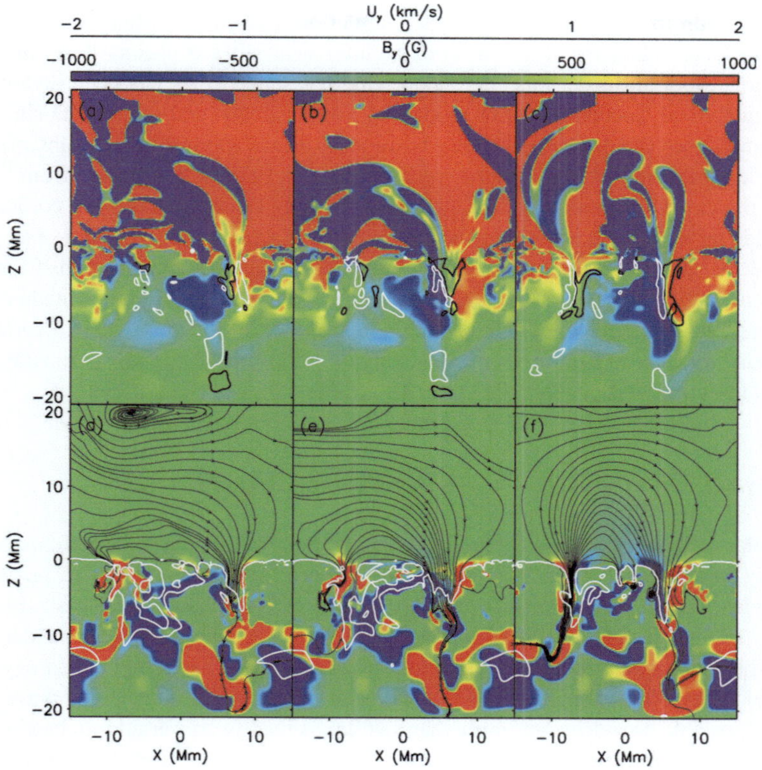

Fig. 8 The structure of flow component u_y (*top*) and magnetic field component B_y (*bottom*) in a vertical slice through the emerging flux rope at three different times: $t = 4{:}21{:}00$ ((**a**) and (**d**)), 4:37:00 ((**b**) and (**e**)), and 5:13:00 ((**c**) and (**f**)). It shows coherent rotational motion in the negative polarity root of the flux rope down to a depth of 10 Mm below the photosphere. The *black* and *white contour lines* represent regions with Lorentz force density $f_y = -20$ and 20 dyne cm^{-3} (*top*). The *black lines with arrows* (*bottom*) indicate streamlines of (B_x, B_z), *white lines* outline Alfvénic Mach number $u/u_A = 1$. From Fang et al. (2012b) reproduced by the permission of the AAS

The coherent rotational motion reaches a speed of about 2 km/s and extends to a depth of about 10 Mm below the photosphere. Shear flows and significant flux cancelation are also found at the polarity inversion line of the emerging region (Fang et al. 2012a). The shear flows at the polarity inversion line and the sunspot rotation build up magnetic shear and twist in the corona, leading to the formation of sigmoid shaped fields. However the formation of a coherent flux rope structure with dipped fields as was found in the laminar models has not been clearly seen in these simulations.

Cheung et al. (2010) have carried out a radiative MHD simulation of the emergence and formation of a solar active region on the solar surface, starting with a twisted magnetic half-torus transported through the lower boundary (at 7.5 Mm below the photosphere) into the convecting domain, rising buoyantly towards the photosphere. It is found that small-scale magnetic elements at the surface first appear, followed by their gradual coalescence into a pair of opposite polarity sunspots. The pair of sunspots both exhibit coherent rotational motions at the photosphere. Although this simulation only extends up to about 700 km above the photosphere, the more accurate radiative transfer calculation in the magnetoconvection allows for a realistic description of the p-mode wave excitation and propagation in these layers. Thus the simulation data can be used as synthetic data for testing local helioseismology methods to infer near surface plasma flows and detect the rotational motion in the flux tube underlying the sunspot. As has been demonstrated both by analytical calculations (Longcope and Welsch 2000) and also MHD simulations of flux emergence (as described above), rotational/vortical motions in the subsurface flux tubes underlying sunspots are a major means by which twist is transported from the interior into the corona in emerging active regions, building up sigmoid-shaped coronal magnetic fields capable of driving solar eruptions (e.g. Manchester et al. 2004; Fan 2009b; Fang et al. 2012a, 2012b). Helioseismology may provide an independent means for measuring such near surface flows under the active regions, complementing other methods using time sequences of vector magnetograms (e.g. Liu et al. 2012), and thus help to determine the lower boundary electric field (and hence the Poynting flux) that drives the coronal magnetic field evolution (e.g. Fisher et al. 2012).

4 Energy Flow and the Role of Waves in the Solar Atmosphere

As the solar atmospheric magnetic field is anchored deep in the Sun's convection zone, it is constantly buffeted and moved around by the subsurface flows and motions. This continuous buffeting leads to the generation of a variety of waves and oscillations, potentially providing a flow of energy through the solar atmosphere. Indeed, Narain and Ulmschneider (1996) estimated that solar convection generates upward propagating waves with an energy budget of several times $10^7\ \mathrm{erg\,cm^{-2}\,s^{-1}}$. However, even though an abundance of wave energy might be available, this does not imply that this flux is readily converted into heating of the solar atmosphere. In this section, we review some of the recent observations of waves and oscillations in the solar atmosphere, in light of their potential contribution to heating of the atmosphere, as well as some of the ongoing theoretical efforts.

4.1 Recent Observations of Waves in the Solar Atmosphere

During the last one to two decades, substantial improvements in spatial and temporal resolution of both ground and space-based observing have resulted in, first of all, the unambiguous detection (imaging) of waves and oscillations in the Sun's atmosphere and, secondly, observational confirmation that waves and oscillations are in fact abundant, if not omnipresent in the solar atmosphere.

It is not possible within the limits of this review to give a comprehensive overview of the potential role of (MHD) waves and oscillations in coronal heating and the rapidly-developing field of coronal seismology. Instead, we will just give a flavour of some of the recent observational discoveries and theoretical developments, mainly focussing on the renewed interest in the potential role of waves as a heating mechanism. For reviews on the subject of coronal heating, we refer the interested reader to, e.g. Walsh and Ireland (2003), Ofman (2005), Klimchuk (2006) or Parnell and De Moortel (2012). Atmospheric seismology has been reviewed by e.g. Nakariakov and Verwichte (2005), Erdélyi (2006) or De Moortel and Nakariakov (2012). Furthermore, two relatively recent volumes of Space Science Reviews were dedicated to more detailed reviews of various aspects of MHD waves and oscillations (Nakariakov and Erdélyi 2009; Erdélyi and Goossens 2011).

Several different types of periodic disturbance which can be interpreted as MHD waves and oscillations have recently been observed. For example, imaging observations have confirmed the existence of propagating, transverse perturbations in a wide variety of structures

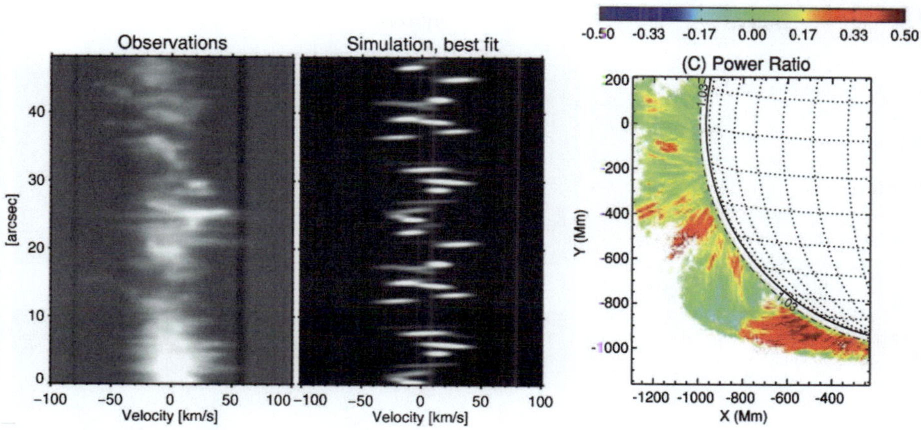

Fig. 9 *Left*: A comparison of λ–y plots of observations in Ca II H of large torsional motion in off-limb spicules and Monte Carlo simulations, modified from De Pontieu et al. (2012). This best-fit simulation has field-aligned velocities of 60 $\mathrm{km\,s^{-1}}$, torsional velocities of 30 $\mathrm{km\,s^{-1}}$, and swaying motions of 15 $\mathrm{km\,s^{-1}}$ with periods of 100–300 s. *Right*: A power ratio map (outward/inward power) of CoMP Doppler shift observations, modified from Tomczyk and McIntosh (2009)

such as prominences (Okamoto et al. 2007; Hershaw et al. 2011), magnetic bright points and inter granular lanes (Jess et al. 2012a), chromospheric spicules (De Pontieu et al. 2007; He et al. 2009; Jess et al. 2012b) and mottles (Kuridze et al. 2012), coronal loops (McIntosh et al. 2011), coronal rain (Antolin and Verwichte 2011) and jets (Cirtain et al. 2007; Morton et al. 2012a). Similar transverse perturbations were reported in Doppler shift observations of large, off-limb, coronal loops by Tomczyk et al. (2007) and Tomczyk and McIntosh (2009). Other observations show evidence for torsional waves in e.g. chromospheric bright points (Jess et al. 2009), spicules (De Pontieu et al. 2012) and coronal holes (Banerjee et al. 2009) and the first, tantalising evidence for sausage modes (Morton et al. 2011a).

All these observed waves and oscillations show periods in the range of a few minutes, reminiscent of the solar surface p-mode spectrum (Tomczyk et al. 2007). As an example, a λ–y plot (wavelength vs distance along the slit) of Ca II H observations (taken by the Swedish Solar Telescope) of off-limb spicules (De Pontieu et al. 2012) is shown on the left hand side of Fig. 9. Using Monte Carlo simulations, these authors demonstrated that a best visual fit is obtained using periods of the order of 100–300 seconds and a combination of field-aligned velocities of 50–100 $\mathrm{km\,s^{-1}}$, swaying motions of 15–20 $\mathrm{km\,s^{-1}}$ and torsional velocities of 25–30 $\mathrm{km\,s^{-1}}$. The right hand side of Fig. 9 shows a power ratio map of the Doppler shift oscillations observed by the Coronal Multi-channel Polarimeter (CoMP) instrument, taken from Tomczyk and McIntosh (2009). Focussing on oscillations in the 3.5 mHz range, this map shows that, apart from in shorter loops, most of the power is directed outwards. In other words, Doppler shift oscillations are seen to propagate up from loop footpoints but only occasionally to reach the other footpoint, implying that significant damping of the observed perturbations must take place as they travel along the loop.

Although this Review does not focus on coronal seismology, it is worthwhile mentioning that this area is developing rapidly in response to the observations of waves and oscillations in such a wide range of structures. To give but a few recent examples, seismology studies have been proposed involving torsional Alfvén waves (Morton et al.

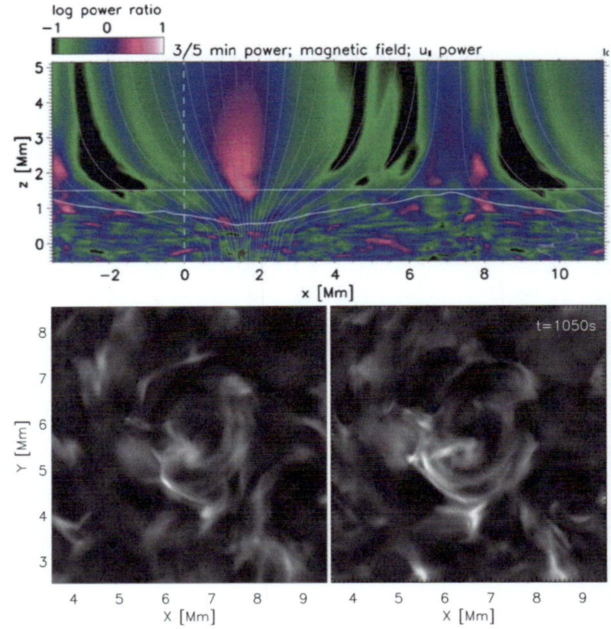

Fig. 10 *Top*: A snapshot of the ratio of 3-minute to 5-minute power (with time-averaged $\beta = 1$ surface (*white line*) and magnetic field lines (*gray lines*) superimposed, adapted from Heggland et al. (2011). *Bottom*: Synthesized narrowband images (forward modelled from numerical simulations at the core of the chromospheric $\lambda = 854.2$ nm line from singly ionized calcium at two times showing swirling motion, taken from Carlsson et al. (2010)

2011b), spicules (Verth et al. 2011), transverse loop oscillations (Terradas et al. 2011; White et al. 2012), propagation waves (Goossens et al. 2012), jets (Morton et al. 2012b), prominences (Foullon et al. 2009) and quasi-periodic pulsations (QPP's—see Sect. 5 in the recent review of De Moortel and Nakariakov (2012)). Again, we refer the reader to the many recent reviews of this subject listed at the start of this section.

4.2 Generation & Propagation of (Alfvénic) Waves in the Solar Atmosphere

As the solar surface motions constantly buffet the footpoints of the magnetic field, different types of MHD waves and oscillations are initiated. Given the recent boom in observations, theoretical efforts are again focussing on modelling wave propagation through the complex and highly dynamical solar atmosphere. Although in a 'theoretical' atmosphere, with a vertical magnetic field, p-modes with periods on the order of five minutes or more are supposed to be trapped, it has long been known that an inclined magnetic field decreases the cut-off frequency sufficiently to allow lower frequency p-modes to leak into the overlying atmosphere (see Bel and Leroy 1977; De Pontieu et al. 2005; Bogdan and Judge 2006; Hansteen et al. 2006; Fedun et al. 2009, 2011, to name but a few). Observational evidence has also confirmed that such 'magnetic portals' do indeed allow the global surface oscillations to propagate into the atmosphere (e.g. Jefferies et al. 2006; Bloomfield et al. 2006; Vecchio et al. 2007). A more detailed description of wave propagation and mode coupling in the solar atmosphere will be given in Sect. 5.

As an example, Fig. 10 (upper panel) shows a snapshot of numerical simulations performed by Heggland et al. (2011) to study wave propagation and jet formation in the solar atmosphere. Looking at different magnetic field configurations, these authors confirm the importance of field inclination for low-frequency wave propagation; 5-minute periods are seen to dominate in regions with strong, inclined fields. In weak or near-vertical magnetic fields, on the other hand, the 3-minute period is dominant. Figure 10 shows one of the configurations considered by Heggland et al. (2011) where the colours represent the ratio of the 3-minute (Fourier) power to the 5-minute power. It is clear that the 3 minute waves are concentrated in the centre of the flux tube, where the field is indeed near-vertical, whereas the 5-minute power is mainly found at the edges of strong flux tubes where the field expands considerably.

Footpoint motions that have been attracting considerable attention in the recent literature are 'swirling' motions, which could potentially propagate into the atmosphere as torsional Alfvén waves. Although predicted by theoretical modelling of solar convection, their observational discovery remained elusive. However, exploiting the extremely high spatial resolution of the Swedish Solar Telescope (SST), Bonet et al. (2008) reported on the discovery of spiral motions of magnetic bright points "on the way to being engulfed by a downdraft". Similar chromospheric swirling motions and associated bright point motions (also observed using the SST), were interpreted by Wedemeyer-Böhm and Rouppe van der Voort (2009) as evidence of twisting and braiding of the atmospheric magnetic field by the convective buffeting of magnetic footprints. Numerical simulations have indeed confirmed that convection naturally leads to vortex motions of magnetic flux elements (e.g. Vögler et al. 2005; Carlsson et al. 2010; Shelyag et al. 2010), which can travel into the overlying atmosphere. An example, taken from a numerical simulation by Carlsson et al. (2010) is shown in the lower panel of Fig. 10 and clearly demonstrates an apparent swirling motions of the magnetic structure (see also Wedemeyer-Böhm et al. 2012).

4.3 Wave Heating

Historically, waves were the first mechanism to be suggested as a potential heating mechanism and hence, there is a vast literature on this subject. Especially throughout the 1980's and 1990's, many wave-based heating theories were proposed (see e.g. Goedbloed and Poedts 2004; Goossens et al. 2011, for reviews). Following the recent observations of low-frequency waves which contain a substantial amount of energy, wave-based heating mechanisms have regained the attention of the Solar Physics community. Two types of observation in particular have come to the forefront, namely, quasi-periodic, propagating disturbances and propagating, transverse oscillations.

The solar corona not only has to be heated but has also to be (continually) supplied with mass. Hence, the energy and mass cycles of the solar atmosphere (and the solar wind) are intrinsically linked. Whether the supplied mass is heated in-situ or not is still under considerable debate. Analysing Hinode and SDO data, De Pontieu et al. (2009, 2011), McIntosh and De Pontieu (2009) report that a significant fraction of the hot coronal mass can be supplied by "Type II" spicules. McIntosh et al. (2012) observed slow, counterstreaming downflows ($\sim$10 km/s), which could be the cool return component of the coronal material, at the end of the mass cycle. Although the role of spicules and propagating, coronal disturbances is under considerable debate (see e.g., De Pontieu and McIntosh 2010; Verwichte et al. 2010; Klimchuk 2012), discussion has re-enforced that the solar corona cannot be treated in isolation.

Transverse, oscillatory loop displacements, most likely driven by footpoint motions have now also been observed, in a wide range of structures (see Sect. 4.1). Some of these observed oscillations are reported to contain substantial energy, comparable to the requirements of, at least, the quiet Sun. For example, McIntosh et al. (2011) derive an energy budget of $\sim$1–2 $\times$ 10^5 $\mathrm{erg\,cm^{-2}\,s^{-1}}$ in quiet Sun and coronal holes, which is indeed on the order of the estimated requirements (Withbroe and Noyes 1977). However, although the observed waves contain a substantial amount of energy, one has to remember that this does not necessarily imply a solution to the coronal heating problem. Theoretical modelling will have to establish whether the observed waves can lead to heating on the appropriate timescales, in the right places. For example, numerical simulations suggest that the damping of the periodic Doppler shift oscillations observed by CoMP (Tomczyk et al. 2007; Tomczyk and McIntosh 2009) can be explained due to the energy transfer from the footpoint driven observed transverse oscillation to an azimuthal Alfvén wave in the tube boundary (see e.g. Pascoe et al. 2012 or Sect. 3 in De Moortel and Nakariakov 2012). Phase mixing of the Alfvén waves in the inhomogeneous tube boundaries could lead to enhanced dissipation and hence, local heating, but the viability of this remains to be demonstrated. For example, using a 1.5D model of heating based on torsional Alfvén waves, Antolin and Shibata (2010) showed that this mechanism is only viable for long and thick loops.

So far, relatively little work has been done on trying to establish unique, observational characteristics of wave-based heating mechanisms (see, Taroyan and Erdélyi 2009, for a review of heating diagnostics with MHD). In particular, it is not a priori obvious whether a wave-based heating mechanism would be near-continuos in nature or episodic and "bursty", and, hence, hard to distinguish from reconnection-based nanoflare models (e.g. Ofman et al. 1998; De Groof and Goossens 2002; Mendoza-Briceño et al. 2005; Klimchuk 2006; Antolin and Shibata 2010). Thus from 3D reduced MHD modelling of Alfvén wave propagation, van Ballegooijen et al. (2011) equally found a heating profile similar, in spatial and temporal distribution of heating events, to the profile of a nanoflare storm. However, looking at the dependence of the coronal heating rate on the loop parameters, they suggest that this heating rate increases with coronal field strength and decreases with loop length.

The recent observations described above have re-opened the debate on the potential role of waves and oscillations as a coronal (and chromospheric!) heating mechanism. However, wave propagation from the solar surface into the complex and dynamical atmosphere is a highly non-trivial topic, as described in the following section.

5 Wave Propagation and Mode Coupling in the Sun's Magnetic Atmosphere

Magnetic fields at the solar surface profoundly affect acoustic waves. We first consider a few basic effects on the basis of ray theory. Then we discuss results from numerical simulations in comparison with measurements of wave travel-times and Doppler velocities.

5.1 Modifications of Acoustic Waves by Magnetic Fields

Text-book illustrations on helioseismology typically display acoustic rays traveling around the Sun in a waveguide between an internal reflection point given by the spherical harmonic degree l, and the upper reflection at the solar surface, provided by the atmospheric acoustic cutoff frequency $\omega_{\mathrm{A}} \approx 5.5$ mHz. Only acoustic waves with a frequency $\omega < \omega_{\mathrm{A}}$ are reflected at the surface—waves with a frequency $\omega > \omega_{\mathrm{A}}$, so called "high-frequency waves", are not subjected to this surface reflection and can freely propagate through the photosphere into the chromosphere. Wave paths, reflection point, and cutoff frequency, all become modified in the presence of a strong magnetic field, i.e., when its energy density $p_{\mathrm{mag}} = B^2/(8\pi)$ is larger than or similar to the thermal energy density p_{gas}. Typically, $\beta = p_{\mathrm{gas}}/p_{\mathrm{mag}} \lesssim 1$ occurs in the photosphere and chromosphere, while in the solar interior $\beta \gg 1$ so that a direct, substantial modification of wave propagation by the magnetic field occurs at the solar surface only. The basic modifications are illustrated by the two cases shown in Fig. 11 (from Cally 2007).

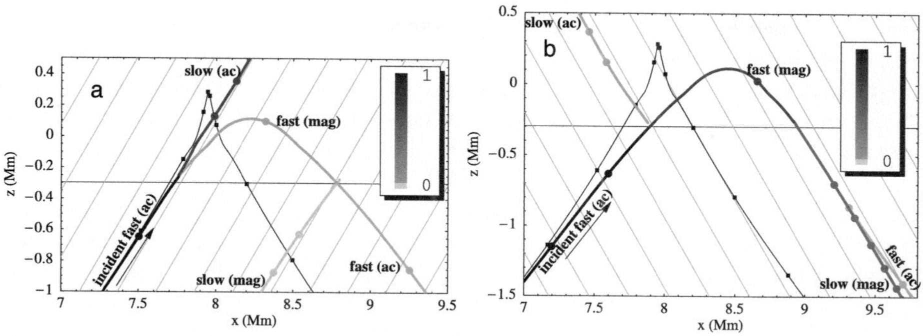

Fig. 11 Ray paths in the x–z vertical plane of a surface-layer model with a 2 kG uniform magnetic field inclined at (**a**) $+30°$ and (**b**) $-30°$ to the vertical. The incoming 5 mHz rays have lower turning points at $z = -5$ Mm. The *horizontal grey line* indicates where the sound and Alfvén speeds coincide, which is approximately where mode conversion happens. The fractional energy remaining in each resulting ray is indicated by the gray-scale legend. The *dots* on the ray paths indicate 1 min group travel-time intervals. The *thin black curve* represents the acoustic ray that would be there in the absence of the magnetic field. Note that the fast ray travels faster through the surface layers than the nonmagnetic ray. From Cally (2007)

The Ramp Effect In Fig. 11a, the wave vector of the incident wave is close to parallel to the magnetic field. In this case, the energetically largest part of the acoustic wave continues to travel along the magnetic field (heavy black curve) beyond $z = 0.2$ Mm across the temperature minimum region into the chromosphere, despite its frequency being below the nominal cutoff frequency. The magnetic field acts for the wave like a ramp for lifting a weight, where the force to be applied is reduced by the factor $\cos\theta$. Here, the 'magnetic ramp' lowers the cutoff frequency by the same factor (Cally 2007). Thus, the upper reflection point for helioseismic waves is absent even for waves below the nominal cutoff frequency in the presence of inclined, strong surface magnetic fields when the magnetic field and incident wave vector happen to be close to parallel. In this case, most wave energy leaks to the outer atmosphere and only a small fraction may return back into the interior (Michalitsanos 1973; Bel and Leroy 1977; Suematsu 1990). The significance of the ramp effect for chromospheric physics was only recently fully appreciated. Jefferies et al. (2006) show that inclined magnetic fields at the boundaries of supergranular cells provide "portals" through which low-frequency magnetoacoustic waves can propagate into the solar chromosphere. They concluded that these waves may provide a significant source of energy for balancing the radiative losses in the solar chromosphere. De Pontieu et al. (2004) found that the leakage of the global 5-minute p-mode oscillation along inclined magnetic flux tubes, again, due to the ramp effect, provides sufficient energy to power shocks that drive upward flows, forming type I spicules. The ramp effect is also manifest in the numerical simulations of Heggland et al. (2011), as discussed in Sect. 4.2, Fig. 10.

Mode Transmission Besides lowering the cutoff frequency, the inclined magnetic field of Fig. 11a also acts as a wave guide for the incident acoustic wave. Thus, strictly speaking, this wave is not purely acoustic but of magneto-acoustic nature. Below the horizontal gray line, which marks the level where sound speed c and Alfvén speed a are of equal magnitude, i.e., in the region where $\beta \gg 1$, the incident wave is hardly affected by the magnetic field and is predominantly an acoustic wave. It travels with the speed of sound, which here is much larger than the Alfvén speed, and therefore is called a fast magneto acoustic wave. As it travels further up, it remains a predominantly acoustic wave, the magnetic field merely acting as a guide. As the wave crosses the horizontal gray line (in the following called the equipartition level), it enters the region where the Alfvén speed quickly surpasses the sound speed. Henceforth, it is called a slow magneto acoustic wave. Note that the physical nature of the wave does not change: it remains predominantly acoustic but changes in relative speed from fast to slow when transmitted from the high to the low β region (Cally 2007).

Mode Conversion In Fig. 11a, a less energetic (light grey) ray splits off the principal ray, at about the position where the incident ray crosses the equipartition level. It quickly turns to the right, in the horizontal direction, and back again into the Sun. Different from the reflective, cusp shaped ray path in the absence of the magnetic field, this curved ray path is due to refraction. This wave undergoes refraction due to the Alfvén speed steeply increasing with height, in combination with the fact that it is magnetic in nature, the predominant restoring force being magnetic pressure. Hence, its characteristic speed is the Alfvén speed, which is much larger than the sound speed in this region and it is therefore called a fast magneto acoustic wave. For this branch of the wave path, the incident, predominantly acoustic fast wave gets converted to a predominantly magnetic fast wave (Cally 2007). The partitioning of the initial wave energy into these two branches depends on the angle between the incident wave vector and the magnetic field vector at the location of the equipartition level (the attack angle), on the wave number, and on the width of the equipartition level, i.e., the region where $a \approx c$. In Fig. 11a, the attack angle is small and consequently, most of the incident wave energy stays in the acoustic branch that leaves the domain through the upper boundary. In Fig. 11b on the other hand, the attack angle is large and consequently, most energy goes into the magnetic branch, which gets refracted and turns back into the Sun. After refraction of the fast magneto acoustic wave, this branch crosses the equipartition level again, now in reverse direction. Again, the ray path splits; this time into a fast acoustic and into a slow magnetic branch.

Wave Travel-Times Both cases in Fig. 11 start with a fast incident acoustic ray and both cases also show a fast outgoing acoustic ray traveling back into the Sun again, although of considerable less energy. In the intermediate case, when the magnetic field is vertical, the energy of the downward traveling fast acoustic wave may be larger than in the cases shown on Fig. 11. Considering only the incident upward traveling and the outgoing downward traveling fast acoustic ray, we have a situation equivalent to the purely acoustic ray of the magnetic field-free atmosphere. The difference however is that the travel distance between upstroke and downstroke (skip distance) is larger in the former case suggesting a longer travel time. However, because the wave converts to a fast magnetoacoustic one in the intermediate stretch above the equipartition level, the longer distance is more than compensated for by the higher speed of this wave compared to the very low speed of the acoustic wave near the cusp-shaped reflection point in the field-free atmosphere. The difference (skip-time perturbation) in case of Figures 11a and b is about 3 min. Thus, discarding the magnetic field, one would from wave-travel times derive a higher than the true speed of sound, or a shorter path than the true wave travel-distance, i.e., a lower reflection point.

5.2 Development of the Theory

Theory and theoretical aspects of magneto-acoustic waves in a gravitationally stratified stellar atmosphere, sometimes called magneto-acoustic-gravity (MAG) waves or magneto-atmospheric waves, have developed over many decades and received increasing attention in recent years (e.g. Osterbrock 1961; Thomas 1983; Shibata 1983; Zhugzhda and Dzhalilov

1984a, 1984b). Several numerical experiments with non-uniform magnetic field equilibria in a two-dimensional, stratified atmosphere, are due to Rosenthal et al. (2002) and Bogdan et al. (2003). These authors recognized and highlighted the role of refraction of the fast magneto-acoustic wave in the low β regime (first demonstrated by Osterbrock 1961) and the role of the equipartition surface as a wave conversion zone. Aiming at applications in local helio-seismology, Cally (2005, 2006), and Schunker and Cally (2006) derive gravito-magneto-acoustic dispersion relations and then use these to examine how acoustic rays split into fast and slow components when entering regions of strong field and the subsequent fates of each as discussed in Sect. 5.1. Thus, they developed an "extended ray theory" for stellar atmospheric application, which includes tunneling between the fast and slow branches, which goes beyond the eikonal assumption upon which it is based. With numerical simulations of MAG-wave propagation in three-dimensional space, Cally and Goossens (2008), Cally and Hansen (2011), and Hansen and Cally (2012) show that the intermediate Alfvén wave may couple to the fast and slow magnetoacoustic waves with important implications for energy loss from helioseismic modes. They found that energy branching to upward travelling Alfvén waves can be substantial, in many cases exceeding the loss to the slow, acoustic waves.

Applications of the extended ray theory and direct numerical simulations of magneto-acoustic wave propagation have largely focussed on sunspots, where the magnetic field is strong enough for mode conversion to occur already in the photosphere (Cally and Goossens 2008; Moradi and Cally 2008; Moradi et al. 2009). Khomenko and Collados (2006) demonstrated refraction of the fast wave due to the vertical and horizontal gradients of the Alfvén speed in sunspots. For small-scale flux-tubes, Hasan et al. (2005) and Khomenko et al. (2008) find that deep horizontal motions of the flux tube initially generate a slow (magnetic) mode and a surface mode that are efficiently transformed into a slow (acoustic) mode when the magnetic field starts to dominate. This slow mode propagates vertically along the magnetic field remaining always within the flux tube, where it steepens to a shock. Only a small part of the driver energy is returned to the photosphere by the fast magneto-acoustic mode (see also Vigeesh et al. 2009). These works suggest that vertical magnetic field concentrations play an essential role in coupling the dynamics of the photosphere to the chromosphere through efficient channeling and conversion of magneto-acoustic waves.

5.3 Atmospheric Seismology

Travel-times of high-frequency waves between different height levels in the solar atmosphere can become drastically reduced in the presence of magnetic fields. This effect can be exploited to map the three-dimensional topography of the equipartition surface, sometimes referred to as the 'magnetic canopy', or the '$\beta = 1$ surface', in and around active regions. Finsterle et al. (2004) analyzed Doppler data of two chromospheric spectral lines obtained with the MOTH instrument (Magneto-Optical filters at Two Heights, Cacciani et al. 2003) from which they determined travel-times of 7 mHz waves. In combination with the photospheric Ni-line from SOHO/MDI, they were able to map the funnel shaped magnetic canopies of a bipolar active region from the photosphere to the chromosphere. The wave travel-time can even vanish between the two chromospheric lines in locations of strong magnetic fields, which Nutto et al. (2012b) attribute to the refraction of the fast magneto acoustic wave. Depending on whether the spectral lines form below or above the equipartition level, they either can catch the fast acoustic or the fast magnetic mode, respectively. In the latter case, the travel-time becomes reduced due to the high characteristic speed in the region of low β, and it may even vanish or become negative because of wave refraction: when the

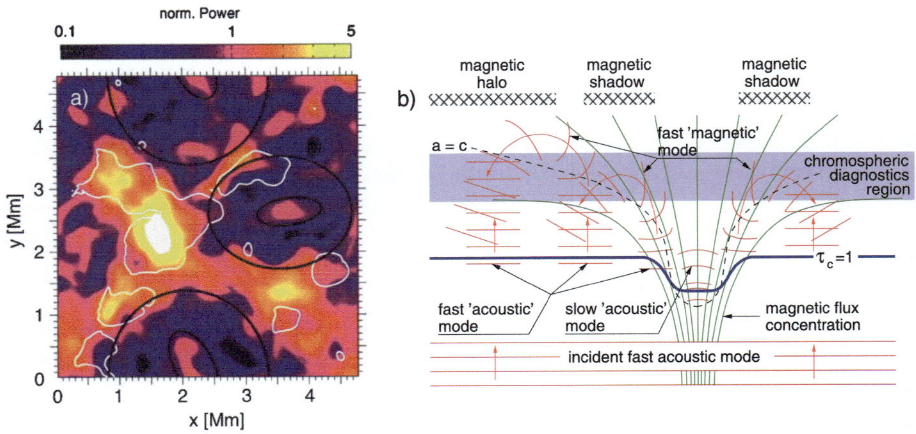

Fig. 12 (**a**) Power map of the vertical velocity perturbation, δv_z, at a continuum optical depth layer τ in the lower chromosphere of a numerical simulation. An artificial acoustic source, placed at the bottom boundary at a depth of 1.4 Mm, generates outward travelling plane parallel acoustic waves of frequency 10 mHz. The *white contours* indicate the intersection of the equipartition level with the τ-level. The *three inner ellipses* mark regions of magnetic flux concentrations. Between the large and the small ellipses, the characteristic annulus of suppressed power of the magnetic shadow is apparent. Adapted from Nutto et al. (2012a). (**b**) Sketch of the different wave modes arising from the incident plane parallel acoustic wave front. See text for explanations

wave path bends into the horizontal direction, the phase speed becomes infinitely large, and when turning back again, downwards into the Sun, the phase speed becomes negative.

Another two phenomena, the *magnetic halo* and the *magnetic shadow*, are thought to result from mode conversion and refraction. The magnetic halo is an enhancement in the high-frequency (non-trapped) wave power in an area surrounding active regions both in the photosphere (Brown et al. 1992) and in the chromosphere (Braun et al. 1992; Toner and Labonte 1993); its properties are summarized in Khomenko and Collados (2009), Rajaguru et al. (2013), and Schunker and Braun (2011). The magnetic shadow, on the other hand, is a narrow seam of reduced power surrounding active regions, seen in high-frequency chromospheric intensity and velocity oscillations (Judge et al. 2001; Vecchio et al. 2007; Kontogiannis et al. 2010a, 2010b; Rajaguru et al. 2013).

The magnetic halo was reproduced with numerical simulations, by Khomenko and Collados (2009), featuring a sunspot that shows a clear enhancement of the photospheric vertical velocity power in an annular region around and outside the location where the equipartition surface dips into the photosphere.[1] The power enhancement comes about because of waves that reflect at the equipartition surface and subsequently interfere with the preexisting waves from the acoustic source when traveling back into the Sun (see also Fig. 12b). The reflection is due to the conversion of the originally acoustic waves into fast magneto-acoustic waves that undergo refraction above the equipartition surface until traveling back into the Sun and converting back to acoustic waves again when crossing the equipartition surface from above. Waves below the photospheric cutoff frequency become evanescent before reaching the equipartition layer and neither refract nor produce interference, which explains the absence of the halo for low frequency oscillations.

[1] As the magnetic field strength increases towards the sunspot axis, the equipartition surface lowers and even reaches below the optical depth surface $\tau = 1$ at the center of the sunspot.

In a slightly different vein, the 3-D radiation MHD simulations of a magnetic network element of Nutto et al. (2012a) show instead magnetic shadows; quite similar to those observed by Vecchio et al. (2007). Figure 12a shows the power of the vertical velocity perturbation at photospheric height levels, due to the incident plane parallel wave from the artificial source. There are three locations of magnetic flux concentrations, which approximately coincide with the three inner ellipses indicated in the panel. Between the inner (small) and the outer (large) ellipses, the characteristic annulus of suppressed power of the magnetic shadow is apparent. Above the magnetic flux concentrations proper, there is some power left (inner ellipses) and further away from the ellipses, in regions which are arched by horizontal magnetic field, the power is enhanced.

Figure 12b sketches the physics behind this peculiar power pattern. It can be thought to represent a vertical section at $y = 2.6$ Mm through the computational box of panel (a). In the center of the magnetic flux concentration, the wave vector of the incident acoustic wave is parallel to the magnetic field (vanishing attack angle) so that it is transmitted across the equipartition level, $a = c$, (dashed curve) to a slow acoustic mode, which gives rise to the power within the small ellipses. In the peripheral region of the flux concentration, acoustic waves that travel across the equipartition level are partially converted to fast, predominantly magnetic waves, which refract above this level (see also Sect. 5.1) and travel back into the Sun. Two such refractive wave paths are indicated to the left of the magnetic flux concentration in Fig. 12b, another one to the right of it. The apex of the inner wave train falls into the chromospheric diagnostics region, and because its wave vector is close to horizontal there, the vertical velocity perturbation vanishes and with it the power in the corresponding power map, hence the magnetic shadow. Note, that at this same location, the horizontal velocity power would be large and also the intensity power: the latter was actually observed by Rajaguru et al. (2013) in the core of Fe I 617.33 nm and in the SDO/AIA channel at 170 nm. Further out, the downward traveling refracted wave interferes with the continuing incident fast acoustic wave already within the chromospheric diagnostics region, which leads to excess power in the region outside the large ellipse, which can be identified with the excess power in the magnetic halo (Khomenko and Collados 2009). This interference continues to exist down into the photosphere, which explains the occurrence of the halo in both the photosphere and the chromosphere.

In reality, details will differ from the sketch of Fig. 12b, depending on the actual magnetic field configuration and field strength, on the relative position between the equipartition surface and the spectral line forming region (diagnostics region), and depending on the extent and location of the wave source. Also the chromospheric diagnostics region must be expected to be a corrugated, rather than the straight horizontal layer as indicated in Fig. 12.

6 Future Developments

Recent observations have shown that subsurface flows of active regions are clearly linked with the conditions of the magnetic activity in the photosphere and above. With the availability of SDO/HMI data, it is now possible to measure simultaneously subsurface flows and photospheric vector magnetic fields, which will help to address the question how close the link is between them. Local helioseismology provides estimates of the subsurface dynamics and thermal and magnetic structure. However, reliable inferences are difficult in the presence of strong magnetic fields especially for structure inversions (e.g. Gizon et al. 2009). To improve the reliability, the techniques are being tested with numerical simulations (e.g. Braun et al. 2012; Parchevsky et al. 2012). A way to substantially improve helioseismic inferences would be to build a network of instruments capable of multi-wavelength observations, as discussed in another chapter. Such measurements could also probe the structure of the solar atmosphere.

Realistic radiative MHD simulations of magneto-convection, flux emergence and sunspot formation (e.g. the type of simulations by Cheung et al. 2010; Stein and Nordlund 2012) that spans from the interior ($\approx$30 Mm depth) to the solar atmosphere on realistic active region size scales are becoming practical. These simulations not only provide fundamental insight into the magnetic connection between the interior and the corona in solar active regions, but also can be used as synthetic data for testing local helioseismology techniques for detecting emerging active regions and for inferring the near surface dynamics. Given the continuous observation by SDO/HMI, the latter provides important information for determining the Poynting flux and the helicity flux that build up the free magnetic energy in active region coronal magnetic fields.

Recent observations have shown that an abundance of waves and oscillations is present in the solar atmosphere and, that these perturbations contain a substantial amount of energy, potentially sufficient to account for e.g. the quiet Sun requirements. These recent observational findings have placed wave-based heating mechanisms back at the forefront of the coronal heating debate. However, both the propagation of wave-like perturbations through the solar atmosphere and the efficiency of wave-dissipation mechanisms need to be (re-)examined using realistic numerical simulations before an accurate assessment can be made of the contribution of these observed waves and oscillations to the energy budget of the chromosphere, corona and/or solar wind. In particular, observational signatures of wave-based heating mechanisms need to be synthesised from the theoretical/numerical models as it is currently not clear how to distinguish such mechanisms observationally (or indeed, whether this is even possible).

Realistic numerical simulations of large-scale active regions including sunspots, plage, and network magnetic fields are also needed for completing the physical picture regarding halos and magnetic shadows of power maps of photospheric and chromospheric intensities and velocities. These simulations should reach deep enough to self-consistently include the acoustic sources and run long enough for a seismological analysis equivalent to that of real data. Such three-dimensional simulations should also shed light on the coupling of the fast and slow magneto-acoustic to Alfvén waves, which typically occurs around and above the apex of the fast wave path, depending on the local angle between the background magnetic field and the wave vector (Khomenko and Cally 2012; Felipe 2012). Consequently, one can also expect halos of Alfvén energy flux to the corona in active regions and plages. Rather idealistic sunspot models (Schunker et al. 2011) may be favored in simulations of wave propagation for the interpretation of local helioseismic measurements within and around sunspots. Finally, global models including active region magnetic fields should be constructed, possibly in the form of a patchwork of local models, in view of global helioseismic simulations. Such simulations should improve our understanding of the influence of magnetic fields on global modes and their dependency on the solar cycle.

Acknowledgements RK acknowledges support from NSF/SHINE Award No. 1062054 and NASA grant NNX11AQ57G to the National Solar Observatory, which is operated by AURA, Inc. under a cooperative agreement with the National Science Foundation. IDM acknowledges support from a Royal Society University Research Fellowship. YF acknowledges support from NASA LWS grant NNX09AJ89G to NCAR. NCAR is sponsored by the National Science Foundation. SI acknowledges support from NASA contract NAS5-02139 to Stanford University.

References

W.P. Abbett, The magnetic connection between the convection zone and corona in the quiet Sun. Astrophys. J. **665**, 1469–1488 (2007). doi:10.1086/519788

P. Antolin, K. Shibata, The role of torsional Alfvén waves in coronal heating. Astrophys. J. **712**, 494–510 (2010). doi:10.1088/0004-637X/712/1/494

P. Antolin, E. Verwichte, Transverse oscillations of loops with coronal rain observed by Hinode/Solar optical telescope. Astrophys. J. **736**, 121 (2011). doi:10.1088/0004-637X/736/2/121

V. Archontis, T. Török, Eruption of magnetic flux ropes during flux emergence. Astron. Astrophys. **492**, 35–38 (2008). doi:10.1051/0004-6361:200811131

V. Archontis, F. Moreno-Insertis, K. Galsgaard, A. Hood, E. O'Shea, Emergence of magnetic flux from the convection zone into the corona. Astron. Astrophys. **426**, 1047–1063 (2004). doi:10.1051/0004-6361:20035934

V. Archontis, A.W. Hood, A. Savcheva, L. Golub, E. Deluca, On the structure and evolution of complexity in sigmoids: a flux emergence model. Astrophys. J. **691**, 1276–1291 (2009). doi:10.1088/0004-637X/691/2/1276

D. Banerjee, D. Pérez-Suárez, J.G. Doyle, Signatures of Alfvén waves in the polar coronal holes as seen by EIS/Hinode. Astron. Astrophys. **501**, 15–18 (2009). doi:10.1051/0004-6361/200912242

S. Bao, H. Zhang, Patterns of current helicity for the twenty-second solar cycle. Astrophys. J. Lett. **496**, 43 (1998). doi:10.1086/311232

G. Barnes, K.D. Leka, Evaluating the performance of solar flare forecasting methods. Astrophys. J. Lett. **688**, 107–110 (2008). doi:10.1086/595550

N. Bel, B. Leroy, Analytical study of magnetoacoustic gravity waves. Astron. Astrophys. **55**, 239 (1977)

M.A. Berger, G.B. Field, The topological properties of magnetic helicity. J. Fluid Mech. **147**, 133–148 (1984). doi:10.1017/S0022112084002019

A.C. Birch, D.C. Braun, Y. Fan, An estimate of the detectability of rising flux tubes. Astrophys. J. Lett. **723**, 190–194 (2010). doi:10.1088/2041-8205/723/2/L190

A.C. Birch, D.C. Braun, K.D. Leka, G. Barnes, B. Javornik, Helioseismology of pre-emerging active regions. II. Average emergence properties. Astrophys. J. **762**, 131 (2013). doi:10.1088/0004-637X/762/2/131

D.S. Bloomfield, R.T.J. McAteer, M. Mathioudakis, F.P. Keenan, The influence of magnetic field on oscillations in the solar chromosphere. Astrophys. J. **652**, 812–819 (2006). doi:10.1086/508409

T.J. Bogdan, P.G. Judge, Observational aspects of sunspot oscillations. Philos. Trans. R. Soc. Lond. Ser. A, Math. Phys. Sci. **364**, 313–331 (2006). doi:10.1098/rsta.2005.1701

T.J. Bogdan, M. Carlsson, V.H. Hansteen, A. McMurry, C.S. Rosenthal, M. Johnson, S. Petty-Powell, E.J. Zita, R.F. Stein, S.W. McIntosh, Å. Nordlund, Waves in the magnetized solar atmosphere. II. Waves from localized sources in magnetic flux concentrations. Astrophys. J. **599**, 626–660 (2003). doi:10.1086/378512

J.A. Bonet, I. Márquez, J. Sánchez Almeida, I. Cabello, V. Domingo, Convectively driven vortex flows in the Sun. Astrophys. J. Lett. **687**, 131–134 (2008). doi:10.1086/593329

D.C. Braun, Sunspot seismology: new observations and prospects, in *GONG 1994. Helio- and Astro-Seismology from the Earth and Space*, ed. by R.K. Ulrich, E.J. Rhodes Jr., W. Dappen. Astronomical Society of the Pacific Conference Series, vol. 76, 1995, p. 250

D.C. Braun, Helioseismic holography of an artificial submerged sound speed perturbation and implications for the detection of pre-emergence signatures of active regions. Sol. Phys. (2012a). doi:10.1007/s11207-012-0185-4

D.C. Braun, Comment on "Detection of emerging sunspot regions in the solar interior". Science **336**, 296 (2012b). doi:10.1126/science.1215425

D.C. Braun, C. Lindsey, Helioseismic imaging of the farside and the interior, in *GONG+ 2002. Local and Global Helioseismology: The Present and Future*, ed. by H. Sawaya-Lacoste. ESA Special Publication, vol. 517, 2003, pp. 15–22

D.C. Braun, C. Lindsey, Y. Fan, S.M. Jefferies, Local acoustic diagnostics of the solar interior. Astrophys. J. **392**, 739–745 (1992). doi:10.1086/171477

D.C. Braun, A.C. Birch, C. Lindsey, Local helioseismology of near-surface flows, in *SOHO 14 Helio- and Asteroseismology: Towards a Golden Future*, ed. by D. Danesy. ESA Special Publication, vol. 559, 2004, p. 337

D.C. Braun, A.C. Birch, M. Rempel, T.L. Duvall, Helioseismology of a realistic magnetoconvective sunspot simulation. Astrophys. J. **744**, 77 (2012). doi:10.1088/0004-637X/744/1/77

T.M. Brown, T.J. Bogdan, B.W. Lites, J.H. Thomas, Localized sources of propagating acoustic waves in the solar photosphere. Astrophys. J. Lett. **394**, 65–68 (1992). doi:10.1086/186473

D.S. Brown, R.W. Nightingale, D. Alexander, C.J. Schrijver, T.R. Metcalf, R.A. Shine, A.M. Title, C.J. Wolfson, Observations of rotating sunspots from TRACE. Sol. Phys. **216**, 79–108 (2003). doi:10.1023/A:1026138413791

A. Cacciani, S.M. Jefferies, W. Finsterle, P. Rapex, A. Knox, C. Giebink, V. di Martino, A new instrument for sounding the solar atmosphere, in *GONG+ 2002. Local and Global Helioseismology: The Present and Future*, ed. by H. Sawaya-Lacoste. ESA Special Publication, vol. 517, 2003, pp. 243–245
P.S. Cally, Local magnetohelioseismology of active regions. Mon. Not. R. Astron. Soc. **358**, 353–362 (2005). doi:10.1111/j.1365-2966.2005.08742.x
P.S. Cally, Dispersion relations, rays and ray splitting in magnetohelioseismology. Philos. Trans. R. Soc. Lond. Ser. A, Math. Phys. Sci. **364**, 333–349 (2006). doi:10.1098/rsta.2005.1702
P.S. Cally, What to look for in the seismology of solar active regions. Astron. Nachr. **328**, 286 (2007). doi:10.1002/asna.200610731
P.S. Cally, M. Goossens, Three-dimensional MHD wave propagation and conversion to Alfvén waves near the solar surface. I. Direct numerical solution. Sol. Phys. **251**, 251–265 (2008). doi:10.1007/s11207-008-9142-7
P.S. Cally, S.C. Hansen, Benchmarking fast-to-Alfvén mode conversion in a cold magnetohydrodynamic plasma. Astrophys. J. **738**, 119 (2011). doi:10.1088/0004-637X/738/2/119
A. Canou, T. Amari, V. Bommier, B. Schmieder, G. Aulanier, H. Li, Evidence for a pre-eruptive twisted flux rope using the themis vector magnetograph. Astrophys. J. Lett. **693**, 27–30 (2009). doi:10.1088/0004-637X/693/1/L27
M. Carlsson, V.H. Hansteen, B.V. Gudiksen, Chromospheric heating and structure as determined from high resolution 3D simulations. Mem. Soc. Astron. Ital. **81**, 582 (2010)
H.-K. Chang, D.-Y. Chou, M.-T. Sun, In search of emerging magnetic flux underneath the solar surface with acoustic imaging. Astrophys. J. Lett. **526**, 53–56 (1999). doi:10.1086/312366
M.C.M. Cheung, M. Rempel, A.M. Title, M. Schüssler, Simulation of the formation of a solar active region. Astrophys. J. **720**, 233–244 (2010). doi:10.1088/0004-637X/720/1/233
J.W. Cirtain, L. Golub, L. Lundquist, A. van Ballegooijen, A. Savcheva, M. Shimojo, E. DeLuca, S. Tsuneta, T. Sakao, K. Reeves, M. Weber, R. Kano, N. Narukage, K. Shibasaki, Evidence for Alfvén waves in solar x-ray jets. Science **318**, 1580 (2007). doi:10.1126/science.1147050
A. De Groof, M. Goossens, Fast and Alfvén waves driven by azimuthal footpoint motions. II. Random driver. Astron. Astrophys. **386**, 691–698 (2002). doi:10.1051/0004-6361:20020204
I. De Moortel, V.M. Nakariakov, Magnetohydrodynamic waves and coronal seismology: an overview of recent results. Philos. Trans. R. Soc. Lond. Ser. A, Math. Phys. Sci. **370**, 3193–3216 (2012). doi:10.1098/rsta.2011.0640
B. De Pontieu, S.W. McIntosh, Quasi-periodic propagating signals in the solar corona: the signature of magnetoacoustic waves or high-velocity upflows? Astrophys. J. **722**, 1013–1029 (2010). doi:10.1088/0004-637X/722/2/1013
B. De Pontieu, R. Erdélyi, S.P. James, Solar chromospheric spicules from the leakage of photospheric oscillations and flows. Nature **430**, 536–539 (2004). doi:10.1038/nature02749
B. De Pontieu, R. Erdélyi, I. De Moortel, How to channel photospheric oscillations into the corona. Astrophys. J. **624**, 61–64 (2005). doi:10.1086/430345
B. De Pontieu, S.W. McIntosh, M. Carlsson, V.H. Hansteen, T.D. Tarbell, C.J. Schrijver, A.M. Title, R.A. Shine, S. Tsuneta, Y. Katsukawa, K. Ichimoto, Y. Suematsu, T. Shimizu, S. Nagata, Chromospheric Alfvénic waves strong enough to power the solar wind. Science **318**, 1574 (2007). doi:10.1126/science.1151747
B. De Pontieu, S.W. McIntosh, V.H. Hansteen, C.J. Schrijver, Observing the roots of solar coronal heating in the chromosphere. Astrophys. J. Lett. **701**, 1–6 (2009). doi:10.1088/0004-637X/701/1/L1
B. De Pontieu, S.W. McIntosh, M. Carlsson, V.H. Hansteen, T.D. Tarbell, P. Boerner, J. Martinez-Sykora, C.J. Schrijver, A.M. Title, The origins of hot plasma in the solar corona. Science **331**, 55 (2011). doi:10.1126/science.1197738
B. De Pontieu, M. Carlsson, L.H.M. Rouppe van der Voort, R.J. Rutten, V.H. Hansteen, H. Watanabe, Ubiquitous torsional motions in type II spicules. Astrophys. J. Lett. **752**, 12 (2012). doi:10.1088/2041-8205/752/1/L12
T.L. Duvall Jr., S.M. Jefferies, J.W. Harvey, M.A. Pomerantz, Time-distance helioseismology. Nature **362**, 430–432 (1993). doi:10.1038/362430a0
R. Erdélyi, Magnetic coupling of waves and oscillations in the lower solar atmosphere: can the tail wag the dog? Philos. Trans. R. Soc. Lond. Ser. A, Math. Phys. Sci. **364**, 351–381 (2006). doi:10.1098/rsta.2005.1703
R. Erdélyi, M. Goossens, Magnetohydrodynamic waves and seismology of the solar atmosphere. Space Sci. Rev. **158**, 167–168 (2011). doi:10.1007/s11214-011-9800-1
Y. Fan, The emergence of a twisted Ω-tube into the solar atmosphere. Astrophys. J. Lett. **554**, 111–114 (2001). doi:10.1086/320935
Y. Fan, The three-dimensional evolution of buoyant magnetic flux tubes in a model solar convective envelope. Astrophys. J. **676**, 680–697 (2008). doi:10.1086/527317

Y. Fan, Magnetic fields in the solar convection zone. Living Rev. Sol. Phys. **6**, 4 (2009a)

Y. Fan, The emergence of a twisted flux tube into the solar atmosphere: sunspot rotations and the formation of a coronal flux rope. Astrophys. J. **697**, 1529–1542 (2009b). doi:10.1088/0004-637X/697/2/1529

F. Fang, W. Manchester, W.P. Abbett, B. van der Holst, Simulation of flux emergence from the convection zone to the corona. Astrophys. J. **714**, 1649–1657 (2010). doi:10.1088/0004-637X/714/2/1649

F. Fang, W. Manchester IV, W.P. Abbett, B. van der Holst, Buildup of magnetic shear and free energy during flux emergence and cancellation. Astrophys. J. **754**, 15 (2012a). doi:10.1088/0004-637X/754/1/15

F. Fang, W. Manchester IV, W.P. Abbett, B. van der Holst, Dynamic coupling of convective flows and magnetic field during flux emergence. Astrophys. J. **745**, 37 (2012b). doi:10.1088/0004-637X/745/1/37

V. Fedun, R. Erdélyi, S. Shelyag, Oscillatory response of the 3D solar atmosphere to the leakage of photospheric motion. Sol. Phys. **258**, 219–241 (2009). doi:10.1007/s11207-009-9407-9

V. Fedun, S. Shelyag, R. Erdélyi, Numerical modeling of footpoint-driven magneto-acoustic wave propagation in a localized solar flux tube. Astrophys. J. **727**, 17 (2011). doi:10.1088/0004-637X/727/1/17

T. Felipe, Three-dimensional numerical simulations of fast-to-Alfvén conversion in sunspots. Astrophys. J. **758**, 96 (2012). doi:10.1088/0004-637X/758/2/96

W. Finsterle, S.M. Jefferies, A. Cacciani, P. Rapex, S.W. McIntosh, Helioseismic mapping of the magnetic canopy in the solar chromosphere. Astrophys. J. Lett. **613**, 185–188 (2004). doi:10.1086/424996

G.H. Fisher, B.T. Welsch, W.P. Abbett, Can we determine electric fields and Poynting fluxes from vector magnetograms and Doppler measurements? Sol. Phys. **277**, 153–163 (2012). doi:10.1007/s11207-011-9816-4

C. Foullon, E. Verwichte, V.M. Nakariakov, Ultra-long-period oscillations in EUV filaments near to eruption: two-wavelength correlation and seismology. Astrophys. J. **700**, 1658–1665 (2009). doi:10.1088/0004-637X/700/2/1658

Y. Gao, J. Zhao, H. Zhang, Analysis on correlations between subsurface kinetic helicity and photospheric current helicity in active regions. Astrophys. J. Lett. **761**, 9 (2012). doi:10.1088/2041-8205/761/1/L9

S.E. Gibson, L. Fletcher, G. Del Zanna, C.D. Pike, H.E. Mason, C.H. Mandrini, P. Démoulin, H. Gilbert, J. Burkepile, T. Holzer, D. Alexander, Y. Liu, N. Nitta, J. Qiu, B. Schmieder, B.J. Thompson, The structure and evolution of a sigmoidal active region. Astrophys. J. **574**, 1021–1038 (2002). doi:10.1086/341090

L. Gizon, A.C. Birch, Local helioseismology. Living Rev. Sol. Phys. **2**, 6 (2005)

L. Gizon, T.L. Duvall Jr., R.M. Larsen, Seismic tomography of the near solar surface. J. Astrophys. Astron. **21**, 339 (2000). doi:10.1007/BF02702420

L. Gizon, H. Schunker, C.S. Baldner, S. Basu, A.C. Birch, R.S. Bogart, D.C. Braun, R. Cameron, T.L. Duvall, S.M. Hanasoge, J. Jackiewicz, M. Roth, T. Stahn, M.J. Thompson, S. Zharkov, Helioseismology of sunspots: a case study of NOAA region 9787. Space Sci. Rev. **144**, 249–273 (2009). doi:10.1007/s11214-008-9466-5

J.P.H. Goedbloed, S. Poedts, *Principles of Magnetohydrodynamics* (Cambridge University Press, Cambridge, 2004)

I. González Hernández, R. Komm, L. van Driel-Gesztelyi, L. Harra, L. Baker, R. Howe, Subsurface flows associated with non-Joy oriented active regions: a case study. J. Phys. Conf. Ser. **440**, 012050 (2013). doi:10.1088/1742-6596/440/1/012050

M. Goossens, R. Erdélyi, M.S. Ruderman, Resonant MHD waves in the solar atmosphere. Space Sci. Rev. **158**, 289–338 (2011). doi:10.1007/s11214-010-9702-7

M. Goossens, R. Soler, I. Arregui, J. Terradas, Analytic approximate seismology of propagating magnetohydrodynamic waves in the solar corona. Astrophys. J. **760**, 98 (2012). doi:10.1088/0004-637X/760/2/98

O.S. Gopasyuk, A.G. Kosovichev, Analysis of SOHO/MDI and TRACE observations of sunspot torsional oscillation in AR10421. Astrophys. J. **729**, 95 (2011). doi:10.1088/0004-637X/729/2/95

D.A. Haber, B.W. Hindman, J. Toomre, M.J. Thompson, Organized subsurface flows near active regions. Sol. Phys. **220**, 371–380 (2004). doi:10.1023/B:SOLA.0000031405.52911.08

M. Hagino, T. Sakurai, Solar-cycle variation of magnetic helicity in active regions. Publ. Astron. Soc. Jpn. **57**, 481–485 (2005)

S.C. Hansen, P.S. Cally, Benchmarking fast-to-Alfvén mode conversion in a cold MHD plasma. II. How to get Alfvén waves through the solar transition region? Astrophys. J. **751**, 31 (2012). doi:10.1088/0004-637X/751/1/31

V.H. Hansteen, B. De Pontieu, L. Rouppe van der Voort, M. van Noort, M. Carlsson, Dynamic fibrils are driven by magnetoacoustic shocks. Astrophys. J. Lett. **647**, 73–76 (2006). doi:10.1086/507452

T. Hartlep, A.G. Kosovichev, J. Zhao, N.N. Mansour, Signatures of emerging subsurface structures in acoustic power maps of the sun. Sol. Phys. **268**, 321–327 (2011). doi:10.1007/s11207-010-9544-1

J.W. Harvey, F. Hill, R.P. Hubbard, J.R. Kennedy, J.W. Leibacher, J.A. Pintar, P.A. Gilman, R.W. Noyes, A.M. Title, J. Toomre, R.K. Ulrich, A. Bhatnagar, J.A. Kennewell, W. Marquette, J. Patron, O. Saa, E. Yasukawa, The global oscillation network group (GONG) project. Science **272**, 1284–1286 (1996). doi:10.1126/science.272.5266.1284

S.S. Hasan, A.A. van Ballegooijen, W. Kalkofen, O. Steiner, Dynamics of the solar magnetic network: two-dimensional MHD simulations. Astrophys. J. **631**, 1270–1280 (2005). doi:10.1086/432655
J.-S. He, C.-Y. Tu, E. Marsch, L.-J. Guo, S. Yao, H. Tian, Upward propagating high-frequency Alfvén waves as identified from dynamic wave-like spicules observed by SOT on Hinode. Astron. Astrophys. **497**, 525–535 (2009). doi:10.1051/0004-6361/200810777
L. Heggland, V.H. Hansteen, B. De Pontieu, M. Carlsson, Wave propagation and jet formation in the chromosphere. Astrophys. J. **743**, 142 (2011). doi:10.1088/0004-637X/743/2/142
J. Hershaw, C. Foullon, V.M. Nakariakov, E. Verwichte, Damped large amplitude transverse oscillations in an EUV solar prominence, triggered by large-scale transient coronal waves. Astron. Astrophys. **531**, 53 (2011). doi:10.1051/0004-6361/201116750
F. Hill, Rings and trumpets—three-dimensional power spectra of solar oscillations. Astrophys. J. **333**, 996–1013 (1988). doi:10.1086/166807
B.W. Hindman, L. Gizon, T.L. Duvall Jr., D.A. Haber, J. Toomre, Comparison of solar subsurface flows assessed by ring and time-distance analyses. Astrophys. J. **613**, 1253–1262 (2004). doi:10.1086/423263
B.W. Hindman, D.A. Haber, J. Toomre, Helioseismically determined near-surface flows underlying a quiescent filament. Astrophys. J. **653**, 725–732 (2006). doi:10.1086/508603
B.W. Hindman, D.A. Haber, J. Toomre, Subsurface circulations within active regions. Astrophys. J. **698**, 1749–1760 (2009). doi:10.1088/0004-637X/698/2/1749
R.F. Howard, Solar active regions as diagnostics of subsurface conditions. Annu. Rev. Astron. Astrophys. **34**, 75–110 (1996). doi:10.1146/annurev.astro.34.1.75
S. Ilonidis, J. Zhao, A. Kosovichev, Detection of emerging sunspot regions in the solar interior. Science **333**, 993 (2011). doi:10.1126/science.1206253
S. Ilonidis, J. Zhao, A. Kosovichev, Response to comment on "Detection of emerging sunspot regions in the solar interior". Science **336**, 296 (2012a). doi:10.1126/science.1215539
S. Ilonidis, J. Zhao, A.G. Kosovichev, Helioseismic detection of emerging magnetic flux, in *Progress in Solar/Stellar Physics with Helio- and Asteroseismology*, ed. by H. Shibahashi, M. Takata, A.E. Lynas-Gray. Astronomical Society of the Pacific Conference Series, vol. 462, 2012b, p. 283
S. Ilonidis, J. Zhao, T. Hartlep, Helioseismic investigation of emerging magnetic flux in the solar convection zone. Astrophys. J. (2013, in press)
K. Jain, R.W. Komm, I. González Hernández, S.C. Tripathy, F. Hill, Subsurface flows in and around active regions with rotating and non-rotating sunspots. Sol. Phys. **279**, 349–363 (2012). doi:10.1007/s11207-012-0039-0
S.M. Jefferies, S.W. McIntosh, J.D. Armstrong, T.J. Bogdan, A. Cacciani, B. Fleck, Magnetoacoustic portals and the basal heating of the solar chromosphere. Astrophys. J. Lett. **648**, 151–155 (2006). doi:10.1086/508165
J.M. Jensen, T.L. Duvall Jr., B.H. Jacobsen, J. Christensen-Dalsgaard, Imaging an emerging active region with helioseismic tomography. Astrophys. J. Lett. **553**, 193–196 (2001). doi:10.1086/320677
D.B. Jess, M. Mathioudakis, R. Erdélyi, P.J. Crockett, F.P. Keenan, D.J. Christian, Alfvén waves in the lower solar atmosphere. Science **323**, 1582 (2009). doi:10.1126/science.1168680
D.B. Jess, S. Shelyag, M. Mathioudakis, P.H. Keys, D.J. Christian, F.P. Keenan, Propagating wave phenomena detected in observations and simulations of the lower solar atmosphere. Astrophys. J. **746**, 183 (2012a). doi:10.1088/0004-637X/746/2/183
D.B. Jess, D.J. Pascoe, D.J. Christian, M. Mathioudakis, P.H. Keys, F.P. Keenan, The origin of type I spicule oscillations. Astrophys. J. Lett. **744**, 5 (2012b). doi:10.1088/2041-8205/744/1/L5
Y. Jiang, R. Zheng, J. Yang, J. Hong, B. Yi, D. Yang, Rapid sunspot rotation associated with the X2.2 flare on 2011 February 15. Astrophys. J. **744**, 50 (2012). doi:10.1088/0004-637X/744/1/50
P.G. Judge, T.D. Tarbell, K. Wilhelm, A study of chromospheric oscillations using the SOHO and TRACE spacecraft. Astrophys. J. **554**, 424–444 (2001)
M.D. Kazachenko, R.C. Canfield, D.W. Longcope, J. Qiu, Predictions of energy and helicity in four major eruptive solar flares. Sol. Phys. **277**, 165–183 (2012). doi:10.1007/s11207-011-9786-6
S. Kholikov, A search for helioseismic signature of emerging active regions. Sol. Phys. (2013). doi:10.1007/s11207-013-0321-9
E. Khomenko, P.S. Cally, Numerical simulations of conversion to Alfvén waves in sunspots. Astrophys. J. **746**, 68 (2012). doi:10.1088/0004-637X/746/1/68
E. Khomenko, M. Collados, Numerical modeling of magnetohydrodynamic wave propagation and refraction in sunspots. Astrophys. J. **653**, 739–755 (2006). doi:10.1086/507760
E. Khomenko, M. Collados, Sunspot seismic halos generated by fast MHD wave refraction. Astron. Astrophys. **506**, 5–8 (2009). doi:10.1051/0004-6361/200913030
E. Khomenko, M. Collados, T. Felipe, Nonlinear numerical simulations of magneto-acoustic wave propagation in small-scale flux tubes. Sol. Phys. **251**, 589–611 (2008). doi:10.1007/s11207-008-9133-8

J.A. Klimchuk, On solving the coronal heating problem. Sol. Phys. **234**, 41–77 (2006). doi:10.1007/s11207-006-0055-z
J.A. Klimchuk, The role of type II spicules in the upper solar atmosphere. J. Geophys. Res. **117**(A16), 12102 (2012). doi:10.1029/2012JA018170
R. Komm, F. Hill, Solar flares and solar subphotospheric vorticity. J. Geophys. Res. **114**(A13), 6105 (2009). doi:10.1029/2008JA013977
R. Komm, T. Corbard, B.R. Durney, I. González Hernández, F. Hill, R. Howe, C. Toner, Solar subsurface fluid dynamics descriptors derived from global oscillation network group and Michelson-Doppler imager data. Astrophys. J. **605**, 554–567 (2004). doi:10.1086/382187
R. Komm, R. Howe, F. Hill, M. Miesch, D. Haber, B. Hindman, Divergence and vorticity of solar subsurface flows derived from ring-diagram analysis of MDI and GONG data. Astrophys. J. **667**, 571–584 (2007). doi:10.1086/520765
R. Komm, S. Morita, R. Howe, F. Hill, Emerging active regions studied with ring-diagram analysis. Astrophys. J. **672**, 1254–1265 (2008). doi:10.1086/523998
R. Komm, R. Howe, F. Hill, Emerging and decaying magnetic flux and subsurface flows. Sol. Phys. **258**, 13–30 (2009a). doi:10.1007/s11207-009-9398-6
R. Komm, R. Howe, F. Hill, I. González Hernández, Subsurface zonal flows. Sol. Phys. **254**, 1–15 (2009b). doi:10.1007/s11207-008-9285-6
R. Komm, R. Ferguson, F. Hill, G. Barnes, K.D. Leka, Subsurface vorticity of flaring versus flare-quiet active regions. Sol. Phys. **268**, 389–406 (2011a). doi:10.1007/s11207-010-9552-1
R. Komm, R. Howe, F. Hill, Subsurface velocity of emerging and decaying active regions. Sol. Phys. **268**, 407–428 (2011b). doi:10.1007/s11207-010-9692-3
R. Komm, R. Howe, F. Hill, Vorticity of subsurface flows of emerging and decaying active regions. Sol. Phys. **277**, 205–226 (2012). doi:10.1007/s11207-011-9920-5
R. Komm, S. Gosain, A. Pevtsov, Active regions with superpenumbral whirls and their subsurface kinetic helicity. Sol. Phys. (2013a). doi:10.1007/s11207-012-0218-z
R. Komm, R. Howe, I. González Hernández, L. Harra, L. Baker, L. van Driel-Gesztelyi, Are subsurface flows and coronal holes related? J. Phys. Conf. Ser. **440**, 012022 (2013b). doi:10.1088/1742-6596/440/1/012022
I. Kontogiannis, G. Tsiropoula, K. Tziotziou, Power halo and magnetic shadow in a solar quiet region observed in the Hα line. Astron. Astrophys. **510**, 41 (2010a). doi:10.1051/0004-6361/200912841
I. Kontogiannis, G. Tsiropoula, K. Tziotziou, M.K. Georgoulis, Oscillations in a network region observed in the Hα line and their relation to the magnetic field. Astron. Astrophys. **524**, 12 (2010b). doi:10.1051/0004-6361/201015066
A.G. Kosovichev, Local helioseismology of sunspots: current status and perspectives. Sol. Phys. **279**, 323–348 (2012). doi:10.1007/s11207-012-9996-6
A.G. Kosovichev, T.L. Duvall Jr., Local helioseismology and magnetic flux emergence, in *Subsurface and Atmospheric Influences on Solar Activity*, ed. by R. Howe, R.W. Komm, K.S. Balasubramaniam, G.J.D. Petrie. Astronomical Society of the Pacific Conference Series, vol. 383, 2008, p. 59
A.G. Kosovichev, T.L.. Duvall Jr., P.H. Scherrer, Time-distance inversion methods and results (Invited Review). Sol. Phys. **192**, 159–176 (2000)
F. Krause, K.-H. Raedler, *Mean-Field Magnetohydrodynamics and Dynamo Theory* (Pergamon Press, Oxford, 1980)
D. Kuridze, R.J. Morton, R. Erdélyi, G.D. Dorrian, M. Mathioudakis, D.B. Jess, F.P. Keenan, Transverse oscillations in chromospheric mottles. Astrophys. J. **750**, 51 (2012). doi:10.1088/0004-637X/750/1/51
K.D. Leka, G. Barnes, A.C. Birch, I. Gonzalez-Hernandez, T. Dunn, B. Javornik, D.C. Braun, Helioseismology of pre-emerging active regions. I. Overview, data, and target selection criteria. Astrophys. J. **762**, 130 (2013). doi:10.1088/0004-637X/762/2/130
C. Lindsey, D.C. Braun, Helioseismic imaging of sunspots at their antipodes. Sol. Phys. **126**, 101–115 (1990). doi:10.1007/BF00158301
Y. Liu, J. Zhao, P.W. Schuck, Horizontal flows in the photosphere and subphotosphere of two active regions. Sol. Phys. **287**(1–2), 279–291 (2012). doi:10.1007/s11207-012-0089-3
D.W. Longcope, B.T. Welsch, A model for the emergence of a twisted magnetic flux tube. Astrophys. J. **545**, 1089–1100 (2000). doi:10.1086/317846
D.W. Longcope, G.H. Fisher, A.A. Pevtsov, Flux-tube twist resulting from helical turbulence: the sigma-effect. Astrophys. J. **507**, 417–432 (1998). doi:10.1086/306312
M.L. Luoni, P. Démoulin, C.H. Mandrini, L. van Driel-Gesztelyi, Twisted flux tube emergence evidenced in longitudinal magnetograms: magnetic tongues. Sol. Phys. **270**, 45–74 (2011). doi:10.1007/s11207-011-9731-8
T. Magara, A model for dynamic evolution of emerging magnetic fields in the sun. Astrophys. J. **605**, 480–492 (2004). doi:10.1086/382148

T. Magara, Dynamic and topological features of photospheric and coronal activities produced by flux emergence in the Sun. Astrophys. J. **653**, 1499–1509 (2006). doi:10.1086/508926
T. Magara, D.W. Longcope, Sigmoid structure of an emerging flux tube. Astrophys. J. Lett. **559**, 55–59 (2001). doi:10.1086/323635
W. Manchester IV, T. Gombosi, D. DeZeeuw, Y. Fan, Eruption of a buoyantly emerging magnetic flux rope. Astrophys. J. **610**, 588–596 (2004). doi:10.1086/421516
D. Mason, R. Komm, F. Hill, R. Howe, D. Haber, B.W. Hindman, Flares, magnetic fields, and subsurface vorticity: a survey of GONG and MDI data. Astrophys. J. **645**, 1543–1553 (2006). doi:10.1086/503761
R.A. Maurya, A. Ambastha, V. Reddy, Kinetic and magnetic helicities in solar active regions. J. Phys. Conf. Ser. **271**(1), 012003 (2011). doi:10.1088/1742-6596/271/1/012003
S.W. McIntosh, B. De Pontieu, Observing episodic coronal heating events rooted in chromospheric activity. Astrophys. J. Lett. **706**, 80–85 (2009). doi:10.1088/0004-637X/706/1/L80
S.W. McIntosh, B. de Pontieu, M. Carlsson, V. Hansteen, P. Boerner, M. Goossens, Alfvénic waves with sufficient energy to power the quiet solar corona and fast solar wind. Nature **475**, 477–480 (2011). doi:10.1038/nature10235
S.W. McIntosh, H. Tian, M. Sechler, B. De Pontieu, On the Doppler velocity of emission line profiles formed in the "Coronal contraflow" that is the chromosphere-corona mass cycle. Astrophys. J. **749**, 60 (2012). doi:10.1088/0004-637X/749/1/60
C.A. Mendoza-Briceño, L.D.G. Sigalotti, R. Erdélyi, Catastrophic cooling of impulsively heated coronal loops. Astrophys. J. **624**, 1080–1092 (2005). doi:10.1086/429249
A.G. Michalitsanos, The five minute period oscillation in magnetically active regions. Sol. Phys. **30**, 47–61 (1973). doi:10.1007/BF00156172
S. Min, J. Chae, The rotating sunspot in AR 10930. Sol. Phys. **258**, 203–217 (2009). doi:10.1007/s11207-009-9425-7
H.K. Moffatt, A. Tsinober, Helicity in laminar and turbulent flow. Annu. Rev. Fluid Mech. **24**, 281–312 (1992). doi:10.1146/annurev.fl.24.010192.001433
H. Moradi, P.S. Cally, Time-distance modelling in a simulated sunspot atmosphere. Sol. Phys. **251**, 309–327 (2008). doi:10.1007/s11207-008-9190-z
H. Moradi, S.M. Hanasoge, P.S. Cally, Numerical models of travel-time inhomogeneities in sunspots. Astrophys. J. Lett. **690**, 72–75 (2009). doi:10.1088/0004-637X/690/1/L72
H. Moradi, C. Baldner, A.C. Birch, D.C. Braun, R.H. Cameron, T.L. Duvall, L. Gizon, D. Haber, S.M. Hanasoge, B.W. Hindman, J. Jackiewicz, E. Khomenko, R. Komm, P. Rajaguru, M. Rempel, M. Roth, R. Schlichenmaier, H. Schunker, H.C. Spruit, K.G. Strassmeier, M.J. Thompson, S. Zharkov, Modeling the subsurface structure of sunspots. Sol. Phys. **267**, 1–62 (2010). doi:10.1007/s11207-010-9630-4
R.J. Morton, R. Erdélyi, D.B. Jess, M. Mathioudakis, Observations of sausage modes in magnetic pores. Astrophys. J. Lett. **729**, 18 (2011a). doi:10.1088/2041-8205/729/2/L18
R.J. Morton, M.S. Ruderman, R. Erdélyi, Torsional Alfvén waves: magneto-seismology in static and dynamic coronal plasmas. Astron. Astrophys. **534**, 27 (2011b). doi:10.1051/0004-6361/201117020
R.J. Morton, A.K. Srivastava, R. Erdélyi, Observations of quasi-periodic phenomena associated with a large blowout solar jet. Astron. Astrophys. **542**, 70 (2012a). doi:10.1051/0004-6361/201117218
R.J. Morton, G. Verth, J.A. McLaughlin, R. Erdélyi, Determination of sub-resolution structure of a jet by solar magnetoseismology. Astrophys. J. **744**, 5 (2012b). doi:10.1088/0004-637X/744/1/5
M.J. Murray, A.W. Hood, F. Moreno-Insertis, K. Galsgaard, V. Archontis, 3D simulations identifying the effects of varying the twist and field strength of an emerging flux tube. Astron. Astrophys. **460**, 909–923 (2006). doi:10.1051/0004-6361:20065950
V.M. Nakariakov, R. Erdélyi, Foreword. Space Sci. Rev. **149**, 1–2 (2009). doi:10.1007/s11214-009-9620-8
V.M. Nakariakov, E. Verwichte, Coronal waves and oscillations. Living Rev. Sol. Phys. **2**, 3 (2005)
U. Narain, P. Ulmschneider, Chromospheric and coronal heating mechanisms II. Space Sci. Rev. **75**, 453–509 (1996). doi:10.1007/BF00833341
C. Nutto, O. Steiner, M. Roth, Revealing the nature of magnetic shadows with numerical 3D-MHD simulations. Astron. Astrophys. **542**, 30 (2012a). doi:10.1051/0004-6361/201218856
C. Nutto, O. Steiner, W. Schaffenberger, M. Roth, Modification of wave propagation and wave travel-time by the presence of magnetic fields in the solar network atmosphere. Astron. Astrophys. **538**, 79 (2012b). doi:10.1051/0004-6361/201016083
L. Ofman, MHD waves and heating in coronal holes. Space Sci. Rev. **120**, 67–94 (2005). doi:10.1007/s11214-005-5098-1
L. Ofman, J.A. Klimchuk, J.M. Davila, A self-consistent model for the resonant heating of coronal loops: the effects of coupling with the chromosphere. Astrophys. J. **493**, 474 (1998). doi:10.1086/305109
T.J. Okamoto, S. Tsuneta, T.E. Berger, K. Ichimoto, Y. Katsukawa, B.W. Lites, S. Nagata, K. Shibata, T. Shimizu, R.A. Shine, Y. Suematsu, T.D. Tarbell, A.M. Title, Coronal transverse magnetohydrodynamic waves in a solar prominence. Science **318**, 1577 (2007). doi:10.1126/science.1145447

D.E. Osterbrock, The heating of the solar chromosphere, plages, and corona by magnetohydrodynamic waves. Astrophys. J. **134**, 347 (1961). doi:10.1086/147165

K.V. Parchevsky, J. Zhao, T. Hartlep, A.G. Kosovichev, Verification of the travel time measurement technique and the helioseismic inversion procedure for sound speed using artificial data. ArXiv e-prints (2012)

C.E. Parnell, I. De Moortel, A contemporary view of coronal heating. Philos. Trans. R. Soc. Lond. Ser. A, Math. Phys. Sci. **370**, 3217–3240 (2012). doi:10.1098/rsta.2012.0113

D.J. Pascoe, A.W. Hood, I. de Moortel, A.N. Wright, Spatial damping of propagating kink waves due to mode coupling. Astron. Astrophys. **539**, 37 (2012). doi:10.1051/0004-6361/201117979

A.A. Pevtsov, R.C. Canfield, T.R. Metcalf, Patterns of helicity in solar active regions. Astrophys. J. Lett. **425**, 117–119 (1994). doi:10.1086/187324

A.A. Pevtsov, R.C. Canfield, T.R. Metcalf, Latitudinal variation of helicity of photospheric magnetic fields. Astrophys. J. Lett. **440**, 109–112 (1995). doi:10.1086/187773

A.A. Pevtsov, R.C. Canfield, S.M. Latushko, Hemispheric helicity trend for solar cycle 23. Astrophys. J. Lett. **549**, 261–263 (2001). doi:10.1086/319179

S.P. Rajaguru, S. Couvidat, X. Sun, K. Hayashi, H. Schunker, Properties of high-frequency wave power halos around active regions: an analysis of multi-height data from HMI and AIA onboard SDO. Sol. Phys. **287**, 107–127 (2013). doi:10.1007/s11207-012-0180-9

A.A. Reinard, J. Henthorn, R. Komm, F. Hill, Evidence that temporal changes in solar subsurface helicity precede active region flaring. Astrophys. J. Lett. **710**, 121–125 (2010). doi:10.1088/2041-8205/710/2/L121

C.S. Rosenthal, T.J. Bogdan, M. Carlsson, S.B.F. Dorch, V. Hansteen, S.W. McIntosh, A. McMurry, Å. Nordlund, R.F. Stein, Waves in the magnetized solar atmosphere. I. Basic processes and internetwork oscillations. Astrophys. J. **564**, 508–524 (2002). doi:10.1086/324214

G. Rüdiger, V.V. Pipin, G. Belvedère, Alpha-effect, helicity and angular momentum transport for a magnetically driven turbulence in the solar convection zone. Sol. Phys. **198**, 241–251 (2001). doi:10.1023/A:1005217606269

P.H. Scherrer, R.S. Bogart, R.I. Bush, J.T. Hoeksema, A.G. Kosovichev, J. Schou, W. Rosenberg, L. Springer, T.D. Tarbell, A. Title, C.J. Wolfson, I. Zayer, MDI Engineering Team, The solar oscillations investigation—Michelson-Doppler imager. Sol. Phys. **162**, 129–188 (1995). doi:10.1007/BF00733429

P.H. Scherrer, J. Schou, R.I. Bush, A.G. Kosovichev, R.S. Bogart, J.T. Hoeksema, Y. Liu, T.L. Duvall, J. Zhao, A.M. Title, C.J. Schrijver, T.D. Tarbell, S. Tomczyk, The helioseismic and magnetic imager (HMI) investigation for the solar dynamics observatory (SDO). Sol. Phys. **275**, 207–227 (2012). doi:10.1007/s11207-011-9834-2

C.J. Schrijver, Driving major solar flares and eruptions: a review. Adv. Space Res. **43**, 739–755 (2009). doi:10.1016/j.asr.2008.11.004

C.J. Schrijver, A.M. Title, Active regions losing their moorings by subsurface reconnection. Sol. Phys. **188**, 331–344 (1999)

H. Schunker, D.C. Braun, Newly identified properties of surface acoustic power. Sol. Phys. **268**, 349–362 (2011). doi:10.1007/s11207-010-9550-3

H. Schunker, P.S. Cally, Magnetic field inclination and atmospheric oscillations above solar active regions. Mon. Not. R. Astron. Soc. **372**, 551–564 (2006). doi:10.1111/j.1365-2966.2006.10855.x

H. Schunker, R.H. Cameron, L. Gizon, H. Moradi, Constructing and characterising solar structure models for computational helioseismology. Sol. Phys. **271**, 1–26 (2011). doi:10.1007/s11207-011-9790-x

M. Schüssler, M. Rempel, The dynamical disconnection of sunspots from their magnetic roots. Astron. Astrophys. **441**, 337–346 (2005). doi:10.1051/0004-6361:20052962

S. Shelyag, M. Mathioudakis, F.P. Keenan, D.B. Jess, A photospheric bright point model. Astron. Astrophys. **515**, 107 (2010). doi:10.1051/0004-6361/200913846

K. Shibata, Nonlinear MHD wave propagation in the solar chromosphere. I. The case of a uniform vertical magnetic field. Publ. Astron. Soc. Jpn. **35**, 263–284 (1983)

K. Shibata, T. Tajima, R.S. Steinolfson, R. Matsumoto, Two-dimensional magnetohydrodynamic model of emerging magnetic flux in the solar atmosphere. Astrophys. J. **345**, 584–596 (1989). doi:10.1086/167932

R.F. Stein, Å. Nordlund, On the formation of active regions. Astrophys. J. Lett. **753**, 13 (2012). doi:10.1088/2041-8205/753/1/L13

Y. Suematsu, Influence of photospheric 5-minute oscillations on the formation of chromospheric fine structures, in *Progress of Seismology of the Sun and Stars*, ed. by Y. Osaki, H. Shibahashi. Lecture Notes in Physics, vol. 367 (Springer, Berlin, 1990), p. 211. doi:10.1007/3-540-53091-6-83

Y. Taroyan, R. Erdélyi, Heating diagnostics with MHD waves. Space Sci. Rev. **149**, 229–254 (2009). doi:10.1007/s11214-009-9506-9

J. Terradas, I. Arregui, G. Verth, M. Goossens, Seismology of transversely oscillating coronal loops with siphon flows. Astrophys. J. Lett. **729**, 22 (2011). doi:10.1088/2041-8205/729/2/L22

J.H. Thomas, Magneto-atmospheric waves. Annu. Rev. Fluid Mech. **15**, 321–343 (1983). doi:10.1146/annurev.fl.15.010183.001541

L. Tian, D. Alexander, Role of sunspot and sunspot-group rotation in driving sigmoidal active region eruptions. Sol. Phys. **233**, 29–43 (2006). doi:10.1007/s11207-006-2505-z

S. Tomczyk, S.W. McIntosh, Time-distance seismology of the solar corona with CoMP. Astrophys. J. **697**, 1384–1391 (2009). doi:10.1088/0004-637X/697/2/1384

S. Tomczyk, S.W. McIntosh, S.L. Keil, P.G. Judge, T. Schad, D.H. Seeley, J. Edmondson, Alfvén waves in the solar corona. Science **317**, 1192 (2007). doi:10.1126/science.1143304

C.G. Toner, B.J. Labonte, Direct mapping of solar acoustic power. Astrophys. J. **415**, 847 (1993). doi:10.1086/173206

S. Toriumi, K. Hayashi, T. Yokoyama, Detection of the horizontal divergent flow prior to the solar flux emergence. Astrophys. J. **751**, 154 (2012). doi:10.1088/0004-637X/751/2/154

S. Toriumi, S. Ilonidis, T. Sekii, T. Yokoyama, Probing the shallow convection zone: rising motion of subsurface magnetic fields in the solar active region. Astrophys. J. Lett. **770**, 11 (2013). doi:10.1088/2041-8205/770/1/L11

A.A. van Ballegooijen, D.H. Mackay, Model for the coupled evolution of subsurface and coronal magnetic fields in solar active regions. Astrophys. J. **659**, 1713–1725 (2007). doi:10.1086/512849

A.A. van Ballegooijen, M. Asgari-Targhi, S.R. Cranmer, E.E. DeLuca, Heating of the solar chromosphere and corona by Alfvén wave turbulence. Astrophys. J. **736**, 3 (2011). doi:10.1088/0004-637X/736/1/3

A. Vecchio, G. Cauzzi, K.P. Reardon, K. Janssen, T. Rimmele, Solar atmospheric oscillations and the chromospheric magnetic topology. Astron. Astrophys. **461**, 1–4 (2007)

G. Verth, M. Goossens, J.-S. He, Magnetoseismological determination of magnetic field and plasma density height variation in a solar spicule. Astrophys. J. Lett. **733**, 15 (2011). doi:10.1088/2041-8205/733/1/L15

E. Verwichte, M. Marsh, C. Foullon, T. Van Doorsselaere, I. De Moortel, A.W. Hood, V.M. Nakariakov, Periodic spectral line asymmetries in solar coronal structures from slow magnetoacoustic waves. Astrophys. J. Lett. **724**, 194–198 (2010). doi:10.1088/2041-8205/724/2/L194

G. Vigeesh, S.S. Hasan, O. Steiner, Wave propagation and energy transport in the magnetic network of the sun. Astron. Astrophys. **508**, 951–962 (2009). doi:10.1051/0004-6361/200912450

A. Vögler, S. Shelyag, M. Schüssler, F. Cattaneo, T. Emonet, T. Linde, Simulations of magneto-convection in the solar photosphere. Equations, methods, and results of the MURaM code. Astron. Astrophys. **429**, 335–351 (2005). doi:10.1051/0004-6361:20041507

R.W. Walsh, J. Ireland, The heating of the solar corona. Astron. Astrophys. Rev. **12**, 1–41 (2003). doi:10.1007/s00159-003-0021-9

S. Wedemeyer-Böhm, L. Rouppe van der Voort, Small-scale swirl events in the quiet Sun chromosphere. Astron. Astrophys. **507**, 9–12 (2009). doi:10.1051/0004-6361/200913380

S. Wedemeyer-Böhm, E. Scullion, O. Steiner, L. Rouppe van der Voort, J. de La Cruz Rodriguez, V. Fedun, R. Erdélyi, Magnetic tornadoes as energy channels into the solar corona. Nature **486**, 505–508 (2012). doi:10.1038/nature11202

R.S. White, E. Verwichte, C. Foullon, First observation of a transverse vertical oscillation during the formation of a hot post-flare loop. Astron. Astrophys. **545**, 129 (2012). doi:10.1051/0004-6361/201219856

G.L. Withbroe, R.W. Noyes, Mass and energy flow in the solar chromosphere and corona. Annu. Rev. Astron. Astrophys. **15**, 363–387 (1977). doi:10.1146/annurev.aa.15.090177.002051

X.L. Yan, Z.Q. Qu, D.F. Kong, C.L. Xu, Sunspot rotation, sigmoidal filament, flare, and coronal mass ejection: the event on 2000 February 10. Astrophys. J. **754**, 16 (2012). doi:10.1088/0004-637X/754/1/16

M. Zhang, Helicity observations of weak and strong fields. Astrophys. J. Lett. **646**, 85–88 (2006). doi:10.1086/506560

Y. Zhang, J. Liu, H. Zhang, Relationship between rotating sunspots and flares. Sol. Phys. **247**, 39–52 (2008). doi:10.1007/s11207-007-9089-0

J. Zhao, A.G. Kosovichev, Helioseismic observation of the structure and dynamics of a rotating sunspot beneath the solar surface. Astrophys. J. **591**, 446–453 (2003). doi:10.1086/375343

J. Zhao, A.G. Kosovichev, Torsional oscillation, meridional flows, and vorticity inferred in the upper convection zone of the Sun by time-distance helioseismology. Astrophys. J. **603**, 776–784 (2004). doi:10.1086/381489

J. Zhao, A.G. Kosovichev, T.L. Duvall Jr., Investigation of mass flows beneath a sunspot by time-distance helioseismology. Astrophys. J. **557**, 384–388 (2001). doi:10.1086/321491

J. Zhao, A.G. Kosovichev, T.L. Duvall Jr., On the relationship between the rotational velocity and the field strength of solar magnetic elements. Astrophys. J. Lett. **607**, 135–138 (2004). doi:10.1086/421974

J. Zhao, A.G. Kosovichev, T. Sekii, High-resolution helioseismic imaging of subsurface structures and flows of a solar active region observed by Hinode. Astrophys. J. **708**, 304–313 (2010). doi:10.1088/0004-637X/708/1/304

S. Zharkov, M.J. Thompson, Time distance analysis of the emerging active region NOAA 10790. Sol. Phys. **251**, 369–380 (2008). doi:10.1007/s11207-008-9239-z

Y.D. Zhugzhda, N.S. Dzhalilov, Magneto-acoustic-gravity waves on the Sun. I—Exact solution for an oblique magnetic field. Astron. Astrophys. **132**, 45–51 (1984a)

Y.D. Zhugzhda, N.S. Dzhalilov, Magneto-acoustic-gravity waves on the Sun. II—Transformation and propagation. Astron. Astrophys. **132**, 52–57 (1984b)

Space Sci Rev (2015) 196:201–219
DOI 10.1007/s11214-015-0136-0

Interpretation of Helioseismic Travel Times

Sensitivity to Sound Speed, Pressure, Density, and Flows

Raymond Burston[1] · **Laurent Gizon**[1,2] · **Aaron C. Birch**[1]

Received: 8 February 2014 / Accepted: 27 January 2015 / Published online: 20 March 2015

Abstract Time-distance helioseismology uses cross-covariances of wave motions on the solar surface to determine the travel times of wave packets moving from one surface location to another. We review the methodology to interpret travel-time measurements in terms of small, localised perturbations to a horizontally homogeneous reference solar model. Using the first Born approximation, we derive and compute 3D travel-time sensitivity (Fréchet) kernels for perturbations in sound-speed, density, pressure, and vector flows. While kernels for sound speed and flows had been computed previously, here we extend the calculation to kernels for density and pressure, hence providing a complete description of the effects of solar dynamics and structure on travel times. We treat three thermodynamic quantities as independent and do not assume hydrostatic equilibrium. We present a convenient approach to computing damped Green's functions using a normal-mode summation. The Green's function must be computed on a wavenumber grid that has sufficient resolution to resolve the longest lived modes. The typical kernel calculations used in this paper are computer intensive and require on the order of 600 CPU hours per kernel. Kernels are validated by computing the travel-time perturbation that results from horizontally-invariant perturbations using two independent approaches. At fixed sound-speed, the density and pressure kernels are approximately related through a negative multiplicative factor, therefore implying that perturbations in density and pressure are difficult to disentangle. Mean travel-times are not only sensitive to sound-speed, density and pressure perturbations, but also to flows, especially vertical flows. Accurate sensitivity kernels are needed to interpret complex flow patterns such as convection.

Keywords Helioseismology · Solar interior · Wave propagation · Scattering

✉ L. Gizon
gizon@mps.mpg.de

[1] Max-Planck-Institut für Sonnensystemforschung, Justus-von-Liebig-Weg 3, 37077 Göttingen, Germany

[2] Institut für Astrophysik, Georg-August-Universität Göttingen, Friedrich-Hund-Platz 1, 37077 Göttingen, Germany

1 The Forward Problem in Time-Distance Helioseismology

There are two general approaches to helioseismology. Global helioseismology consists of interpreting the frequencies of the normal modes of oscillations, whereas local helioseismology uses spatial-temporal correlations of the wave field as the input data (time-distance helioseismology, acoustic holography, direct modelling) or local power spectra of solar oscillations (ring-diagram analysis). For overviews of local helioseismology techniques, see reviews by Christensen-Dalsgaard (2002), Gizon and Birch (2005), and Gizon et al. (2010).

In this paper, we focus on the forward problem of time-distance helioseismology, i.e. how to compute the effects of subsurface heterogeneities on wave travel times. The problem of inferring subsurface heterogeneities from observed travel times (the inverse problem) was first formalised and solved by Kosovichev (1996) and is also discussed in the reviews cited above. Throughout this paper we consider travel times measured between two points on the solar surface using a cross-covariance technique (Duvall et al. 1993). The purpose of the paper is to review and extend the computation of the sensitivity of wave travel times to small-amplitude changes to the local physical conditions of the solar interior. By a small-amplitude change to a reference solar model we mean a perturbation that causes a phase shift in the observed wave field that is much less than a radian. These sensitivity functions depend on the 3D position in the solar interior (the coordinates of local scatterers) and are known as Fréchet kernels, or simply kernels, in Earth seismology (Marquering et al. 1998, 1999; Dahlen et al. 2000; Hung et al. 2000). The solar problem differs from the earthquake seismology problem because solar oscillations are continuously excited by a random process, turbulent solar convection (granulation).

The first kernel calculations in time-distance helioseismology employed the ray approximation, in which the wavelength is assumed to be small compared to the other length scales in the problem (Kosovichev 1996; Kosovichev and Duvall 1997). Birch and Kosovichev (2000) used the first Born approximation (single-scattering approximation) to compute 3D kernels for sound-speed perturbations using the single-source approximation. Subsequently, Gizon and Birch (2002) introduced a general framework for Born kernel calculations including a treatment of wave excitation by near-surface convection, this framework is the basis for the calculations presented here. In particular, Gizon and Birch (2002) showed that spatially distributed wave sources alter the sensitivity kernels: scattering far away from the observation points can be important. They also showed that the details of the wave power spectrum, wave attenuation, and the details of the measurement procedure must all be included in the kernel computations. They proposed a measurement procedure that enables a consistent computation of kernels by relating small-amplitude changes in the cross-covariance to travel-time shifts. Using this approach kernels have been computed for source amplitude and wave attenuation (Gizon and Birch 2002), sound-speed perturbations (Birch et al. 2004), and vector flows (Gizon et al. 2000; Jackiewicz et al. 2006, 2007; Birch and Gizon 2007; Birch et al. 2011). We note that the same approach was used to compute kernels for ring-diagram analysis (Birch et al. 2007). All of these calculations assume that the helioseismic observable is the vertical wave velocity at the solar surface. Away from disk center, the horizontal velocity contributes to the Doppler velocity and short wavelengths are less visible towards the limb (Jackiewicz et al. 2006, 2007). All of these calculations employed a plane-parallel approximation for the background solar model (used for the study of small patches on the Sun).

More recent approaches based on the adjoint method are particularly useful for solving problems with 3D heterogeneous background models (Hanasoge et al. 2011). This method has been applied to a background model containing a magnetic field and 3D kernels for sound-speed, density, 3D flows, and magnetic fields were computed (Hanasoge et al. 2012). Mathematical expressions for density kernels, as well as sound-speed kernels, have also been presented by Schlottmann and Kosovichev (2011) using "functional analytic methods". The contribution of the lowest-order non-zero perturbations of magnetic field, to a background containing no magnetic field, is discussed in Gizon and Thompson (2007).

Other approaches than the Born approximation exist. For example, Jensen and Pijpers (2003) applied the Rytov approximation, which is closely related to the Born approximation, to compute kernels for sound-speed and flows in the single-source approximation.

In this paper, we work within the framework of Gizon and Birch (2002) and extend the computation of travel-time kernels to treat density and pressure perturbations, in addition to sound-speed and flow perturbations. For the sake of computational simplicity, we work in Cartesian geometry starting from a 1D static and non-magnetic reference solar model. Furthermore, we do not assume *a priori* hydrostatic equilibrium in the perturbed model to allow for contributions to the force balance from flows and magnetic fields and thus we retain three thermodynamic variables (sound speed, density, and pressure in the perturbed model).

2 Equivalent-Source Description of Wave Interaction

Linear adiabatic stellar oscillations (e.g. Lynden-Bell and Ostriker 1967) are typically governed by equations of the form $\mathcal{L}\boldsymbol{\xi}(\mathbf{r},t) = \mathbf{S}(\mathbf{r},t)$, where $\mathcal{L}$ is a linear differential operator, $\boldsymbol{\xi}(\mathbf{r},t)$ is the vector-valued wave displacement, and $\mathbf{S}(\mathbf{r},t)$ is the vector-valued source function responsible for wave excitation. We use $\mathbf{r}$ for the three dimensional (3D) position vector and t as time.

In the first-order Born approximation (e.g. Gizon and Birch 2002), the standard coupled system of linear differential equations is

$$\mathcal{L}_0\boldsymbol{\xi}_0 = \mathbf{S}_0, \tag{1}$$

$$\mathcal{L}_0\delta\boldsymbol{\xi} = -\delta\mathcal{L}\boldsymbol{\xi}_0 + \delta\mathbf{S}. \tag{2}$$

Equation (1) describes the propagation of small amplitude waves through a background solar model. The operator $\mathcal{L}_0$ is the reference linear wave operator, $\boldsymbol{\xi}_0$ the displacement field in the reference model and $\mathbf{S}_0$ is the reference source field. Equation (2) governs the perturbed displacement field, $\delta\boldsymbol{\xi}$. The first-order linear operator $\delta\mathcal{L}$ encapsulates perturbations to the solar thermodynamic structure and the introduction of flows, and $\delta\mathbf{S}$ is the perturbation to the source field. In this paper, we ignore $\delta\mathbf{S}$ (see e.g. Gizon and Birch 2002, for a discussion).

For a reference solar model with vanishing flows and magnetic fields, and also using the Cowling approximation, the reference linear operator governing the reference displacement field is (see e.g. Cameron and Gizon 2006)

$$\mathcal{L}_0\boldsymbol{\xi}_0 := \rho_0(\partial_t + \Gamma)^2\boldsymbol{\xi}_0 - \nabla\left(c_0^2\rho_0\nabla\cdot\boldsymbol{\xi}_0 + \boldsymbol{\xi}_0\cdot\nabla P_0\right) + \mathbf{g}_0\nabla\cdot(\rho_0\boldsymbol{\xi}_0). \tag{3}$$

The background solar model variables, which depend on the distance from the center of the Sun, have their usual symbols: c_0 is the sound-speed, ρ_0 is the density, P_0 is the pressure, and $\mathbf{g}_0$ is the acceleration due to gravity. The partial derivative with subscript t denotes differentiation with respect to time and the vector ∇ is the 3 dimensional spatial gradient operator. The phenomenological operator, Γ, is responsible for damping the oscillations. The random source function $\mathbf{S}_0$ is assumed to be statistically homogeneous in the horizontal direction and statistically stationary in time and represents the driving of oscillations by near-surface turbulent convection.

The first-order linear operator $\delta\mathcal{L}$ due to first-order perturbations in sound-speed squared (δc^2), density ($\delta\rho$), pressure (δP) and 3D mass flows ($\mathbf{v}$) is,

$$\begin{aligned}\delta\mathcal{L}\boldsymbol{\xi}_0 := &-\nabla\big(\delta c^2\rho_0\nabla\cdot\boldsymbol{\xi}_0\big)+2\rho_0(\mathbf{v}\cdot\nabla)\partial_t\boldsymbol{\xi}_0-\nabla(\boldsymbol{\xi}_0\cdot\nabla\delta P)\\ &+\delta\rho(\partial_t+\Gamma)^2\boldsymbol{\xi}_0-\nabla\big(c_0^2\delta\rho\nabla\cdot\boldsymbol{\xi}_0\big)+\mathbf{g}_0\nabla\cdot(\delta\rho\boldsymbol{\xi}_0)+\delta\mathbf{g}\nabla\cdot(\rho_0\boldsymbol{\xi}_0).\end{aligned}\tag{4}$$

The first-order changes in gravity $\delta\mathbf{g}$ are related to first-order changes in density and are expected to be very small near the solar surface.

We use sound-speed, density, and pressure as our three independent thermodynamic variables. It is possible to make other choices; e.g. the adiabatic exponent γ may be introduced at the expense of the others by using the relationship $\rho c^2 = P\gamma$. We do not assume hydrostatic equilibrium for the thermodynamic perturbations. If hydrostatic equilibrium was assumed, then it would be possible to eliminate the gradient of the pressure perturbation (in Eq. (4)) with a contribution from the density perturbation (or vice-versa), and consequently, reduce the number of independent perturbation variables to two (e.g. sound-speed and density) as is done in global helioseismology (e.g. Christensen-Dalsgaard 2002).

3 Computing Travel-Time Sensitivity Kernels for Structure and Flows

In time-distance helioseismology wave travel times (τ) are measured from the temporal cross-covariance of the observed signal, $\phi(\mathbf{x}, t)$:

$$C(t;\mathbf{x}_1,\mathbf{x}_2)=\frac{1}{T}\int_0^T dt'\phi\big(\mathbf{x}_1,t'\big)\phi\big(\mathbf{x}_2,t'+t\big),\tag{5}$$

where t is the time lag, $\mathbf{x}_1$ and $\mathbf{x}_2$ are two spatial locations on the solar surface and T is the duration of the time series. In the temporal Fourier domain it becomes

$$C(\omega;\mathbf{x}_1,\mathbf{x}_2)=h_\omega\phi^*(\mathbf{x}_1,\omega)\phi(\mathbf{x}_2,\omega),\tag{6}$$

where ω is the angular frequency and the superscript $*$ denotes complex conjugate (throughout this paper we use the Fourier conventions from Gizon and Birch 2002), $C(\omega;\mathbf{x}_1,\mathbf{x}_2)$ is the temporal Fourier transform of $C(t;\mathbf{x}_1,\mathbf{x}_2)$ and $\phi(\mathbf{x},\omega)$ is the temporal Fourier transform of $\phi(\mathbf{x},t)$. The sampling in frequency space is inversely proportional to the duration $h_\omega = 2\pi/T$.

The observable ϕ is often a line-of-sight Doppler velocity; it is connected to the oscillations of the atmosphere caused by seismic waves. There exists a non-trivial relationship between $\boldsymbol{\xi}$ and ϕ (see e.g. Nagashima et al. 2014) which can be expressed in general as

$$\phi(\mathbf{x},\omega)=O\big\{\boldsymbol{\xi}(\mathbf{r},\omega)\big\},\tag{7}$$

where O is an operator that acts on the 3D displacement and maps it onto the CCD detector and $\mathbf{r} = (\mathbf{x}, z)$. The operator includes the Point Spread Function (PSF) of the instrument plus any filters that were applied to the data.

Then using the Born approximation (Sect. 2) the first-order change to the cross-covariance is given by

$$\delta C(\omega; \mathbf{x}_2, \mathbf{x}_1) = -(2\pi)^4 \sum_\alpha \int_\odot d\mathbf{r} \sum_j \mathcal{G}^j(\mathbf{x}_2, \omega, \mathbf{r})\hat{\mathbf{e}}_j \cdot \delta\mathcal{L}^\alpha_{\mathbf{r},\omega}\mathbf{X}(\mathbf{r}, \omega, \mathbf{x}_1) + (1 \leftrightarrow 2)^* \quad (8)$$

$$= \sum_\alpha \int_\odot d\mathbf{r}\mathcal{C}_\alpha(\mathbf{r}, \omega; \mathbf{x}_2, \mathbf{x}_1)\delta q_\alpha(\mathbf{r}) + (1 \leftrightarrow 2)^*, \quad (9)$$

$$\text{where } \mathbf{X}(\mathbf{r}, \omega, \mathbf{x}_1) := \sum_{i,i'} \int_\odot d\mathbf{r}_s \int_\odot d\mathbf{r}'_s \mathcal{G}^{i*}(\mathbf{x}_1, \omega, \mathbf{r}_s)\mathbf{G}^{i'}(\mathbf{r}, \omega, \mathbf{r}'_s) M_{ii'}(\mathbf{r}_s, \mathbf{r}'_s, \omega), \quad (10)$$

$$\text{and } M_{ii'}(\mathbf{r}_s, \mathbf{r}'_s, \omega) := E\left[S^*_i(\mathbf{r}_s, \omega) S_{i'}(\mathbf{r}'_s, \omega)\right]. \quad (11)$$

The Green's tensor $\mathbf{G}^i$ is defined by

$$\mathcal{L}_0\mathbf{G}^i(\mathbf{r}, t - t_s, \mathbf{r}_s) = \delta(\mathbf{r} - \mathbf{r}_s)\delta(t - t_s)\hat{\mathbf{e}}_i, \quad (12)$$

and for convenience we also defined

$$\mathcal{G}^i = O\{\mathbf{G}^i\}. \quad (13)$$

The operator $\delta\mathcal{L}^\alpha_{\mathbf{r},\omega}$ is the component of the perturbed wave operator ($\delta\mathcal{L}$) that is due to the change in the physical quantity $\delta q_\alpha(\mathbf{r})$. The quantity $M_{ii'}$ is the source covariance tensor. The symbol $(1 \leftrightarrow 2)^*$ is a short notation to mean switch $\mathbf{x}_1$ and $\mathbf{x}_2$ and take the complex conjugate of all terms in the expression. The scattering physics is included in the functions $\mathcal{C}_\alpha$ that give the linear sensitivity of the expectation value of the cross-covariance to changes in the solar model. In order to arrive at (9), it is often required to use integration by parts to separate the perturbation variables from spatial derivatives, and thus in those cases it is been assumed that relevant quantities vanish on the boundaries.

Using (9), $\mathcal{C}_\alpha$ can be expressed for each of the perturbations

$$\mathcal{C}_{c^2}(\mathbf{r}; \mathbf{x}_2, \mathbf{x}_1) = -(2\pi)^4\rho_0(\mathbf{r})c_0^2(\mathbf{r})\left[\sum_j \partial_{r_j}\mathcal{G}^j(\mathbf{x}_2, \mathbf{r})\right]\nabla_\mathbf{r} \cdot \mathbf{X}(\mathbf{r}, \mathbf{x}_1) + (1 \leftrightarrow 2)^*, \quad (14)$$

$$\begin{aligned}\mathcal{C}_\rho(\mathbf{r}; \mathbf{x}_2, \mathbf{x}_1) = -(2\pi)^4\rho_0(\mathbf{r})\Big[&(\omega + \mathrm{i}\gamma)^2\sum_j \mathcal{G}^j(\mathbf{x}_2, \mathbf{r})X_j(\mathbf{r}, \mathbf{x}_1) \\ &+ c_0^2(\mathbf{r})\nabla_\mathbf{r} \cdot \mathbf{X}(\mathbf{r}, \mathbf{x}_1)\sum_j \partial_{r_j}\mathcal{G}^j(\mathbf{x}_2, \mathbf{r}) \\ &- \mathbf{X}(\mathbf{r}, \mathbf{x}_1) \cdot \nabla_\mathbf{r}\sum_j \mathcal{G}^j(\mathbf{x}_2, \mathbf{r})g_j(\mathbf{r})\Big] \\ &- (2\pi)^4\delta g_i(\mathbf{r})\frac{\rho_0}{\delta\rho}(\mathbf{r})\mathcal{G}^i(\mathbf{x}_2, \mathbf{r})\nabla_\mathbf{r} \cdot \left[\rho_0\mathbf{X}(\mathbf{r}, \mathbf{x}_1)\right] + (1 \leftrightarrow 2)^*,\end{aligned} \quad (15)$$

$$\mathcal{C}_P(\mathbf{r}; \mathbf{x}_2, \mathbf{x}_1) = (2\pi)^4 P_0(\mathbf{r})\nabla \cdot \left[\mathbf{X}(\mathbf{r}, \mathbf{x}_1)\sum_j \partial_{r_j}\mathcal{G}^j(\mathbf{x}_2, \mathbf{r})\right] + (1 \leftrightarrow 2)^*, \quad (16)$$

$$\mathcal{C}_{\mathbf{v}_i}(\mathbf{r}; \mathbf{x}_2, \mathbf{x}_1) = (2\pi)^4 2\mathrm{i}\omega\rho_0(\mathbf{r})\mathcal{G}^j(\mathbf{x}_2, \mathbf{r})\partial_{r_i}X_j(\mathbf{r}, \mathbf{x}_1) + (1 \leftrightarrow 2)^*, \quad (17)$$

where i is the imaginary number. Note that the first-order perturbation to gravity can be expressed in terms of first-order density according to

$$\nabla \cdot \delta \mathbf{g} = -4\pi G \delta \rho \tag{18}$$

where G is the gravitational constant.

The integrals in (8)–(10) are 3D and over the volume of the Sun. The 3D flow vector ($\mathbf{v}$) has been decomposed into its three components denoted v_i. The sensitivity kernel K_α for a given physical quantity q_α is computed with all other physical quantities held fixed. Each kernel $K_\alpha(\mathbf{r})$ due to a point scatterer at $\mathbf{r}$ in the interior is computed according to

$$K_\alpha(\mathbf{r}; \mathbf{x}_1, \mathbf{x}_2) = 4\pi \mathrm{Re} \int_0^\infty W^*(\omega; \mathbf{x}_1, \mathbf{x}_2) \mathcal{C}_\alpha(\mathbf{r}, \omega; \mathbf{x}_1, \mathbf{x}_2) d\omega, \tag{19}$$

where Re takes the real part of the following expression and the function W gives the linear dependence of the travel-time shifts on the cross-covariance. The functions W incorporate all the details of the travel-time measurement procedure (Gizon and Birch 2002). One-way travel-times are sensitive to all types of perturbations. Duvall et al. (1997) introduced travel-time differences $\delta\tau_{\mathrm{diff}} = \delta\tau(\mathbf{x}_1, \mathbf{x}_2) - \delta\tau(\mathbf{x}_2, \mathbf{x}_1)$ to partially isolate the contribution from flows and mean travel times $\delta\tau_{\mathrm{mean}} = [\delta\tau(\mathbf{x}_1, \mathbf{x}_2) + \delta\tau(\mathbf{x}_2, \mathbf{x}_1)]/2$ to measure the other contributions. The corresponding W-functions are

$$W_{\mathrm{diff}}(\omega; \mathbf{x}_1, \mathbf{x}_2) = W(\omega; \mathbf{x}_1, \mathbf{x}_2) - W(\omega; \mathbf{x}_2, \mathbf{x}_1), \tag{20}$$

$$W_{\mathrm{mean}}(\omega; \mathbf{x}_1, \mathbf{x}_2) = \big[W(\omega; \mathbf{x}_1, \mathbf{x}_2) + W(\omega; \mathbf{x}_2, \mathbf{x}_1)\big]/2. \tag{21}$$

The forward problem consists of modelling the chain of perturbations $\delta\boldsymbol{\xi} \to \delta\phi \to \delta C \to \delta\tau$ caused by changes in the structure and dynamics of the solar interior. Travel-time shifts ($\delta\tau$) due to first-order Born perturbations in sound-speed squared, density, pressure and 3D vector flows are given by

$$\begin{aligned} \delta\tau(\mathbf{x}_1, \mathbf{x}_2) = \int_\odot d\mathbf{r} \bigg[& K_{c^2}(\mathbf{r}; \mathbf{x}_1, \mathbf{x}_2) \frac{\delta c^2}{c_0^2}(\mathbf{r}) + K_\rho(\mathbf{r}; \mathbf{x}_1, \mathbf{x}_2) \frac{\delta\rho}{\rho_0}(\mathbf{r}) \\ & + K_P(\mathbf{r}; \mathbf{x}_1, \mathbf{x}_2) \frac{\delta P}{P_0}(\mathbf{r}) + \sum_{i=1}^{3} K_{v_i}(\mathbf{r}; \mathbf{x}_1, \mathbf{x}_2) v_i(\mathbf{r}) \bigg]. \end{aligned} \tag{22}$$

4 Plane Parallel Example

Here we compute example kernels assuming a plane-parallel model which is appropriate for studying small areas on the Sun. We assume the observable is given by

$$\phi(\mathbf{k}, \omega) = -\mathrm{i}\omega F(\mathbf{k}, \omega) \hat{\boldsymbol{\ell}} \cdot \boldsymbol{\xi}(\mathbf{k}, z_{\mathrm{obs}}, \omega), \tag{23}$$

where $\mathbf{k}$ is a horizontal wave vector, $\hat{\boldsymbol{\ell}}$ is the line-of-sight unit vector that we assume points in the vertical direction ($\hat{\boldsymbol{\ell}} = \hat{\mathbf{z}}$). We used an observation height of $z_{\mathrm{obs}} = 300$ km.

The function F is a combination of both the instrumental PSF and filters. In this paper, we focus on modelling observations of waves with the same radial order. The solar power spectrum shows ridges of enhanced power. Each of these ridges corresponds to waves with a particular radial order. To isolate a single radial order, the power spectrum is filtered using a ridge filter that suppress all waves at other radial orders (e.g. see Jackiewicz et al. 2008). In this paper we consider both an f-mode ridge filter and p_1-mode ridge filter.

We assume spatially uncorrelated sources, such that the example source covariance tensor is

$$M_{ii'}\big(\mathbf{r}_s, \mathbf{r}'_s, \omega\big) = (2\pi)^2 \delta_{i,z}\delta_{i',z}\delta\big(\mathbf{x}_s - \mathbf{x}'_s\big) m(\omega)\partial_{z_s}\delta(z_s - z_{\mathrm{src}})\partial_{z'_s}\delta\big(z'_s - z_{\mathrm{src}}\big), \tag{24}$$

where $\mathbf{r}_s = (\mathbf{x}_s, z_s)$ and $\mathbf{r}'_s = (\mathbf{x}'_s, z'_s)$. This is the source covariance described by Birch et al. (2004). The depth dependence is a vertical derivative of a vertical momentum source. This form of the source covariance gives a good match to the observed power spectrum for the p modes of low radial order. The frequency dependence here is Gaussian as in Birch et al. (2004) however, there is an additional factor of $1/\omega^2$ since we are solving for displacement whereas Birch et al. (2004) was solving for velocity. Here, the source auto-correlation function is $m(\omega) := \omega^{-2} \exp[-(\omega T_{\mathrm{src}})^2]$, with a correlation time $T_{\mathrm{src}} = 48$ s. We note that M is defined to within an overall scale factor, which changes the amplitude of the C_q but does not change the kernels K_q since the W-functions are inversely proportional to this scale factor. We used a source height of $z_{\mathrm{src}} = -100$ km (Birch et al. 2004).

4.1 Convenient Mode-Summation Formula for the Green's Tensor

The solutions to (1) and (2) can be written in terms of Green's functions (Gizon and Birch 2002). We use normal-mode summation to compute Green's functions, however, the approach presented here is different to what was implemented in Birch et al. (2004). Here we write the Green's functions for the damped problem in terms of normal modes of the undamped problem, and consequently, the new approach is less mathematically and computationally demanding. The equation governing the reference solar oscillations can be expressed as

$$\mathcal{L}\boldsymbol{\xi}_0 = \rho_0\big[(\partial_t + \Gamma)^2 + \mathcal{H}\big]\boldsymbol{\xi}_0 = \mathbf{S}_0, \tag{25}$$

where $\mathcal{H}$ can be identified from Sect. 2. Normal-mode eigensolutions (undamped) thus must satisfy

$$\mathcal{H}\boldsymbol{\xi}^n(\mathbf{r}, \mathbf{k}) = \omega_n^2(k)\boldsymbol{\xi}^n(\mathbf{r}, \mathbf{k}), \tag{26}$$

where $\boldsymbol{\xi}^n(\mathbf{r}, \mathbf{k})$ is the eigenfunction with radial order n and horizontal wave vector $\mathbf{k}$, $\omega_n(k)$ is the associated (real) eigenfrequency, and $k := \|\mathbf{k}\|$ is the wavenumber. Notice that because of the assumed horizontal isotropy of the reference solar model the normal-mode frequencies are independent of the direction of $\mathbf{k}$. The displacement eigenfunctions may be decomposed into vertical and horizontal components according to

$$\boldsymbol{\xi}^n(\mathbf{r}, \mathbf{k}) = \big[\hat{\mathbf{z}}\xi_z^n(z, k) + \mathrm{i}\hat{\mathbf{k}}\xi_h^n(z, k)\big]e^{\mathrm{i}\mathbf{k}\cdot\mathbf{x}}, \tag{27}$$

where $\mathbf{x}$ is a horizontal position vector and z denotes height. Because the only restoring forces are pressure and buoyancy there is no motion in the direction $\hat{\mathbf{k}} \times \hat{\mathbf{z}}$. We choose to normalise the eigenfunctions according to

$$\int \rho_0\boldsymbol{\xi}^{n*}(\mathbf{r}, \mathbf{k}) \cdot \boldsymbol{\xi}^{n'}\big(\mathbf{r}, \mathbf{k}'\big)d^3\mathbf{r} = \delta_{n,n'}\delta\big(\mathbf{k} - \mathbf{k}'\big), \tag{28}$$

where $\delta_{n,n'}$ is a Kronecker delta, $\delta(\mathbf{k}-\mathbf{k}')$ is a Dirac delta function. The integral is three dimensional and over all space. For the horizontally homogeneous medium under study, we define the Green's vector $\mathbf{G}^i(\mathbf{x}-\mathbf{s}, z, t-t_s, z_s)$ as the displacement vector at position $\mathbf{r}=(\mathbf{x}, z)$ and time t that results from a delta function source at position $\mathbf{r}_s=(\mathbf{s}, z_s)$ and time t_s pointing in the $\hat{\mathbf{e}}_i$ direction, where $\hat{\mathbf{e}}_i$ is one of the unit vectors $\hat{\mathbf{e}}_x$, $\hat{\mathbf{e}}_y$ or $\hat{\mathbf{e}}_z$. If we take the Fourier transform in time, then we get

$$\rho_0\big[-(\omega+i\gamma)^2+\mathcal{H}\big]\mathbf{G}^i(\mathbf{x}-\mathbf{s}, z, \omega, z_s)=\frac{1}{2\pi}\delta(\mathbf{r}-\mathbf{r}_s)\hat{\mathbf{e}}_i. \tag{29}$$

The function γ damps the waves so their lifetimes are finite. Typically it is chosen such that the subsequent line-widths of the model reproduce the line-widths observed in the Sun.

We now expand the $\mathbf{G}^i$ on a basis of undamped normal-mode solutions according to

$$\mathbf{G}^i(\mathbf{x}-\mathbf{s}, z, \omega, z_s)=\sum_n\int d^2\mathbf{k}c_n^i(\omega, \mathbf{r}_s, \mathbf{k})\boldsymbol{\xi}^n(\mathbf{r}, \mathbf{k}), \tag{30}$$

where $c_n^i(\omega, \mathbf{r}_s, \mathbf{k})$ are coefficients to be determined. The sum is over all radial orders n and the integral is two dimensional over all horizontal wave numbers. After taking the dot product of (29) with $\boldsymbol{\xi}^{n'*}(\mathbf{r}, \mathbf{k}')$ and integrating over $\mathbf{r}$, we find

$$\big[-\big(\omega+\mathrm{i}\gamma_{n'}(k)\big)^2+\omega_{n'}^2(k)\big]c_{n'}^i(\omega, \mathbf{r}_s, \mathbf{k})=\frac{1}{2\pi}\xi_i^{n'*}(\mathbf{r}_s, \mathbf{k}), \tag{31}$$

where $\gamma(k)\boldsymbol{\xi}^n(\mathbf{r}, \mathbf{k})=\gamma_n(k)\boldsymbol{\xi}^n(\mathbf{r}, \mathbf{k})$. Finally, after some additional manipulation we have

$$\mathbf{G}^i(\mathbf{k}, z, \omega, z_s)=\frac{1}{2\pi}\sum_n\frac{\boldsymbol{\zeta}^n(z, \mathbf{k})\zeta_i^{n*}(z_s, \mathbf{k})}{-[\omega+\mathrm{i}\gamma_n(k)]^2+\omega_n^2(k)}, \tag{32}$$

where we introduce the convenient notation for the eigenfunction

$$\boldsymbol{\zeta}^n(z, \mathbf{k}):=\hat{\mathbf{z}}\xi_z^n(z, k)+\mathrm{i}\hat{\mathbf{k}}\xi_h^n(z, k). \tag{33}$$

Equations (32) and (33) give a convenient set of expressions for computing the Green's functions for the damped problem in terms of the modes of the undamped problem. They are, in practice, easier to use than the expressions of Birch et al. (2004), which involve the modes of the damped problem; the modes of the undamped problem are easier to compute than the modes of the damped problem. In addition, the damping model (i.e., the choice of the γ_n) can be updated independently of the computation of the normal modes.

Figure 1 shows how the complex Green's function presented here compares with two computations from Birch et al. (2004) versus acoustic depth, at fixed ω and k. Birch et al. (2004) used a slightly different definition of the Green's function: $(\partial_t+\Gamma)^2$ is replaced by $\partial_t^2+2\Gamma\partial_t$ in the wave operator $\mathcal{L}_0$ and $\Gamma=\Gamma(z)$. The problem was solved by Birch et al. (2004) by summation over the modes of the damped problem (unlike in the present paper) and also by direct numerical solution of the differential equations. The source depth is $z_s=3.7$ Mm and the horizontal wavenumber is $k=1$ rad Mm^{-1}. For the 2004 paper, the cyclic frequency is $\omega/2\pi=3.92$ mHz, a frequency just above the $n=2$ resonance. The normal mode eigenfrequencies from the model used in this paper are slightly different from those in Birch et al. (2004); on the order of a few μHz. Thus in order to make a good comparison the new Green's function was evaluated a few μHz higher at $\omega/2\pi=3.92265$ mHz, in order to be a similar distance from the $n=2$ mode. We find that the two mode-summation

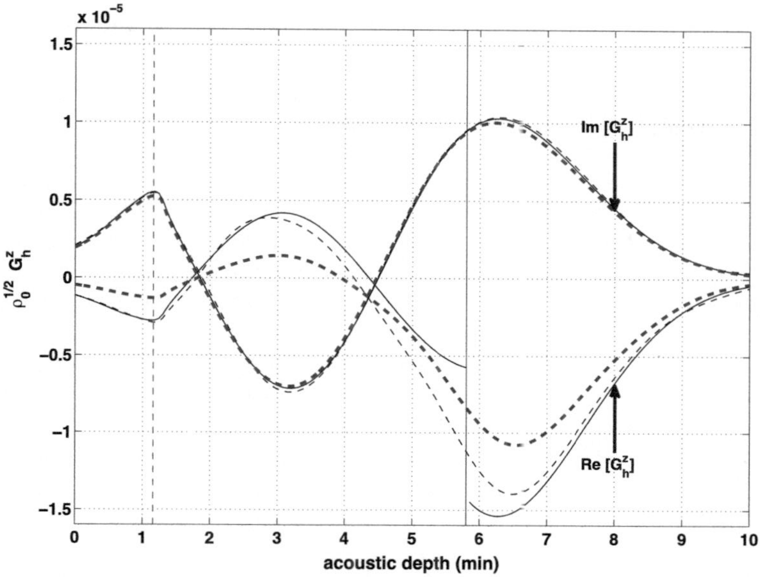

Fig. 1 Horizontal component of the Green's function $\mathbf{G}^z$ from this study (*blue dashed curves*) compared with normal-mode summation Green's function from Birch et al. (2004) (*dashed black curves*) and the numerical solution from Birch et al. (2004) (*solid black curves*). There are two curves for each Green's function to show the real and imaginary parts. The source depth is $z_s = 3.7$ Mm and the horizontal wavenumber is $k = 1\ \mathrm{rad\,Mm^{-1}}$. The mode summation formulae uses modes from $n = 0$ up to $n = 14$. *The vertical black dashed line* shows the observation height and *the vertical solid black line* shows the source depth

solutions (both used radial orders $0 \leq n \leq 14$) give similar answers (dashed curves in Fig. 1): The imaginary parts of the Green's functions match very well, the real parts less so. The difference in the real parts is due to the non-vanishing imaginary component of the damped eigenfunctions from Birch et al. (2004). The real part of the numerical Green's function (solid curves in Fig. 1) has a discontinuity at the source depth that cannot be reproduced by a sum over only 15 modes. Whether the above differences are important for the interpretation of helioseismic travel times deserves to be studied.

4.2 Background Solar Model and Zero Order Wavefield

We use model S (Christensen-Dalsgaard et al. 1996) for the background model of the Sun, where sound-speed, density, pressure, and gravity are functions of height only. There are no flows or magnetic fields in this background model.

The reference displacement field for the undamped problem (i.e. $\boldsymbol{\xi}_0$) is computed by solving (1). We follow the work of Gizon and Birch (2002) and Birch et al. (2004) and use a plane-parallel (Cartesian) approximation. The Cowling approximation is also enforced. This reduces to an eigenvalue problem similar to that encountered in Birch et al. (2004), however here we solve for $\boldsymbol{\xi}_0$ rather than $\partial_t^2 \boldsymbol{\xi}_0$. It was particularly important to ensure the eigenfunction solutions were well behaved under double numerical differentiation over height, since some of the features of the pressure kernels are sensitive to small errors. We use the same source covariance as in Birch et al. (2004).

The undamped reference displacement field is used to compute the attenuated Green's function through (32).

4.3 Numerical Resolution in Fourier Space and Spatial Computational Domain

The kernels were computed using the Model-S non-uniform height grid that is provided as part of the Aarhus adiabatic pulsation package. We selected to use a subset of the height domain from $z = 0.4959$ Mm to $z = -20.9066$ Mm. We use a uniform horizontal grid ($\Delta x = \Delta y = 96$ km) spanning 96 Mm in each direction. The size of the computational box was selected to ensure the kernels essentially vanish on the horizontal boundaries and also the bottom boundary. In particular, the amplitudes of the hyperbolic features of the kernels are notorious for decaying slowly with distance. We conducted resolution tests to confirm the example kernels presented in this paper did not change significantly with higher resolution.

Following Birch et al. (2004), we used the line-width measurements from Korzennik et al. (2004) as a basis for the empirical damping model. These line widths set the numerical resolution in Fourier space that is required in the calculation of the kernels. For the p_1-mode examples shown in this paper, we used frequency and wavenumber resolutions of $\Delta\nu = 3.8$ μHz and $\Delta k = 1.59 \times 10^{-4}$ rad Mm^{-1} respectively. The frequency domain ranged from $f = 1.5$ mHz to $f = 5.3$ mHz. The wavenumber domain was from $k = 0.11$ rad Mm^{-1} to $k = 2.5$ rad Mm^{-1}. We computed higher resolution tests to confirm that the resolution used here was sufficient.

To obtain a good representation of p_1 modes (including the line asymmetry), normal modes from $n = 0$ to $n = 14$ were used in the summation for the Green's functions. A filter was then applied to isolate the p_1 ridge. We used a Modulation Transfer Function (MTF) of the form $e^{-\beta k}$ where $\beta = 1.15$ Mm.

4.4 Example Kernels

The code used to compute travel-time sensitivity kernels for this paper was written in MATLAB and is a modified and extended version of the code used by Birch et al. (2004) to compute mean sound-speed kernels and by Birch and Gizon (2007) to compute difference velocity kernels. The main modifications are the computation of density and pressure kernels and the new method to compute Green's functions (Sect. 4.1). For simplicity, when computing the density kernel we neglected the perturbation the gravitational field ($\delta\mathbf{g}$) and γ in (15); both these effects are small.

Figure 2 shows horizontal and vertical slices through kernels for the changes in the mean travel-time due to fractional perturbations to the squared sound-speed, density, and pressure (for p_1-modes) and a separation of $\Delta = \mathbf{x}_2 - \mathbf{x}_1 = 16.0347$ Mm $\hat{\mathbf{x}}$ between the observation points. All the kernels are symmetric in both horizontal coordinates and show the same general features as the sound-speed kernels shown by Birch et al. (2004). The kernels have complicated features and are largest in amplitude near the two observation points. The kernels extend roughly over the first few Mm below the photosphere; this is the region where the p_1 kinetic energy density is significant for frequencies 1.5–5.3 mHz. The total integral of the sound-speed kernel is negative (increased sound-speed leads to a reduction in $\delta\tau_{\rm mn}$). The total integral of the density kernel is also negative implying a uniform increase in density would also lead to a reduction in $\delta\tau_{\rm mn}$. Conversely, the total integral of the pressure kernel is positive, which implies a uniform increase in pressure would lead to an increase in $\delta\tau_{\rm mn}$. The pressure and density kernels are strongly anti-correlated.

Figure 3 shows slices through the kernels for the sensitivity of the travel-time differences to changes in sound speed, density, and pressure, for the same geometry and filter as the kernels shown in Fig. 2. These kernels, unlike in Fig. 2, are anti-symmetric in x, but still symmetric in y. This is a consequence of the definition of the travel-time difference (the

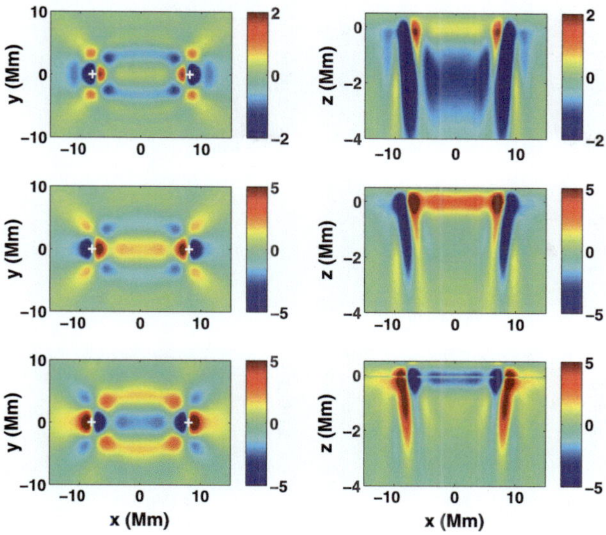

Fig. 2 Mean point-to-point travel-time sensitivity kernels for relative changes in sound-speed squared (*top panels*), density (*middle panels*), and pressure (*bottom panels*). Kernels have units of $\mathrm{s\,Mm^{-3}}$. Normal modes $n = 0$ to $n = 14$ were used to compute the Green's function and a filter was applied to isolate the p_1-mode ridge. *The left column* displays horizontal cuts at a height of $z = -0.3$ Mm. *The right column* displays vertical slices along the line $y = 0$ Mm. *All colour scales* are symmetric about zero (*green*) with *red* positive and *blue* negative. *The crosses in the left panels* represent the two points $\mathbf{x}_1$ and $\mathbf{x}_2$; the distance between the two points is 16.0347 Mm

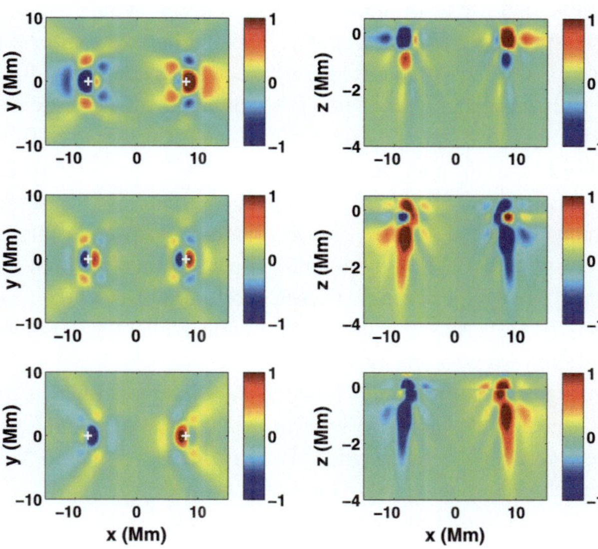

Fig. 3 Difference point-to-point travel-time sensitivity kernels for relative changes in sound-speed squared (*top panels*), density (*middle panels*), and pressure (*bottom panels*). Kernels have units of $\mathrm{s\,Mm^{-3}}$. Normal modes $n = 0$ to $n = 14$ were used to compute the Green's function and a filter was applied to isolate the p_1-mode ridge. *The left column* displays horizontal cuts at a height of $z = -0.3$ Mm. *The right column* displays vertical slices along the line $y = 0$ Mm. *The crosses in the left panels* represent the two points $\mathbf{x}_1$ and $\mathbf{x}_2$; the distance between the two points is 16.0347 Mm

difference between the two one-way travel times). The anti-symmetry in x is important as it implies that the total integral of these kernels vanishes, i.e., a spatially uniform change in solar structure does not produce a change in the travel-time differences.

Figure 4 shows slices through the kernels for the sensitivity of the travel-time differences to changes in v_x, v_y, and v_z, for the same geometry and filter as the kernels shown in Fig. 2.

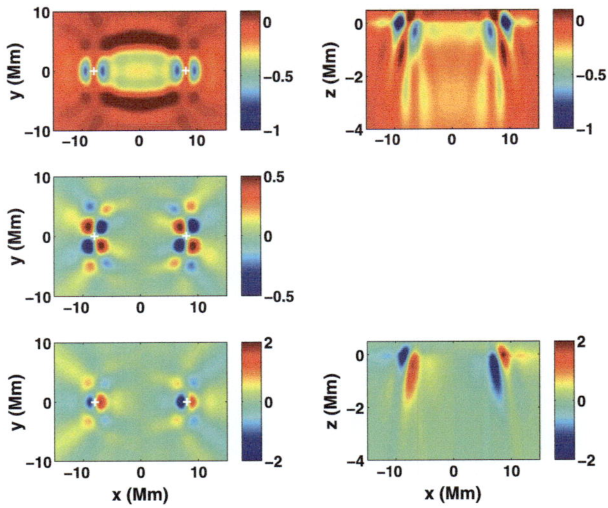

Fig. 4 Difference point-to-point travel-time sensitivity kernels for v_x (*top panels*), v_y (*middle panels*), and v_z (*bottom panels*). Kernels have units of $\mathrm{s\,Mm^{-3}/(km/s)}$. Normal modes $n = 0$ to $n = 14$ were used to compute the Green's function and a filter was applied to isolate the p_1-mode ridge. *The left column* displays horizontal cuts at a height of $z = -0.3$ Mm. *The right column* displays vertical slices along the line $y = 0$ Mm. *All colour scales* are symmetric about zero (*green*) with *red* positive and *blue* negative. *The crosses in the left panels* represent the two points $\mathbf{x}_1$ and $\mathbf{x}_2$; the distance between the two points is 16.0347 Mm

The kernel for v_z is anti-symmetric in x and symmetric in y, whereas the kernel for v_y is anti-symmetric in both x and y. This implies that uniform perturbations in v_z and v_y will not effect travel-time differences. The difference kernel for v_x is symmetric in both x and y and consequently has a non-zero total integral. 3D difference kernels for the f-mode were presented in Birch and Gizon (2007) and Birch et al. (2011). The present calculations are fully consistent with Birch and Gizon (2007) and Birch et al. (2011).

Figure 5 shows slices through the kernels for the sensitivity of mean travel-times to changes in v_x, v_y, and v_z, for the same geometry and filter as the kernels shown in Fig. 2. The symmetries imply that uniform perturbations in both v_x and v_y would not affect mean travel times. Equivalent 2D kernels for v_x and v_y, for the f-mode, were shown in Jackiewicz et al. (2007). We show here that the mean kernel for v_z is symmetric in both x and y and has a significant non-zero total integral.

Figure 6 demonstrates how the total integrals of the kernels converge as a function of distance from the observation points. The total integrals for the structure and horizontal flow kernels have approximately converged over the 96 Mm^2 domain, however, the mean kernel for vertical velocity has not yet converged.

Table 1 shows the mean and difference total integrals for p_1 modes. This gives an indication of how travel-time shifts respond to uniform perturbations. The largest contribution to the mean travel-times is from a uniform change in relative sound-speed. The contributions to mean travel-times from relative density and pressure perturbations are almost equal but opposite in sign. For the flows, the only non-zero contribution to the mean travel-times would be a uniform vertical flow, the horizontal components vanish due to their symmetries. The only contribution to the difference travel times due to uniform perturbations is due to flows along the two observation points.

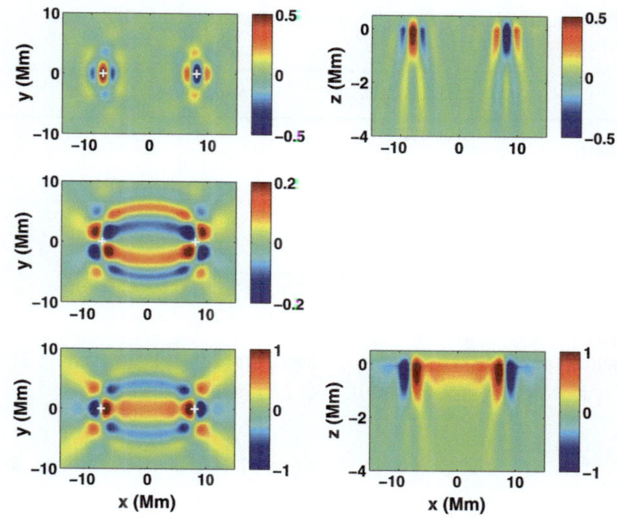

Fig. 5 Mean point-to-point travel-time sensitivity kernels for v_x (*top panels*), v_y (*middle panels*), and v_z (*bottom panels*). Kernels have units of $\mathrm{s\,Mm^{-3}/(km/s)}$. Normal modes $n = 0$ to $n = 14$ were used to compute the Green's function and a filter was applied to isolate the p_1-mode ridge. *The left column* displays horizontal cuts at a height of $z = -0.3$ Mm. *The right column* displays vertical slices along the line $y = 0$ Mm. *All colour scales* are symmetric about zero (*green*) with *red* positive and *blue* negative. *The crosses in the left panels* represent the two points $\mathbf{x}_1$ and $\mathbf{x}_2$; the distance between the two points is 16.0347 Mm

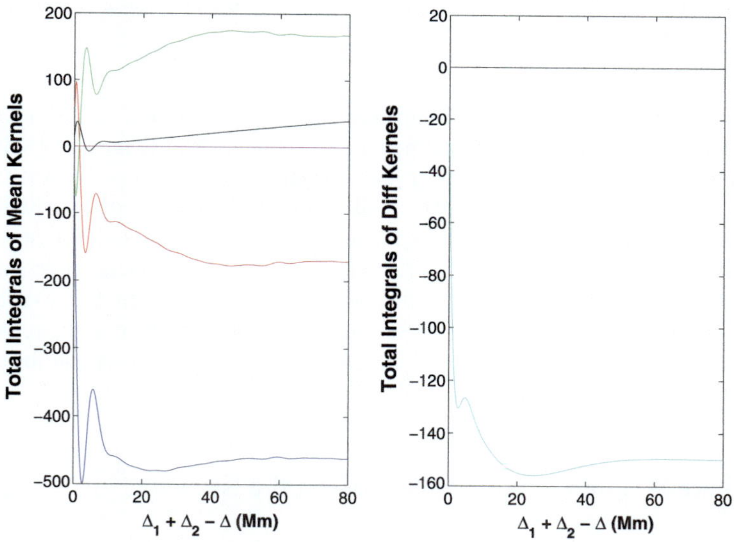

Fig. 6 Total 3D integrals of the kernels as a function of $\Delta_1 + \Delta_2 - \Delta$ (Mm), where Δ is the distance between the two points. The flow kernels were scaled by a uniform flow of 1 km/s such that all the total integrals have units of seconds. *The left panel* is for mean kernels and *the right panel* for difference kernels. The integrals for sound-speed squared are *blue*, density are *red*, pressure are *green*, v_x are *cyan*, v_y are *magenta*, and v_z are *black*. The total integrals for the mean kernels for sound-speed squared, density and pressure are close to converged at large distances from the two points however, for the vertical velocity a larger domain is necessary to ensure convergence. The total integrals of the mean kernels for horizontal flows vanish. The only non-zero total integral for the difference kernels is the one for v_x (*cyan*) and its total integral is close to converged with the domain of 96 Mm

5 A Consistency Check

In this section, we compute the change in the cross-covariance (e.g. see (5)), ΔC, caused by a simple sound-speed perturbation of the form

Table 1 Total integrals of the mean and difference kernels for sound-speed, density, pressure and flows for p_1 modes

	$\int K_{c^2} d\mathbf{r}$	$\int K_\rho d\mathbf{r}$	$\int K_P d\mathbf{r}$	$\int K_{v_x} d\mathbf{r}$	$\int K_{v_y} d\mathbf{r}$	$\int K_{v_z} d\mathbf{r}$
mean	−460 s	−170 s	168 s	0 s/(km/s)	0 s/(km/s)	54 s/(km/s)
diff	0 s	0 s	0 s	−152 s/(km/s)	0 s/(km/s)	0 s/(km/s)

$$\frac{\Delta c^2}{c_0^2} = \varepsilon \exp\left[\frac{-(z-z_0)^2}{2\sigma^2}\right], \tag{34}$$

where $\varepsilon = 0.0001$, $z_0 = -1.5$ Mm, and $\sigma = (1\ \text{Mm})/(8\ln 2)^{1/2}$.

Because ε is very small, ΔC can be approximated by

$$\delta C(\omega; \mathbf{x}_1, \mathbf{x}_2) = \int d\mathbf{x}dz \mathcal{C}_{c^2}(\mathbf{x}, z, \omega; \mathbf{x}_1, \mathbf{x}_2)\frac{\delta c^2}{c_0^2}(z), \tag{35}$$

where the two points are separated by a distance of 16 Mm. The function $\delta C(\omega)$ is real.

Because Δc^2 depends on height only, we can also compute ΔC using

$$\Delta C(\omega) = C(\omega) - C_0(\omega), \tag{36}$$

where $C(\omega)$ is the cross-covariance computed using the eigenmodes of the solar model with sound speed $c_0 + \Delta c$. To compute the new eigenmodes we use the same solver as in Sect. 4. Here we used a different damping model where each mode was damped according to $\gamma_n(k) = \tilde{\gamma}|\omega_0(k)/\tilde{\omega}|^{4.4}$, where $\tilde{\omega}/(2\pi) = 3$ mHz and $\tilde{\gamma}/(2\pi) = 100$ μHz. In addition, for wave numbers less than 0.8 Mm we set the damping to remain constant. The resolution in wavenumber was the same as earlier, however the resolution in frequency was changed to 19.1 mHz. In Fig. 7, we find that there is a very good agreement between $\Delta C(\omega)$ and $\delta C(\omega)$, which is expected in the limit $\varepsilon \to 0$. These results give us confidence in the reliability of the computations. Furthermore, it appears that the numerical resolutions and the extent of the horizontal domain are sufficient.

6 Are There Simple Relationships Between the Kernels?

In order to produce reliable time-distance inversions of the solar interior it is helpful if the kernels have different structures and features. Figures 8 and 9 display cuts through the mean and difference kernels at $z = -0.3$ Mm along the x and y axes. We have plotted the negative values of the pressure kernel (green) in order to better illustrate the strong anti-correlation with the kernel for density (red). It is evident that kernels for sound-speed squared (blue), density (red) and negative pressure (green) have similar features.

Travel-time shifts due to sound-speed perturbations are directly related to the divergence of the eigenfunctions. Therefore, travel-times shifts for f-modes are not sensitive to sound-speed perturbations since the divergences of the f-mode eigenfunctions vanish. However, the power spectrum in the neighbourhood of the f-mode ridge has a small contribution from the wings of the p modes and thus f-mode mean travel-time measurements are weakly sensitive to sound-speed perturbations. For example, we find that the total integral of K_{c^2} for the f-mode ridge at $\Delta = 16.0347$ Mm is approximately 3 % of the total integral of K_{c^2} for the p_1 ridge.

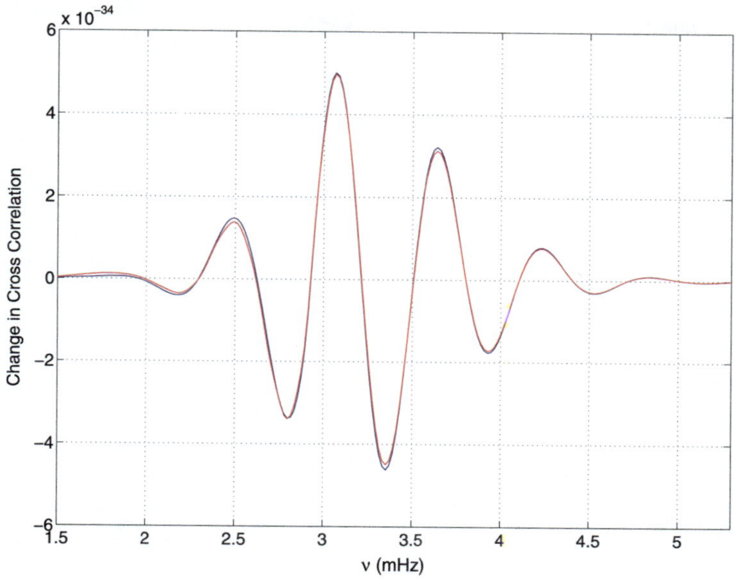

Fig. 7 Change in the cross-covariance (at a fixed point-to-point distance of $\Delta = 16.0347$ Mm) for the case of the Gaussian perturbation to the sound-speed of Model S. *The red curve* is δC and *the blue curve* ΔC. The results show very good agreement

7 Specific Case of Hydrostatic Equilibrium

In the fully general non-hydrodynamic case, where we have three independent thermodynamic perturbation variables, the travel-time shifts (ignoring flows) are given by

$$\delta\tau = \int_{\odot} d\mathbf{r}\left[K_{c^2,\rho P}(\mathbf{r})\frac{\delta c^2}{c_0^2}(\mathbf{r}) + K_{\rho,c^2P}(\mathbf{r})\frac{\delta\rho}{\rho_0}(\mathbf{r}) + K_{P,c^2\rho}(\mathbf{r})\frac{\delta P}{P_0}(\mathbf{r})\right]. \tag{37}$$

We have now explicitly expressed which quantities are held fixed. The kernel $K_{c^2,\rho P}$ is due to perturbations in sound-speed at fixed density and pressure, similarly K_{ρ,c^2P} is the density kernel at constant sound-speed and pressure and finally $K_{P,c^2\rho}$ is the pressure kernel at constant sound-speed and density.

In global helioseismology, it is typically assumed that the density and pressure perturbations are in hydrostatic equilibrium (e.g Gough and Thompson 1991)

$$\nabla\delta P = \mathbf{g}_0\delta\rho, \tag{38}$$

where we assume the perturbation to the gravitational field is negligible for the sake of simplicity.

Equation (38) may be manipulated to express the pressure perturbation entirely in terms of a density perturbation (or vice-versa) and thus the number of independent thermodynamic variables is reduced from three to two.

The horizontal components of (38) implies that the pressure perturbation is at most a function of height, i.e. $\delta P = \delta P(z)$. The vertical component implies that the density perturbation is also at most a function of height i.e $\delta\rho = \delta\rho(z)$, and hence

$$\frac{d\delta P}{dz} = -g_0\delta\rho. \tag{39}$$

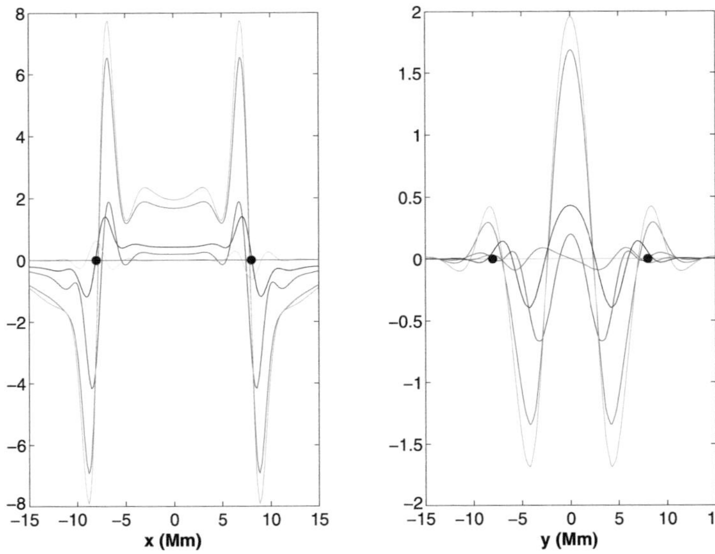

Fig. 8 Cuts through the mean point-to-point travel-time sensitivity kernels for sound-speed squared (*blue*), density (*red*), negative pressure (*green*), v_x (*cyan*), v_y (*magenta*), and v_z (*black*). The sound-speed, density and pressure kernels have units of $\mathrm{s\,Mm^{-3}}$, whereas the velocity kernels have units $\mathrm{Mm^{-3}/(km/s)}$. Normal modes $n=0$ to $n=14$ were used to compute the Green's function and a filter was applied to isolate the p_1-mode ridge. The cuts are at taken at a height of $z=-0.3$ Mm. *The left panel* is a cut along the x axis, whereas *the right panel* is a cut along the y axis. The distance between the two points is 16.0347 Mm

It is possible to implement the assumption of hydrostatic equilibrium, for example by eliminating pressure in terms of density. To achieve this it is necessary to manipulate the last term on the right-hand-side of (37) as a gradient of the pressure. So since the perturbations are all functions of height only, we require

$$\frac{\partial A(\mathbf{r})}{\partial z} = \frac{K_{P,c^2\rho}(\mathbf{r})}{P_0(z)} \tag{40}$$

with boundary condition $A(\mathbf{x}, z_t) = 0$. This could be solved for each $\mathbf{x}$. Provided a solution exists, after integration by parts once, then, with the assumptions that the boundary terms vanish, the travel time shifts become

$$\delta\tau = \int_\odot d\mathbf{r}\left[K_{c^2,\rho P}(\mathbf{r})\frac{\delta c^2}{c_0^2}(z) + \big(K_{\rho,c^2P}(\mathbf{r}) + \rho_0(z)g_0(z)A(\mathbf{r})\big)\frac{\delta\rho}{\rho_0}(\mathbf{r})\right]. \tag{41}$$

However, since we know we have two independent thermodynamic quantities then it is also possible to express the travel time shifts in terms of a different set of kernels according to

$$\delta\tau = \int_\odot d\mathbf{r}\left[K^H_{c^2,\rho}(\mathbf{r})\frac{\delta c^2}{c_0^2}(z) + K^H_{\rho,c^2}(\mathbf{r})\frac{\delta\rho}{\rho_0}(z)\right], \tag{42}$$

where the superscript H is to remind the reader that the assumption of hydrodynamic balance has been asserted. Then by equating coefficients we get

$$K^H_{c^2,\rho}(\mathbf{r}) = K_{c^2,\rho P}(\mathbf{r}), \tag{43}$$

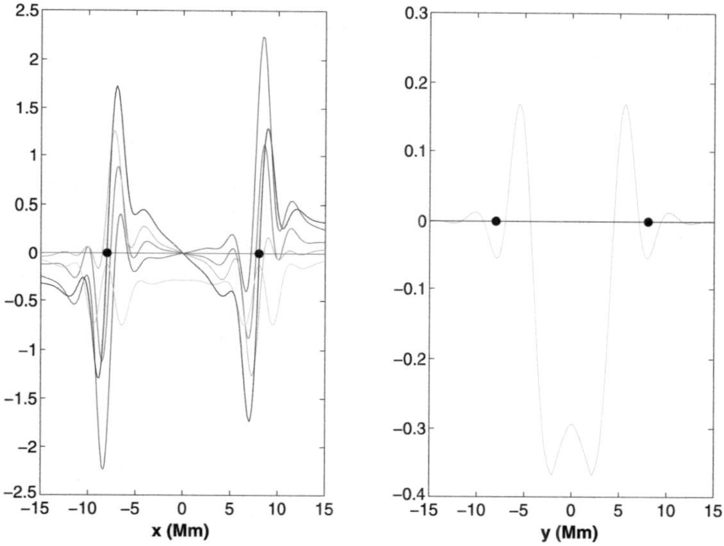

Fig. 9 Cuts through the difference point-to-point travel-time sensitivity kernels for sound-speed squared (*blue*), density (*red*), negative pressure (*green*), v_x (*cyan*), v_y (*magenta*), and v_z (*black*). The sound-speed, density and pressure kernels have units of $\mathrm{s\,Mm^{-3}}$, whereas the velocity kernels have units $\mathrm{Mm^{-3}/(km/s)}$. Normal modes $n = 0$ to $n = 14$ were used to compute the Green's function and a filter was applied to isolate the p_1-mode ridge. The cuts are at taken at a height of $z = -0.3$ Mm. *The left panel* is a cut along the x axis, whereas *the right panel* is a cut along the y axis. The distance between the two points is 16.0347 Mm

$$K^H_{\rho,c^2}(\mathbf{r}) = K_{\rho,c^2 P}(\mathbf{r}) + \rho_0(z) g_0(z) A(\mathbf{r}). \tag{44}$$

Equation (43) implies that the sound-speed kernel at constant density and pressure in the general problem without hydrostatic equilibrium is equal to the sound-speed kernel at constant density in the hydrostatic case. Equation (44) demonstrates that more work is required to obtain the density kernel at constant sound-speed in the hydrostatic case. It is related to the density kernel at constant sound-speed and pressure in the non-hydrodynamic case plus an additional term involving the function $A(\mathbf{r})$. Most of the work in going from one set of kernels to another involves computing $A(\mathbf{r})$.

8 Discussion

We have reviewed the sensitivity of time-distance travel times to small-amplitude steady changes in the structure of a solar model as well as their sensitivity to weak and steady flows. In the course of this paper, we introduced a new convenient method for computing Green's functions based on normal-mode summation using the normal modes of the adiabatic wave equation (in contrast with the approach of Birch et al. 2004, based on the modes of the damped problem). The sensitivity kernels we computed are essential for time-distance inversions of the solar interior.

It is difficult to disentangle density and pressure perturbations and this will complicate time-distance inversions. Our calculations have shown that, to a good approximation, there exists a constant $\alpha > 0$, such that $K_{\rho,cP} \simeq -\alpha K_{P,c\rho}$. We have

$$\delta\tau = \int dV \left(K_{c^2,P\rho} \frac{\delta c^2}{c^2} + K_{P,c\rho} \frac{\delta P}{P} + K_{\rho,cP} \frac{\delta\rho}{\rho} \right) \simeq \int dV \left(K_{c^2,u} \frac{\delta c^2}{c^2} + K_{u,c} \frac{\delta u}{u} \right) \tag{45}$$

where $u \equiv P/\rho^\alpha$, $K_{c^2,u} \equiv K_{c^2,P\rho}$, and $K_{u,c} \equiv K_{P,c\rho}$. Thus, to a good approximation, two thermodynamic quantities are enough to describe the medium. If the density and pressure perturbations are in hydrostatic equilibrium, then two thermodynamic quantities are exactly enough (see Sect. 7).

The correct treatment of instrumental, geometrical, and mode physics effects is an important future goal for accurately imaging the solar interior. Correctly including line-of-sight and foreshortening effects in the computation of the kernels, will allow accurate applications using large line-of-sight angles. Jackiewicz et al. (2007) studied line-of-sight effects for 2D kernels (for surface waves) and found striking changes to the appearance of kernels that will be important for accurately inverting for small-scale flows. Computing travel-time sensitivity kernels in full spherical geometry, rather than using the Cartesian approximation, is important for larger distances between the observation points.

The kernels discussed in this review assume that the Lagrangian pressure perturbation vanishes at the top boundary. This is not suitable for frequencies above the acoustic cut-off frequency and thus needs to be improved. Another important question is how much do time-distance inversion results depend on accurate descriptions of input physics, such as source models or damping.

Most cases assume a laterally homogeneous reference (non-magnetic) model. To study more complex features on the Sun (e.g. sunspots), kernels with laterally heterogeneous reference models are important. Initial steps towards this goal has already been achieved by Hanasoge et al. (2011, 2012).

Acknowledgements This work was carried out in the framework of Collaborative Research Center SFB 963 "Astrophysical Flow Instabilities and Turbulence" (Project A1) from the German Science Foundation (DFG). L.G. acknowledges support from EU FP7 Collaborative Project "Exploitation of Space Data for Innovative Helio- and Asteroseismology" (SPACEINN) and ERC Starting Grant "Seismic Imaging of the Solar Interior".

References

A.C. Birch, L. Gizon, Linear sensitivity of helioseismic travel times to local flows. Astron. Nachr. **328**, 228 (2007). doi:10.1002/asna.200610724

A.C. Birch, A.G. Kosovichev, Travel time sensitivity kernels. Sol. Phys. **192**, 193–201 (2000)

A.C. Birch, A.G. Kosovichev, T.L. Duvall Jr., Sensitivity of acoustic wave travel times to sound-speed perturbations in the solar interior. Astrophys. J. **608**, 580–600 (2004). doi:10.1086/386361

A.C. Birch, L. Gizon, B.W. Hindman, D.A. Haber, The linear sensitivity of helioseismic ring diagrams to local flows. Astrophys. J. **662**, 730–737 (2007). doi:10.1086/513683

A.C. Birch, L. Gizon, R. Burston, Erratum: "Linear sensitivity of helioseismic travel times to local flows". Astron. Nachr. **332**, 658 (2011). doi:10.1002/asna.201111557

R. Cameron, L. Gizon, Three-dimensional numerical simulation of wave propagation through a model sunspot, in *Proceedings of SOHO 18/GONG 2006/HELAS I, Beyond the Spherical Sun*, vol. 624 (2006)

J. Christensen-Dalsgaard, Helioseismology. Rev. Mod. Phys. **74**, 1073–1129 (2002). doi:10.1103/RevModPhys.74.1073

J. Christensen-Dalsgaard, W. Dappen, S.V. Ajukov, E.R. Anderson, H.M. Antia, S. Basu, V.A. Baturin, G. Berthomieu, B. Chaboyer, S.M. Chitre, A.N. Cox, P. Demarque, J. Donatowicz, W.A. Dziembowski, M. Gabriel, D.O. Gough, D.B. Guenther, J.A. Guzik, J.W. Harvey, F. Hill, G. Houdek, C.A. Iglesias, A.G. Kosovichev, J.W. Leibacher, P. Morel, C.R. Proffitt, J. Provost, J. Reiter, E.J. Rhodes Jr., F.J. Rogers, I.W. Roxburgh, M.J. Thompson, R.K. Ulrich, The current state of solar modeling. Science **272**, 1286–1292 (1996). doi:10.1126/science.272.5266.1286

F.A. Dahlen, S.-H. Hung, G. Nolet, Fréchet kernels for finite-frequency traveltimes-I. Theory. Geophys. J. Int. **141**, 157–174 (2000). doi:10.1046/j.1365-246X.2000.00070.x

T.L. Duvall Jr., S.M. Jefferies, J.W. Harvey, M.A. Pomerantz, Time-distance helioseismology. Nature **362**, 430–432 (1993). doi:10.1038/362430a0

T.L. Duvall Jr., A.G. Kosovichev, P.H. Scherrer, R.S. Bogart, R.I. Bush, C. de Forest, J.T. Hoeksema, J. Schou, J.L.R. Saba, T.D. Tarbell, A.M. Title, C.J. Wolfson, P.N. Milford, Time-distance helioseismology with the MDI instrument: initial results. Sol. Phys. **170**, 63–73 (1997)

L. Gizon, A.C. Birch, Time-distance helioseismology: the forward problem for random distributed sources. Astrophys. J. **571**, 966–986 (2002). doi:10.1086/340015

L. Gizon, A.C. Birch, Local helioseismology. Living Rev. Sol. Phys. **2**, 6 (2005)

L. Gizon, M.J. Thompson, Outstanding problems in local helioseismology. Astron. Nachr. **328**, 204 (2007). doi:10.1002/asna.200610720

L. Gizon, T.L. Duvall Jr., R.M. Larsen, Seismic tomography of the near solar surface. J. Astrophys. Astron. **21**, 339 (2000). doi:10.1007/BF02702420

L. Gizon, A.C. Birch, H.C. Spruit, Local helioseismology: three-dimensional imaging of the solar interior. Annu. Rev. Astron. Astrophys. **48**, 289–338 (2010). doi:10.1146/annurev-astro-082708-101722

D.O. Gough, M.J. Thompson, The inversion problem, in *Solar Interior and Atmosphere*, ed. by A.N. Cox, W.C. Livingston, M.S. Matthews (1991), pp. 519–561. http://adsabs.harvard.edu/abs/1991sia..book..519G

S.M. Hanasoge, A. Birch, L. Gizon, J. Tromp, The adjoint method applied to time-distance helioseismology. Astrophys. J. **738**, 100 (2011). doi:10.1088/0004-637X/738/1/100

S. Hanasoge, A. Birch, L. Gizon, J. Tromp, Seismic probes of solar interior magnetic structure. Phys. Rev. Lett. **109**(1), 101101 (2012)

S.-H. Hung, F.A. Dahlen, G. Nolet, Fréchet kernels for finite-frequency traveltimes-II. Examples. Geophys. J. Int. **141**, 175–203 (2000). doi:10.1046/j.1365-246X.2000.00072.x

J. Jackiewicz, L. Gizon, A.C. Birch, Sensitivity of solar f-mode travel times to internal flows, in *SOHO-17, 10 Years of SOHO and Beyond*. ESA Special Publication, vol. 617 (2006)

J. Jackiewicz, L. Gizon, A.C. Birch, T.L. Duvall Jr., Time-distance helioseismology: sensitivity of f-mode travel times to flows. Astrophys. J. **671**, 1051–1064 (2007). doi:10.1086/522914

J. Jackiewicz, L. Gizon, A.C. Birch, The forward and inverse problems in time-distance helioseismology. J. Phys. Conf. Ser. **118**, 1–11 (2008). doi:10.1088/1742-6596/118/1/012033

J.M. Jensen, F.P. Pijpers, Sensitivity kernels for time-distance inversion based on the Rytov approximation. Astron. Astrophys. **412**, 257–265 (2003). doi:10.1051/0004-6361:20031361

S.G. Korzennik, M.C. Rabello-Soares, J. Schou, On the determination of Michelson Doppler imager high-degree mode frequencies. Astrophys. J. **602**, 481–516 (2004). doi:10.1086/381021

A.G. Kosovichev, Tomographic imaging of the Sun's interior. Astrophys. J. **461**, 55 (1996). doi:10.1086/309989

A.G. Kosovichev, T.L. Duvall Jr., Acoustic tomography of solar convective flows and structures, in *SCORe'96: Solar Convection and Oscillations and Their Relationship*, ed. by F.P. Pijpers, J. Christensen-Dalsgaard, C.S. Rosenthal. Astrophysics and Space Science Library, vol. 225 (1997), pp. 241–260

D. Lynden-Bell, J.P. Ostriker, On the stability of differentially rotating bodies. Mon. Not. R. Astron. Soc. **136**, 293 (1967)

H. Marquering, G. Nolet, F.A. Dahlen, Three-dimensional waveform sensitivity kernels. Geophys. J. Int. **132**, 521–534 (1998). doi:10.1046/j.1365-246X.1998.00426.x

H. Marquering, F.A. Dahlen, G. Nolet, Three-dimensional sensitivity kernels for finite-frequency traveltimes: the banana-doughnut paradox. Geophys. J. Int. **137**, 805–815 (1999). doi:10.1046/j.1365-246X.1999.00837.x

K. Nagashima, L. Gizon, A. Birch, B. Löptien, S. Couvidat, B. Fleck, Helioseismic and magnetic imager multi-height Dopplergrams. ArXiv e-prints, 2014

R.B. Schlottmann, A.G. Kosovichev, Helioseismic Fréchet traveltime kernels in spherical coordinates. ArXiv e-prints, 2011

Space Sci Rev (2015) 196:221–249
DOI 10.1007/s11214-015-0199-y

Recent Developments in Helioseismic Analysis Methods and Solar Data Assimilation

A. Schad[1] · L. Jouve[2] · T.L. Duvall Jr.[3] · M. Roth[1] · S. Vorontsov[4,5]

Received: 4 March 2015 / Accepted: 11 August 2015 / Published online: 11 September 2015

Abstract We review recent advances and results in enhancing and developing helioseismic analysis methods and in solar data assimilation. In the first part of this paper we will focus on selected developments in time-distance and global helioseismology. In the second part, we review the application of data assimilation methods on solar data. Relating solar surface observations as well as helioseismic proxies with solar dynamo models by means of the techniques from data assimilation is a promising new approach to explore and to predict the magnetic activity cycle of the Sun.

Keywords Sun · Helioseismology · Data analysis

1 Introduction

Helioseismic inferences are subject to statistical and systematic errors of different origin. Part of the errors may be reduced by analyzing longer observations or averaging over observations. But especially systematic errors often result from the neglect of inevitable instrumental effects on the observations or from model misspecifications in the data analysis and cannot be diminished for example by longer observations. Consequently, more accurate models and a better account of instrumental effects are essential for the enhancement of local and global helioseismic analysis methods. This will improve the accuracy and reliability of estimates of helioseismic parameters like mode frequencies and wave travel times.

✉ A. Schad
schad@kis.uni-freiburg.de

1 Kiepenheuer-Institut für Sonnenphysik, Schöneckstr. 6, 79104 Freiburg, Germany

2 UPS-OMP, Institut de Recherche en Astrophysique et Planétologie, Université de Toulouse, 31028 Toulouse Cedex 4, France

3 Max-Planck-Institut für Sonnensystemforschung, Justus-von-Liebig-Weg 3, 37077 Göttingen, Germany

4 Astronomy Unit, School of Physics and Astronomy, Queen Mary University of London, Mile End Road, London E1 4NS, UK

5 Institute of Physics of the Earth, B. Gruzinskaya 10, Moscow 123810, Russia

Just recently, for example, progress was made in measuring the meridional flow in the deep layers with the time-distance technique not only by the availability of high-resolution data from the HMI instrument (Schou et al. 2012) but also after becoming aware and removal of a systematic center-to-limb effect in the travel-time estimates (Zhao et al. 2012, 2013). In order to detect weak processes like the meridional flow in the deeper interior or highly dynamic processes like supergranular convective motions with the time-distance technique, travel-time measurements of high signal-to-noise ratio are needed. The development of spatial averaging strategies and optimized filters can help here. Regarding global helioseismic investigations, different models and estimation schemes are used to extract the mode parameters of modes of medium and large harmonic degree from the data of GONG, MDI or HMI (e.g., Anderson et al. 1990; Schou 1992, 1998; Hill et al. 1996; Korzennik 2005; Reiter et al. 2015). Recent advances are made on identification and eradication of systematic influences of different origin on the mode parameter estimates (Larson and Schou 2008; Rabello-Soares et al. 2008; Vorontsov et al. 2009; Korzennik et al. 2013). Further advances in global helioseismology are made by the development of analysis methods that exploit the perturbation of mode eigenfunctions due to the advection of the acoustic waves by flows. This kind of perturbation is also described as mode coupling. It manifests in correlations that can be investigated by cross-spectral analysis of spherical harmonic decomposed global oscillation data (Woodard 2000; Schad et al. 2013). Such an approach is promising for inferences on the meridional flow inside the Sun since its influence on mode frequencies is only of second order and very small (Roth and Stix 2008; Gough and Hindman 2010; Schad et al. 2011; Vorontsov 2011).

One essential purpose of helioseismic investigations of dynamic processes within the solar interior is to explore and better understand the solar dynamo and the related activity cycle. In particular, some dynamo models predict a strong link between the meridional flow speed and the magnetic cycle period. Enabling observations as well as helioseismic constraints and models to work hand in hand could definitely help to progress on our understanding of the solar cycle and possibly produce reliable forecasts of solar phenomena and activity. The concepts of data assimilation are promising for solving these issues. Data assimilation has been extensively used for decades to predict the weather on Earth. Moreover, by combining physically-based models and well-chosen observables, it allows to constrain processes unaccessible to any measurement through their indirect impact on observations. With the increasing collection of solar data, this technique has become more and more attractive in solar physics and starts to be adapted to this field.

In the following we review some of the limitations related with the estimation of the helioseismic quantities in time-distance and global helioseismology and ideas and current attempts to accomplish them. For the time-distance analysis two recent advances, ensemble averaging and optimal filtering, will be discussed in detail. The inclusion of enhanced models of the solar oscillation spectrum to improve global helioseismic parameter estimates is addressed by the "Global Helioseismic Metrology" project. As another advancement in global helioseismology we review the mode eigenfunction perturbation analyses that uses cross-spectra of solar oscillations. In the second part of this paper we overview the latest developments and results from application of data assimilation methods in solar physics with regard to activity cycle predictions.

2 Developments in Local Helioseismology

One recent challenge in local helioseismology is to obtain precise three-dimensional maps of the temporal evolution of the solar sub-surface layers. Here we focus on the time-distance method which is one tool to obtain such measurements and promises for example the detection of the emergence of magnetic flux or the analysis of convective motions like supergranulation.

2.1 Large and Small Separations

In time-distance helioseismology, travel times are measured between pairs of surface locations with angular heliocentric separation Δ. A problem in the analysis of travel times is that information from larger depths is only obtained from signals measured at locations with larger separations Δ and these are adversely affected by larger noise levels due to the geometrical spreading of the wavefronts (Gizon and Birch 2004, Eq. (31)). In recent years much of time-distance helioseismology has been carried out using what we will call small separations, like $\Delta < 2.4^\circ$ (29.2 Mm) (Jackiewicz et al. 2008) and $\Delta < 5^\circ$ (61 Mm) (Zhao and Kosovichev 2003). In particular for the analysis time periods appropriate for studying near-surface features, say 8 hours, and to get sufficient signal-to-noise ratio to easily see perturbation signals, analyses have been restricted to relatively small separations Δ. One of the difficulties with the restriction to small separations Δ is the inability to separate horizontal and vertical flow signals near the surface. The travel-time signal from a flow is measured from the time difference between counterpropagating waves traveling between two points. The ray theory predicts that the time difference $\delta\tau$ is given by the integral of the flow along the ray path Γ, $\delta\tau = -2\int_\Gamma \mathbf{v}\cdot\mathbf{ds}/c^2$, where $\mathbf{v}$ is the flow velocity, $\mathbf{ds}$ is the element of length along the ray path, and c is the sound speed. When studying near-surface phenomena, like supergranulation or sunspots, the restriction to short distances means that horizontal and vertical flows are not cleanly separated (Zhao and Kosovichev 2003). However, for large distances, the separation is much cleaner. To illustrate this point, Fig. 1b shows the travel-time contributions of the vertical and horizontal flow components as well as the sum for the shallow supergranular flow model shown in Fig. 1a. For such a shallow flow model, the horizontal component contributes almost nothing for $\Delta > 6$ deg. But the vertical contribution of ≈ 5 [s] at large Δ can be used to define a constant of the model, namely an integral over depth of the vertical flow at cell center, $\delta\tau^z = -2\int v_z \mathrm{d}z/c^2$.

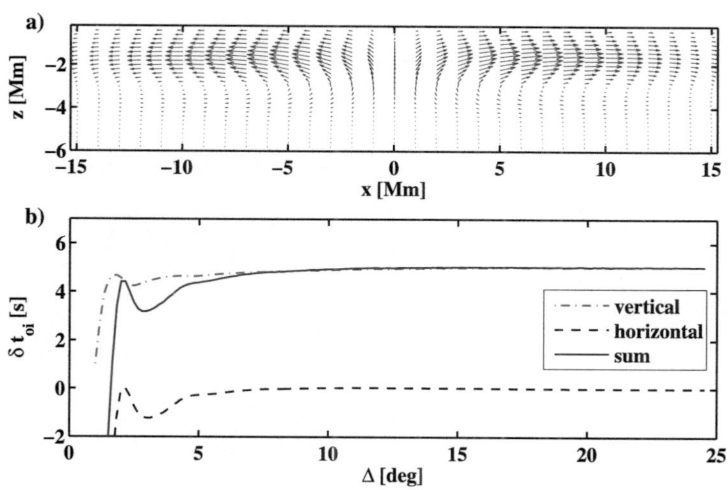

Fig. 1 (**a**) Flow vectors of a shallow supergranular model (Duvall and Hanasoge 2013), and (**b**) Vertical flow and horizontal flow contributions to the center-annulus travel-time difference δt_{oi} [s] and their sum

To make full use of large travel distances, the signal-to-noise ratio in the travel-time maps needs to be increased. This requires several steps: optimal choice of phase-speed filters and spatial averaging schemes. For example, Ilonidis et al. (2013) used a broad phase speed filter with a non-Gaussian shape and particular averaging schemes with multiple arc configurations to obtain their results on the detection of emerging magnetic flux (Ilonidis et al. 2011). However, the reliability of their findings still needs to be validated (Braun 2012).

2.2 Averaging over Features

Different averaging schemes are applied in time-distance helioseismology to get a significant signature of a particular feature in the wave travel times. As an example Birch et al. (2010) employed numerical models to estimate that a rising flux tube causes travel-time shifts in the order of 1 s. A signal which could be certainly detected only by averaging over 150 of such events.

A powerful method to increase the signal-to-noise ratio and to obtain average properties directly is to average either cross correlations or travel times about the locations of features detected near the surface. The basic assumptions of this feature averaging are that there is an underlying linear process that can be averaged and that the systematic errors are sufficiently understood. This averaging technique has been used with small magnetic features (Duvall et al. 2006; Felipe et al. 2012) and to supergranulation (Birch et al. 2006; Hirzberger et al. 2008; Duvall and Birch 2010; Duvall and Hanasoge 2013, and Švanda 2012). An example of features corresponding to the supergranulation cell centers is shown in Fig. 2 from Duvall and Birch (2010). To derive these feature locations, a map of the center-annulus travel-time differences of f-mode waves is used to approximate the horizontal divergence of the flow. Local maxima of a smoothed version of this map are located and only features farther apart than 22 Mm are accepted. The smaller (in terms of travel-time difference) of a pair closer than 22 Mm is rejected. This results in a useful choice of supergranules, although it is likely biased towards larger than average cells (Švanda 2012).

From the feature analysis of Duvall and Birch (2010), only a single parameter is derived from each cell, namely the strength of the f-mode divergence signal. Travel times could be binned based on the value of this parameter, enabling some discrimination of different strength or size cells. No one has published such results yet, but it is an obvious extension of the present work. However, it is likely that an additional parameter (at least one) would be required to describe supergranular cells. Cell size comes to mind as a candidate. However, it

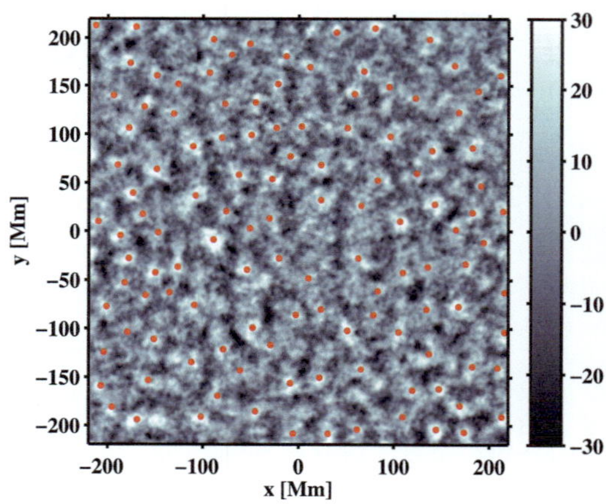

Fig. 2 Feature locations (*red*) determined by the local peaks in the horizontal divergence signal (*gray scale*) derived as the annulus minus center travel-time difference for the f mode (Figure adapted from Duvall and Birch 2010)

is very likely that the peak f-mode divergence signal will be highly correlated with the size. If the cells were to be described by two parameters, it would be useful to have parameters that are orthogonal. A way to analyze the horizontal divergence map to derive additional cell information is the Fourier segmentation procedure developed by Hirzberger et al. (2008).

2.3 Optimized Filters

An essential element of time-distance helioseismic methods is the construction and usage of optimized filters. They have been used in time-distance helioseismology since its inception (Duvall et al. 1993), where it was shown that filtering the data in horizontal phase speed ω/k (ω circular frequency of a wave and k its horizontal wavenumber) leads to isolated features in a time-distance correlation function. This is important because waves with the same horizontal phase speed travel to the same depth in the Sun (Duvall 1982). In the early work, it was considered that by measuring travel-time differences between surface points that the effect of perturbations along the ray path could be measured just by considering waves with the particular phase speed corresponding to that depth. However, it was shown by Woodard (1998) that perturbations, such as supergranulation, spread power in the $k-\omega$ power spectrum by an amount corresponding to their inverse size. Supergranulation, with a spectrum peaking at spherical harmonic degree $\ell = kR = 120$, spreads signal considerably in the power spectrum.

To study features such as supergranulation, it is useful to have a filter broad enough to admit most or all of the signal. This issue has been examined by Duvall and Hanasoge (2013). They have constructed filters with a central phase speed but a full width at half maximum (FWHM) Γ_ℓ that is independent of ω. For a range of large separations of $\Delta = 19$–22 deg, they have measured the center-annulus travel-time difference [δt_{oi}] for the average supergranules versus filter width (Fig. 3a). At small widths, the signal strength is approximately linear with filter width until most of the supergranular signal is captured for $\Gamma_\ell > 240$. Another important parameter of the filter is the subsequent signal to noise ratio. This ratio is shown in Fig. 3b. It is found that narrow filters, in addition to not capturing most of the signal, do not yield the best signal to noise ratio. From Fig. 3b a filter width of $\Gamma_\ell = 400$ was chosen for further study.

2.4 Discussion

In this section we have touched on two relatively new advances in time-distance helioseismology: ensemble or feature averaging and optimal filtering. Both strategies improve the

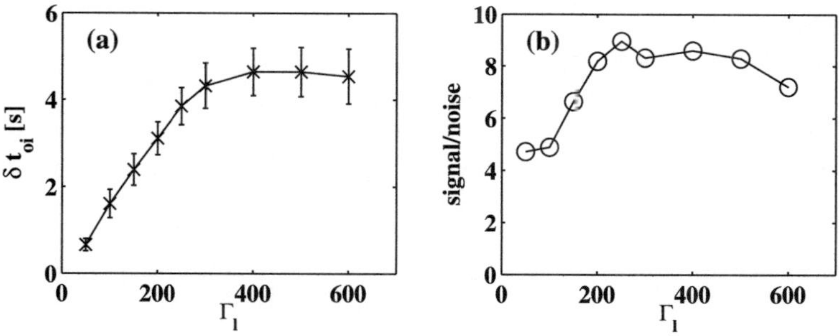

Fig. 3 (**a**) Travel-time difference [δt_{oi}] versus filter FWHM [Γ_ℓ]. The unfiltered case has 5.3 ± 1.2 seconds. (**b**) The travel-time difference from (**a**) divided by the size of the *error bar* from (**a**) *versus* the filter FWHM Γ_ℓ. The value for the unfiltered case is 4.6. Figure from Duvall and Hanasoge (2013)

signal-to-noise ratio of travel-time measurements that is necessary to measure the solar subsurface velocities. Of course other methods such as the ring-diagram analysis (Hill 1988), helioseismic holography (Lindsey and Braun 1990), and Fourier-Hankel analysis (Braun et al. 1987) can provide independent views on the processes insides the Sun. However, all local helioseismic techniques require a certain strategy of averaging or filtering to increase the signal-to-noise ratio. Another issue of time-distance analysis concerns its reliability and ability to retrieve subsurface structures, especially in the presence of magnetic fields (Gizon et al. 2009; Moradi et al. 2010), but also of subsurface supergranular-like flows as shown by DeGrave et al. (2014). The authors of that simulation study showed that current time-distance techniques were not able to adequately retrieve the supergranular flow pattern and suggested that averaging schemes, as proposed by Duvall and Birch (2010) and Švanda et al. (2011), might help to obtain reliable results in this case.

3 Developments in Global Helioseismology

Global helioseismic analyses of mode frequencies were very fruitful in the past for our current picture of the Sun's internal structure and dynamics. The fidelity and resolution of these findings is restricted by the quality of the observations but also by systematic errors entering the analysis methods used to determine the mode parameters from the data. Especially the systematic errors are crucial for global helioseismic inferences and cannot be overcome by improving the quality or length of observations. In the following we focus on some of the difficulties in estimating global mode parameters and the efforts on identifying and eradicating associated systematic errors by means of the "Global Helioseismic Metrology" project.

3.1 Measurement of Solar Oscillation Frequencies: Uncertainties and Limitations

A significant part of the uncertainty in global helioseismic measurements has been identified as systematic in nature. The evidence for these errors comes both from the direct comparison of the results provided by different data-analysis techniques applied to the same data (Larson and Schou 2008; Vorontsov et al. 2009), and from the helioseismic inversions, where they are revealed as an internal inconsistency in the input data set. Figure 4 illustrates systematic errors in published SOHO MDI rotational splitting coefficients, revealed by the inversion in an attempt to implement the splitting measurements, accumulated over five years of observations and corrected for temporal variations in the solar internal rotation ("torsional oscillations"), to improve the measurement of the time-independent component of the rotation (Vorontsov et al. 2002a). The prominent horn-like structures in the mismatch between the data and the inverted model do not allow the measurement to benefit from the prolonged observation. Figure 5 illustrates a simpler example with centroid (m-averaged) frequencies. Here, the centroid frequencies tabulated over the entire span of MDI observations were averaged without any corrections for variations with solar activity, for addressing the equation of state of solar plasma (Vorontsov et al. 2013a). The difference between the averaged frequencies and those measured in the first year of SOHO mission (low solar activity) demonstrates a well-known frequency dependence, but only on average: the obvious outliers are the higher-degree modes, the most precious part of data for this particular measurement (modes with turning points in the vicinity of HeII ionization region). Again, attempting to implement a prolonged observation brings no benefit, but only makes systematic errors more evident.

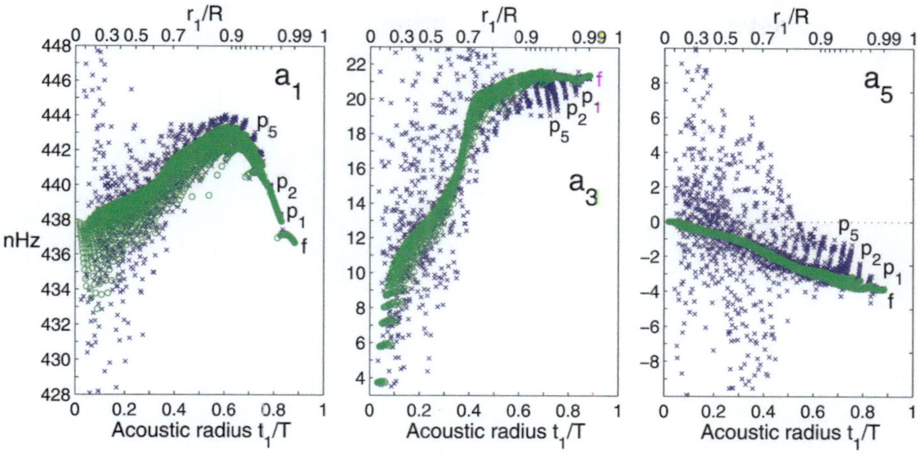

Fig. 4 First three rotational splitting coefficients a_1, a_3, a_5 of solar p- and f modes, plotted versus the position of their inner turning points. *Blue crosses*: average of splitting coefficients measured from 27 consecutive 72d slots of SOHO MDI observations and corrected for the temporal variation of the solar rotation. *Green circles*: the same splitting coefficients but calculated from the inverted rotating model

Fig. 5 Difference between centroid frequencies averaged over 15 years of SOHO MDI observations and frequencies of the 1-yr measurement at solar activity minimum (Vorontsov et al. 2013a). The error bars are 1-σ errors of the 1-yr data set

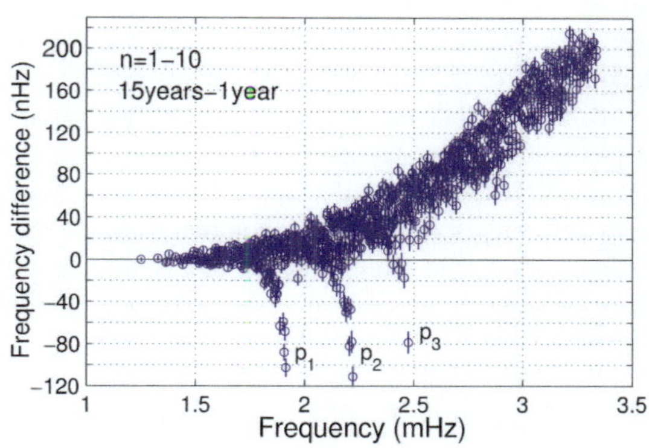

The challenges to the accurate measurement of the solar oscillation frequencies are illustrated by two examples of Doppler-velocity power spectra shown in Fig. 6. The difficulties originate largely from the spatial leaks coming from modes of neighboring values of degree l (Fig. 6a) and azimuthal order m (Fig. 6b). The accurate modeling of the spatial leaks, which may blend into the target peak, puts very tough requirements on the accuracy of the leakage matrix. Further, the asymmetry of the line profiles (Fig. 6a) needs to be properly accounted for. Finally, deviations from the spherical symmetry of the equilibrium solar configuration lead, in general, to mode coupling—instead of individual modes described by a particular pair of (l, m) values, we have the coherent composite states. The biggest effect comes from the mode coupling by differential rotation; an example of how the mode coupling affects the power spectra is the asymmetry of the amplitudes of $m-2$ and $m+2$ leaks, clearly seen in Fig. 6b. The biggest challenge is data analysis at high degree l, where modes of individual degree can no longer be identified and the modal analysis has to be replaced by "ridge-fitting" techniques (Rabello-Soares et al. 2008; Korzennik et al. 2013; Reiter et al. 2015). Better spectral modeling in the intermediate-degree range, where systematic errors of frequency measurements are clearly seen and their origin can be identi-

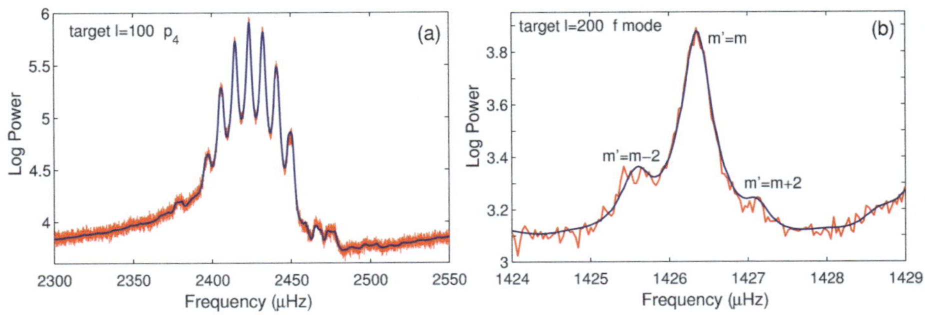

Fig. 6 1-year SOHO MDI power spectra (*red*) and their models (*blue*) for p_4 mode of degree $l = 100$ (**a**) and f mode of $l = 200$ (**b**)

fied, will also bring better confidence to the analysis of lower-degree (Korzennik 2005) and higher-frequency (Rhodes et al. 2011) domains of the solar-oscillation power spectra.

3.2 Global Helioseismic Metrology

Significant efforts are now invested in enhancing the data analysis in global solar seismology. The standard SOHO MDI data analysis pipeline is largely redeveloped, from better accounting for instrumental effects when mapping Dopplergrams to spherical harmonics to more elaborated techniques of frequency fitting (Larson and Schou 2008), and extended for implementation to SDO HMI data (Larson and Schou 2011). Below, we address in more detail the prospects of the "Global Helioseismic Metrology" project, based on modeling the solar acoustic oscillation in the continuous spectrum.

For each pair of target degree l and azimuthal order m, the observed power spectrum is approximated by the spectral model defined as (Jefferies et al. 2006; Vorontsov and Jefferies 2013b)

$$P_{\text{obs}} = |L_U + hL_V|^2 \left\{ \left[\frac{A\cos(\varphi - S)}{1 - R^2} \right]^2 + B^2 \right\}, \quad \tan\varphi = \frac{1+R}{1-R}\tan\theta. \tag{1}$$

In this model, θ is the phase integral of the trapped acoustic wave—a continuous function of frequency ω with $\theta(\omega) = \pi n$ at frequencies of acoustic resonances. The energy losses are assumed to be localized in the near-surface layers and described by the surface acoustic reflectivity R. The average strength of the stochastic excitation is described by the excitation amplitude A. The composite background B is contributed by both the coherent and incoherent components of the solar noise. Parameter S describes line asymmetry, which is governed by the depth and parity of the excitation source and by the coherent component of the solar noise. L_U and L_V are two components of the leakage matrix, which account for vertical and horizontal velocities on the solar surface, h is the ratio of the magnitudes of the two velocity components. The leakage matrix is calculated using a computationally efficient semi-analytic approach described in Vorontsov and Jefferies (2005) and extended later to account for non-zero B-angle and for the discrete bin sampling implemented in the SOHO MDI "medium-l" program. Mode coupling by differential rotation is treated as suggested in Vorontsov (2007).

The parameters of the spectral model, resulted from fitting SOHO MDI power spectra obtained from the first year of observations (at low solar activity) are illustrated by Fig. 7. The results are shown for p modes of radial order n from 1 to 10 and for f modes, in the degree range limited by $l = 200$. The maximum-likelihood solution was obtained by an iterative improvement of the spectral parameters (A, R, S, B) and of the resonant frequencies and frequency splittings. The resulted agreement between the model and the data is almost adequate, as can be judged by comparing the m-averaged power spectra (Fig. 6). A small systematic inaccuracy in the predicted amplitudes of spatial leaks remains, however. The origin of this mismatch is not yet properly understood; it may be related, in part, with asymmetric distortion of the point-spread function of the MDI instrument (Rabello-Soares et al. 2008).

An important property of this model, which is behind its diagnostic potential, is that the spectral parameters of individual modes do not depend on the degree and collapse to slowly-varying functions of frequency only when the degree l is not too high (less than about 100). The composite background B^2 (Fig. 7d) may look as an exception. However, an accurate measurement of the background at frequencies higher than about 2 mHz is a difficult task, because the background level appears to be significantly smaller than the resonant signals coming from the spatial leaks. As a result, the measurement of B^2 can be distorted by small inaccuracies in the leakage matrix. The fitted background is also significantly higher than the average at lowest values of target degree l (p_{10} mode of $l = 2$ in Fig. 7d). A possible explanation of this excess is the contribution of the instrumental noise, which can probably be modeled by adding an $l = 0$ component to the (otherwise degree-independent) solar $B^2(\omega)$.

Interestingly, the analysis of solar f modes reveal the same values of the excitation amplitude A and "acoustic reflectivity" R as solar p modes of similar frequencies (Fig. 7a,b).

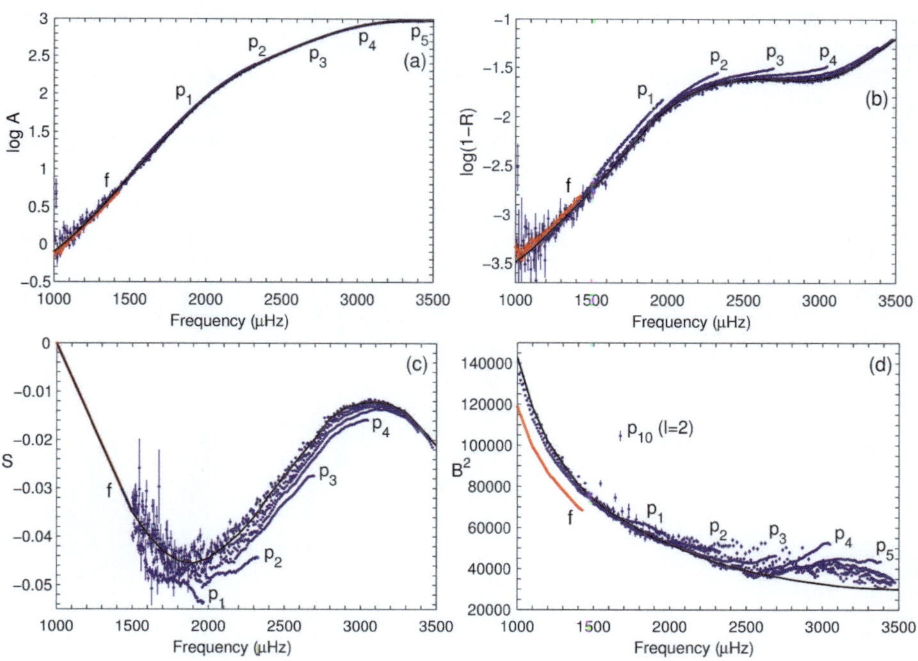

Fig. 7 Excitation amplitude (**a**), acoustic reflectivity (**b**), asymmetry parameter (**c**), and composite background (**d**) obtained from a 1-year SOHO MDI Doppler-velocity power spectra (adapted from Vorontsov and Jefferies 2013b). *Solid lines* show approximations by slowly-varying functions of frequency for lower-degree modes. The results obtained for f modes are shown in *red*. The asymmetry parameter S is extrapolated by a straight line at frequencies below 1500 μHz, where its measurement suffers from uncertainties coming from smaller signal-to-noise ratio and narrow line profiles

It indicates that excitation and damping mechanisms do not distinguish between p- and f modes, despite the difference in their physical nature (the f modes are incompressible waves). The composite background of f modes appears to be smaller than that of p modes (Fig. 7d). Since at low frequencies the fitted background B^2 is dominated by the granulation noise, one possible explanation is that its contribution to the observational power shall be modeled with a smaller (or zero) value of h, the ratio of horizontal and vertical velocities. In data processing, a simple theoretical value $h = GM_\odot/(R_\odot^3\omega^2)$ has been used, which corresponds to incompressible motion: this approximation is hardly relevant to the granulation noise.

The major benefit to the p-mode data analysis, expected from the global description of spectral parameters, is that the global spectral variables (A, R, S, B as functions of frequency), inferred from the large volume of high-quality intermediate-degree data, can be used in measurements at lower degree l. Reducing the number of free parameters in low-degree measurements will bring significant improvement to the accuracy and precision of frequencies (and frequency splittings) of the p modes which penetrate into the deepest solar interior.

Figure 8 shows the rotational splitting coefficients resulting from the measurement which is described above, in comparison with published splitting coefficients. The horn-like structures, which signify systematic errors (cf. Fig. 4) are now eliminated. As indicated by a detailed analysis (Vorontsov et al. 2009), the dominant part of the systematic errors came from discarding the effects of mode coupling by differential rotation in the original version of the SOHO MDI data analysis pipeline (which was later improved to include these effects, among others. Systematic errors in centroid frequencies, illustrated by Fig. 5, are due apparently to the temporal variation of the plate-scale error).

The differences between the centroid frequencies and their published values are shown in Fig. 9. The major part of the discrepancies is due to the line asymmetries, discarded in the original version of the MDI data analysis pipeline (cf. Fig. 7c). Smaller-scale features

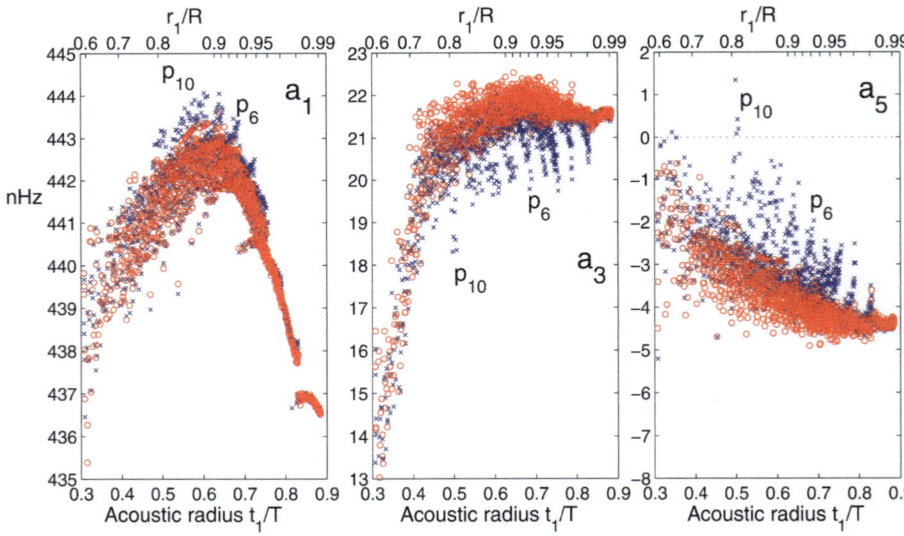

Fig. 8 First rotational splitting coefficients a_1, a_3, a_5, inferred from the first year of SOHO MDI measurements. *Blue crosses*: published coefficients. *Red circles*: the result obtained in the "Global Helioseismic Metrology" project (Vorontsov and Jefferies 2013b)

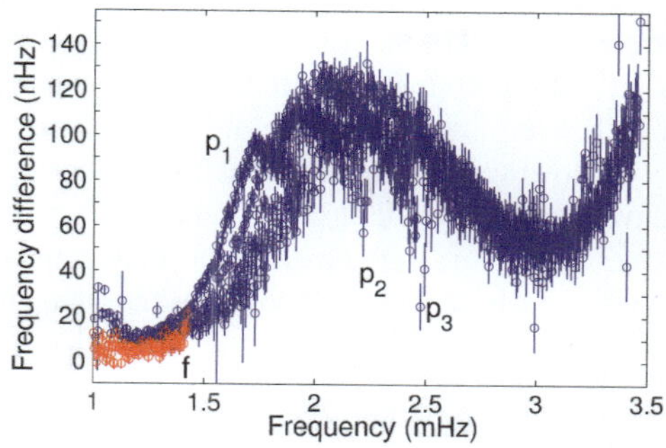

Fig. 9 Difference between the centroid frequencies measured in the "Global Helioseismic Metrology" project from the 1-year SOHO MDI power spectra and their published values (Vorontsov and Jefferies 2013b)

are due apparently to the combination of the mode-coupling effects with the plate-scale error. The centroid frequencies measured in the "Global Helioseismic Metrology" project have been used in a recent study targeted at the seismic diagnostics of the equation of state (Vorontsov et al. 2013a). Interestingly, it was found that these frequencies allow the achievement of a significantly better agreement with solar models. This is quite an unusual finding: better accuracy of observational data brings better agreement with theoretical models.

The final goal of the project is a new approach to helioseismic inversion, where frequency measurements will be eliminated from the analysis, and the parameters of the rotating solar model will be matched directly with p-mode power spectra. This approach will bring benefits of streamlined regularization, first of all—by eliminating problems with error correlation and possible mode misidentification in frequency measurements.

3.3 Global Helioseismology from Cross-Spectral Analysis

In the past, global helioseismology was very successful in measuring the differential rotation inside the Sun from the splitting of the p-mode frequencies. But the meridional flow in the deeper interior is hardly accessible from mode frequencies since the flow is small in amplitude and its influence on the frequencies is only of second order in the flow. Numerical simulations suggest frequency shifts of the order of a few nHz due to the flow (Roth and Stix 2008; Chatterjee and Antia 2009; Schad et al. 2011; Vorontsov 2011). The perturbation of another characteristic of resonant waves by the meridional flow, the mode eigenfunction, and its potential for helioseismic analyses was early recognized, but largely neglected, likely due the difficulty to measure this kind of perturbation and to discriminate it from systematic instrumental effects in the observations. Below we address the recent advances in using cross-spectral analysis for a global helioseismic measurement of the meridional flow from the perturbation of mode eigenfunctions.

3.3.1 Perturbation of Mode Eigenfunctions by Advective Mode Coupling

Expressions for the eigenfunctions perturbed by the meridional flow were derived by several authors (Woodard 2000; Schad et al. 2011; Vorontsov 2011), who used however partly different assumptions and approximations. The perturbations are considered with respect to a flow-free, purely hydrodynamic reference model of the Sun from which the frequencies and eigenfunctions, $(\omega_{nlm}, \boldsymbol{\xi}^0_{nlm})$, of the unperturbed p-modes are obtained. A flow $\boldsymbol{u}$ leads to an advection of the acoustic waves, which causes a perturbation of eigenfunctions and eigenfrequencies of the p-modes with respect to the reference model. If the flow amplitude is small compared to the speed of sound, the perturbed eigenfunctions can be expanded linearly in terms of the unperturbed eigenfunctions: $\boldsymbol{\xi}_k = \sum_{k'} c_{kk'} \boldsymbol{\xi}^0_{k'}$, where k substitutes the triplet (n, l, m) of the wave numbers of a mode. The perturbations are often considered as a kind of mode coupling and the expansion coefficients $\{c_{kk'}\}$ specify the coupling coefficients which determine the strength of the coupling.

Following the general mathematical framework for the perturbation of p-modes from Lavely and Ritzwoller (1992), Schad et al. (2011) have shown that the coupling coefficients of a mode k due to the meridional flow are approximated in first order by a linear integral equation, whose integrand specifies the advection kernel,

$$c_{kk'} \approx 2\mathrm{i}\,\frac{\omega_k}{\omega_k^2 - \omega_{k'}^2} \int \rho_0\, \overline{\boldsymbol{\xi}_{k'}^0} \cdot \left(\boldsymbol{u} \cdot \nabla \boldsymbol{\xi}_k^0\right) d^3\boldsymbol{r} \quad \text{for } k' \neq k\,, \tag{2}$$

where $c_{kk} = 1$. For the meridional flow, the coupling coefficients $c_{kk'}$ are purely imaginary. They are expected to be largest for modes of similar frequency and wave numbers (n, l). The meridional flow does not contribute to a self-coupling of modes. As a consequence, the shift of the mode frequencies due to the meridional flow is only of second order in $\boldsymbol{u}$. Assuming azimuthal symmetry for the meridional flow, coupling is only possible between modes of equal harmonic degree m and the coupling coefficient for fixed pairs of (n, l) and (n', l') may be considered as a function of m: $c_{kk'} = c_{nl,n'l'}(m)$. This expression can be expanded in terms of polynomials in azimuthal order, where the polynomial order is given by the harmonic degree s of the individual meridional flow components when expanded in spherical harmonics Y_s^0. Different polynomial expansions are suggested. Schad et al. (2011) uses a complete set of orthogonal polynomials $\{\mathcal{P}_{l',l}^s(m)\}$ that are based on the Clebsch–Gordan coefficients. In the asymptotic case of high degree l and for coupling modes of equal radial order $n' = n$, Vorontsov (2011) uses associated Legendre functions which are however not perfectly orthogonal on a discrete grid in m. The expression in Eq. (2) defines a linear inversion problem. The polynomial expansion allows to simplify the expression of the coupling coefficients in terms of one dimensional integral equations and to investigate the harmonic components of the flow separately.

3.3.2 *Cross-Spectral Analysis of Solar Oscillations*

The oscillating amplitude $\alpha_k(t)$ of individual global modes of harmonic degree l' and azimuthal order m' can be extracted by a spherical harmonic transformation (SHT) from sequences of Dopplergrams. Each of these spherical harmonic (SH) coefficients is a weighted sum over the amplitudes of several modes: $o_{l'm'}(t) = \sum_k \alpha_k(t) \sum_{k''} c_{kk''} L_{l'm',k''} U_{k''}(R)$. The weighting is given by the coupling coefficients $c_{kk''}$, the amplitude of the radial eigenfunction component $U_{k''}(R)$ of mode k'' observed at radius R, where the respective absorption line is formed, and the leakage matrix elements $\{L_{k'k''}\}$ of the observing instrument. The leakage origins from the imperfect orthogonality of the spherical harmonic functions when not integrated over the complete solar sphere. The leakage matrix elements are further largely influenced by the apodization mask and the projection of the solar velocity field onto the line-of-sight axis. As a consequence of leakage and mode coupling, modes of a specific target degree l and order m are not perfectly separated from modes of neighboring degree l' and order m' by the SHT. This results in a cross-talk between the SH coefficients and the power of a mode is spread amongst the spectra of spherical harmonic coefficients of neighboring (l', m'), where it shows up as sidelobes at the respective mode frequency.

Woodard (2000) pointed out the sensitivity of the cross-spectrum of the SH coefficients to mode coupling. The cross-spectrum of two SH coefficients o_{lm} and $o_{l'm'}$ is defined by $CS_{lm,l'm'}(\omega) = \langle \tilde{o}_{lm}^*(\omega)\tilde{o}_{l'm'}(\omega)\rangle$, where $\tilde{o}_{..}(\omega)$ is the Fourier transform of $o_{..}(t)$ and $\langle X\rangle$

denotes the statistical expectation value of the random variable X. It is related to the coupling coefficients by

$$CS_{lm,l'm}(\omega) = \sum_{k} S_{\alpha,k}(\omega) \sum_{k',k''} c^*_{kk'} c_{kk''} L^*_{lm,k'} L_{l'm,k''} U_{k'}(R) U_{k''}(R) , \tag{3}$$

where it is assumed that the solar modes are excited independently and $S_{\alpha,k} = \langle|\alpha_k(\omega)|^2\rangle$ is the auto-spectrum of mode k.

Schad et al. (2011) introduced a slightly different quantity to relate the coupling-coefficients with observations: the amplitude ratio $y_{lm,l'm}(\omega_{nlm}) := \tilde{o}_{l'm}(\omega_{nlm})/\tilde{o}_{lm}(\omega_{nlm})$. It measures the relative cross-talk of power of a reference mode with frequency ω_{nlm} between the SH coefficients due to mode coupling and leakage. Given certain assumptions on the separability of the mode frequencies in the solar oscillation spectrum, which are met only by modes of low and medium harmonic degree l, the amplitude ratio is independent of the amplitude α_k and can be approximated in first order by (Schad et al. 2011)

$$y_{lm,l'm}(\omega_{nlm}) \approx \frac{\sum_{k''} c_{kk''} L_{k'k''} U_{k''}(R)}{\sum_{k''} c_{kk''} L_{kk''} U_{k''}(R)} \in \mathbb{C} . \tag{4}$$

Its expectation value is in leading order determined by the cross-spectrum, since $\langle y_{lm,l'm}(\omega_{nlm})\rangle \sim CS_{lm,l'm}(\omega_{nlm})/S_{nlm}(\omega_{nlm})$. This quantity, denoted as the complex gain, is related to the gain in linear filter theory. The asymptotic statistical distribution of the estimator of the complex gain can be expressed analytically (Schad et al. 2013).

3.3.3 Influence of Differential Rotation

The toroidal velocity field of differential rotation also leads to a coupling of modes, which cannot be neglected in the eigenfunction perturbation analysis. The perturbation of eigenfunctions due to rotation was investigated by Vorontsov (2007, 2011) for the asymptotic case of large degree l. The case of low and medium degree l was investigated by Schad (2013). Formally rotation leads to an additional term to the coupling coefficient: $c_{k'k} = c^{(rot)}_{k'k} + c^{(poloidal)}_{k'k}$. Again, rotation couples only modes of similar frequency and of equal azimuthal order and the coupling coefficients can be expanded by polynomials in m which are equal to the ones used for meridional flow. But in contrast to the coupling coefficients due to the meridional flow, the rotational coupling coefficients $c^{(rot)}_{k'k}$ are real valued and anti-symmetric with respect to the azimuthal order. The different symmetry properties can be exploited to compensate approximately the influence of rotation on the amplitude ratios in analyses for the meridional flow. In this case, the amplitude ratios $y_{lm,l'm}$ are symmetrized with respect to azimuthal order (Schad et al. 2013). Exemplary amplitude ratios estimated from about 6 years of MDI data for ($n = 1, l = 180$) and ($n = 1, l = 182$), as well as simulated amplitude ratios from numerical forward computations of simple flow profiles, are shown in Fig. 10. The influence of rotation is clearly visible. Both the real and the imaginary part of the amplitude ratios deviate significantly from azimuthal symmetry if rotation is present. The compensation of the rotational influences by azimuthal symmetrization is also illustrated using simulated amplitude ratios. The symmetrized amplitude ratios matches very well to the amplitude ratios obtained for a velocity field without solar rotation.

3.3.4 Application to Data: Measurement of the Meridional Flow

First measurements of the meridional flow from analysis of the perturbation of p-mode eigenfunctions were given by Schad et al. (2012, 2013), Woodard et al. (2013).

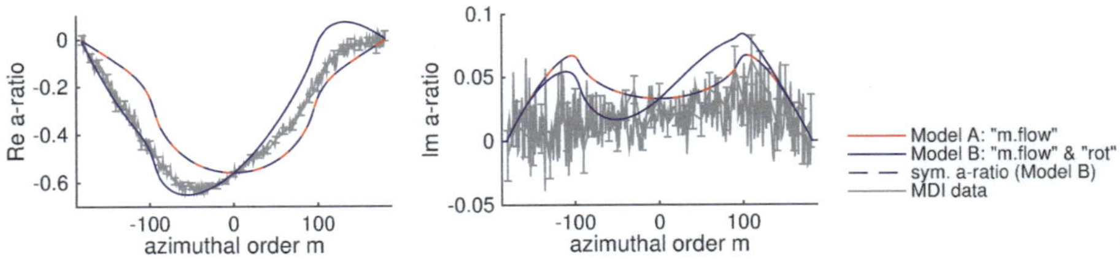

Fig. 10 Real and imaginary part of amplitude ratios for $(n = 1, l = 180)$ and $(n' = 1, l' = 182)$ as a function of azimuthal order m (Schad 2013). The amplitude ratios are estimated from SH coefficients from MDI data covering 2004–2010. The ratios are compared to theoretical amplitude ratios computed for simple flow models: Model A—two-cell meridional flow; Model B—two-cell meridional flow and differential rotation adapted to observed rotation rates. The real part essentially exhibits features from leakage and rotation, the latter is responsible for an antisymmetry in azimuthal order. The imaginary part is influenced by leakage, rotation, and meridional flow. Theoretical amplitude ratios for Model B after compensation for rotation by azimuthal symmetrization are shown by a *dashed blue line*

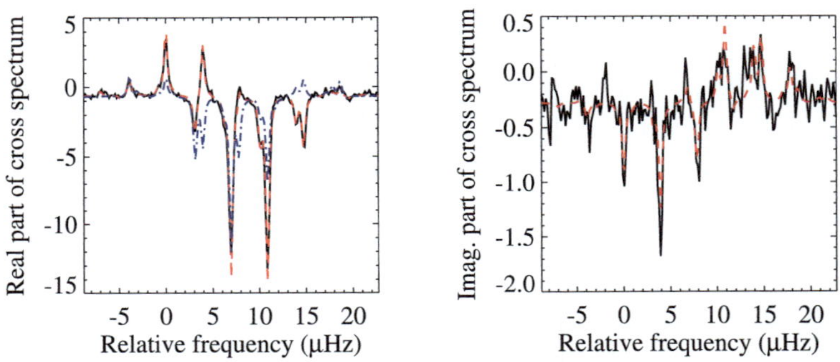

Fig. 11 De-rotated cross-spectra for $l = 200$ and $l' = 202$ averaged over positive order m as a function of ν relative to the frequency of the f-mode multiplet of degree l (Figures adapted from Woodard et al. (2013)). *Left*: Real part. *Right*: Imaginary part. The *black curves* depict the cross-spectrum estimated from 500 days of HMI data. The *blue dash-dotted curve* is the fitted cross-spectrum when ignoring the effect of flows on mode eigenfunctions. The *red dashed curves* are fitted cross-spectra including the effect of differential rotation and meridional flow

Woodard et al. (2013) fitted a model to cross-spectra estimated from HMI data to measure the horizontal component of the meridional flow. The model for the cross-spectrum incorporates instrumental leakage, differential rotation, and the solar background. They analyzed HMI data with a length of 500 days and investigated couplings between modes of the same radial order n for the harmonic degrees $20 \leq l \leq 300$. Cross-spectra estimated from the HMI data and averaged over azimuthal order as well as cross-spectra after fitting different cross-spectral models to the data are depicted in Fig. 11. The comparison illustrates the improvement of the cross-spectral model by inclusion of flow-dependent eigenfunction perturbations (Woodard et al. 2013).

The estimated peak velocities of the horizontal flow component with harmonic degree 2 ($V2$) are shown in Fig. 12 as a function of ν/L that is related to the lower turning point of the acoustic waves (Woodard et al. 2013). Near the surface, the horizontal peak velocities are of the order of about 20 m/s as expected from local helioseismic measurements. Toward the interior, the velocities exhibit an unexpected large increase and the authors assume that their measurements are likely affected by a systematic effect (Woodard et al. 2013).

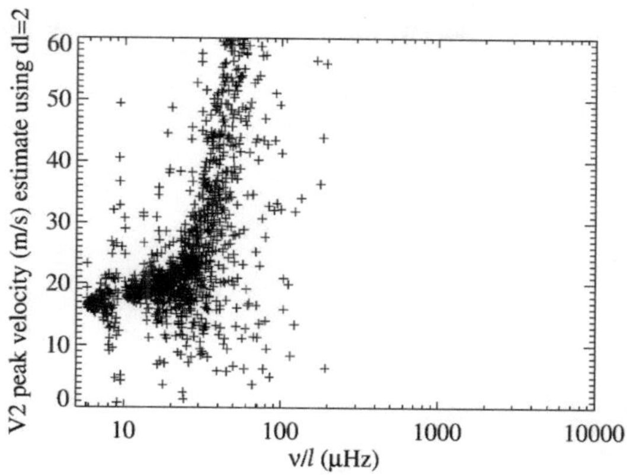

Fig. 12 Surface peak velocities for the horizontal flow component $V2$ estimated from a cross-spectral analysis of HMI data as a function of ν/L as found by Woodard et al. (2013). The steep increase of the peak velocity with depth toward speeds > 30 m/s is suspected to be due to systematic errors (Woodard et al. 2013)

Schad et al. (2012, 2013) set up a global helioseismic estimation scheme for the meridional flow based on the analysis of amplitude ratios of coupling modes. It takes into account leakage and the influence of rotation is compensated by symmetrization of amplitude ratios. They applied their method to MDI data covering 2004–2010 for modes with $0 \le l \le 200$ and investigated couplings between modes of different radial order and for flow components of harmonic degree $s = 1, \dots, 10$. They found two flow components of harmonic degree $s = 2$ and $s = 8$, which differ significantly from zero. Individual components, like the $s = 8$ flow component were measured deep down to 0.5 R. The radial component of the flow $U(r,\theta) = \sum_s u_s(r) Y_s^0(\theta,\phi)$ is estimated from a composite of the individual flow components. The horizontal flow component $V(r,\theta) = \sum_s v_s(r)\partial_\theta Y_s^0(\theta,\phi)$ is reconstructed from the radial components $\{u_s\}$ assuming mass conservation. The composite of flow components of even degree $s = 2, \dots, 8$ are depicted in Fig. 13 as a function of radius and latitude. The meridional flow exhibits a multi-cellular pattern over latitude and depth. Near the solar surface, the horizontal flow is consistent with subsurface flow measurements from ring-diagram analyses indicating a poleward directed flow on each hemisphere with a small-scale latitudinal modulation and a speed of about 20 m/s at mid-latitudes. Their findings substantiate the assumption that the flow is confined between the tachocline region and the solar surface.

3.4 Discussion

The systematic features of the estimates of global mode parameters considered in Sect. 3.1 were found from analyses of medium-l MDI data and can be reduced by the approach used by the global metrology project. Parameters estimated from data from the same instrument but of different preprocessing or from data from other instruments, e.g., HMI or GONG, may show other systematic features. But independent of the instrument, one cannot avoid for example leakage of mode power due to the observational restrictions or mode coupling dominated by rotation.

The development of global analysis methods based on mode eigenfunction perturbations and the analysis of cross-spectra of spherical harmonic Doppler velocity coefficients is a promising emerging field for inferences on the meridional flow in the deep solar interior. In this approach, care must be taken of mode eigenfunction perturbations from other sources, for example from differential rotation. Vice versa, analyses of mode eigenfunction perturbations may also be used to investigate rotation. Next to the solar subsurface, the meridional flow measured by Schad et al. (2013) seems to be consistent with flow measurements from a ring-diagram analysis. But, the recent measurement of the meridional flow from a deep

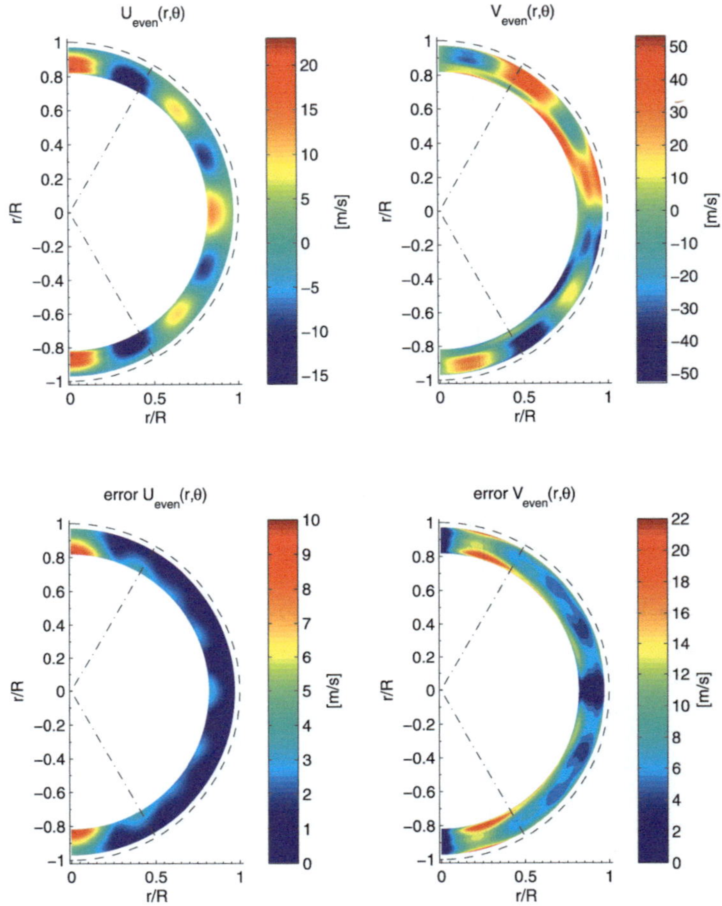

Fig. 13 *Top row*: The meridional flow estimated from MDI data covering 2004–2010 as a function of radius r/R and latitude (Schad et al. 2013). The profile is obtained from a composite of flow components of the harmonic degrees $s = 2, 4, 6, 8$. A *dashed line* indicates the solar surface. *Left*: Radial flow component. Positive (negative) values of U refer to an outward (inward) directed radial flow. *Right*: Horizontal flow component. Positive (negative) values of V refer to a northward (southward) directed horizontal flow. *Bottom row*: 1σ-error on the respective flow estimates

focusing time-distance helioseismic analysis of HMI data by Zhao et al. (2013) seems to disagree with their findings. However, the investigated HMI data cover a later observation time period. Further investigations are necessary to test the reliability of these results.

The success of each of the global helioseismic methods considered here strongly depends on the accurate knowledge of the leakage matrix. The inferences from global helioseismic analyses suffer from systematic effects if the leakage matrix is inaccurate or the models used to estimate the helioseismic parameters from spectra or cross-spectra are misspecified. For example, a comparison study of splitting coefficients and rotation profiles obtained from medium-l and full-disk MDI data indicate that the there used leakage matrix used for the medium-l MDI data probably does not perfectly account for the apodization function and that might be responsible for a spurious polar jet in the solar rotation profile from the medium-l MDI data (Larson and Schou 2009).

The results obtained so far clearly illustrate the necessity of putting efforts into enhancing the models of the solar oscillation spectrum and of improving the accuracy of the leakage matrix of the respective instruments. A difficulty in the computation of the leakage matrix comes along with taking into account systematics that change over time, like the B-angle, or from not properly known instrument characteristics, like the plate-scale error or the point spread function.

4 Data Assimilation in Solar Physics

Understanding and predicting the solar activity cycle poses one of the main problems in solar physics and comes along with questions about the timing, amplitude, and shape of the currently evolving and following cycle. Convection, rotation and the mean meridional flow are thought to be key ingredients that drive the generation and evolution of the solar magnetic field. In the previous sections it was shown that new helioseismic techniques could provide estimates of some of these ingredients. We will see in this section how observations and helioseismic measurements could be combined with physical models to improve our knowledge of the solar activity cycle.

We have entered an era where extremely large amounts of data are available concerning the Sun, thanks to the high-resolution observations of satellites like Hinode or more recently SDO. At the same time, considerable progress has been made on the multi-dimensional numerical simulations of highly non-linear physical processes interacting in the solar interior and in its atmosphere. Although still far from realistic values of the parameters, several impressive local and global computations now seem to reach high levels of turbulence and capture relevant physical processes occurring in our star. Various 3D MHD codes are used for that purpose, among which the ASH code (e.g. Miesch et al. 2008; Brun et al. 2011; Nelson et al. 2013; Alvan et al. 2014), the PENCIL code (e.g. Käpylä et al. 2012; Warnecke et al. 2014), the MURaM code (e.g. Cheung et al. 2010; Rempel and Cheung 2014) or the STAGGER code (e.g. Stein and Nordlund 2012). This improvement is intimately linked to the recent fast development of high performance computing, several computers in the world now reaching the performance of tens of PetaFlops. Finally, we live today in a technological society in which strong solar flares, CMEs or any violent events linked to solar activity could cause significant damage to satellites, air traffic or telecommunication networks. That is why a solar cycle panel, whose role is to produce predictions of solar activity, was created in 1997 and has provided us with estimates of the sunspot number for Cycle 23 and current Cycle 24.

As Cycle 24 progressed, it became clearer and clearer that it would be a weak cycle. A quantitative estimate of the cycle strength can be given by the Wolf number, defined as $R = k(10g + s)$, with g the number of sunspot groups, s the total number of individual sunspots in all groups and k a variable scaling factor that accounts for instruments or observation conditions. We now know that for Cycle 24, the monthly smoothed Wolf number reached a peak of about 82 in April 2014. This will probably become the official maximum. This second peak surpassed the level of the first peak (about 67 in February 2012). Many cycles are double peaked but this is the first in which the second peak in sunspot number was larger than the first. These features make Cycle 24 the weakest cycle since Cycle 14, which peaked in 1906. Among the predictions, less than 25 % had anticipated such a small number. It thus still seems extremely difficult to provide reliable predictions of the long-term solar activity with the techniques that have been used so far (mostly relying on geomagnetic precursors or other statistical estimates—see Hathaway 2009 for a review on the subject). In meteorology, data assimilation which cleverly combines observational data and numerical models has been used for decades and now routinely in weather-forecasting. Considering the high-quality observations at our disposal, the recent progress in numerical simulations and the necessity to produce predictions of the amplitude, timing and shape of the next solar cycle, it seems rather reasonable to try to apply data assimilation to solar physics (Brun 2007).

4.1 First Attempts to Introduce Data into Models

There have been first attempts to connect models and data, not exactly through data assimilation but rather by driving models with a time series of well-chosen data. This procedure has been implemented for flux-transport dynamo models and for photospheric and heliospheric magnetic field evolutions. In their mean-field dynamo model, Dikpati and Gilman (2006) have introduced a surface source of magnetic field that depended upon the sunspot areas observed since 1874, when Choudhuri et al. (2007) chose the surface field at minimum to be their driving observations. Both models produced good agreement with previous observed cycles but differed completely on the predictions for Cycle 24. One of the reasons was that one model (Dikpati's) was dominated by the advection process while the other (Choudhuri's) was dominated by diffusion, changing drastically the characteristics of the memory of the system, as was shown by Yeates et al. (2008). It should be noted that both predictions happen to be quite far from reality, either on the timing of the cycle, or its amplitude. Data-driven models for the solar atmosphere now also tend to develop. A first attempt was made by Schrijver and DeRosa (2003) when they introduced SOHO/MDI magnetograms into a flux-dispersal model to see the influence on the coronal reconfigurations. More recently, Cheung and DeRosa (2012) simulated the evolution of the active region coronal field driven by temporal sequences of photospheric magnetograms from SDO/HMI. Under certain conditions, they found that data-driven simulations could produce flux ropes that were ejected from the modeled active region due to loss of equilibrium, possibly showing a way to predict violent events through simulations.

In the future, we would like to make observations and models really work together and feed the simulations with data so that the model improves itself and allows to make reliable predictions. To do so, data assimilation techniques are exactly what we are looking for (see Kalnay 2003 for a general introduction related to atmospheric sciences and Fournier et al. 2010 for applications to geophysics). Indeed, they consist in combining observational data and numerical models to produce what is called an analysis: the new information provided by the observational data is taken into account in order to advance in time the "background" state that the numerical code has predicted. The increment is obtained by taking the difference, or innovation, between the observational data and the observation operator. More specifically, let $\mathbf{x^b}$ be the background vector state characterizing the current state of the model, H the observational operator and $\mathbf{y^o}$ the observational data to be assimilated in the model, then one can show that the analysis $\mathbf{x^a}$ is:

$$\mathbf{x^a} = \mathbf{x^b} + W\big(\mathbf{y^o} - H\big(\mathbf{x^b}\big)\big), \tag{5}$$

where W represents weights whose exact determination will differ from one assimilation technique to another. Two approaches and their applications to solar physics will now be briefly presented and discussed.

4.2 Sequential Assimilation and Application to Solar Physics

In the sequential assimilation technique, the background state is advanced in time thanks to the numerical model and corrected (i.e. an analysis is performed) each time an observation is available. The analysis is performed using Eq. (5), where W is given by the Kalman gain matrix which is a combination of the covariance matrices of the forecast and of the observational errors.

The analysis at time k thus provides the new state vector x_k^a from which the forecast is calculated by running the numerical model. This forecast step thus produces the new background x_{k+1}^b which in turn will be corrected by the observation at time $k+1$. This sequential process is illustrated in Fig. 14. This technique thus propagates information forward in time and also gives an estimate of the forecast errors and of their evolution.

As far as predictions of future solar activity is concerned, sequential assimilation has been implemented for the first time by Kitiashvili and Kosovichev (2008) in a 1D mean-field dynamo model evolving jointly the three components of the magnetic field and a measure of the magnetic helicity. The observations used were the annually smoothed Wolf sunspot number for the period 1857–2007. To derive the observational operator H, the following relationship between the Wolf number and the toroidal (in the azimuthal direction) magnetic field was used: $|B_\phi| \propto R^{2/3}$. The model used was a classical $\alpha\Omega$ dynamo model in which the toroidal field owes its origin to the differential rotation shearing the poloidal (in the meridian plane) field lines (the Ω-effect) and where the poloidal field is due to helical turbulence within the solar convection zone acting on the toroidal field (the α-effect). This model contains a number of simplifications (see Brun et al. 2013 and Charbonneau 2010), most notably a rather crude parameterization of the effects of turbulence on the large-scale magnetic field but has the advantage of producing a cyclic variation of **B** and an exponential growth of the magnetic energy (saturated here by the back-reaction of the magnetic field on the α-effect).

Figure 15 shows the result of their forecasting step for Cycle 24, after they applied their sequential assimilation to get the analysis at the time they wrote the paper (2008). Their predictions were shown for three different estimates of the Wolf number in 2008 since they did not have yet the real data. What we see on this figure is actually quite a good agreement with what Cycle 24 looks like now, reaching its maximum of around 80 in 2013 (a bit early then). However, we should keep in mind that the model was here very simple and the relationship between the observational data and the outputs of the model rather uncertain, leading

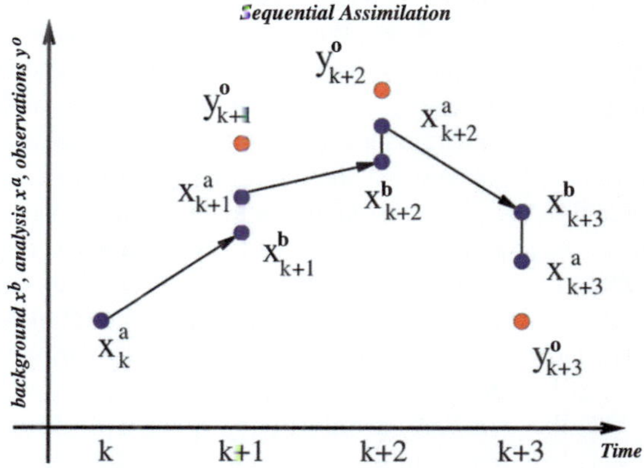

Fig. 14 Schematic representation of the sequential data assimilation method used in weather forecast (adapted from Bocquet 2015). The background state (x^b) is updated every time an observation is available and the model evolves the state until the next step (following the arrows), at which observational data (y^o) are again assimilated to produce the analysis (x^a)

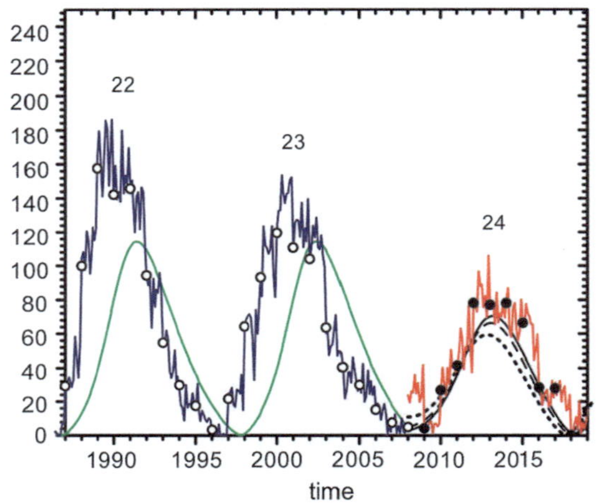

Fig. 15 Prediction of the sunspot number for Cycle 24 from Kitiashvili and Kosovichev (2008). The *green curve* shows the model reference solution. The *blue curve* shows the best estimate of the sunspot number using the observational data (*open circles*) and the model outputs (*black dots*), for the previous cycles. The *black curve* the model solution according to the initial conditions of the last measurement. The *red curve* shows the prediction results. The model solution is shown for three different estimates of the sunspot number for 2008: 3 (*black curve*), 5 (*dashed curve*), and 10 (*dotted curve*)

them to state in the conclusion that additional observations like the latitudinal distribution of sunspots should probably also be taken into account to make progress in the subject.

First steps towards using a more complete model including a large-scale meridional circulation were undertaken recently by Dikpati and Anderson (2012) and Dikpati et al. (2014). Their idea is to use a 2D dynamo model in which the poloidal field is generated by the decay of active regions emerging at the solar surface (Babcock 1961; Leighton 1969) and not by small-scale turbulence in the convection zone like in classical $\alpha\Omega$ dynamo models. The data they are planning to use is similar to what Kitiashvili and Kosovichev (2008) chose, namely the monthly smoothed sunspot number data from the Royal Observatory of Belgium but they also intend to make efficient use of data concerning the solar meridional circulation. Their first step was thus to determine the response time of the whole system to perturbations of the meridional flow, which is not very well measured below the first few Mm of the Sun and which is known to produce large changes in the timing and possibly the shape of the magnetic cycles in those types of flux-transport dynamo models (Dikpati and Charbonneau 1999; Jouve and Brun 2007). Dikpati and Anderson found that in the advection-dominated regimes they considered, the model's time of peak response to a change in flow speed was rather short compared to the full circulation time and thus that a modification of the amplitude of the flow would quickly show large changes in the evolution of the magnetic field. In a recent subsequent paper (Dikpati et al. 2014), they applied a sequential data assimilation technique to reconstruct the meridional flow speed at the solar surface (fixing the meridional flow profile to one large circulation cell per hemisphere) from synthetic observations of the magnetic field. The synthetic observations were produced by running the model with a fixed meridional flow profile and speed and then noised to mimic observational errors. They found that the best reconstruction of meridional flow-speed could be obtained when 10 or more observations were used with an up-dating time of 15 days and an observational error of less than 10 %.

Another relevant quantity to assess when dealing with forecasting in such dynamical systems is what Lhuillier et al. (2011) call the forecast horizon or the time interval over which reliable predictions can be achieved. They made a detailed analysis of the growth rate of perturbations applied to the magnetic, velocity or temperature field for geodynamo simulations and found that the limit of predictability would be a combination of this growth time (estimated to be of about 30 yr), of all types of errors affecting the initial conditions and of the limited numerical resolution. Recently, Sanchez et al. (2014) performed the same kind of analysis for a flux-transport mean-field dynamo model, similar to the one considered by Dikpati et al. (2014). They measured the rate associated with the exponential growth of an initial perturbation of the model trajectory, and found a characteristic e-folding time of 2.76 solar cycle durations. These results are quite promising for possible future predictions of solar activity. However, thorough studies of the sensitivity to all model parameters and on the predictability skills of these models will have to be undertaken before being able to apply more complete sequential data assimilation and to rely on predictions coming from simplified solar dynamo models.

4.3 Variational Assimilation and Applications to Solar Physics

As opposed to sequential assimilation, the variational technique consists in adjusting the trajectory of the model through observations over a significant time interval. This is illustrated in Fig. 16. To be more formal, the analysis x^a is found here by minimizing a cost function J, defined over the entire time window where observations are available. This cost (or objective) function is the sum of two terms. The first one measures the distance between the outputs of the model and the observations and the other one measures the distance to an a priori estimate of the background state, if there is any. This second term is called the background term, it can be used to provide the system with information about regularity of the solutions we are looking for, approximations that our flows need to satisfy or typical values we expect at certain points in the domain. After the analysis is found, the forecast step is similar to the one performed in the sequential technique, it is calculated by applying the numerical model to the analysis which has just been determined. This technique thus propagates information both forward and backward in time since the present state is estimated using the past and future observations available over the entire time window. This is important if we wish to reassimilate past data or in other words propagate backward in time the current quality of observational data. This advantage of variational assimilation might be of great use for solar physics for example if we wish to have a better insight on the state of the system at periods where observations were missing because of a lack of instruments or

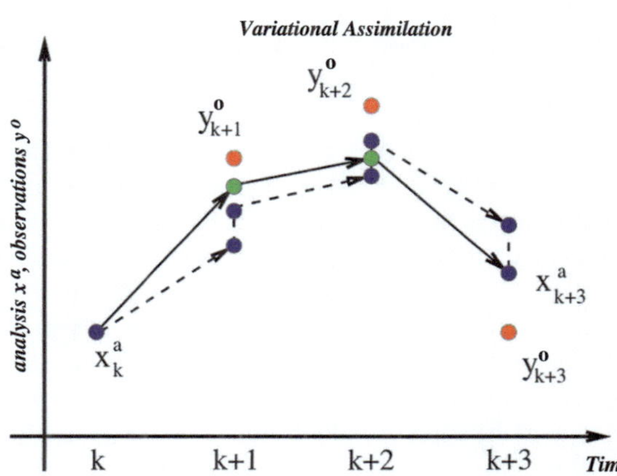

Fig. 16 4D variational method and comparison with sequential assimilation (adapted from Bocquet 2015). In the 4D variational method, within a time interval the model and the observations are taken into account in the cost function that needs to be "minimized". The minimization of this cost function results in a best trajectory (plain arrows) across the observations

of a lack of surface events, like in the periods of grand minima. However, as stated before, we should of course keep in mind that the quality of the forecast will be strongly limited by the range of predictability in such dynamical systems (Lhuillier et al. 2011).

The main drawback however of using variational assimilation is that it requires the development of an adjoint model which will be used to provide the gradient of the cost function with respect to all input variables. Both the values of the function itself and of its gradient are then combined in a minimization algorithm to produce the analysis. Simple recipes can be used to write the adjoint code of a numerical program by hand (e.g. Talagrand 1991; Giering and Kaminski 1998) but for very large codes, it can be tempting to resort to automatic differentiation (AD) algorithms. AD is becoming a very efficient and powerful tool used to produce the adjoint code of general circulation models in meteorology (among many others, we can quote the on-line tool *TAPENADE*, developed by the TROPICS team in INRIA Sophia Antipolis, France).

Variational assimilation (or 3D/4D-VAR) was used recently in astrophysics for the problem of 2D stratified convection (Svedin et al. 2013) and in the context of solar physics for two main applications as of today: models of solar flares and dynamo models. In both studies, only synthetic observations were used, i.e. produced by a model and not by nature. Nevertheless, those are first steps towards understanding how variational assimilation may help us understand the current state of our Sun and hopefully predict its magnetic activity. In 2007, Bélanger et al. used a phenomenological model, called the avalanche model, which, although a priori far removed from the physics of magnetic reconnection and magnetohydrodynamical evolution of coronal structures, nonetheless reproduces quite well the observed statistical distribution of flare characteristics. This model is the continuous analogous to one of a sandpile where grains are dropped one by one until the pile reaches an equilibrium conical shape. Addition of more grains will then sometimes produce small to large avalanches or may have no consequences at all, leading to a strongly intermittent unloading while the loading remains slow and gradual. The cost function which is minimized here is the misfit between the outputs of the model and the synthetic observations (produced with a known set of parameters), in terms of an amount of energy released, averaged in time. They show that, despite the unpredictable (and unobservable) stochastic nature of the driving/triggering mechanism within the avalanche model, 4D-VAR succeeds in producing optimal initial conditions that reproduce adequately the time series of energy released by avalanches and flares. More recent works about the predictability of solar flares with the avalanche model have however shown that only a modified version of it, purely deterministically driven, could produce reliable predictions for large eruptive events (Strugarek and Charbonneau 2014).

Variational data assimilation was also applied in the context of solar dynamo models. Since the idea is to produce predictions of future solar activity, trying to use a technique which is now used routinely in weather forecasting on Earth sounds like a reasonable way to go. In Jouve et al. (2011), a variational data assimilation technique was developed, using a 2D mean-field $\alpha\Omega$ dynamo model. As we said before, synthetic data were used, consisting of outputs from the model produced by a particular set of parameters and more specifically a particular choice of α as a function of latitude (designated by the coordinate y here). After the observations were produced, a cost function was chosen to be the misfit between the model toroidal field and the observations of the same quantity. The result of the minimization was then supposed to reproduce the α function used to calculate the observations. The authors performed an analysis of the dependence of the quality of this recovery as a function of the number and location of observations. In Fig. 17, an example of such a calculation is shown, where data were only present in one hemisphere (from $y=-1$ to $y=0$).

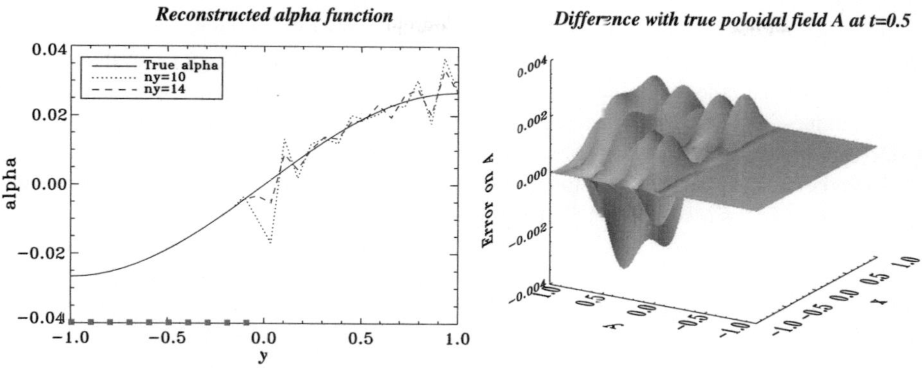

Fig. 17 Adapted from Jouve et al. (2011). *Left panel*: α function reconstructed after assimilation of data in the Southern hemisphere only, with 2 different sets of observations, superimposed with the true state. The *red dots* indicate the locations of the observations in the case were we have 10 observations in latitude ($ny = 10$). *Right panel*: difference between the poloidal potential produced by the reconstructed α and the true state at $t = 0.5$ (in the middle of the time interval)

The recovery of the α function was very good in the hemisphere where data were present but the difference with the real solution (called the true state) was also found to be reduced in the other hemisphere when more observations were assimilated. This indicates that the cost function (and thus the values of the magnetic field) in one hemisphere was sensitive to the α-effect in the other hemisphere, which is the kind of insight that can be gained from variational data assimilation techniques.

One of the challenges in solar physics that can be tackled under the angle of data assimilation is to get a better knowledge of the meridional flow and its role in the dynamo process. Hung et al. (2015) are now developing a variational data assimilation technique applied to a 2D spherical flux-transport dynamo model. This will allow consideration of various possible observations beyond the basic averaged sunspot number. For the magnetic field, one could use the amount of poloidal field at the poles, the timing of its reversal (that we can get from Hinode or SDO for example), the components of the multipolar expansion of the magnetic field (accessible of course on the Sun, DeRosa et al. 2012, but also now for other stars through spectropolarimetry, see Petit et al. 2008). For the velocity field, use could be made of the amplitude and profile of the meridional flow (both at the surface and deeper down thanks to new helioseismic techniques, as discussed in this paper) and possibly of torsional oscillations (Vorontsov et al. 2002b; Spruit 2003). By applying data assimilation with those various sources of observations and an a priori knowledge of part of the meridional flow, the first results of this study indicate that it is possible to reconstruct not only its amplitude as in the work of Dikpati et al. (2014) but also its profile in the solar interior and its structure close to the base of the convection zone, which are of prime interest for dynamo modelers.

4.4 Perspectives for Data Assimilation in Solar Physics

The applications presented here are of course very preliminary in terms of models and observations. Firstly, they do not yet use real data but were only tested on synthetic observations or on very smoothed proxies. Secondly, the models are extremely simple compared to the most up-to-date 3D models evolving the full set of magneto-hydrodynamics equations in spherical geometry. However, and that is maybe what constitutes a big difference with meteorology, we have limited access to observations within the solar interior (with the notable exception of what is learned thanks to helioseismology and particularly new global and local techniques, as is shown in this paper) and moreover we do not have yet at our disposal a self-consistent 3D dynamo model reproducing the main characteristics of the large-scale solar magnetic field. We thus have to move step by step towards this goal by first considering surface observations assimilated in simplified solar models.

In this context, it could be useful to work on ensemble forecasting, similar to what is used today in weather predictions on Earth. In the case of weather forecasting, the idea is to make several predictions starting with slightly different initial conditions and to take an average of those predictions to get the actual forecast. This has mainly three goals: improving the forecast thanks to the averaging, providing an indication of the reliability of the prediction and giving a quantitative way to assess the quality of each individual forecast. In solar physics, we could think of forming an ensemble not by perturbing the initial conditions for the same model but rather by considering different models, where the key physical processes may have more or less impact on the evolution of the magnetic field. Each model would then provide its own forecast. Eventually, the averaging process would give us a way to distinguish between those various models and would indicate which characteristics of the cycle we are the most likely to correctly anticipate. That is probably a future development to be considered on the way towards better and more reliable long-term predictions of solar activity.

5 Summary

In the first part of this paper we reviewed some recent developments of local and global helioseismic data analysis methods. For the local helioseismic analysis we focused on the time-distance method as used for the estimation of velocity fields in the solar subsurface, e.g., supergranulation. Such analyses of rather rapidly evolving processes benefit from sophisticated averaging schemes and filtering methods that help to increase the signal-to-noise ratio and to retrieve signatures from travel-time shifts.

Regarding global helioseismology, we considered two advances, the metrology project and the mode eigenfunction perturbation analysis for inferences of the meridional flow. Global helioseismic inferences rely on the accuracy and reliability of estimated global seismic mode parameters, like the mode frequencies. The global helioseismic metrology project aims to improve accuracy and reliability of the parameter estimates by better incorporation of systematic influences in the parameter estimation scheme. The systematic influences cannot be overcome by longer observations and are of different origin, for example from leakage of mode power in the l–ω-diagram that comes essentially from the technical restrictions in observing the solar velocity field in the photosphere and mode coupling dominated by rotation. The method used by the global helioseismic metrology project is able to reduce some of the systematic errors of the estimated parameters which provide a better agreement with solar models constructed with recent versions of the equation of state.

One systematic influence, the mode coupling, actually results from a perturbation of the mode eigenfunctions by flows in the solar interior. These perturbations lead to a cross-talk between the spherical harmonic coefficients of the Doppler velocity measurements that manifests as leakage in the power spectra. This phenomenon was recently exploited to develop global helioseismic methods to infer the meridional flow in the deeper interior. Here, the characteristic cross-talk due to the meridional flow is investigated by a cross-spectral analysis of time series of these spherical harmonic velocity coefficients. The meridional flow measured by this approach shows a complex flow pattern over latitude and depth and extends from the surface down to the base of the convection zone. Since mode eigenfunctions are also perturbed by other kinds of disturbance, e.g., rotation, their analysis may be also of interest for studies on these perturbations.

One important objective of helioseismic investigations is to reveal the dynamic processes in the interior associated with the solar dynamo, e.g., the meridional flow, in order to better understand this mechanism. Another approach to explore this mechanism and moreover to predict the related solar magnetic activity cycle was recently made by the use of data assimilation methods. This subject was reviewed in detail in the second part of this paper. We discussed two approaches, sequential and variational data assimilation, and their application to solar data. So far, different kinds of observational quantities, for example the sunspot numbers, the speed of the meridional flow, the polar magnetic field as well as synthetic data, and rather simple models relating observations to the solar dynamo were combined to afford forecasts of the magnetic solar cycle or flares. The results of these investigations and the increasing amount of solar data of high quality make data assimilation a promising approach for forecasts of the solar activity cycle as well as magnetic activities relevant for space weather.

Acknowledgements LJ would like to thank Sacha Brun for helpful comments on the data assimilation part in this paper. MR and AS have received funding from the European Research Council under the European Union's Seventh Framework Program (FP/2007-2013)/ERC Grant Agreement no. 307117. The authors thank M.F. Woodard for providing Fig. 11.

Conflict of interest The authors declare that they have no conflict of interest.

References

L. Alvan, A.S. Brun, S. Mathis, Theoretical seismology in 3D: nonlinear simulations of internal gravity waves in solar-like stars. Astron. Astrophys. **565**, 42 (2014)

E.R. Anderson, T.L. Duvall Jr., S.M. Jeffries, Modelling of solar oscillation power spectra. Astrophys. J. **364**, 699–705 (1990)

H.W. Babcock, The topology of the Sun's magnetic field and the 22-year cycle. Astrophys. J. **133**, 572 (1961)

E. Bélanger, A. Vincent, P. Charbonneau, Predicting solar flares by data assimilation in avalanche models. I. Model design and validation. Sol. Phys. **245**, 141–165 (2007)

A. Birch, T.L. Duvall Jr., L. Gizon, J. Jackiewicz, Helioseismology of the "average" supergranule. Bull. Am. Astron. Soc. **38**, 224 (2006)

A. Birch, D.C. Braun, Y. Fan, An estimate of the detectability of rising flux tubes. Astrophys. J. Lett. **723**, L190–L194 (2010)

M. Bocquet, in M2 OACOS/WAPE/ParisTech lecture notes (2015), pp. 2014–2015. http://cerea.enpc.fr/HomePages/bocquet/teaching.html

D.C. Braun, T.L. Duvall Jr., B.J. LaBonte, Acoustic absorption by sunspots. Astrophys. J. Lett. **319**, L27–L31 (1987)

D.C. Braun, Comment on "Detection of emerging sunspot regions in the solar interior". Science **336**, 296 (2012)

A.S. Brun, Towards using modern data assimilation and weather forecasting methods in solar physics. Astron. Nachr. **328**, 329–338 (2007)

A.S. Brun, M.S. Miesch, J. Toomre, Modeling the dynamical coupling of solar convection with the radiative interior. Astrophys. J. **742**, 79 (2011)

A.S. Brun, M.K. Browning, M. Dikpati, H. Hotta, A. Strugarek, Recent advances on solar global magnetism and variability. Space Sci. Rev. (2013). doi:10.1007/s11214-013-0028-0

P. Charbonneau, Dynamo models of the solar cycle. Living Rev. Sol. Phys. **7**, 3 (2010)

P. Chatterjee, H.M. Antia, Solar flows and their effect on frequencies of acoustic modes. Astrophys. J. **707**, 208–217 (2009)
M.C.M. Cheung, M. Rempel, A.M. Title, M. Schüssler, Simulation of the formation of a solar active region. Astrophys. J. **720**, 233–244 (2010)
M.C.M. Cheung, M.L. DeRosa, A method for data-driven simulations of evolving solar active regions. Astrophys. J. **757**, 147 (2012)
A.R. Choudhuri, P. Chatterjee, J. Jiang, Predicting solar cycle 24 with a solar dynamo model. Phys. Rev. Lett. **98**, 13 (2007)
K. DeGrave, J. Jackiewicz, M. Rempel, Validating time-distance helioseismology with realistic quiet-Sun simulations. Astrophys. J. **127**, 15 (2014)
M.L. DeRosa, A.S. Brun, J.T. Hoeksema, Solar magnetic field reversals and the role of dynamo families. Astrophys. J. **757**, 96 (2012)
M. Dikpati, P. Charbonneau, A Babcock-Leighton flux transport dynamo with solar-like differential rotation. Astrophys. J. **518**, 508–520 (1999)
M. Dikpati, P.A. Gilman, Simulating and predicting solar cycles using a flux-transport dynamo. Astrophys. J. **649**, 498–514 (2006)
M. Dikpati, J.L. Anderson, Evaluating potential for data assimilation in a flux-transport dynamo model by assessing sensitivity and response to meridional flow variation. Astrophys. J. **756**, 20 (2012)
M. Dikpati, J.L. Anderson, D. Mitra, Ensemble Kalman filter data assimilation in a Babcock-Leighton solar dynamo model: an observation system simulation experiment for reconstructing meridional flow-speed. Geophys. Res. Lett. **41**, 5361 (2014)
T.L. Duvall Jr., A dispersion law for solar oscillations. Nature **300**, 242–243 (1982)
T.L. Duvall Jr., S.M. Jefferies, J.W. Harvey, M.A. Pomerantz, Time-Distance Helioseismology. Nature **362**, 430–432 (1993)
T.L. Duvall Jr., A.C. Birch, L. Gizon, Direct measurement of travel-time kernels for helioseismology. Astrophys. J. **646**, 553–559 (2006)
T.L. Duvall Jr., A.C. Birch, The vertical component of the supergranular motion. Astrophys. J. **725**, L47–L51 (2010)
T.L. Duvall Jr., S.M. Hanasoge, Subsurface supergranular vertical flows as measured using large distance separations in time-distance helioseismology. Sol. Phys. **287**, 71–83 (2013)
T. Felipe, D. Braun, A. Crouch, A. Birch, Scattering of the f-mode by small magnetic flux elements from observations and numerical simulations. Astrophys. J. **757**, 148–160 (2012)
A. Fournier, G. Hulot, D. Jault, W. Kuang, A. Tangborn, N. Gillet, E. Canet, J. Aubert, F. Lhuillier, An introduction to data assimilation and predictability in geomagnetism. Space Sci. Rev. **155**, 247–291 (2010)
L. Gizon, A.C. Birch, Time-distance helioseismology: noise estimation. Astrophys. J. **614**, 472–489 (2004)
L. Gizon, H. Schunker, C.S. Baldner, S. Basu, A.C. Birch, R.S. Bogart, D.C. Braun, R. Cameron, T.L. Duvall Jr., S.M. Hanasoge, J. Jackiewicz, M. Roth, T. Stahn, M.J. Thompson, S. Zharkov, Helioseismology of sunspots: a case study of NOAA region 9787. Space Sci. Rev. **114**, 249–273 (2009)
R. Giering, T. Kaminski, Recipes for adjoint code construction. ACM Trans. Math. Softw. **24**, 437–474 (1998)
D. Gough, B.W. Hindman, Helioseismic detection of deep meridional flow. Astrophys. J. **714**, 960–970 (2010)
D.H. Hathaway, Solar cycle forecasting. Space Sci. Rev. **144**, 401–412 (2009)
F. Hill, Rings and trumpets – three-dimensional power spectra of solar oscillations. Astrophys. J. **333**, 996–1013 (1988)
F. Hill, P.B. Stark, R.T. Stebbins, E.R. Anderson, H.M. Antia, T.M. Brown, T.L. Duvall Jr., D.A. Haber, J.W. Harvey, D.H. Hathaway, R. Howe, R. Hubbard, H.P. Jones, J.R. Kennedy, S.G. Korzennik, A. Kosovichev, J.W. Leibacher, K.G. Libbrecht, J.A. Pintar, E.J. Rhodes Jr., J. Schou, M.J. Thompson, S. Tomczyk, C.G. Toner, R. Toussaint, W.E. Williams, The solar acoustic spectrum and eigenmode parameters. Science **272**, 1292–1295 (1996)
J. Hirzberger, L. Gizon, S.K. Solanki, T.L. Duvall Jr., Structure and evolution of supergranulation from local helioseismology. Sol. Phys. **251**, 417–437 (2008)
C.P. Hung, L. Jouve, A.S. Brun, A. Fournier, O. Talagrand, Estimating the deep solar meridional circulation using magnetic observations and a dynamo model: a variational approach, Astrophys. J. (2015) (to appear)
S. Ilonidis, J. Zhao, A. Kosovichev, Detection of emerging sunspot regions in the solar interior. Science **333**, 993–996 (2011)
S. Ilonidis, J. Zhao, T. Hartlep, Helioseismic investigation of emerging magnetic flux in the solar convection zone. Astrophys. J. **777**, 138 (2013)

J. Jackiewicz, L. Gizon, A.C. Birch, High-resolution mapping of flows in the solar interior: fully consistent OLA inversion of helioseismic travel times. Sol. Phys. **251**, 381–415 (2008)

S.M. Jefferies, S.V. Vorontsov, C. Giebink, Toward improving the seismic visibility of the solar tachocline, in *Proc. SOHO 18/GONG 2006/HELAS I: Beyond the Spherical Sun*, ed. by K. Fletcher, M. Thompson (ESA, Noordwijk, 2006), ESA SP-624

L. Jouve, A.S. Brun, On the role of meridional flows in flux transport dynamo models. Astron. Astrophys. **474**, 239–250 (2007)

L. Jouve, A.S. Brun, O. Talagrand, Assimilating data into an $\alpha\Omega$ dynamo model of the Sun: a variational approach. Astrophys. J. **735**, 31 (2011)

E. Kalnay, *Atmospheric Modeling, Data Assimilation and Predictability* (Cambridge University Press, Cambridge, 2003)

P.J. Käpylä, M.J. Mantere, A. Brandenburg, Cyclic magnetic activity due to turbulent convection in spherical wedge geometry. Astrophys. J. Lett. **755**, L22 (2012)

I. Kitiashvili, A.G. Kosovichev, Application of data assimilation method for predicting solar cycles. Astrophys. J. Lett. **688**, L49 (2008)

S.G. Korzennik, A mode-fitting methodology optimized for very long helioseismic time series. Astrophys. J. **626**, 585–615 (2005)

S.G. Korzennik, M.C. Rabello-Soares, J. Schou, T.P. Larson, Accurate characterization of high-degree modes using MDI observations. Astrophys. J. **772**, 87 (2013), 28 pp

T.P. Larson, J. Schou, Improvements in global mode analysis. J. Phys. Conf. Ser. **118**, 012083 (2008)

T.P. Larson, J. Schou, Variations in global mode analysis, in *Solar-Stellar Dynamos as Revealed by Helio- and Asteroseismology*, ed. by M. Dikpati, T. Arentoft, I. Gonzalez Hernandez, C. Lindsey, F. Hill. Astron. Soc. Pacific Conf. Series, vol. 416 (2009), pp. 311–314

T.P. Larson, J. Schou, HMI global helioseismology data analysis pipeline. J. Phys. Conf. Ser. **271**, 012062 (2011)

E.M. Lavely, M.H. Ritzwoller, The effect of global-scale, steady-state convection and elastic-gravitational asphericities on helioseismic oscillations. Philos. Trans. R. Soc. Lond. A **339**, 431–496 (1992)

R.B. Leighton, A magneto-kinematic model of the solar cycle. Astrophys. J. **156**, 1 (1969)

F. Lhuillier, J. Aubert, G. Hulot, Earth's dynamo limit of predictability controlled by magnetic dissipation. Geophys. J. Int. **186**, 492–508 (2011)

C. Lindsey, D.C. Braun, Helioseismic imaging of sunspots at their antipodes. Sol. Phys. **126**, 101–115 (1990)

M.S. Miesch, A.S. Brun, M.L. DeRosa, J. Toomre, Structure and evolution of giant cells in global models of solar convection. Astrophys. J. **673**, 557 (2008)

H. Moradi, C. Baldner, A.C. Birch, D.C. Braun, R.H. Cameron, T.L. Duvall Jr., L. Gizon, D. Haber, S.M. Hanasoge, B.W. Hindman, J. Jackiewicz, E. Khomenko, R. Komm, P. Rajaguru, M. Rempel, M. Roth, R. Schlichenmaier, H.J. Schunker, H.C. Spruit, K.G. Strassmeier, M.J. Thompson, S. Zharkov, Modeling the subsurface structure of sunspots. Sol. Phys. **267**, 1–62 (2010)

N.J. Nelson, B.P. Brown, A.S. Brun, M.S. Miesch, J. Toomre, Magnetic wreaths and cycles in convective dynamos. Astrophys. J. **762**, 73 (2013)

P. Petit, B. Dintrans, S.K. Solanki, M. Aurière, F. Lignières, J. Morin, F. Paletou, J. Ramirez, C. Catala, R. Fares, Toroidal vs. poloidal magnetic fields in sun-like stars: a rotation threshold. Mon. Not. R. Astron. Soc. **388**, 80–88 (2008)

M.C. Rabello-Soares, S.G. Korzennik, J. Schou, Analysis of MDI high-degree mode frequencies and their rotational splittings. Sol. Phys. **251**, 197 (2008)

J. Reiter, E.J. Rhodes Jr., A.G. Kosovichev, J. Schou, P.H. Scherer, T.P. Larson, A method for the estimation of p-mode parameters from averaged solar oscillation power spectra. Astrophys. J. **803**, 92 (2015), 42 pp

M. Rempel, M.C.M. Cheung, Numerical simulations of active region scale flux emergence: from spot formation to decay. Astrophys. J. **785**, 90 (2014)

E.J. Rhodes Jr., J. Reiter, J. Schou, T. Larson, P. Scherrer, J. Brooks, P. McFaddin, B. Miller, J. Rodriguez, J. Yoo, Temporal changes in the frequencies and widths of the solar p-mode oscillations. J. Phys. Conf. Ser. **271**, 012029 (2011)

M. Roth, M. Stix, Meridional circulation and global solar oscillations. Sol. Phys. **251**, 77–89 (2008)

S. Sanchez, A. Fournier, J. Aubert, The predictability of advection-dominated flux-transport solar dynamo models. Astrophys. J. **781**, 8 (2014)

A. Schad, J. Timmer, M. Roth, A unified approach to the helioseismic inversion problem of the solar meridional flow from global oscillations. Astrophys. J. **734**, 97–105 (2011)

A. Schad, J. Timmer, M. Roth, Measuring the solar meridional flow from perturbations of eigenfunctions of global oscillation. Astron. Nachr. **333**, 991–994 (2012)

A. Schad, J. Timmer, M. Roth, Global helioseismic evidence for a deeply penetrating meridional flow consisting of multiple flow cells. Astrophys. J. Lett. **778**, L38–L44 (2013)

A. Schad, A new approach for the global helioseismic investigation of the solar meridional flow. Doctoral dissertation, Albert-Ludwig, University of Freiburg, Germany (2013)

J. Schou, On the analysis of helioseismic data. Doctoral Dissertation, Aarhus University, Denmark (1992)

J. Schou, Observations of medium- and high-degree modes: methods and sandtraps, in *Structure and Dynamics of the Interior of the Sun and Sun-Like Stars*, ed. by S.G. Korzennik, A. Wilson (ESA, Noordwijk, (1998), pp. 47–52, ESA SP-418

J. Schou, P.H. Scherrer, R.I. Bush, R. Wachter, S. Couvidat, M.C. Rabello-Soares, R.S. Bogart, J.T. Hoeksema, Y. Liu, T.L. Duvall Jr., D.J. Akin, B.A. Allard, J.W. Miles, R. Rairden, R.A. Shine, T.D. Tarbell, A.M. Title, C.J. Wolfson, D.F. Elmore, A.A. Norton, S. Tomczyk, Design and ground calibration of the Helioseismic and Magnetic Imager (HMI) instrument on the Solar Dynamics Observatory (SDO). Sol. Phys. **275**, 229–259 (2012)

C.J. Schrijver, M.L. DeRosa, Photospheric and heliospheric magnetic fields. Sol. Phys. **212**, 165–200 (2003)

H.C. Spruit, Origin of the torsional oscillation pattern of solar rotation. Sol. Phys. **213**, 1–21 (2003)

R.F. Stein, Å. Nordlund, On the formation of active regions. Astrophys. J. Lett. **753**, L13 (2012)

A. Strugarek, P. Charbonneau, Predictive capabilities of avalanche models for solar flares. Sol. Phys. **289**, 4137 (2014)

M. Švanda, L. Gizon, S.M. Hanasoge, S.D. Ustyugov, Validated helioseismic inversions for 3D vector flows. Astron. Astrophys. **530**, 148 (2011)

M. Švanda, Inversions for average supergranular flows using finite-frequency kernels. Astrophys. J. Lett. **759**, 29–33 (2012)

A. Svedin, M.C. Cuéllar, A. Brandenburg, Data assimilation for stratified convection. Mon. Not. R. Astron. Soc. **433**, 2278 (2013)

O. Talagrand, The use of adjoint equations in numerical modeling of the atmospheric circulation, in *Automatic Differentiation of Algorithms: Theory, Implementation, and Application*, ed. by A. Griewank, G.F. Corliss (Society for Industrial and Applied Mathematics, Philadelphia, 1991), pp. 169–180

S.V. Vorontsov, J. Christensen-Dalsgaard, J. Schou, V.N. Strakhov, M.J. Thompson, Solar internal rotation as seen from SOHO MDI data, in *From Solar Min to Max: Half a Solar Cycle with SOHO*, ed. by A. Wilson (2002a), pp. 111–114, ESA SP-508

S.V. Vorontsov, J. Christensen-Dalsgaard, J. Schou, V.N. Strakhov, M.J. Thompson, Helioseismic measurement of solar torsional oscillations. Science **296**, 101–103 (2002b)

S.V. Vorontsov, S.M. Jefferies, Modeling solar oscillation power spectra. I. Adaptive response function for Doppler velocity measurements. Astrophys. J. **623**, 1202–1214 (2005)

S.V. Vorontsov, Solar p modes of high degree l: Coupling by differential rotation. Mon. Not. R. Astron. Soc. **378**, 1499–1506 (2007)

S.V. Vorontsov, S.M. Jefferies, C. Giebink, J. Schou, Toward eliminating systematic errors in intermediate-degree p-mode measurements, in *Solar-Stellar Dynamos as Revealed by Helio- and Asteroseismology*, ed. by M. Dikpati, T. Arentoft, I. Gonzalez Hernandez, C. Lindsey, F. Hill. Astron. Soc. Pacific Conf. Series, vol. 416 (2009), pp. 301–305

S.V. Vorontsov, Effects of differential rotation and meridional circulation in solar oscillations of high degree l. Mon. Not. R. Astron. Soc. **418**, 1146–1155 (2011)

S.V. Vorontsov, V.A. Baturin, S.V. Ayukov, V.K. Gryaznov, Helioseismic calibration of the equation of state and chemical composition in the solar convective envelope. Mon. Not. R. Astron. Soc. **430**, 1636–1652 (2013a)

S.V. Vorontsov, S.M. Jefferies, Modeling solar oscillation power spectra. II. Parametric model of spectral lines observed in Doppler-velocity measurements. Astrophys. J. **778**, 75 (2013b) 10 pp

J. Warnecke, P.J. Käpylä, M.J. Käpylä, A. Brandenburg, On the cause of solar-like equatorward migration in global convective dynamo simulations. Astrophys. J. Lett. **796**, L12 (2014)

M.F. Woodard, Effect of subsurface inhomogeneities on the statistics of solar oscillation power spectra. Sol. Phys. **180**, 19–28 (1998)

M.F. Woodard, Theoretical signature of solar meridional flow in global seismic data. Sol. Phys. **197**, 11–20 (2000)

M.F. Woodard, J. Schou, A.C. Birch, T.P. Larson, Global-oscillation eigenfunction measurements of solar meridional flow. Sol. Phys. **287**, 129–147 (2013)

A.R. Yeates, D. Nandy, D.H. Mackay, Exploring the physical basis of solar cycle predictions: flux transport dynamics and persistence of memory in advection versus diffusion-dominated solar convection zones. Astrophys. J. **673**, 544–556 (2008)

J. Zhao, A.G. Kosovichev, On the inference of supergranular flows by time-distance helioseismology, in *GONG+ 2002. Local and Global Helioseismology: The Present and Future*, ed. by H. Sawaya-Lacoste, (ESA, Noordwijk 2003), pp. 417–420, ESA SP-517

J. Zhao, K. Nagashima, R.S. Bogart, A.G. Kosovichev, T.L. Duvall Jr., Systematic center-to-limb variation in measured helioseismic travel times and its effect on inferences of solar interior meridional flows. Astrophys. J. Lett. **749**, 5 (2012)

J. Zhao, R.S. Bogart, A.G. Kosovichev, T.L. Duvall Jr., T. Hartlep, Detection of equatorward meridional flow and evidence of double-cell meridional circulation. Astrophys. J. Lett. **774**, 29 (2013)

Space Sci Rev (2015) 196:251–283
DOI 10.1007/s11214-014-0065-3

Helioseismology with *Solar Orbiter*

Björn Löptien · Aaron C. Birch · Laurent Gizon · Jesper Schou · Thierry Appourchaux · Julián Blanco Rodríguez · Paul S. Cally · Carlos Dominguez-Tagle · Achim Gandorfer · Frank Hill · Johann Hirzberger · Philip H. Scherrer · Sami K. Solanki

Received: 24 February 2014 / Accepted: 20 June 2014 / Published online: 15 July 2014

Abstract The *Solar Orbiter* mission, to be launched in July 2017, will carry a suite of remote sensing and in-situ instruments, including the *Polarimetric and Helioseismic Imager* (PHI). PHI will deliver high-cadence images of the Sun in intensity and Doppler velocity suitable for carrying out novel helioseismic studies. The orbit of the *Solar Orbiter* spacecraft will reach a solar latitude of up to 21° (up to 34° by the end of the extended mission) and thus will enable the first local helioseismology studies of the polar regions. Here we consider an array of science objectives to be addressed by helioseismology within the baseline telemetry allocation (51 Gbit per orbit, current baseline) and within the science observing windows (baseline 3 × 10 days per orbit). A particularly important objective is the measurement of large-scale flows at high latitudes (rotation and meridional flow), which are largely unknown but play an important role in flux transport dynamos. For both helioseismology and feature tracking methods convection is a source of noise in the measurement

B. Löptien · L. Gizon (✉)
Institut für Astrophysik, Georg-August Universität Göttingen, 37077 Göttingen, Germany
e-mail: gizon@mps.mpg.de

A.C. Birch · L. Gizon · J. Schou · A. Gandorfer · J. Hirzberger · S.K. Solanki
Max-Planck-Institut für Sonnensystemforschung, Justus-von-Liebig-Weg 3, 37077 Göttingen, Germany

T. Appourchaux · C. Dominguez-Tagle
Institut d'Astrophysique Spatiale, CNRS, Université Paris Sud, 91405 Orsay, France

J. Blanco Rodríguez
GACE/IPL, Universidad de Valencia, 46980 Valencia, Spain

P.S. Cally
Monash Centre for Astrophysics and School of Mathematical Sciences, Monash University, Clayton, Victoria 3800, Australia

F. Hill
National Solar Observatory, Tucson, AZ 85719, USA

P.H. Scherrer
Hansen Experimental Physics Laboratory, Stanford University, Stanford, CA 94305-4085, USA

of longitudinally averaged large-scale flows, which decreases as $T^{-1/2}$ where T is the total duration of the observations. Therefore, the detection of small amplitude signals (e.g., meridional circulation, flows in the deep solar interior) requires long observation times. As an example, one hundred days of observations at lower spatial resolution would provide a noise level of about three m/s on the meridional flow at 80° latitude. Longer time-series are also needed to study temporal variations with the solar cycle. The full range of Earth-Sun-spacecraft angles provided by the orbit will enable helioseismology from two vantage points by combining PHI with another instrument: stereoscopic helioseismology will allow the study of the deep solar interior and a better understanding of the physics of solar oscillations in both quiet Sun and sunspots. We have used a model of the PHI instrument to study its performance for helioseismology applications. As input we used a 6 hr time-series of realistic solar magneto-convection simulation (Stagger code) and the SPINOR radiative transfer code to synthesize the observables. The simulated power spectra of solar oscillations show that the instrument is suitable for helioseismology. In particular, the specified point spread function, image jitter, and photon noise are no obstacle to a successful mission.

Keywords Helioseismology · Space missions: *Solar Orbiter* · Solar physics · Solar dynamo

1 Introduction: Solar Orbiter

Solar Orbiter[1] is an ESA M-Class mission and Europe's follow-up to the highly successful SOHO mission of ESA and NASA. It was selected in 2011 with a baseline launch date of July 2017. An important feature of *Solar Orbiter* is its orbit; the spacecraft will approach the Sun as closely as 0.28 AU and reach heliographic latitudes of up to 34°, which will allow *Solar Orbiter* to directly observe the solar poles at a much lower angle than is possible from Earth. The spacecraft will host a suite of in-situ and remote-sensing instruments. Direct measurements within the inner heliosphere can be related directly to observations of the different layers of the solar atmosphere, from the photosphere through the chromosphere, the corona, and the inner heliosphere. This will provide unique opportunities for studying the connection between the Sun and the heliosphere. As described in the *Solar Orbiter Definition Study Report* (Marsden and Müller 2011) and in Müller et al. (2013), the central science goal of the mission is:

How does the Sun create and control the heliosphere?

In order to answer this question, *Solar Orbiter* will address four separate science questions:

- How and where do the solar wind plasma and magnetic field originate in the corona?
- How do solar transients drive heliospheric variability?
- How do solar eruptions produce the energetic particle radiation that fills the heliosphere?
- How does the solar dynamo work and drive the connections between the Sun and the heliosphere?

One key aspect for addressing these science questions, particularly the last one, will be probing the solar interior by using helioseismology. Helioseismology is expected to contribute significantly to the determination of subsurface flows in the upper solar convection zone,

[1]The web page of the mission is located here: http://sci.esa.int/solar-orbiter/.

especially in the polar regions. This will offer unique opportunities to study the dynamics of plasma flows and the solar dynamo. By combining observations from other instruments in space or on the ground, it will also be possible to experiment with stereoscopic helioseismology to probe the deep solar interior.

2 Mission Profile

2.1 Orbit Design

Solar Orbiter draws its unique capabilities from its special orbit characteristics. After separation from the launch vehicle, *Solar Orbiter* will start its 3 year cruise phase. Subject to several Gravity-Assist-Maneuvers (GAMs) at Venus and Earth, the spacecraft will lose orbital energy, which will allow *Solar Orbiter* to reduce the distance to the Sun. After a second GAM at Venus, *Solar Orbiter* begins its operational phase. From then on its orbit is in resonance with Venus with an initial period of about 180 days, such that the inclination of the orbital plane with respect to the ecliptical plane can be increased by Venus gravity assists. This particular and unique feature gives *Solar Orbiter* access to the high latitude regions of the Sun (see Fig. 1 and Table 1). The final orbit will be highly eccentric with a perihelion distance of 0.28 AU. At perihelion, *Solar Orbiter* will have a low angular velocity relative to the Sun. This will allow *Solar Orbiter* to follow the evolution of surface structures and solar features for a longer time than possible from Earth (19 days with a viewing angle between $\pm 75^\circ$). The nominal mission phase lasts for about 4.5 years. Afterward, the mission could be extended for another 3.5 years.

2.2 Instrument Suite

One of the most important aspects of the *Solar Orbiter* mission is the combination of remote observing with in-situ measurements. Further, the high-resolution instruments are all designed to observe the same target region on the solar surface. This will allow relating observations of different atmospheric layers, which will be seen using the different instruments. The instruments of *Solar Orbiter* are described in detail in the *Solar Orbiter Definition Study Report* (Marsden and Müller 2011). They can be grouped in three major packages, each consisting of several instruments:

- Field Package: *Radio and Plasma Waves Instrument* (RPW) and *Magnetometer* (MAG).
- Particle Package: *Energetic Particle Detector* (EPD) and *Solar Wind Plasma Analyzer* (SWA).
- Solar remote sensing instrumentation: *Polarimetric and Helioseismic Imager* (PHI), *Extreme Ultraviolet Imager* (EUI), *Multi Element Telescope for Imaging and Spectroscopy* (METIS), *Solar Orbiter Heliospheric Imager* (SoloHI), *Spectral Imaging of the Coronal Environment* (SPICE) and *Spectrometer/Telescope for Imaging X-Rays* (STIX).

The *Polarimetric and Helioseismic Imager* (PHI) will provide the vector magnetic field and line-of-sight (LOS) velocity on the visible solar surface at high spatial resolution, thanks to the close-by observing conditions during the perihelion passages. The magnetic field anchored at the solar surface produces most of the structures and energetic events in the upper solar atmosphere and significantly influences the heliosphere. Extrapolations of the magnetic field observed by PHI into the Sun's upper atmosphere and heliosphere will provide the information needed for other optical and in-situ instruments to analyze and understand

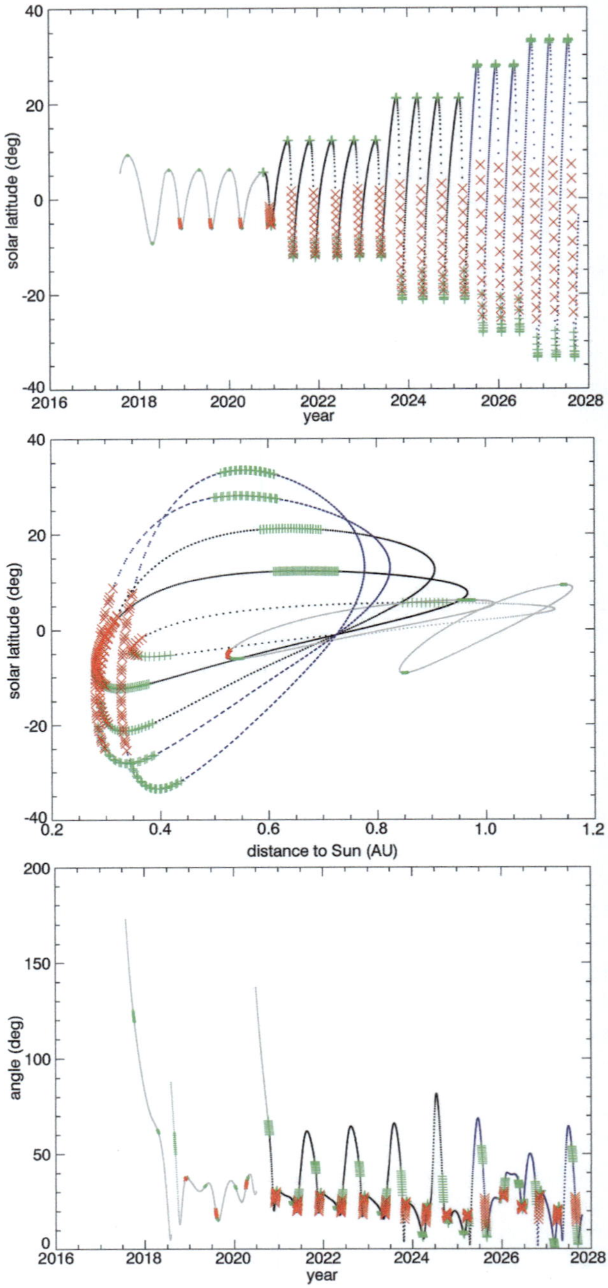

Fig. 1 Orbit of *Solar Orbiter* (July 2017 baseline launch). *From top to bottom:* Solar latitude as a function of time, solar latitude as a function of distance to the Sun and the angle Sun-Earth-spacecraft as a function of time. The *black dots* show the orbit for the science nominal mission and the *gray* and *blue dots* denote the cruise phase and the extended mission phase. The current baseline for the remote-sensing instruments is to observe in only three science windows per orbit (at perihelion and maximum northern/southern heliographic latitude). The science windows are given by the *red X-symbols* (perihelion) and *large green crosses* (maximum latitude). The *small symbols* show the perihelion and the maximum latitude during the cruise phase. The spacecraft will reach a maximum solar latitude of 21° during the nominal mission and up to 34° during the extended mission. The closest perihelion is at 0.28 AU

Table 1 Orbit parameters for a launch in July 2017

	Nominal mission	Extended mission
Starting date	2020	2025
Closest distance [AU]	0.28	0.28
Max. solar latitude [°]	21	34
Max. relative velocity [km/s]	25	23

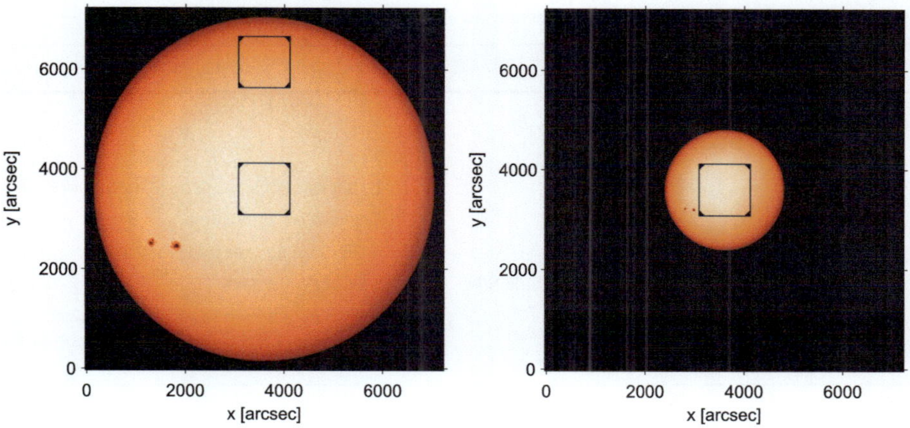

Fig. 2 Field-of-view (FOV) of PHI at 0.28 AU (*left*) and 0.8 AU (*right*) solar distance. The *full images* show the FOV of the FDT and the *small squares* that of the HRT. Although PHI will not have a pointing mechanism of its own, observations close to the limb with the HRT will be possible by changing the pointing of the spacecraft (*upper small square in the image on the left*)

the data recorded by them in a proper physical context. In addition, the Dopplergrams or intensity images obtained with PHI will be used to measure solar oscillations and thereby allow probing of the solar interior using helioseismology.

2.3 PHI: Observables and Operation

PHI is based on a tunable narrow-band filtergraph which is designed to scan the Fe I 6173 Å line (the same as observed by HMI, Schou et al. 2012a) with a planned cadence of 60 s. This line forms in a broad range of heights, ranging from about 100 to 400 km above the photosphere (Fleck et al. 2011). Due to telemetry constraints, the instrument will perform onboard inversion for line-of-sight velocity and magnetic field vector. In addition to the inverted quantities the continuum intensity, which will be measured directly by PHI, will be transferred to the ground.

PHI will consist of two telescopes that cannot be used simultaneously, a Full Disk Telescope (FDT) with a field of view of 2°, sufficient for observing the entire solar disk during the whole orbit with a spatial resolution of about $9''$ ($\sim$1800 km at disk center at perihelion) and a High Resolution Telescope (HRT) that will image a fraction of the Sun with high spatial resolution. The HRT has a field of view of $16.8'$ and a spatial resolution of about $1''$, corresponding to 200 km at disk center at perihelion. As can be seen in Fig. 2, the FOV and the resolution of the HRT and FDT vary significantly over the course of the orbit.

The current baseline for PHI and the other remote sensing instruments is to observe only during three science windows per orbit (see Fig. 1), at perihelion and at maximum northern/southern heliographic latitude. Each science window is planned to last for 10 days. Observations outside of the science windows are in principle possible but they require the spacecraft to provide accurate pointing and thermal stability. Also, operating the remote-sensing instruments might interfere with measurements taken by the in-situ instruments. Helioseismology, however, requires long observing times and is thus a strong argument for increasing the length of the science windows.

PHI is also affected by the low telemetry rate of *Solar Orbiter*. PHI has an allocated telemetry rate of 51 Gbits per science orbit, which corresponds to $\sim$3.3 kbps on average.

Table 2 Choices that have to be made in planning an observing scheme for helioseismology

Parameter	Options	Baseline for helioseismology (this paper)
Observables	Doppler ($V_{\rm los}$), $I_{\rm c}$, $B_{\rm los}$	$V_{\rm los}$
Cadence, dt	>10 s	60 s
Duration, T	Set by telemetry	10 days nominal, hopefully much more
Telescope	FDT, HRT	Depends on application
Spatial sampling	Open	A few samples per min. wavelength
Compression	Set by desired p-mode S/N	5 bit per observable
Vantage point	Set by orbit	High inclination to study polar regions; Full range of Earth-Sun-spacecraft angles for stereoscopy

This is not sufficient for transmitting continuous observations with full resolution to Earth. Telemetry is thus a constraint for helioseismology and motivates a well-planned observing strategy. PHI will perform extensive onboard processing of the data, consisting of image processing and radiative transfer equation (RTE) inversion. For larger data volumes, additional compression will be applied. Since the same computational resources are used for data acquisition and processing, the processing can only be carried out when no observations are being performed. The instrument will host 4 Tbits of flash memory, which will allow storage of raw and processed data. Data obtained within the science windows can be processed during the whole orbit and transmitted to Earth whenever telemetry is available. In addition to the inverted quantities, raw data will be transferred as well from time to time for testing purposes. When obtaining data for helioseismology only, a full inversion is not required. Instead, a simple algorithm for determining the LOS velocity can be used, as was successfully done by MDI (Scherrer et al. 1995).

3 Helioseismology: Science Objectives

The helioseismology science objectives for *Solar Orbiter* include measuring differential rotation, torsional oscillations, meridional flow, and convective flows at high latitudes and, also, in combination with another instrument (e.g., HMI or GONG++, Hill et al. 2003) using stereoscopic helioseismology to probe the deep convection zone and the tachocline. These science objectives are discussed in the *Solar Orbiter Definition Study Report* (Marsden and Müller 2011). Discussions of the science objectives involving helioseismology are also given in Marsch et al. (2000), Gizon et al. (2001) and Woch and Gizon (2007).

Here we present an array of science objectives for helioseismology that could be addressed by PHI. We focus on science targets that we expect will go beyond what is possible, or will be possible, with current or future observations from HMI, MDI, GONG++, or BiSON (Chaplin et al. 1996). For all of the science objectives presented here we make suggestions for an observing strategy, in all cases we give both the observing time and the number of pixels (see Table 2 for a list of parameters). In order to reduce telemetry, we suggest rebinning the data to a given resolution (a certain number of pixels across the solar disk). Most of the time, observations would be performed using the FDT. For some science goals, this resolution cannot be achieved when using the FDT when observing around aphelion. In this case, observations should be performed using the HRT, which will also cover a large fraction of the solar disk around aphelion (see Fig. 2).

The observing strategies for the science goals are summarized in Table 3. More details for the individual science objectives are given in the following subsections. For all science objectives, maps of the magnetic field should be transferred from time to time to provide context information. We expect that a duty cycle greater than 80 % may be acceptable for helioseismology.

Some of the science goals are not within the limitations of the baseline telemetry allocation and observing windows (three science windows per orbit). As discussed in Sect. 4.2, it might be possible to decrease the required telemetry by compressing the data. However, reducing the observing duration would severely affect the signal-to-noise ratio. Probing large-scale flows is affected by the realization noise of the solar oscillations and noise arising from supergranulation. Both decrease with $T^{-1/2}$, where T is the observing time.

In addition to the observing duration and telemetry, many of the science goals described below have some requirements regarding the orbit. This is especially the case for the measurements at high latitudes. These should be performed while the spacecraft is at high latitudes. During the extended mission, *Solar Orbiter* will reach a maximum solar latitude of 34° and will be above 30° for up to 23 consecutive days. This is comparable to the required observing time for high-latitude measurements.

We note that helioseismology of PHI data will require some (potentially complicated) modifications to standard methods to account for the time-varying image geometry.

3.1 Near-Surface Rotation, Meridional Circulation, and Solar-Cycle Variations at High Latitudes

One major science objective of *Solar Orbiter* is to understand large scale flows near the solar poles, which are in turn important for understanding solar dynamics and the solar cycle. Differential rotation and meridional circulation are essential components in dynamo models, especially flux transport dynamos. The importance of these large scale flows at high latitudes has been discussed by, e.g., Jiang et al. (2009), Dikpati and Gilman (2012). While helioseismology has mapped differential rotation in most of the solar interior (e.g., see the review by Howe (2009) and references therein), the cycle dependence of differential rotation ("torsional oscillations", > 20 m/s) is not well measured for the full Hale cycle at latitudes above about sixty degrees. Variations from one sunspot cycle to the next have been seen in the differential rotation at high latitude (Hill et al. 2009). The Sun might also exhibit a polar vortex (Gilman 1979), as already seen on planets (e.g., Fletcher et al. 2008). In some dynamo models, the meridional circulation sets the cycle period and thus is an important target. The meridional circulation has an amplitude of less than fifteen m/s at mid-latitudes (and less at high latitudes) and is thus a challenging objective. Progress has been made using local helioseismology (Zhao et al. 2013), although a lot of uncertainty remains in the deep convection zone and at high latitudes.

Near the surface, the limiting factor in the measurement of longitudinally-averaged flows is the noise introduced by the supergranulation flows (~200 m/s rms). The noise introduced by supergranulation flows goes up with latitude, as the number of supergranules at fixed latitude goes down. This is illustrated in Fig. 3, which is based on a very simple model that does not take account of the inclination of the orbit. Time averaging over long periods is necessary to measure longitudinally-averaged flows at high latitudes (e.g. Howe et al. 2013; Komm et al. 2013). One month of data will result in a noise level of two m/s at a latitude of 75°, using time-distance helioseismology. A noise level of one m/s at 75° would require ~ 120 days of observations. These noise estimates are not specific to the particular method employed (helioseismology or other); it is inherently due to noise from solar convective flows. One m/s at 85° would require continuous coverage for a significant fraction

Table 3 Table of helioseismology science objectives. Some science goals exceed the currently allocated observing time or telemetry (highlighted in red, green colors indicate that the parameter is within the allocation)

Science target	# Spatial points	Observing time	Observables	Approx. telemetry (5 bits/observable)
Near-surface rotation, meridional circulation, and solar-cycle variations at high latitudes				
– Helioseismology	512 × 512	30 days	V_{los} every 60 s	60 Gbit
– Solar-cycle variations from helioseismology	512 × 512	4 × 30 days, 2 years apart	V_{los} every 60 s	4 × 60 Gbit
– Meridional circulation to 3 m/s at 75° (see Fig. 3)	512 × 512	100+ days	V_{los}, I_c & B_{los}	Method dependent
– Granulation and magnetic-feature tracking	2048 × 2048	30 days	I_c & B_{los}, two consecutive images every 8 h	8 Gbit
– Supergranulation tracking	512 × 512	30 days	V_{los} every 60 min	1 Gbit
Deep and large-scale solar dynamics				
– MDI-like medium-*l* program	128 × 128	Continuous	V_{los} every 60 s	40 Gbit/year
– Stereoscopic helioseismology (PHI + other instrument)	128 × 128	Continuous	V_{los} every 60 s	40 Gbit/year
Convection at high latitudes				
– Helioseismology	1024 × 1024	7 days	V_{los} every 60 s	50 Gbit
– Feature tracking	2048 × 2048	7 days	I_c & B_{los}, two consecutive images every 8 h	2 Gbit
Deep convection and giant cells				
– Helioseismology	128 × 128	Continuous	V_{los} every 60 s	40 Gbit/year
– Feature tracking	512 × 512	4 × 60 days	V_{los} every 60 min	4 × 2 Gbit
Active regions and sunspots				
– Active region flows & structure	512 × 512	20 days	V_{los} every 60 s	40 Gbit
– Sunspot oscillations	1024 × 1024	2 days	V_{los}, I_c & $\mathbf{B}$ every 60 s	80 Gbit
– Calibration far-side helioseismology	128 × 128	5 × 2 days	B_{los}, I_c, & V_{los} every 60 s	5 × 0.3 Gbit

Table 3 (*Continued*)

Science target	# Spatial points	Observing time	Observables	Approx. telemetry (5 bits/observable)
Physics of oscillations (stereoscopic obs.)				
– Effect of granulation on oscillations	2048 × 256	1 day	6 filtergrams every 60 s	20 Gbit
– Two components of velocity	512 × 512	10 days	V_{los} every 60 s	20 Gbit
- Magnetic oscillations	2048 × 2048	1 day	V_{los}, I_{c}, B_{los} every 60 s & **B** at max. cadence	100 Gbit
Low resolution observations				
– LOI-like observations (solar-cycle variations, active longitudes)	4 × 4 for V_{los} & I_{c}, 32 × 32 for B_{los}	Continuous	V_{los} & I_{c} every 60 s & B_{los} once per day	0.1 Gbit/year
– Shape of the Sun	10 × 6000	Every few months	I_{c} at 12 angles during rolls	4 Mbit for one roll

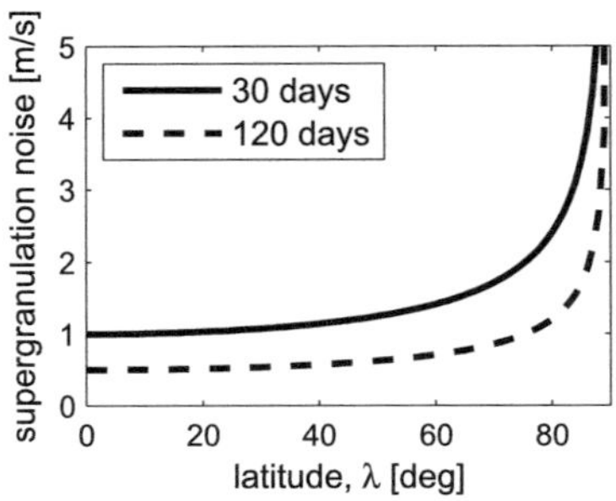

Fig. 3 Noise level due to supergranulation in a surface average of rotation or meridional circulation over longitude in 30 days (*solid curve*) and 120 days (*dashed curve*) of continuous measurements as a function of heliographic latitude, λ. The noise level for an observing time of 30 days scales as $\sim(1\ \mathrm{m/s}) \times (\cos\lambda)^{-1/2}$, where 1 m/s is the typical noise level at the equator (Fig. 6 of Gizon et al. 2001), and $\cos\lambda$ is the ratio of the number of supergranules between the equator and latitude λ. The same estimate follows from simple assumptions for rms amplitude, lifetime, and length scale of the supergranulation pattern. In order to reduce the noise from supergranulation, time averaging (or time sampling) is required. The noise due to solar convective flows is not only important for helioseismology, but must also be considered for correlation tracking and other surface measurements

of the mission. Standard local helioseismology requires spatially resolved observations with resolution of at least 512 × 512 pixels across the solar disk. All helioseismology applications require a cadence of at least 60 s. In order to look for solar-cycle variations of these large-scale flows, we suggest to repeat these observations several times during the mission.

It should be noted that the helioseismology observations (e.g., V_{los} or I_{c}) will be used for non-helioseismic purposes. For example, flows at the surface can also be determined by correlation tracking of granulation or supergranulation. Granulation tracking requires observations at full spatial resolution (e.g. Roudier et al. 2013) and pairs of images taken 60 s apart, with at least a few pairs taken per day.

Another way to determine the flows near the surface is to track the supergranulation in Dopplergrams (e.g. Hathaway et al. 2013). This has the advantage that only temporally averaged Dopplergrams with low cadence (one hour) and modest resolution (512 × 512 is known to be sufficient, but 256 × 256 may be acceptable) are needed. Schou (2003) used 60 days of data to detect the meridional flow up to about ∼75° latitude, but without resolution in depth.

3.2 Deep and Large-Scale Solar Dynamics

One of the main science goals of *Solar Orbiter* is to measure flows in the deep interior of the Sun. These are currently not well characterized. Recently, Zhao et al. (2013) reported the discovery of multiple cells of the meridional circulation throughout the convection zone but high latitudes have not been explored. In addition, the 1.3-year oscillations in the rotation rate near the tachocline (Howe et al. 2000) remain a puzzling result. These oscillations have an amplitude that increases with latitude and could be measured with *Solar Orbiter*.

Only low spatial resolution is needed for deeper flows. On the other hand, the measurement of flows in the deep interior requires long time-series. Braun and Birch (2008) estimated that up to ten years would be required to measure a few m/s return meridional flow at the base of the convection zone. However, a noise level of 5 m/s could already be achieved with 6 months of data.

By combining observations from *Solar Orbiter* with those from space or the ground to maximize the spatial coverage, improvements are expected. Thanks to its unique orbit, *Solar*

Orbiter will be the first mission to study the advantages of stereoscopic helioseismology. For the purpose of stereoscopic helioseismology of the deep interior, observations with a resolution of 128 × 128 should be sufficient. Observing the Sun from opposing angles offers the opportunity to perform time-distance helioseismology using ray-paths probing the deep interior of the Sun. This method has never been used before and may lead to new discoveries. Other than the preliminary estimates of Ruzmaikin and Lindsey (2003), models have not yet been developed to predict the signal-to-noise ratios that will be possible using this method. For this science target, the duty cycle should be above 80 %. The data for this represents only a fraction of the currently allocated telemetry, but does require observations over very long periods of time.

3.3 Convection at High Latitudes

Another major objective is to understand the properties of convection, in particular supergranulation, which almost 60 years after its discovery still remains enigmatic (e.g. preferred length scale of convection, wavelike properties). This in turn will allow us to better understand issues such as flux transport and thus the solar dynamo.

Supergranulation is one of the strongest signals in time-distance helioseismology and will no doubt be a feasible science target for PHI within the baseline resources. There are several reasons to expect an unusual behavior of the supergranulation at high latitudes. According to Nagashima et al. (2011) supergranules may preferentially align in the North-South direction at high latitudes. Furthermore, the effects of the Coriolis force on supergranulation flows (see Gizon et al. 2010) is expected to be strongest in the polar regions. Finally, the wavelike properties reported for supergranulation (Gizon et al. 2003; Schou 2003) show a clear latitude dependence.

A snapshot of supergranulation flows near the surface can be obtained using local helioseismology with a few hours of observations of moderate spatial resolution (1024 × 1024 pixels). The study of the evolution of the supergranulation pattern requires at least a week of data (a few lifetimes). In addition to helioseismology, local correlation tracking with high resolution (2048 × 2048) but low cadence (two consecutive images every 8 h) could be used to study the supergranulation flows at the surface.

3.4 Deep Convection and Giant Cells

Recent helioseismology results call into question current models of deep solar convection. Hanasoge et al. (2012) found much lower flow velocities for deep convection than predicted. Extending this work with *Solar Orbiter* would require continuous observations over a large fraction of the mission, with a spatial resolution of 128 × 128 pixels.

Recently, Hathaway et al. (2013), using correlation tracking, reported the detection of persistent giant cells at the solar surface, but were unable to observe close to the poles. Giant cells may play an important role in transporting angular momentum. This presents a new and exciting opportunity for PHI. For studying giant cells, we will need several months of observations consisting of 512 × 512 pixels with a cadence of 60 min (Hathaway et al. 2013). This could either be achieved using a single but very long time-series (of order a full orbit) or using a few moderately long (2–3 months) time-series with a similar combined length. It may also be desirable to repeat this after a substantial period to look for any solar-cycle dependence in the giant cell pattern. This analysis could also be done with helioseismology over the same observing time, but with a cadence of 60 s.

3.5 Active Regions and Sunspots

Another interesting possibility is to be able to increase the temporal coverage of evolving active regions using both helioseismology and correlation tracking, as well as to observe the flux dispersal over longer times than currently possible. This will require 20 days of data (20 days in combination with, e.g., HMI observations would allow continuous observations of the active region for more than a full solar rotation) with a cadence of 60 s and a resolution of 512×512 pixels.

Stereoscopic observations would also greatly benefit sunspot seismology since it would be possible to observe two components of MHD waves. These observations will likely not require extensive observations. Previous studies (e.g. Schunker et al. 2005; Zhao and Kosovichev 2006) were able to probe sunspots with local helioseismology using MDI data with a length of a few days. We estimate that 2 days of data should be sufficient with a spatial resolution of 1024×1024 pixels. However, in order to study MHD waves, in addition to the Doppler velocity, the vector magnetic field and the continuum intensity will be required. Coordination with an additional instrument will be necessary, meaning that the active regions being investigated will need to be on the nearside of the Sun.

Since *Solar Orbiter* will be able to observe the farside of the Sun, it can be used to calibrate farside helioseismology (Lindsey and Braun 2000). Active regions not visible from the Earth may be imaged with *Solar Orbiter* (B_{los}, I_{c}) and compared with helioseismic farside maps derived from a near-Earth vantage point. In addition, combining *Solar Orbiter* and other helioseismology datasets would improve farside imaging (e.g., using different skip geometries). Since farside helioseismology relies on medium-degree modes, this would be possible with a few days of data obtained for the low resolution helioseismology (128×128 pixels).

3.6 Physics of Oscillations (Stereoscopic Observations)

Vector velocities, utilizing two vantage points to observe the same location, also present exciting possibilities. Recent attempts at measuring deep meridional flows using local helioseismology are affected by apparent center-to-limb effects (Zhao et al. 2012), which are suspected to be caused by asymmetries in granular flows (Baldner and Schou 2012). Stereoscopic observations would allow the evaluation of the dependence of this effect on the observation angle. We suggest to observe a strip with a size of 2048×256 pixels, covering all heliocentric angles, for about a day. In order to study the dependence of this effect with height in the atmosphere, the filtergrams could be transmitted instead of Dopplergrams obtained onboard.

Obtaining vector velocities would allow observations of the relation between radial and horizontal velocities in supergranulation or the phase relationships between different velocity components of helioseismic waves near the solar surface and thus give insight to the physics of p- and f-modes. Observing supergranulation does not require a high spatial resolution (e.g. 512×512 pixels) but it is necessary to observe for several lifetimes of the supergranulation (e.g. 10 days).

In addition, vector velocities in combination with the full magnetic field vector would give insight to MHD waves in active regions (e.g. Norton and Ulrich 2000). This would require stereoscopic observations of an active region with high spatial resolution (2048×2048 pixels). Norton and Ulrich (2000) used a few hours of data, so we estimate that an observing length of the order of one day should be sufficient here.

3.7 Low Resolution Observations

Additional helioseismology observation modes that should be considered are unresolved observations or observations with very low resolution. Sun-as-a-star time-series from SOHO/VIRGO (Fröhlich et al. 1995) are used regularly for testing concepts for asteroseismic data analysis (e.g., García 2009; Sato et al. 2010; Gizon et al. 2013). PHI would provide a unique data set in this regard as it would allow observations from outside of the ecliptic plane. For this purpose (nearly) continuous observations (I_c and V_{los}) with very low resolution (e.g. 4×4 pixels) using the FDT would be appropriate.

Another case for observations using only few pixels is the study of the shape of the Sun (e.g., Kuhn et al. 2012). This would require rolls of the spacecraft, which perhaps would have to be coordinated with calibrations of the in-situ instruments on *Solar Orbiter*. In order to save telemetry, only a band of pixels along the limb would have to be transmitted.

4 Helioseismology: Selected Requirements

4.1 Example Telemetry Requirements

The last column of Table 3 shows a set of example telemetry estimates made using some simple assumptions. Here we have used 5 bits/pixel, regardless of the type of observable or cadence and have not distinguished between the two telescopes (FDT or HRT). For some objectives it is possible to only transmit parts of the images (e.g. the parts that are not seen from the Earth), but this has not been accounted for. In addition, we have not taken into account the number of pixels that are off the solar disk when computing the data rates for full-disk images. Given that the spacecraft-Sun distance will be changing it will be necessary to have variable binning and extraction as a function of time, but here we have simply assumed that the images have the desired size.

Some science objectives in Table 3 are within the baseline of telemetry (51 Gbit/orbit, Marsden and Müller 2011) and others require more telemetry. We are currently investigating ways (compression algorithms) to optimize data rates. As will be discussed in Sect. 4.2 there are many possibilities for reducing the amount of telemetry and a factor of two to five gain in compression efficiency may be possible. These estimates are based on the experience on compression from previous space missions, as for example MDI (Scherrer et al. 1995; Kosovichev et al. 1997).

One promising avenue for refining the very simplistic example estimates shown here is direct numerical simulation of both wave propagation and synthetic PHI observations (see Sects. 6 and 7). In particular for the case of stereoscopic helioseismology of the tachocline, it will be useful to have wave propagation simulations in spherical geometry. These simulations will help in the exploration of methods for getting the most out of stereoscopic or other helioseismic observations.

An important consideration in planning observations is that many of the most important phenomena on the Sun change with time. Snapshots of solar interior structure and dynamics are important but are not as informative as continuous coverage. In the ideal case, we would be able to track the temporal evolution of flows at high latitudes and in the tachocline throughout the *Solar Orbiter* mission. For this reason, it may be important to consider the value of carrying out low data rate but continuous observations in addition to, or in place of, a few short intervals of high data-rate observations. This would, of course, require observations outside of the science windows.

4.2 Thoughts on Compression

The issue of how to best trade off data coverage versus compression is a complex one. Obviously good temporal coverage at high cadence, with good S/N and good resolution is desired, but the telemetry is limited and so something has to give. In some cases, such as reducing the length of observations, it is fairly straightforward to evaluate the impacts. For many other options, it will probably be necessary to simulate data or use existing data from other instruments, which can then be degraded. This study is underway (Löptien et al. submitted).

4.2.1 Binning, Subsampling, and Cropping

These methods are perhaps the simplest ways to reduce the amount of data. Indeed, these were the main methods applied for MDI. Straight binning and subsampling tends to be neither efficient from a compression point of view, nor optimal from a science point of view. A large amount of spatial aliasing is generated leading to artifacts in e.g. power spectra and the data are not efficiently compressed. To overcome these problems, the data should be smoothed with a low-pass filter before subsampling. An example of how to do this is the MDI Medium-l program where the images are convolved with a 2D Gaussian before subsampling, which greatly reduces the amount of aliasing and associated artifacts (Kosovichev et al. 1996, 1997). Note that filtering the data leads to a significant reduction in the compressed size, even in the absence of subsampling, at least for most lossless compression algorithms.

The MDI Medium-l program also crops the images at 0.9 $R_{\odot}$. Even more substantial reductions are also possible, such as was done in some MDI campaigns where only a subsection of the images were downlinked. While the MDI Medium-l program has been extremely successful, it is also clear that a better trade-off could have been made. The images were probably too severely cropped and the anti-aliasing filter could have been better designed. Originally it was planned to do an onboard Spherical Harmonic Transform (SHT) instead of the Medium-l program, but this was not done for a variety of reasons. The advantage of such a scheme (onboard SHT or similar) is that it allows for filtering that is more uniform in physical scale on the Sun than does a simple smoothing and subsampling. While impossible given the MDI hardware, it would also be possible to apply a spatially variable filtering, which would have a similar effect. For PHI it may be worth it to consider some of these options if the hardware and software allow.

4.2.2 Truncation

This is another simple way of reducing the amount of data. The data are divided by a scale factor and then rounded. As long as the scale factor is small compared to the range of the data, this is almost equivalent to adding white noise. Such a rounding comparable to the photon noise is certainly reasonable, but the effect of larger numbers is currently being evaluated.

In the case of intensity data it may make sense to pass the data through a square root lookup table instead. This results in a constant fractional increase in the noise relative to the photon noise. This scheme is used for HMI and MDI.

4.2.3 Lossless Compression

For both MDI and HMI the compression (beyond truncation and sqrt lookup) was lossless and in both cases done by first differencing the numbers in the readout direction (roughly parallel to the equator) and passing the result through a Rice compression. This was, of course, combined with transmitting the first pixel and various data to prevent excessive data loss in case one packet is lost.

A compression scheme like this can be considered to consist of two parts: a prediction of the current pixel (it is the same as the previous one) and an encoding of the difference. This view leads to two obvious directions for improvement: better prediction and better compression of the differences. For the prediction several improvements can be made. Instead of using the preceding value one can use multiple preceding values. Since the data have significant high frequency power, this does not lead to a very large improvement. A further improvement can be made by looking in two dimensions, e.g. by using the average of the nearest preceding pixels in x and y (both of which have already been transmitted). This leads to a modest improvement and is similar to what is done in the standard lossless JPEG algorithm (Wallace 1992). This can also be extended to using the preceding image(s) in time, which in some case can lead to a significant improvement. It may be noted that using a substantial number of images back in time is more or less equivalent to numerically propagating the observed wavefield.

Rice compression has similar performance to a Huffman coding (Huffman 1952), which is in turn optimal if the probabilities are known and we restrict ourselves to encoding one difference at a time. Obvious improvements on a Huffman coding include making it depend on the position in the image, using longer symbol length and going to an arithmetic coding.

Finally it may be possible to use standard lossless compression algorithms, like various LZ algorithms (Ziv and Lempel 1977, 1978), JPEG and so forth.

4.2.4 Lossy Compression

It may also be worth it to test various lossy algorithms, like JPEG. However, there are some serious issues to consider regarding this. First of all the oscillations look essentially like white noise (filtered by the PSF) in a single image. This makes it very difficult to compress away noise while keeping the signal. Another problem is that algorithms such as lossy JPEG divide images into (e.g. 8×8 pixel) blocks. Likely this will cause strong artifacts at the scale of the blocks.

4.2.5 Some Further Thoughts

It may also be possible to employ further tricks to reduce the data volume. One approach is to use more advanced filtering than a simple convolution by a smooth (e.g. Gaussian) kernel. SHTs and variable width Gaussians are some possibilities, but there may be other options. More interesting is to also use the time dimension. One simple possibility is to suppress the signal at low (granulation) and high (above acoustic cutoff) temporal frequencies. But in principle more advanced filtering schemes could be employed on full 3D transforms. Similarly it may be possible to encode the transform directly instead of going back into the space-time domain. Transmitting SHT or temporal Fourier Transform coefficients is one possibility, but others exist.

Clearly some of the various ideas are easy to test. In particular the lossless compression algorithms where the success criteria are straightforward (i.e. the compression efficiency and computational cost). Others are more tricky as they involve evaluating if the damage introduced by various lossy algorithms is acceptable. We are currently working on evaluating some of these methods.

4.3 Other Requirements

In addition to the requirements on duration, resolution, spatial and temporal coverage already mentioned, several other things are needed to make optimal use of the data.

First of all it is important that the image geometry is well known. The image scale and pointing for the FDT telescope should be derived onboard using the solar limb. For the HRT this is more difficult, as the limb is not in general visible. For that it will be necessary to cross-correlate FDT and HRT images. Unfortunately the offset and scale difference is unlikely to be constant. There will be long term drifts due to thermal effects and so measurements will have to be made under various conditions. Also, the offset has a certain random component which will have to be characterized.

Some of the science objectives are achieved with FDT data for which the requirements are easily met. A subset of the rest of the objectives either involve studies where the absolute pointing is not critical (e.g. convection in polar regions, active region science) or ones where we have a simultaneous view from HMI (e.g. observations of the same area from different directions). For some science objectives, it may be necessary to observe the limb in the HRT field of view.

A more difficult requirement is that of knowing the roll angle of the instrument. To avoid that the solar rotation leaks into the measurement of the meridional flow, the absolute roll must be known to about one arcmin. Likely the only way this can be achieved is by cross-correlating with HMI or a ground-based instrument, but this will have to be repeated to check for any thermally induced variations. The roll cannot be determined using the star-trackers of the spacecraft because there might be an offset between the instrument and the star-tracker which needs to be determined. This offset could depend on temperature and, thus, change during the orbit. HMI and Solar Orbiter do not need to have the same viewing angle for this purpose. It will be sufficient if both spacecraft see the same region on the Sun.

Some further thoughts based on lessons learned from MDI and HMI are in Sect. 8.

5 Polarimetric and Helioseismic Imager

5.1 Observables

PHI will obtain two-dimensional intensity images for six wavelength points within the Fe I 6173 Å line, while measuring four polarization states at each wavelength. The current baseline is to measure at five wavelength positions within the line (−160, −80, 0, 80, and 160 mÅ from the line center) and at one continuum point (−400 mÅ from the line center).

5.2 Instrument Concept and Implementation

PHI will consist of two telescopes, which feed one filtergraph and one focal plane array: the High Resolution Telescope (HRT) is designed as an off-axis Ritchey-Chrétien telescope with a decentered pupil of 140 mm diameter. The HRT includes an internal image stabilization system (ISS) based on a fast steerable mirror that reduces residual pointing errors

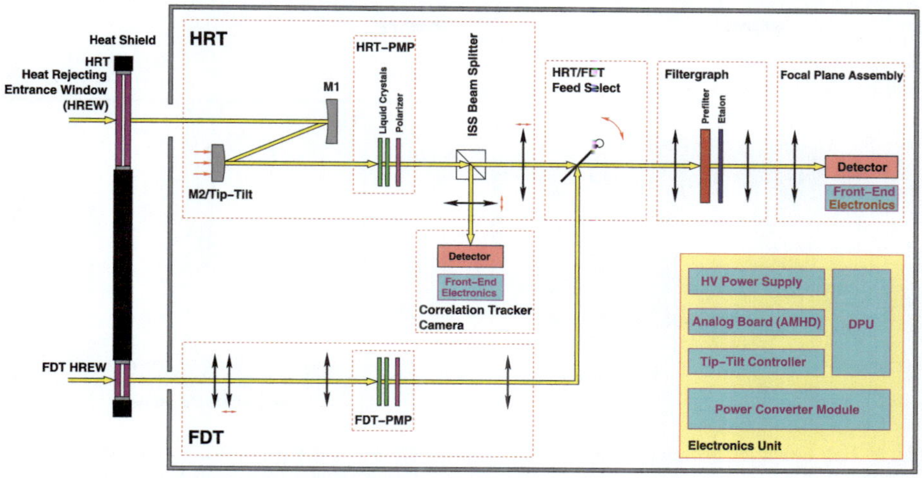

Fig. 4 Schematic drawing of the PHI instrument. PHI will consist of two telescopes, the High Resolution Telescope (HRT, *upper left part of the diagram*) and the Full Disk Telescope (FDT, *lower left part of the diagram*). The HRT will have its own Image Stabilization System (ISS) consisting of a tip-tilt mirror and a correlation tracker. Both telescopes have individual Polarization Modulation Packages (PMPs) but feed the same filtergraph and one detector. For more details about the instrument see text

by the spacecraft to levels compatible with high resolution polarimetry. The error signal is derived from a correlation tracker camera inside the HRT. The Full Disk Telescope (FDT) is designed as a refractive telescope with 17 mm aperture diameter. The two telescopes will be used sequentially. A schematic drawing of PHI is given in Fig. 4.

Both telescope apertures are protected from intense solar flux by special Heat Rejecting Entrance Windows (HREW), which are part of the heat-shield assembly of the spacecraft. They are purely dielectric broad-band reflectors with a narrow notch in the reflectivity curve around the science wavelength of the instruments. With more than 80 % transmittance at the science wavelength, in combination with almost perfect blocking from 200 nm to the far infrared, the heat load into the instrument can be effectively decreased, while preserving the high photometric and polarimetric accuracy of PHI.

The polarimetric analysis is performed by two Polarization Modulation Packages (PMPs), one in each of the telescopes. Each PMP consists of two nematic liquid crystal variable retarders (LCVRs), followed by a linear polarizer as analyzer. The modulation scheme is the same as the one used in the Imaging Magnetograph eXperiment (IMaX, Martínez Pillet et al. 2011) onboard the successful Sunrise balloon-borne observatory (Solanki et al. 2010; Barthol et al. 2011).

The filtergraph (see Fig. 5) for spectral analysis consists of a lithium niobate ($LiNbO_3$) solid state etalon that can scan the line over about 2 Å (the line is 120 mÅ wide). Its optical design is based on heritage from the IMaX instrument, while the filtergraph structure is specifically designed for *Solar Orbiter* by IAS in Orsay. The transmission curve of the etalon exhibits evenly distributed peaks separated by the product of the index of refraction of the lithium niobate and the thickness of the etalon (typically about 250 μm). In order to select the correct transmission peak, the etalon is used in tandem with a set of two prefilters: the wide-band prefilter and the narrow-band prefilter having a Full Width at Half Maximum (FWHM) of 100 Å and 2.7 Å respectively (see right part of Fig. 5). Before the light enters the etalon, the narrow-band prefilter selects the working-order of the etalon. This filter also

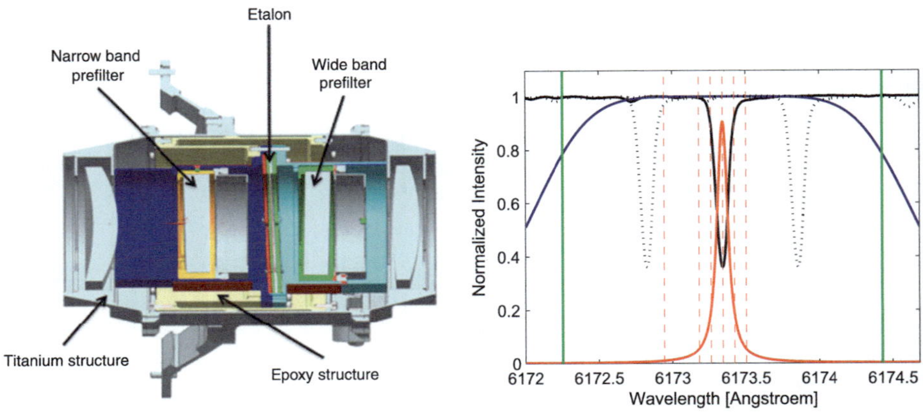

Fig. 5 *Left*: Cut of the filtergraph. The filtergraph consists of a freely tunable etalon in a telecentric mounting and two prefilters on either side of the etalon, a narrow-band prefilter and a wide-band prefilter. *Right*: Schematic representation of the filter curves of PHI. The Fe I 6173 Å line (*solid black line*) will be scanned by using three filters, the wide-band prefilter (not shown here), the order-selecting narrow-band prefilter (*blue curve*) and a freely tunable etalon (example profile shown by the *red curve*). The current baseline is to scan the line at six wavelength positions (*vertical dotted lines*). Due to the large orbital motion of the spacecraft (up to 25 km/s), the line will be subject to a large Dopplershift (up to ±0.5 Å, denoted by the *dotted black lines*). The etalon has a sufficiently large tuning range to account for this (given by the *solid green lines*)

transmits green light that is used for ground-based calibrations. The green light is removed by the narrow-band prefilter, located behind the etalon. The tuning over the line is performed by using the piezoelectric effect of the lithium niobate, affecting both the thickness and the index of refraction. A change of ~6 kV allows tuning the etalon over 2 Å. The width of the narrow-band prefilter and the tuning range of the etalon are sufficient to account for Dopplershifts of the line caused by the orbit (up to ±0.5 Å).

The etalon is used in a telecentric beam, i.e. all the points in the solar image see a cone of light with the same opening impinging on the etalon at normal incidence angle. The stability of the filtergraph is driven by two main design constraints: the thermal stability for periods shorter than 30 min, and by the wavelength repeatability induced by the high voltage power supply. The thermal design of the filtergraph ensures that the inertia is about 7 h, thereby making the instrument extremely stable to variations from the environment. A stability better than 0.01° C peak-to-peak for periods shorter than 30 min was measured during thermal tests of the breadboard corresponding to 20 m/s; this should allow the measurement of modes with degree higher than four. The high voltage repeatability was also measured in the lab to value of 3 m/s, this is 50 times lower than specified.

The focal plane assembly is built around a 2048 × 2048 pixel Active Pixel Sensor (APS), which is especially designed and manufactured for the instrument. It will deliver 10 frames per second which are read out in synchronism with the switching of the polarization modulators. In order to increase the signal-to-noise ratio, PHI will accumulate several images for every wavelength position and polarization state.

6 Simulation Tool: SOPHISM

SOPHISM is a software simulator aimed at a full representation of the PHI instrument (both hardware and software) and is applicable to both telescopes of the instrument. Starting from 2D maps of line profiles for the Fe I 6173 Å line computed from MHD simulations, SOPHISM generates synthetic 2D maps of the observables that will be measured by PHI. This allows estimates of the performance of PHI, which helps optimize the instrument design and the onboard processing. The hardware part of the simulations takes into account all the elements affecting solar light from when it enters the telescope (including some perturbations due to the spacecraft) up to the registering of this light on the detector. The software part deals with the data pipeline and processing that takes place onboard.

The simulator is programmed in the Interactive Data Language[2] (IDL) in a modular structure, each module deals with one aspect of the instrument. The modules are mostly independent from each other and can individually be enabled or disabled. The simulation runs can be saved at each step and made available for subsequent calculations. Also, the code is very flexible, with various parameters that can be modified. Presently, the modules and effects covered are the following:

- Input: this module prepares the input data to be used in the simulation run. If needed, temporal interpolation is performed, as well as spatial operations such as replication of the FOV (for simulation data with periodic boundary conditions) or, considering the Sun-spacecraft distance, scaling from the original spatial resolution to that of the detector.
- Jitter: this module represents the vibrations induced by the spacecraft, including also the correction by the ISS. A random shift of the FOV is generated and then filtered in frequency according to different possibilities, including a jitter model similar to the *Hinode* spacecraft (Katsukawa et al. 2010). Next, the resulting jitter is diminished by means of the ISS attenuation function.
- Polarization: the polarization modulation of the incoming light is parametrized in this module. This comprises the Mueller matrix of the system and the liquid crystal variable retarders (LCVRs) settings, such as orientation and retardances, to produce the desired modulation of the data.
- Spectral Profile: the spectral transmission profile of the instrument is calculated here at the user-defined wavelength positions. The resulting transmission curve, a combination of the prefilter and etalon transmissions, is convolved with the data. An example of a simulated spectral transmission profile can be seen in Fig. 5.
- Optical Aberrations: the MTF of the system with its different aberrations (e.g. defocus, astigmatism, coma, ...) is characterized and convolved with the data.
- Pupil apodization: since the etalon is placed in a telecentric mounting, close to focus, the light is converging over it and enters as a cone. This produces the so-called pupil apodization (Beckers 1998; von der Lühe and Kentischer 2000), which results in a radial gradient of intensity and phase over the pupil. This effect has consequences for the spatial resolution and spectral transmission which are taken into account in this module.
- Focal Plane: all the detector effects are included in this module (e.g. dark current, readout noise, flat-field, shutter, etc.).
- Accumulation: a given number of exposures are added directly on the instrument in order to achieve a better signal-to-noise ratio.
- Demodulation: the Stokes vector is recovered here using the demodulation matrices calculated at the polarization module.
- Inversion: the MILOS inversion code (Orozco Suárez and Del Toro Iniesta 2007), based on a Milne-Eddington approximation, is used to retrieve the full vector magnetic field and LOS velocities maps from the demodulated Stokes vector obtained by the instrument.

[2] IDL is a product of EXELIS Visual Information Solutions, http://www.exelisvis.com/.

Figure 6 shows an example of a simulation run. Although in the following part of this paper we only use Stokes I, we present here results for the magnetic Stokes parameters as well. The left panels correspond to the input data (MURaM MHD simulations with an average magnetic field of 50 G, Vögler et al. 2005), the right panels to the output of SOPHISM. The images presented here are intensity at line center and in the continuum, and $\sqrt{Q^2+U^2}/I$ and V/I at -80 mÅ in the line wing, corresponding to the degree of linear and circular polarization, respectively. The simulations represent an observation at disk center from perihelion (0.28 AU) and have a size of 12×12 Mm (the original simulations have a size of 6×6 Mm, but are replicated in the horizontal spatial dimensions in order to increase the FOV, taking advantage of its periodic boundary conditions). The simulations presented in this section have the following main characteristics: jitter with an RMS normalized to $0.5''$ and attenuated by the ISS; polarization modulation with identity Mueller matrix and ideal modulation; a prefilter with a FWHM of 3 Å and the etalon at the six positions given in Sect. 5 with a resulting transmission FWHM of $\sim$90 mÅ along with pupil apodization considerations; an aberrated wavefront normalized to $\lambda/10$; dark current and photon and readout noises; and 12 accumulations.

In the intensity images presented in Fig. 6, the most evident effect of the simulated instrument is a loss of spatial and spectral resolution. Because of the lower signal of the magnetic Stokes parameters Q, U, and V_{los}, other effects become apparent in the linear and circular polarization images, mainly crosstalk from Stokes I due to spacecraft jitter and noise components. The random parts of the simulation, like the jitter generation, noises, etc., are produced in every run. So, a different simulation run, even with the same settings, will yield different results. This is especially noticeable with the jitter crosstalk in the Q, U, V_{los} parameters, which may show different contributions in the same parameter at the same spectral position because of larger or smaller jitter shifts coinciding in the same data product.

The output of SOPHISM depends significantly on the initial settings mentioned above. Although the configuration used for the results shown in Fig. 6 corresponds to the present assumptions for the behavior of PHI, these simulations will not be identical with the actual instrument. The default settings and characteristics of most elements are taken from other instruments or calculated from a theoretical point of view. For example, the frequency filter for the spacecraft jitter is taken from *Hinode* and the prefilter curve is computed theoretically. As the design of PHI proceeds, more precise settings can be used in the simulator, including parameters derived from measurements in the lab. The development of the code will continue during the following years to represent further realistic characteristics of the instrument, such as temperature dependencies, aging (radiation effects, inefficiencies, …), FOV dependent Mueller matrices, etc.

7 Simulating PHI Time-Series for Helioseismology

We are now at the stage where a detailed strategy has to be developed to maximize the helioseismology output of the mission given the various limitations imposed by the mission (e.g., challenging orbit, and an expected telemetry allocation of 51 Gbit per orbit). In this section we present synthetic data with the same properties as expected from the High Resolution Telescope of the PHI instrument on *Solar Orbiter* and begin characterizing the properties of the data for helioseismic studies (e.g. the expected power spectra). Starting from realistic radiative MHD simulations computed with the Stagger code (Stein and Nordlund 2000), computing line profiles with the SPINOR code (Frutiger et al. 2000), and simulating the PHI instrument using SOPHISM (see Sect. 6), we have generated a time-series of synthetic

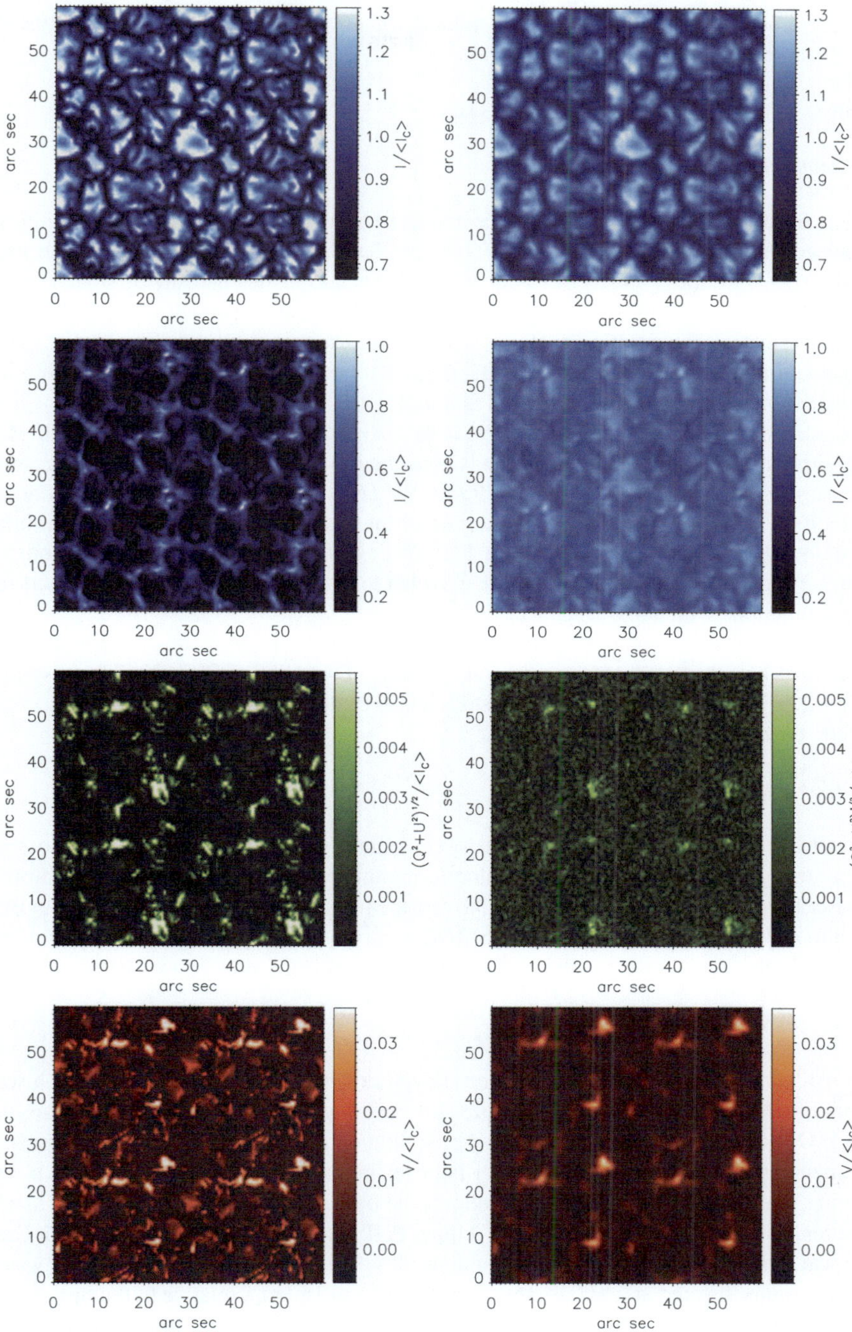

Fig. 6 Example of the influence of SOPHISM on intensity and polarization. The *left column* of panels shows the original data from an MHD simulation, the *right column* of panels show the results after running SOPHISM. *From top to bottom*: Stokes I at -400 mÅ from line center, Stokes I at line center, $\sqrt{Q^2+U^2}/I$ at -80 mÅ and V/I at -80 mÅ from line center. The last two correspond to the degree of linear and circular polarization. Note the change of the colorbars from the original data to the simulation results. The configuration of SOPHISM that was used here and further details about the images are described in the text

Dopplergrams. These Dopplergrams are models for what should be available onboard the *Solar Orbiter* satellite.

7.1 Steps in the Generation of Synthetic Data

7.1.1 Simulations of Solar Convection

We start from simulations of solar surface convection for the quiet Sun computed with the Stagger code. These simulations exhibit solar oscillations and have previously been used in helioseismic studies, mostly for analyzing the excitation mechanism of solar oscillations (e.g. Stein and Nordlund 2001; Nordlund and Stein 2001; Samadi et al. 2003; Stein et al. 2004) but also for testing methods used in helioseismology (e.g. Zhao et al. 2007; Georgobiani et al. 2007; Braun et al. 2007; Couvidat and Birch 2009) or for modeling helioseismic observations (e.g. Baldner and Schou 2012).

We use a simulation run with a size of $96 \times 96 \times 20$ Mm, corresponding to $2016 \times 2016 \times 500$ grid points. The horizontal resolution is constant (about 48 km) and the vertical resolution varies with height between 12 and 79 km. We analyze 359 minutes of the simulation, for which we have snapshots of the entire state of the system with a cadence of one minute corresponding to the planned cadence of PHI. We assume in this work that variations on shorter timescales have a negligible effect. We plan to test this assumption using a short run of the simulations with a high cadence.

In order to reduce computation time we analyze a 48×48 Mm sub-domain from the simulations. This is a small patch of the field of view of the HRT of PHI ($16.8'$, corresponding to $\sim$200 Mm at perihelion) but it is sufficient for studying the solar oscillations in the synthetic data.

7.1.2 Computation of Line Profiles

We synthesize line profiles for the Fe I 6173 Å line for every single pixel of the simulations using the SPINOR code with atomic parameters taken from the Kurucz atomic database ($\log gf = -2.880$, Fuhr et al. 1988) and the iron abundance ($A_{\mathrm{Fe}} = 7.43$) from Bellot Rubio and Borrero (2002). Note that SPINOR allows simulations of observations at heliocentric angles $\rho > 0$ by a synthesis of spectra obtained from inclined ray paths.

7.1.3 Using SOPHISM

Since we want to analyze a relatively long time-series, we configure SOPHISM in a way that minimizes the computation time while still covering all relevant effects for helioseismology. One of the computationally most demanding parts of SOPHISM is the polarimetry. Polarimetry is not of great importance for helioseismology, so we do not include it in the following. We further reduce the computation time by decreasing the number of simulated exposures. When scanning the Fe I 6173 Å line, PHI will make several exposures for each of the wavelength positions over a total time of 60 s. We assume observations at six wavelength positions, located at -400, -160, -80, 0, 80, and 160 mÅ relative to the line center. During this time, the Sun evolves, causing small differences between the individual exposures. This means that the order in which the wavelengths are observed affects the resulting image. The Stagger simulation data we use are sampled at a cadence of 60 s, so we can compute only one exposure. The number of exposures and the order in which they are taken are also important for the impact of spacecraft jitter and photon noise. So, in order to get realistic results when computing only one exposure, in this section we treat the jitter and photon noise in a slightly different way than the standard SOPHISM.

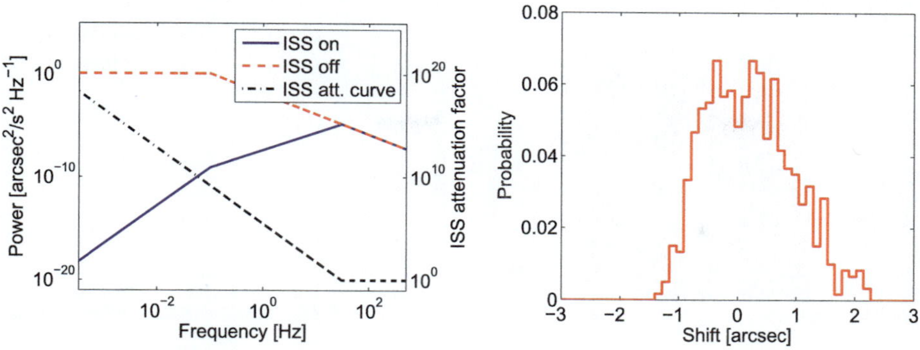

Fig. 7 *Left*: Power spectral density of the shift introduced by the modeled spacecraft jitter. We assume the jitter to be similar to that of the *Hinode* spacecraft and to follow the spacecraft requirement (see text for details). When the ISS is turned off (*dashed curve*), we use a power law to model the power spectral density of the *Hinode* jitter at frequencies above 0.1 Hz ($P(\nu) \propto \nu^{-1.98}$) and a constant power spectral density for frequencies smaller than 0.1 Hz. We model the influence of the ISS by dividing the power spectral density of the jitter (*dashed curve*) by a model for the attenuation curve of the ISS (*dash-dotted curve*). When the ISS is turned on, the jitter is reduced significantly (*solid curve*). The ISS works most efficiently at low frequencies. *Right*: Example of a histogram of the shifts introduced by the jitter (ISS turned off). We used a time-series with a duration of 60 s, which corresponds to the cadence of PHI. The distribution has an RMS of 0.7″

7.1.4 Spacecraft Jitter

The exact behavior of the spacecraft jitter of *Solar Orbiter* is not known yet. When modeling the jitter, we follow the spacecraft requirement that the RMS of the jitter with frequencies above 0.1 Hz must not be higher than 0.5″ (or 0.1″ when the ISS is turned on). We use a model for the jitter that depends on frequency with a power spectral density similar to that of the *Hinode* spacecraft. For frequencies higher than 0.1 Hz, we model the power spectral density with a power law fit of the *Hinode* jitter curve (see left part of Fig. 7) normalized to an RMS of 0.5″. We are not aware of specifications for the jitter at lower frequencies. Here we assume a constant power spectral density of the jitter, corresponding to the power at $\nu = 0.1$ Hz. This low frequency part increases the RMS from 0.5″ to 0.7″. We also model a jitter curve when the ISS is activated by dividing the power spectral density of the jitter by a modeled attenuation curve of the ISS (black curve in Fig. 7). The attenuation is strongest at low frequencies and reduces the RMS from 0.7″ to 0.03″, which is much less than the pixel size (0.5″). We express the jitter in the Fourier domain by using a fixed amplitude for each frequency (derived from the power spectral density) and a random phase. We generate a time-series of the jitter by taking the inverse Fourier transformation.

The influence of the jitter depends on the order in which the filtergrams are measured. We assume the jitter to act on all wavelengths in the same manner by using the same time-series for all wavelengths. Since we compute only one exposure, we do not model the jitter by shifting individual images. Instead, we convolve the image with a distribution corresponding to the accumulated shifts introduced by the jitter within the PHI cadence. The right part of Fig. 7 shows an example for this distribution. Since the jitter is derived from *Hinode*, it does not correspond to the jitter of the final *Solar Orbiter* spacecraft. However, as will be shown in Sect. 7.3, the influence of the jitter on the power spectrum is very small. Hence, we do not expect that our conclusions are affected by selecting a specific model for the jitter.

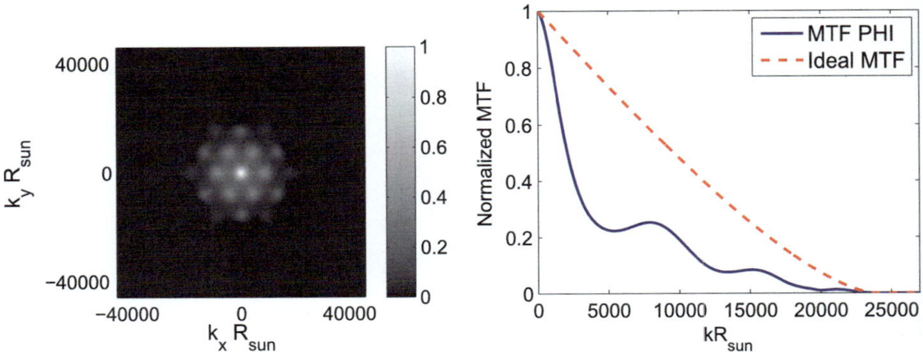

Fig. 8 Modulation Transfer Functions used in this study as a function of wavenumber when observing at perihelion (0.28 AU). *Left*: MTF computed from a modeled instrument, where the entrance window is not designed in the best possible way. *Right*: Azimuthal average of this MTF (*solid curve*) and an MTF corresponding to a diffraction-limited instrument (*dashed curve*)

7.1.5 Point Spread Function

An important characteristic of PHI is the Modulation Transfer Function (MTF). Since the MTF is mostly determined by the optical properties of the HREW when it is exposed to the environmental conditions in the orbit, the final MTF is still subject to tests and simulations. For this work, we use two different Modulation Transfer Functions corresponding to the best and the worst possible outcome. The first MTF represents a diffraction-limited instrument, we assume the Point Spread Function (PSF) to be an Airy function corresponding to the aperture of the High Resolution Telescope (14 cm). The second one is an MTF that was computed from a model of the PHI instrument, where the entrance window was located at a suboptimal position within the heat shield of the spacecraft[3] (see Fig. 8). Compared to the MTF for the diffraction-limited instrument, the MTF computed from the model of PHI strongly decreases with the wavenumber and is also very asymmetric.

7.1.6 Photon Noise

The last step is modeling the detector. Here we resample the data to the detector plate scale (0.5″ per pixel, corresponding to ~100 km at perihelion) and add photon noise to the data.

PHI will accumulate several exposures to generate one final image in order to increase the signal-to-noise ratio. The requirement for PHI for polarimetry is an accuracy of ±10 G per pixel for determining the LOS magnetic field, ±200 G for the transverse component of the magnetic field and ±15 m/s for the LOS velocity. This demands an instrument requirement for the signal-to-noise ratio of the mean continuum of at least 1000. Here we assume the noise to follow the requirement.

Since we simulate only one image, representing the accumulation of many individual exposures, we can add the noise directly to the intensity images using a normal distribution with dispersion

$$\sigma(\lambda, x, y) = 10^{-3}\sqrt{I(\lambda, x, y)\langle I_c\rangle}. \tag{1}$$

[3]Recently, preliminary results from a study for the optimum HREW location have shown that a nearly diffraction-limited telescope under all orbit conditions is feasible.

Here, $I(\lambda, x, y)$ is the intensity at a given wavelength and a given pixel and $\langle I_c \rangle$ is the spatially averaged continuum intensity at disk center. This results in the desired signal-to-noise ratio:

$$\frac{I}{\sigma} = 10^3 \sqrt{\frac{I(\lambda, x, y)}{\langle I_c \rangle}}. \tag{2}$$

Since PHI will perform an onboard correction of the raw images for darks and flat-fields, we do not model their influence here. We also neglect quantization noise. The raw images provided by the detector will have a size of 14 bits per pixel, so the influence of quantization noise is far below the photon noise level. Another source of noise is caused by variations of the exposure time. For degrees above 10 this noise must be less than 7 m/s, corresponding to an accuracy of the exposure time of 450 ppm (Appourchaux 2011).

7.2 Synthetic Intensity and Velocity Maps

Applying the individual steps described in Sect. 7.1 results in a time-series of intensity images at different wavelengths as will be available for onboard processing on *Solar Orbiter*. Here we determined the LOS velocities by computing the barycenter of the intensity at the six wavelength positions that we have selected.

We generated two such time-series with a length of 359 min each, one corresponding to an observation of the disk center and another one for an observation closer to the limb (heliocentric angle $\rho = 60°$). This corresponds to an observation of the solar poles from a solar latitude of 30°, as it will be done by *Solar Orbiter*. In both cases we have assumed the spacecraft to observe from perihelion (0.28 AU). Since the length of the time-series is rather short, we have neglected spacecraft motion. Examples for intensity and velocity maps are shown in Fig. 9 for disk center and in Fig. 10 for $\rho = 60°$.

At disk center, granulation is clearly visible in the intensity images and Dopplergrams. When observing at $\rho = 60°$, the granulation can still be resolved in the intensity images but the contrast between granules and intergranular lanes is reduced. The Dopplergrams at the limb are dominated by horizontal outflows from the granules to the intergranular lanes. After running SOPHISM, the contrast of both the intensity and the velocity is reduced significantly (see Table 4) due to the MTF. There are strong differences between the two MTFs that we use here, the worst-case MTF reduces the RMS both for intensity and velocity by about a factor of two.

7.3 Oscillation Power Spectra

We have a time-series of Dopplergrams and thus we can evaluate the effect of the PHI instrument on the measurement of solar oscillations. The most basic of helioseismic diagnostics is the power spectrum as a function of horizontal wavenumber, k, and frequency, ν. In Fig. 11 we compare two power spectra for an observation at disk center: one generated from synthetic PHI data (i.e. after running SOPHISM) and another one without any influence of the instrument (i.e. for velocity maps derived from the SPINOR line profiles). We see that the power resulting from the synthetic PHI data is lower than the power originating from the line profiles at high-k modes. This is caused by the MTF, which we assume here to correspond to a diffraction-limited instrument.

The impact of the MTF on the power spectrum strongly differs between the two MTFs used in this study (see Sect. 7.1.5). Figure 12 shows the azimuthal averages of power spectra for different setups of the instrument. The worst-case MTF reduces the power significantly

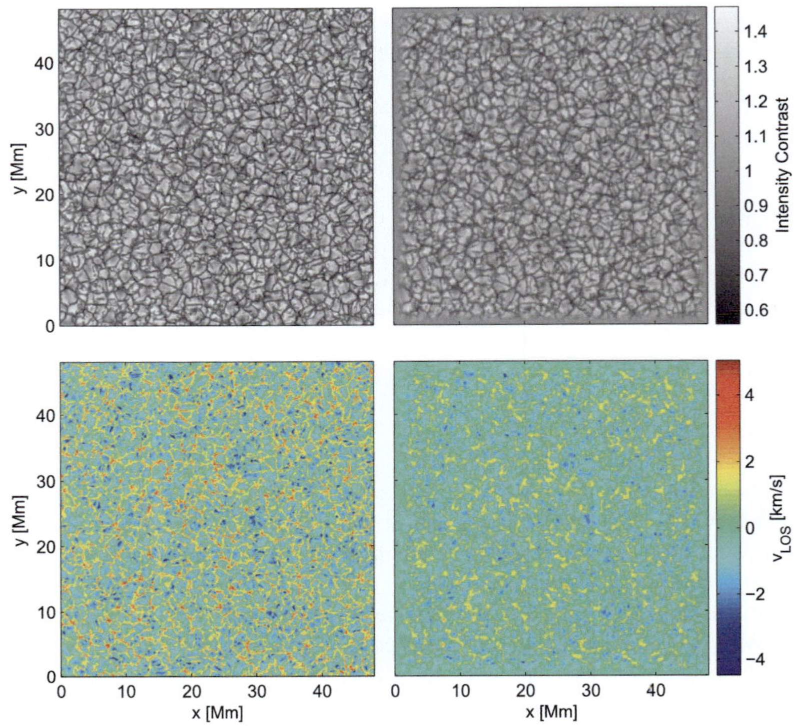

Fig. 9 Influence of the modeled High Resolution Telescope of PHI on intensity (*top*) and line-of-sight velocity (*bottom*) when observing disk center from perihelion (0.28 AU). We compare data from the line profiles (*left*) with our results for the modeled PHI instrument (*right*). The dominant signal in these images is granulation. The Dopplergrams clearly show the upflows in the granules and the downflows in the intergranular lanes. In the modeled PHI data, the spatial resolution is reduced due to the MTF. The edges of the SOPHISM data are smooth, because SOPHISM apodizes the image before applying the MTF in order to avoid artifacts in the Fourier spectrum. Here we use an MTF corresponding to a diffraction-limited instrument. The image is only a fraction of the full FOV for the HRT

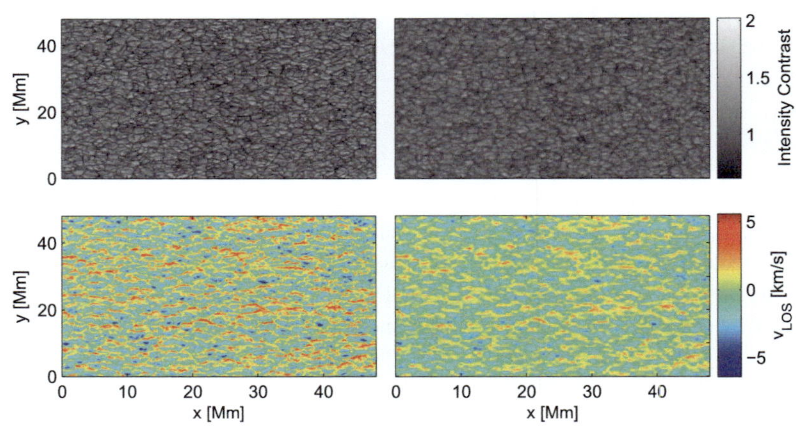

Fig. 10 Same as Fig. 9 for an observation at $\rho = 60^\circ$. Again, we assume an observation from perihelion (0.28 AU). Due to foreshortening the images are squeezed in y-direction. The top of the box is directed towards the limb. While the intensity images show granulation, as for the disk center data, the Dopplergrams are now dominated by flows in the horizontal direction, which originate from outflows from the granules

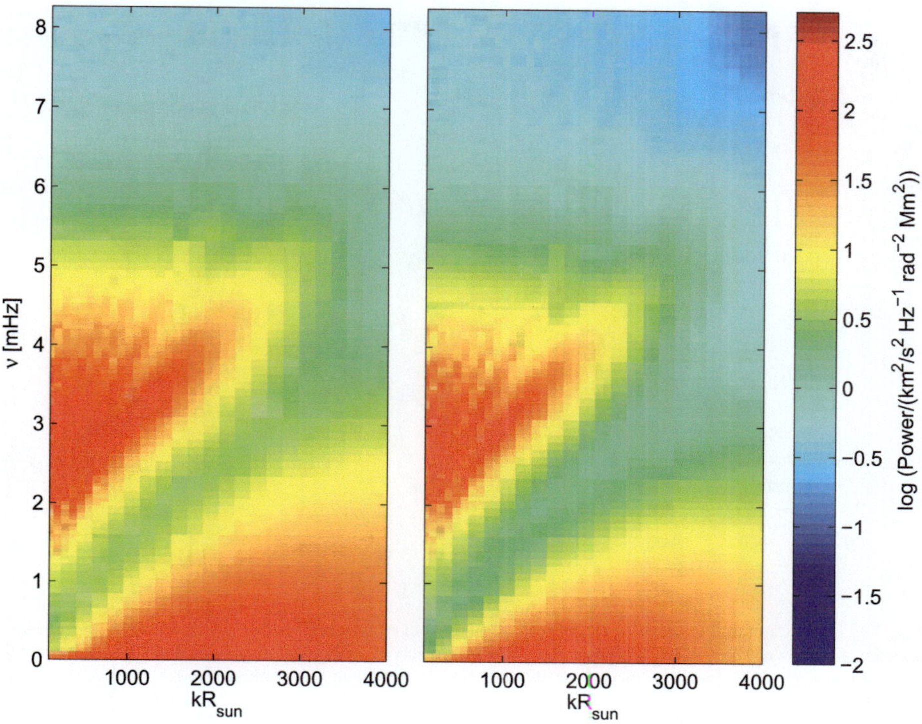

Fig. 11 Azimuthal average of two power spectra at disk center. *Left*: Power for velocities determined directly from the line profiles. *Right*: Power for velocities computed from simulated instrumental data after running SOPHISM. Both images reveal the solar oscillations and granulation. However, the simulated instrument decreases the power, especially for high wavenumbers. This is caused mostly by the Modulation Transfer Function. Here we use an MTF corresponding to a diffraction-limited instrument (the PSF is an Airy function)

Table 4 RMS for continuum intensity contrast and LOS velocity both for disk center and $\rho = 60°$. We compare data from SPINOR with our results from SOPHISM. Both the MTF corresponding to a diffraction-limited instrument and a worst-case MTF are considered

	RMS of intensity contrast		RMS of LOS velocity [km/s]	
	Disk center	$\rho = 60°$	Disk center	$\rho = 60°$
SPINOR	0.17	0.14	1.20	1.54
SOPHISM (diff.-limited MTF)	0.12	0.10	0.72	1.08
SOPHISM (worst-case MTF)	0.06	0.05	0.37	0.62

more at high wavenumbers than the diffraction-limited one. The MTF has a much larger influence on the power spectrum than other instrumental effects, like spacecraft jitter or photon noise. We also show in Fig. 12 power spectra, where the ISS is turned off and the photon noise is increased (S/N = 50). Photon noise and jitter only influence the power spectra if they are much worse than the PHI requirements. The photon noise leads to a constant offset in the power spectrum which is negligible compared to the power of the oscillation modes for any noise level that we expect. The spacecraft jitter reduces the power of the high-k modes, if the ISS is turned off. All these effects barely affect the oscillations.

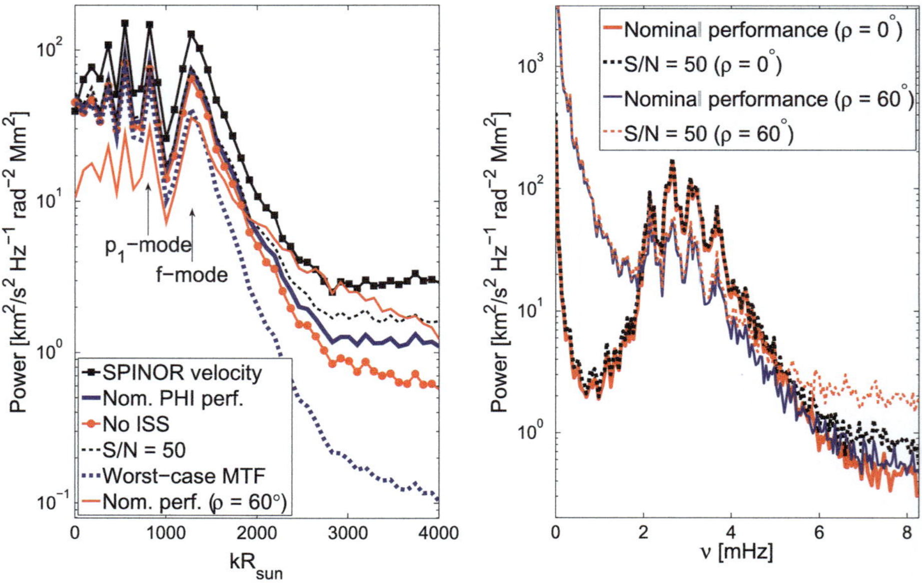

Fig. 12 *Left*: Azimuthal average of the power as a function of wavenumber at $\nu = 3.41$ mHz. Six configurations are shown, *black squares*: Power derived from SPINOR velocities, *blue thick solid curve*: Nominal PHI performance, *red dots*: ISS turned off, *black dashed curve*: Enhanced photon noise (S/N = 50), *blue thick dashed curve*: Worst-case MTF, *red solid curve*: Nominal PHI performance at $\rho = 60°$. *Right*: Power as a function of frequency at $kR_\odot = 456$. *Red thick solid line*: Nominal PHI performance when observing at disk center, *blue solid line*: Nominal PHI performance at $\rho = 60°$, *black thick dashed line*: Enhanced photon noise (S/N = 50) at disk center, *red dashed line*: Enhanced photon noise (S/N = 50) at $\rho = 60°$. We do not show the jitter with the ISS off, since it does not depend on frequency. The photon noise enhances the background power. For the nominal PHI performance, the influence of jitter and photon noise is negligible. If the ISS is turned off, the high-k power is reduced, a higher photon noise level increases the background power. However, a signal-to-noise ratio (S/N) of 50 is much lower than anything we expect

When observing at $\rho = 60°$, the power of the p-modes is lower than at disk center, since the displacement caused by the oscillations is mostly radial. On the other hand, the power resulting from convection is increased. Foreshortening leads to a stronger influence of the MTF and spacecraft jitter. Also, the intensity is lower due to limb darkening, which leads to an increased photon noise level.

In short, the most important instrumental effect is the MTF, which attenuates p-mode power at high spatial frequencies. The two MTFs used in this study lead to strong differences in the power spectrum. By comparison, the uncertainties in the jitter or photon noise level have less of an effect. However, we find that oscillations are still easy to observe even with the worst-case MTF.

8 Lessons Learned from MDI and HMI

While the MDI and HMI instruments and associated spacecraft (SOHO and SDO) are in some way quite different from PHI and *Solar Orbiter*, the experience from those investigations may nonetheless provide some guidance for PHI.

MDI is in many ways similar to PHI. Telemetry is severely restricted, observables computations are done onboard and the onboard software is complex. A major difference is that

Solar Orbiter has significantly more onboard storage, which may permit post-event data selection.

HMI is significantly different in these respects. The telemetry rate is not a limiting factor, (nearly) raw CCD images are downlinked and the onboard software is quite simple. Having said that, PHI is closer to HMI than to MDI in some respects, including thermal perturbations (due to eclipses), rapid LOS velocity changes (due to the orbit), and the ability to make full polarimetric measurements.

One major lesson learned from MDI is that fragmentation of data has to be controlled. Excessive changes between observing modes make it difficult to find good sequences for other uses than those initially envisioned and make it difficult to search through the data. For helioseismology, it is often particularly important that long sequences are available. In the extreme case of global mode analysis, typical lengths are several months and it is essential that the duty cycle is not too low. Duty cycles from existing helioseismic instruments ranges from 80–90 % for GONG and up to well over 99 % for MDI Medium-l for some 72 day periods. As such we expect that a duty cycle greater than 80 % may be acceptable. However, the amount of degradation will depend on the structure of the gaps. Variable duty cycles can lead to variations in the observed quantities which can in turn be confused with solar variations. But even in the case of local helioseismology, time-series of a significant length are needed and for the study of such things as active region evolution it is essential to observe continuously for several days.

Similarly, uncomplicated regular observations from MDI have provided most of the data used for publications. This includes the regular 96 minute cadence magnetograms, the so-called structure program (in particular the Medium-l program, Kosovichev et al. (1996, 1997) and the yearly dynamics programs. The Medium-l program, for example, has probably provided the frequencies and splittings used for most publications in global mode seismology. Open and easy access to data have also been essential to ensure the maximum use of the data.

Well calibrated data are also highly desirable. MDI was not particularly well calibrated from the outset, which combined with the onboard observable calculation led to some issues, especially regarding the magnetograms (Liu et al. 2004, 2007, 2012). Another problem for MDI has been that the noise in the shutter (Liu et al. 2004) also leads to an error in the velocity zero point, which in turn introduces a lot of noise in the low degree modes. HMI, by comparison, was substantially better calibrated from the start (Couvidat et al. 2012; Wachter et al. 2012; Schou et al. 2012b).

The issue of calibrations is particularly difficult for PHI, due to the need to do onboard observables calculations and the massive thermal perturbations. For HMI all the raw data are available on the ground, which means that computations can be done and, if necessary, redone. For PHI this will not be the case and it is thus essential that enough data are downlinked early in the mission to ensure that good calibrations can be derived and adequate onboard algorithms can be developed and tested before regular observations start. It will also be necessary to find some way to ensure that any drifts in the calibrations will be detected and corrected.

Having said that, some of these programs were not designed as well as they could have been, especially in hindsight. In particular, some of the data products could probably have been compressed more (e.g. by truncating more of the bottom bits) and a better trade-off between truncation noise, smoothing, subsampling density and crop radius could probably have been done for the Medium-l program. It might have been better to truncate more bits while keeping a larger FOV. Given the desire to have long and uniform time-series the trade-off was not adjusted during the mission.

A possibility with *Solar Orbiter* is to take data, store it on board and decide later if it should be downlinked. This may be particularly interesting for emerging active regions, flares, and so forth. There is obviously a trade-off involved here, as this will increase operational complexity and fragmentation of the data.

Acknowledgements This work is supported by the European Commission through FP7 project "Exploitation of Space Data for Innovative Helio- and Asteroseismology" (SpaceInn). We are very grateful to R. Stein for making the Stagger simulation data available. We thank A. Lagg for his support with using the SPINOR code. B.L. is a member of the International Max Planck Research School for Solar System Science at the University of Göttingen. B.L. designed and performed the research described in Sect. 7 and wrote this section. B.L. performed research for Sect. 3 and contributed to the writing of Sects. 1, 2, 3 and 5.

References

T. Appourchaux, PHI filtergraph requirements and specifications. Technical report (2011)

C.S. Baldner, J. Schou, Effects of asymmetric flows in solar convection on oscillation modes. Astrophys. J. Lett. **760**, 1 (2012). doi:10.1088/2041-8205/760/1/L1

P. Barthol, A. Gandorfer, S.K. Solanki, M. Schüssler, B. Chares, W. Curdt, W. Deutsch, A. Feller, D. Germerott, B. Grauf, K. Heerlein, J. Hirzberger, M. Kolleck, R. Meller, R. Müller, T.L. Riethmüller, G. Tomasch, M. Knölker, B.W. Lites, G. Card, D. Elmore, J. Fox, A. Lecinski, P. Nelson, R. Summers, A. Watt, V. Martínez Pillet, J.A. Bonet, W. Schmidt, T. Berkefeld, A.M. Title, V. Domingo, J.L. Gasent Blesa, J.C. Del Toro Iniesta, A. López Jiménez, A. Álvarez-Herrero, L. Sabau-Graziati, C. Widani, P. Haberler, K. Härtel, D. Kampf, T. Levin, I. Pérez Grande, A. Sanz-Andrés, E. Schmidt, The Sunrise Mission. Solar Phys. **268**, 1–34 (2011). doi:10.1007/s11207-010-9662-9

J.M. Beckers, On the effect of narrow-band filters on the diffraction limited resolution of astronomical telescopes. Astron. Astrophys. Suppl. **129**, 191–194 (1998). doi:10.1051/aas:1998180

L.R. Bellot Rubio, J.M. Borrero, Iron abundance in the solar photosphere: application of a two-component model atmosphere. Astron. Astrophys. **391**, 331–337 (2002). doi:10.1051/0004-6361:20020656

D.C. Braun, A.C. Birch, Prospects for the detection of the deep solar meridional circulation. Astrophys. J. Lett. **689**, 161–165 (2008). doi:10.1086/595884

D.C. Braun, A.C. Birch, D. Benson, R.F. Stein, Å. Nordlund, Helioseismic holography of simulated solar convection and prospects for the detection of small-scale subsurface flows. Astrophys. J. **669**, 1395–1405 (2007). doi:10.1086/521782

W.J. Chaplin, Y. Elsworth, R. Howe, G.R. Isaak, C.P. McLeod, B.A. Miller, H.B. van der Raay, S.J. Wheeler, R. New, BiSON performance. Solar Phys. **168**, 1–18 (1996). doi:10.1007/BF00145821

S. Couvidat, A.C. Birch, Helioseismic travel-time definitions and sensitivity to horizontal flows obtained from simulations of solar convection. Solar Phys. **257**, 217–235 (2009). doi:10.1007/s11207-009-9371-4

S. Couvidat, J. Schou, R.A. Shine, R.I. Bush, J.W. Miles, P.H. Scherrer, R.L. Rairden, Wavelength dependence of the Helioseismic and Magnetic Imager (HMI) Instrument onboard the Solar Dynamics Observatory (SDO). Solar Phys. **275**, 285–325 (2012). doi:10.1007/s11207-011-9723-8

M. Dikpati, P.A. Gilman, Theory of solar meridional circulation at high latitudes. Astrophys. J. **746**, 65 (2012). doi:10.1088/0004-637X/746/1/65

B. Fleck, S. Couvidat, T. Straus, On the formation height of the SDO/HMI Fe 6173 Å Doppler signal. Solar Phys. **271**, 27–40 (2011). doi:10.1007/s11207-011-9783-9

L.N. Fletcher, P.G.J. Irwin, G.S. Orton, N.A. Teanby, R.K. Achterberg, G.L. Bjoraker, P.L. Read, A.A. Simon-Miller, C. Howett, R. de Kok, N. Bowles, S.B. Calcutt, B. Hesman, F.M. Flasar, Temperature and composition of Saturn's polar hot spots and hexagon. Science **319**, 79 (2008). doi:10.1126/science.1149514

C. Fröhlich, J. Romero, H. Roth, C. Wehrli, B.N. Andersen, T. Appourchaux, V. Domingo, U. Telljohann, G. Berthomieu, P. Delache, J. Provost, T. Toutain, D.A. Crommelynck, A. Chevalier, A. Fichot, W. Däppen, D. Gough, T. Hoeksema, A. Jiménez, M.F. Gómez, J.M. Herreros, T.R. Cortés, A.R. Jones, J.M. Pap, R.C. Willson, VIRGO: experiment for helioseismology and solar irradiance monitoring. Solar Phys. **162**, 101–128 (1995). doi:10.1007/BF00733428

C. Frutiger, S.K. Solanki, M. Fligge, J.H.M.J. Bruls, Properties of the solar granulation obtained from the inversion of low spatial resolution spectra. Astron. Astrophys. **358**, 1109–1121 (2000)

J.R. Fuhr, G.A. Martin, W.L. Wiese, Atomic transition probabilities: iron through nickel. J. Phys. Chem. Ref. Data **17** (1988)

R.A. García, The sun as a star: 13 years of Soho, in *American Institute of Physics Conference Series*, ed. by J.A. Guzik, P.A. Bradley, American Institute of Physics Conference Series, vol. 1170, 2009, pp. 560–565. doi:10.1063/1.3246562

D. Georgobiani, J. Zhao, A.G. Kosovichev, D. Benson, R.F. Stein, Å. Nordlund, Local helioseismology and correlation tracking analysis of surface structures in realistic simulations of solar convection. Astrophys. J. **657**, 1157–1161 (2007). doi:10.1086/511148

P.A. Gilman, Model calculations concerning rotation at high solar latitudes and the depth of the solar convection zone. Astrophys. J. **231**, 284–292 (1979). doi:10.1086/157191

L. Gizon, T.L. Duvall, J. Schou, Wave-like properties of solar supergranulation. Nature **421**, 43–44 (2003). doi:10.1038/nature01287

L. Gizon, A.C. Birch, H.C. Spruit, Local helioseismology: three-dimensional imaging of the solar interior. Ann. Rev. Astron. Astrophys. **48**, 289–338 (2010). doi:10.1146/annurev-astro-082708-101722

L. Gizon, A.C. Birch, R.I. Bush, T.L. Duvall Jr., A.G. Kosovichev, P.H. Scherrer, J. Zhao, Time-distance helioseismology and the Solar Orbiter Mission, in *Solar Encounter: Proceedings of the First Solar Orbiter Workshop*, ed. by B. Battrick, H. Sawaya-Lacoste, E. Marsch, V. Martinez Pillet, B. Fleck, R. Marsden. ESA Special Publication, vol. 493 (2001), pp. 227–231

L. Gizon, J. Ballot, E. Michel, T. Stahn, G. Vauclair, H. Bruntt, P.-O. Quirion, O. Benomar, S. Vauclair, T. Appourchaux, M. Auvergne, A. Baglin, C. Barban, F. Baudin, M. Bazot, T. Campante, C. Catala, W. Chaplin, O. Creevey, S. Deheuvels, N. Dolez, Y. Elsworth, R. Garcia, P. Gaulme, S. Mathis, S. Mathur, B. Mosser, C. Regulo, I. Roxburgh, D. Salabert, R. Samadi, K. Sato, G. Verner, S. Hanasoge, K.R. Sreenivasan, Seismic constraints on rotation of sun-like star and mass of exoplanet. Proc. Natl. Acad. Sci. USA **110**, 13267–13271 (2013). doi:10.1073/pnas.1303291110

S.M. Hanasoge, T.L. Duvall, K.R. Sreenivasan, Anomalously weak solar convection. Proc. Natl. Acad. Sci. USA **109**, 11928–11932 (2012). doi:10.1073/pnas.1206570109

D.H. Hathaway, L. Upton, O. Colegrove, Giant convection cells found on the sun. Science **342**, 1217–1219 (2013). doi:10.1126/science.1244682

F. Hill, J. Bolding, C. Toner, T. Corbard, S. Wampler, B. Goodrich, J. Goodrich, P. Eliason, K.D. Hanna, The Gong++ data processing pipeline, in *GONG+ 2002: Local and Global Helioseismology: The Present and Future*, ed. by H. Sawaya-Lacoste. ESA Special Publication, vol. 517 (2003), pp. 295–298

F. Hill, R. Howe, R. Komm, I. Gonzallez Hernandez, S. Tripathy, K. Jain, A helioseismic comparison of the solar minima preceding cycles 23 and 24, in *AAS/Solar Physics Division Meeting*, vol. 40, No. 24-01 (2009).

R. Howe, Solar interior rotation and its variation. Living Rev. Sol. Phys. **6**, 1 (2009). doi:10.12942/lrsp-2009-1

R. Howe, J. Christensen-Dalsgaard, F. Hill, R.W. Komm, R.M. Larsen, J. Schou, M.J. Thompson, J. Toomre, Dynamic variations at the base of the solar convection zone. Science **287**, 2456–2460 (2000). doi:10.1126/science.287.5462.2456

R. Howe, J. Christensen-Dalsgaard, F. Hill, R. Komm, T.P. Larson, M. Rempel, J. Schou, M.J. Thompson, The high-latitude branch of the solar torsional oscillation in the rising phase of cycle 24. Astrophys. J. Lett. **767**, 20 (2013). doi:10.1088/2041-8205/767/1/L20

D.A. Huffman, A method for the construction of minimum-redundancy codes. Proc. IRE **40**, 1098–1102 (1952). doi:10.1109/JRPROC.1952.273898

J. Jiang, R. Cameron, D. Schmitt, M. Schüssler, Countercell meridional flow and latitudinal distribution of the solar polar magnetic field. Astrophys. J. Lett. **693**, 96–99 (2009). doi:10.1088/0004-637X/693/2/L96

Y. Katsukawa, Y. Masada, T. Shimizu, S. Sakai, K. Ichimoto, Pointing stability of Hinode and requirements for the next solar mission Solar-C, in *International Conference on Space Optics* (2010)

R. Komm, I. González Hernández, F. Hill, R. Bogart, M.C. Rabello-Soares, D. Haber, Subsurface meridional flow from HMI using the ring-diagram pipeline. Solar Phys. **287**, 85–106 (2013). doi:10.1007/s11207-012-0073-y

A.G. Kosovichev, T.L. Duvall, P.H. Scherrer, Simulation of vector-weighted binning for the Medium-l program. SOI technical notes (1996)

A.G. Kosovichev, J. Schou, P.H. Scherrer, R.S. Bogart, R.I. Bush, J.T. Hoeksema, J. Aloise, L. Bacon, A. Burnette, C. de Forest, P.M. Giles, K. Leibrand, R. Nigam, M. Rubin, K. Scott, S.D. Williams, S. Basu, J. Christensen-Dalsgaard, W. Dappen, E.J. Rhodes Jr., T.L. Duvall Jr., R. Howe, M.J. Thompson, D.O. Gough, T. Sekii, J. Toomre, T.D. Tarbell, A.M. Title, D. Mathur, M. Morrison, J.L.R. Saba, C.J. Wolfson, I. Zayer, P.N. Milford, Structure and rotation of the solar interior: initial results from the MDI Medium-l program. Solar Phys. **170**, 43–61 (1997). doi:10.1023/A:1004949311268

J.R. Kuhn, R. Bush, M. Emilio, I.F. Scholl, The precise solar shape and its variability. Science **337**, 1638 (2012). doi:10.1126/science.1223231

C. Lindsey, D.C. Braun, Seismic images of the far side of the sun. Science **287**, 1799–1801 (2000). doi:10.1126/science.287.5459.1799

Y. Liu, X. Zhao, J.T. Hoeksema, Correction of offset in MDI/SOHO magnetograms. Solar Phys. **219**, 39–53 (2004). doi:10.1023/B:SOLA.0000021822.07430.d6

Y. Liu, A.A. Norton, P.H. Scherrer, A note on saturation seen in the MDI/SOHO magnetograms. Solar Phys. **241**, 185–193 (2007). doi:10.1007/s11207-007-0296-5

Y. Liu, J.T. Hoeksema, P.H. Scherrer, J. Schou, S. Couvidat, R.I. Bush, T.L. Duvall, K. Hayashi, X. Sun, X. Zhao, Comparison of line-of-sight magnetograms taken by the Solar Dynamics Observatory/Helioseismic and Magnetic Imager and Solar and Heliospheric Observatory/Michelson Doppler Imager. Solar Phys. **279**, 295–316 (2012). doi:10.1007/s11207-012-9976-x

B. Löptien, A.C. Birch, L. Gizon, J. Schou, Image compression in local helioseismology. Astron. Astrophys. (submitted)

E. Marsch, E. Antonucci, P. Bochsler, J.-L. Bougeret, R. Harrison, R. Schwenn, J.-C. Vial, First Solar Orbiter assessment study report. ESA-SCI (2000) 6

R.G. Marsden, D. Müller, Solar Orbiter Exploring the Sun-Heliosphere Connection. Solar Orbiter Definition Study Report. ESA/SRE (2011) 14

V. Martínez Pillet, J.C. Del Toro Iniesta, A. Álvarez-Herrero, V. Domingo, J.A. Bonet, L. González Fernández, A. López Jiménez, C. Pastor, J.L. Gasent Blesa, P. Mellado, J. Piqueras, B. Aparicio, M. Balaguer, E. Ballesteros, T. Belenguer, L.R. Bellot Rubio, T. Berkefeld, M. Collados, W. Deutsch, A. Feller, F. Girela, B. Grauf, R.L. Heredero, M. Herranz, J.M. Jerónimo, H. Laguna, R. Meller, M. Menéndez, R. Morales, D. Orozco Suárez, G. Ramos, M. Reina, J.L. Ramos, P. Rodríguez, A. Sánchez, N. Uribe-Patarroyo, P. Barthol, A. Gandorfer, M. Knoelker, W. Schmidt, S.K. Solanki, S. Vargas Domínguez, The Imaging Magnetograph eXperiment (IMaX) for the Sunrise Balloon-Borne solar observatory. Solar Phys. **268**, 57–102 (2011). doi:10.1007/s11207-010-9644-y

D. Müller, R.G. Marsden, O.C. St. Cyr, H.R. Gilbert, Solar Orbiter: exploring the sun-heliosphere connection. Solar Phys. **285**, 25–70 (2013). doi:10.1007/s11207-012-0085-7

K. Nagashima, J. Zhao, A.G. Kosovichev, T. Sekii, Detection of supergranulation alignment in polar regions of the sun by helioseismology. Astrophys. J. Lett. **726**, 17 (2011). doi:10.1088/2041-8205/726/2/L17

Å. Nordlund, R.F. Stein, Solar oscillations and convection. I. Formalism for radial oscillations. Astrophys. J. **546**, 576–584 (2001). doi:10.1086/318217

A.A. Norton, R.K. Ulrich, Measuring magnetic oscillations in the solar photosphere: coordinated observations with MDI, ASP and MWO. Solar Phys. **192**, 403–413 (2000). doi:10.1023/A:1005290616918

D. Orozco Suárez, J.C. Del Toro Iniesta, The usefulness of analytic response functions. Astron. Astrophys. **462**, 1137–1145 (2007). doi:10.1051/0004-6361:20066201

T. Roudier, M. Rieutord, V. Prat, J.M. Malherbe, N. Renon, Z. Frank, M. Švanda, T. Berger, R. Burston, L. Gizon, Comparison of solar horizontal velocity fields from SDO/HMI and Hinode data. Astron. Astrophys. **552**, 113 (2013). doi:10.1051/0004-6361/201220867

A. Ruzmaikin, C. Lindsey, Helioseismic probing of the solar dynamo, in *GONG+ 2002: Local and Global Helioseismology: The Present and Future*, ed. by H. Sawaya-Lacoste. ESA Special Publication, vol. 517 (2003), pp. 71–74

R. Samadi, Å. Nordlund, R.F. Stein, M.J. Goupil, I. Roxburgh, Numerical constraints on the model of stochastic excitation of solar-type oscillations. Astron. Astrophys. **403**, 303–312 (2003). doi:10.1051/0004-6361:20030356

K.H. Sato, R.A. Garcia, S. Pires, J. Ballot, S. Mathur, B. Mosser, E. Rodriguez, J.L. Starck, K. Uytterhoeven, Inpainting: a powerful interpolation technique for helio- and asteroseismic data. ArXiv e-prints (2010)

P.H. Scherrer, R.S. Bogart, R.I. Bush, J.T. Hoeksema, A.G. Kosovichev, J. Schou, W. Rosenberg, L. Springer, T.D. Tarbell, A. Title, C.J. Wolfson, I. Zayer, MDI Engineering Team, The solar oscillations investigation—Michelson Doppler imager. Solar Phys. **162**, 129–188 (1995). doi:10.1007/BF00733429

J. Schou, Wavelike properties of solar supergranulation detected in Doppler shift data. Astrophys. J. Lett. **596**, 259–262 (2003). doi:10.1086/379529

J. Schou, P.H. Scherrer, R.I. Bush, R. Wachter, S. Couvidat, M.C. Rabello-Soares, R.S. Bogart, J.T. Hoeksema, Y. Liu, T.L. Duvall, D.J. Akin, B.A. Allard, J.W. Miles, R. Rairden, R.A. Shine, T.D. Tarbell, A.M. Title, C.J. Wolfson, D.F. Elmore, A.A. Norton, S. Tomczyk, Design and ground calibration of the Helioseismic and Magnetic Imager (HMI) instrument on the Solar Dynamics Observatory (SDO). Solar Phys. **275**, 229–259 (2012a). doi:10.1007/s11207-011-9842-2

J. Schou, J.M. Borrero, A.A. Norton, S. Tomczyk, D. Elmore, G.L. Card, Polarization calibration of the Helioseismic and Magnetic Imager (HMI) onboard the Solar Dynamics Observatory (SDO). Solar Phys. **275**, 327–355 (2012b). doi:10.1007/s11207-010-9639-8

H. Schunker, D.C. Braun, P.S. Cally, C. Lindsey, The local helioseismology of inclined magnetic fields and the showerglass effect. Astrophys. J. Lett. **621**, 149–152 (2005). doi:10.1086/429290

S.K. Solanki, P. Barthol, S. Danilovic, A. Feller, A. Gandorfer, J. Hirzberger, T.L. Riethmüller, M. Schüssler, J.A. Bonet, V. Martínez Pillet, J.C. del Toro Iniesta, V. Domingo, J. Palacios, M. Knölker, N. Bello

González, T. Berkefeld, M. Franz, W. Schmidt, A.M. Title, SUNRISE: instrument, mission, data, and first results. Astrophys. J. Lett. **723**, 127–133 (2010). doi:10.1088/2041-8205/723/2/L127

R.F. Stein, Å. Nordlund, Realistic solar convection simulations. Solar Phys. **192**, 91–108 (2000). doi:10.1023/A:1005260918443

R.F. Stein, Å. Nordlund, Solar oscillations and convection. II. Excitation of radial oscillations. Astrophys. J. **546**, 585–603 (2001). doi:10.1086/318218

R. Stein, D. Georgobiani, R. Trampedach, H.-G. Ludwig, Å. Nordlund, Excitation of radial p-modes in the sun and stars. Solar Phys. **220**, 229–242 (2004). doi:10.1023/B:SOLA.0000031458.35121.33

A. Vögler, S. Shelyag, M. Schüssler, F. Cattaneo, T. Emonet, T. Linde, Simulations of magneto-convection in the solar photosphere. equations, methods, and results of the MURaM code. Astron. Astrophys. **429**, 335–351 (2005). doi:10.1051/0004-6361:20041507

O. von der Lühe, T.J. Kentischer, High spatial resolution performance of a triple Fabry-Pérot filtergraph. Astron. Astrophys. Suppl. **146**, 499–506 (2000). doi:10.1051/aas 2000283

R. Wachter, J. Schou, M.C. Rabello-Soares, J.W. Miles, T.L. Duvall, R.I. Bush, Image quality of the Helioseismic and Magnetic Imager (HMI) onboard the Solar Dynamics Observatory (SDO). Solar Phys. **275**, 261–284 (2012). doi:10.1007/s11207-011-9709-6

G.K. Wallace, The JPEG still picture compression standard. IEEE Trans. Consum. Electron. **38**, 18–34 (1992). doi:10.1109/30.125072

J. Woch, L. Gizon, The Solar Orbiter mission and its prospects for helioseismology. Astron. Nachr. **328**, 362 (2007). doi:10.1002/asna.200710743

J. Zhao, A.G. Kosovichev, Surface magnetism effects in time-distance helioseismology. Astrophys. J. **643**, 1317–1324 (2006). doi:10.1086/503248

J. Zhao, D. Georgobiani, A.G. Kosovichev, D. Benson, R.F. Stein, Å. Nordlund, Validation of time-distance helioseismology by use of realistic simulations of solar convection. Astrophys. J. **659**, 848–857 (2007). doi:10.1086/512009

J. Zhao, K. Nagashima, R.S. Bogart, A.G. Kosovichev, T.L. Duvall Jr., Systematic center-to-limb variation in measured helioseismic travel times and its effect on inferences of solar interior meridional flows. Astrophys. J. Lett. **749**, 5 (2012). doi:10.1088/2041-8205/749/1/L5

J. Zhao, R.S. Bogart, A.G. Kosovichev, T.L. Duvall Jr., T. Hartlep, Detection of equatorward meridional flow and evidence of double-cell meridional circulation inside the sun. Astrophys. J. Lett. **774**, 29 (2013). doi:10.1088/2041-8205/774/2/L29

J. Ziv, A. Lempel, A universal algorithm for sequential data compression. IEEE Trans. Inf. Theory **23**, 337–343 (1977). doi:10.1109/TIT.1977.1055714

J. Ziv, A. Lempel, Compression of individual sequences via variable-rate coding. IEEE Trans. Inf. Theory **24**, 530–536 (1978). doi:10.1109/TIT.1978.1055934

Space Sci Rev (2015) 196:285–302
DOI 10.1007/s11214-015-0142-2

Future Mission Concepts for Helioseismology

Takashi Sekii[1] · Thierry Appourchaux[2] · Bernhard Fleck[3] · Sylvaine Turck-Chièze[4]

Received: 17 August 2014 / Accepted: 26 February 2015 / Published online: 17 March 2015

Abstract Future space-mission concepts currently discussed in the helioseismology community are reviewed. One popular idea is to observe the Sun from high latitudes, to explore the polar regions as well as to probe the deep interior using stereoscopic techniques, by combining observations from high latitudes with observations from within the ecliptic plane. Another idea is to stay within the ecliptic plane but still aim for stereoscopic helioseismology for deep layers. A new instrument and a novel mission concept for studying the solar core regions are also discussed.

Keywords Sun · Helioseismology · Space missions

1 Introduction

Before discussing concepts for future helioseismology missions we thought it would be useful to briefly look back and recall how the field was started and how far it has come in

✉ T. Sekii
Fax: +81-422-34-3700
sekii@solar.mtk.nao.ac.jp

T. Appourchaux
Thierry.Appourchaux@ias.u-psud.fr
Fax: +33-169858701

B. Fleck
bfleck@esa.nascom.nasa.gov

S. Turck-Chièze
Sylvaine.Turck-Chieze@cea.fr
Fax: +33-169086577

1 Division of Solar and Plasma Astrophysics, NAOJ, Mitaka, Tokyo 181-8588, Japan

2 Institut d'Astrophysique Spatiale, Université Paris 11 – CNRS (UMR8617), Bâtiment 121, Orsay, France

3 ESA Science Operations Dept., c/o NASA GSFC, Mail Code 671.1, Greenbelt, MD 20771, USA

4 CEA/DSM/IRFU, CE Saclay, 91191 Gif-sur-Yvette, France

the last twenty years. We therefore start with a brief historical review of the early days of helioseismology from space, followed by a condensed account of current missions and their main achievements and then present various new mission concepts that have been proposed for the future.

The first space mission concept that included helioseismic measurements among its scientific objectives was the Solar Cycle and Dynamics Mission (SCADM). The goal of this mission was to reach an understanding of one of the most fundamental processes of nature—the operation of the solar cycle. For an excellent account of the key science questions and the measurement requirements identified at that time the reader is referred to the Proceedings of the symposium on "Study of the Solar Cycle from Space" that was held in June 1979 in Wellesley, Massachusetts (NASA Conference Publication CP-2098, 1979).[1] Interestingly, even as the low-degree unresolved modes were just being discovered, spatially resolved observations for seismology were already recognized as being needed from space.

Around the same time a group of French and Belgian scientists studied a mission dedicated to total and spectral irradiance variations of the Sun. In 1980 they proposed this mission concept to ESA as "DISCO" (Dual Spectral Irradiance and Solar Constant Orbiter, Domingo and Wyn-Roberts 1982). Around the same time the Birmingham group (Claverie et al. 1979) and the Nice Observatory group (Grec et al. 1980) published their pioneering measurements of global solar oscillations using optical resonant scattering photometers. These measurements demonstrated the enormous potential of high-quality global solar oscillation measurements for the study of the solar interior. It was therefore decided to add helioseismology as the primary scientific objective of DISCO and include global oscillation measurements, in velocity as well as in luminosity. DISCO was conceived as a fairly small and cheap spin-stabilized spacecraft, with a spin rate of 5 rpm. It was planned to be launched in 1988 and to be placed into an orbit around the L_1 Sun-Earth Lagrangian point. This orbit, which was later also chosen for SOHO, was selected because it provides extremely good observing conditions for the detection of global velocity fluctuations with high accuracy and sensitivity, as the variations in radial velocity are relatively small and smooth (Domingo and Wyn-Roberts 1984). In addition to the helioseismic and radiometric instruments, the strawman payload of DISCO included a far-UV spectroheliograph to study coronal holes, a magnetometer, a radio wave and plasma wave analyzer, a solar wind instrument and several energetic particle detectors. In January 1983, after completion of a Phase A study (Domingo and Wyn-Roberts 1982), ESA's Solar System Working Group ranked DISCO above "Kepler", a competing Mars mission, but in a final evaluation by ESA's Space Science Advisory Committee DISCO lost out against ISO (Infrared Space Observatory), which was approved by ESA's Science Programme Committee in March 1983 and eventually was launched in November 1995.

In the late seventies ESA had studied another solar mission, GRIST (Grazing Incidence Solar Telescope), which in early 1981 became a victim of ESA's response to NASA's unilateral cancellation of the US contribution to the International Solar Polar Mission (ISPM), the ESA component of which was later called "Ulysses". Following the non-selection of DISCO in early 1983 the solar community discussed a new mission that would combine the science objectives of both GRIST and DISCO. That mission was dubbed "SOHO" (Solar High-Resolution Observatory, later renamed to Solar and Heliospheric Observatory). After getting NASA onboard, which in the early eighties studied a mission to investigate the interior of the Sun (Solar Internal Dynamics Mission), SOHO was eventually approved in early 1986 by ESA's Science Programme Committee as one of the two elements that formed the

[1] http://babel.hathitrust.org/cgi/pt?id=coo.31924004316570;view=1up;seq=1.

Solar-Terrestrial Science Programme (STSP), the first cornerstone of ESA's long-term program "Horizon 2000". STSP in turn was part of the International Solar-Terrestrial Physics Programme (ISTP), which was agreed upon in September 1983 by NASA, ESA, and ISAS (Institute of Space and Astronautical Science of Japan). For a detailed account of the history of SOHO the reader is referred to Huber et al. (1996).

SOHO, which was approved in 1986 and launched on 2 December 1995, is a mission of international cooperation between ESA and NASA to study the Sun, from its deep core to the outer corona, the solar wind, and the interaction with the interstellar medium (Domingo et al. 1995; Fleck et al. 1995). It is probably not an exaggeration to say that it revolutionized solar physics, including the field of helioseismology. SOHO carries two instruments to measure global velocity and luminosity oscillations, similar to those originally considered for DISCO. Those are GOLF (Global Oscillations at Low Frequency, Gabriel et al. 1995) and VIRGO (Variability of solar IRradiance and Gravity Oscillations, Fröhlich et al. 1995). In addition it carries a NASA provided instrument for spatially resolved helioseismology, MDI (Michelson Doppler Imager, Scherrer et al. 1995), which produced data that form the basis of more than 1800 papers in the refereed literature, making it one of the most successful instruments in space-mission history.

GOLF uses a resonant scattering spectrometer technique on sodium D1 and D2 lines to measure the low-degree ($l \leq 3$) solar velocity oscillations. It has been built to detect the low frequency part of the solar spectrum that probes the solar core, in measuring amplitude signals down to 2 m/s^2/Hz above 10^{-4} Hz. The frequency range was chosen to cover the gravity-mode region, which is not too polluted by solar background noise, and the acoustic-mode region. The statistics were adapted to detect a velocity signal of 1 mm/s in 20 days, and consequently a special care had been taken in the detector unit to avoid any electronic perturbations above the statistical noise in the required frequency range (Global Oscillations at Low Frequency, Gabriel et al. 1995). It was operated in its nominal mode (i.e. recording data in both wings of the sodium D lines) for only about 6 weeks during the commissioning phase of the spacecraft in early 1996. Because of concerns about the reliability of the polarizer mechanism, the GOLF team decided to operate the instrument with this mechanism halted, recording data first in the blue wing of the line profile, then in the red wing. After SOHO's recovery in October 1998 (Vandenbussche 1999) GOLF has been tuned to the blue wing of the sodium D1 and D2 profiles which has been considered the most promising one. Excluding the long gaps due to the SOHO mission interruptions in the summer of 1998 and early 1999, GOLF has achieved a duty cycle greater than 98 % (Garcia et al. 2005).

VIRGO uses two different types of absolute radiometers (PMO6-V and DIARAD), a 3-channel sun photometer operating in the blue, red and green, and a twelve-resolution element Luminosity Oscillations Imager (LOI) ($l \leq 10$) to measure low degree brightness oscillations with a duty cycle of about 98 %.

The MDI instrument on SOHO is fed by a 12.5-cm refracting telescope, which images the Sun on a 1024 × 1024 pixels CCD detector through a series of increasingly narrow spectral filters. The final elements are a pair of tunable Michelson interferometers, which allow MDI to take filtergrams of the Ni I 6768 Å line with a spectral resolution of 94 mÅ. From these filtergrams, Dopplergrams, intensitygrams and magnetograms are constructed on-board. MDI can be operated in a full disk mode (2" equivalent pixel size) and in a high resolution mode (0.65" equivalent pixel size).

Data from the three helioseismology instruments on SOHO have led to major advances in the field of helioseismology over the past 17 years, many of which are described elsewhere in this volume. They include reliable measurements of the Sun's differential rotation in the solar interior down well into the radiative core and the confirmation of a prominent shear layer just beneath the solar surface (Schou et al. 1998; Couvidat et al. 2003); measurements of the meridional flow (Giles et al. 1997) and its variations with depth and the solar cycle (Haber et al. 2002; Hathaway and Rightmire 2010), measurements of zonal flow variations from the near surface layers down to the bottom of the convection zone (Kosovichev and Schou 1997; Howe et al. 2000; Vorontsov et al. 2002); ever more reliable measurements of the solar interior structure putting ever more stringent constraints on the standard model hypotheses and the neutrino flux emitted by nuclear reactions in the core and thus helping to definitively solve the solar neutrino puzzle (Turck-Chièze et al. 2001; Turck-Chièze and Couvidat 2011); the first ever images of a star's convection zone and of structures and flows below the Sun's surface (Duvall et al. 1997; Zhao et al. 2001); imaging of activity on the Sun's far side (Lindsey and Braun 2000) and the early detection of rising sunspots in the solar interior (Ilonidis et al. 2011).

On 12 April 2011 at 23:22:31 UT, after successful completion of the cross-calibration with the Helioseismic and Magnetic Imager (HMI, Scherrer et al. 2012; Schou et al. 2012) aboard the Solar Dynamics Observatory (SDO, Pesnell et al. 2012), MDI was commanded to stop taking science data. HMI, which supersedes MDI, started nominal science observations on 1 May 2010. It has a similar design and observing strategy as MDI, with several major improvements: HMI offers significantly improved spatial resolution (4 k $\times$ 4 k detectors *vs.* a 1 k $\times$ 1 k detector on MDI), increased image cadence (45 *vs.* 60 seconds on MDI), full Stokes capabilities (MDI measured only the longitudinal component, or Stokes V), and all observables are transmitted to the ground without any further onboard processing, which allows for a much better correction of instrumental effects than would be possible with more limited on-board resources. Another significant change is the use of a different spectral line. While MDI observed the Ni I 6768 Å line, which is also used by the GONG project, for HMI the Fe I line at 6173 Å was selected. The reason for this switch is that the Fe I line is magnetically more sensitive and offers better performance for vector magnetic-field measurements than the Ni I line (Norton et al. 2006). Acoustic modes have their maximum of amplitude at the surface, so they are now also used as solar activity proxies of the subsurface layers. In addition to the 11 year variability of their frequencies, a two year period has been put in evidence and attributed to these layers. The acoustic mode frequency variabilities depend on the degree and represent a valuable information for the configuration of the magnetic field near the surface (Simoniello et al. 2012).

SOHO and SDO are not the only space missions that provide measurements for helioseismic studies of the Sun. Observations obtained with the 50 cm aperture Solar Optical Telescope (SOT, Tsuneta et al. 2008; Suematsu et al. 2008) on Hinode (Kosugi et al. 2007) can be used for that as well (Sekii et al. 2007). Hinode/SOT offers a particular advantage over observations by SOHO/MDI or SDO/HMI. That is its high spatial resolution of 0.2", which is approximately 4 times better than that of HMI on SDO or MDI's high resolution mode. Of course there is a price to be paid for such a high spatial resolution: In particular the field of view of the Broadband FIlter Imager (BFI), which provides high resolution intensity images in the G-band (CH) and Ca II H line (Sekii et al. 2007), is rather small (218" $\times$ 109"). The field of view of the Narrowband Filter Image (NFI) is about 50 % bigger (328" $\times$ 164"). Nevertheless, in areas where high spatial resolution is essential, such as time-distance measurements of active regions or flows in polar regions, the high resolution data from Hinode can provide new and unique information (see e.g. (Mitra-Kraev et al. 2008; Zhao et al. 2010; Nagashima et al. 2011)). Unfortunately, the partial failure of the X-band transmitter of Hinode led to a significant reduction of the available daily telemetry volume. As a consequence the duration of Hinode/SOT time series is limited (typically not more than 12–16 hours).

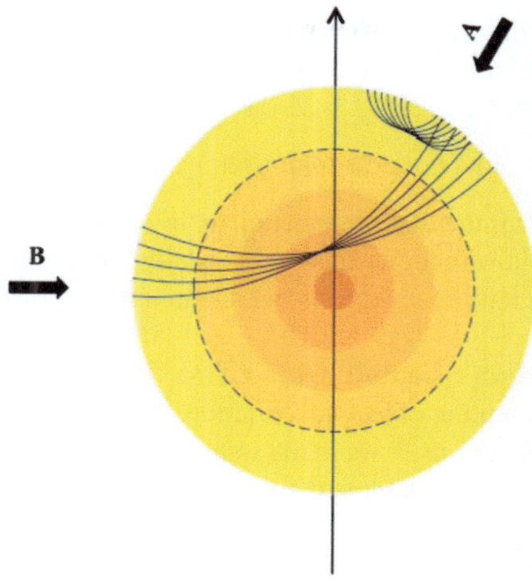

Fig. 1 Sketch illustrating deep-focusing schemes for stereoscopic local helioseismology and local helioseismology of polar regions. An out-of-ecliptic mission provides the vantage point A, from which local helioseismic investigation of the polar region is carried out (indicated by schematic p-mode ray paths). If the vantage point B is provided either by an existing ecliptic mission or a ground-based telescope, stereoscopic helioseismology is done for probing deep interior

Helioseismology naturally requires long and continuous datasets. Issues of telemetry have always been crucial, and they will be more so in future missions, in our attempts to further push the boundary of our knowledge. On the other hand, such datasets are useful not just for helioseismology. Studies of surface fluid motions by correlation tracking, for example, would benefit greatly from these as well.

In the following, we will discuss several concepts for future space missions. The two common themes of most of these are: (i) out-of-ecliptic (polar) measurements and (ii) views from a different angle to perform stereoscopic helioseismic measurements and achieve full sun coverage (Fig. 1). Both themes are also among the scientific objectives of the Polarimetric and Helioseismic Imager (PHI) on Solar Orbiter, which is described elsewhere in this volume (Löptien et al. 2014, this issue). We first discuss out-of-ecliptic missions, which have a common set of primary scientific goals (Solar-D, Solar Polar Imager/POLARIS, Solaris and 4π). Then SAFARI and L5, with stereoscopic helioseismology as one of the objectives but staying within the ecliptic plane, are discussed. Finally, we discuss the GOLD instrument, for probing the deep interior (and atmospheric layers) and ASTROD, a g-mode detection mission based on gravitational wave measurement, before we conclude.

2 Out-of-Ecliptic Missions

2.1 Main Scientific Objectives and Observational Strategies

Main Scientific Questions The main scientific questions that are commonly addressed by out-of-ecliptic missions are as follows:

- How is the global, cyclic, solar magnetic field generated, leading to the quasi-periodic 11/22-yr cycle?
- What is the three-dimensional structure of the solar magnetic field, including the north-south asymmetry, and how does it vary over a solar cycle?
- What is the spatio-temporal structure of the solar internal flows, including those in the polar regions, and how do they interact with magnetism?

These arise from our desire to understand the solar cycle, the greatest of the remaining mysteries around the Sun. There is also an additional question of a great astrophysical importance:

– How does the total solar irradiance and its solar-cycle variation vary with latitude?

Its immediate connection to the solar-cycle puzzle aside, this question can be answered by an out-of-ecliptic mission so easily without much additional cost. These questions lead to the scientific objectives, which are more immediate goals of these missions, discussed below.

Main Scientific Objectives: Observational Study of the Solar Dynamo The primary scientific objective, common to all out-of-ecliptic missions is the helioseismic measurement of various flows in the solar interior. As was discussed in the preceding chapters, these flows are key components in understanding the process of the solar dynamo. We would like to know as much as possible about these flows either as inputs to kinematical dynamo models, or as observational constraints to dynamical dynamo theory.

Another primary scientific objective is the measurement of the surface magnetic field. The flow-field measurements can be cross-correlated with surface magnetic field measurement, to reveal how magnetic flux is transported by flow and by diffusion. In the polar region, how does the polarity reversal take place as interplay between plasma flows and magnetic fields? The predominantly vertical kG-field patches that Hinode has found in the polar regions are large enough to be observed by moderate-sized telescopes considered for these out-of-ecliptic missions from sufficiently high latitudes.

From helioseismic observations, we now have a very good idea about the solar flows, not only the axisymmetric component (rotation) but also the meridional component, in the convection zone in low- to mid-latitudes. Beyond 60° latitude, however, the current measurements become unreliable. Another region of importance that has been elusive is the tachocline region, which is, according to the prevailing idea, playing a crucial role in the solar dynamo.

Let us turn to discussion of observational strategies for such scientific objectives.

Observational Strategy 1: High-Latitude Local Helioseismology The solar rotation is inferred mostly by a global technique, by inverting rotationally induced frequency shifts. The low number of modes that have sensitivity at high latitudes is the reason for our poor knowledge of the high-latitude solar rotation. There is no obvious means to get around this problem and therefore we may have to rely on general horizontal flow measurements using local helioseismology techniques.

Local techniques at high latitudes, however, suffer from the lack of spatial resolution as well as the projection effect, since the solar p-mode velocity field at the solar surface is dominantly radial. For better local helioseismology observations at high latitudes, we need to be able to observe these regions from above i.e. from high latitudes. From 40° latitude, a good coverage of the polar region, with latitude greater than 60°, is provided (Fig. 2). Going up even higher in latitude will permit us increasingly better view of the polar region, by extending the time a spacecraft spends above a certain latitude, and also by providing more longitudinal range, over which one might integrate the quantities of interest, around the pole visible.

Observational Strategy 2: Stereoscopic Helioseismology Imaging the solar tachocline is more challenging, particularly if one's main interest is in the direct magnetic signature of a flux tube stored in the tachocline region and not the proposed fast flow within the tube

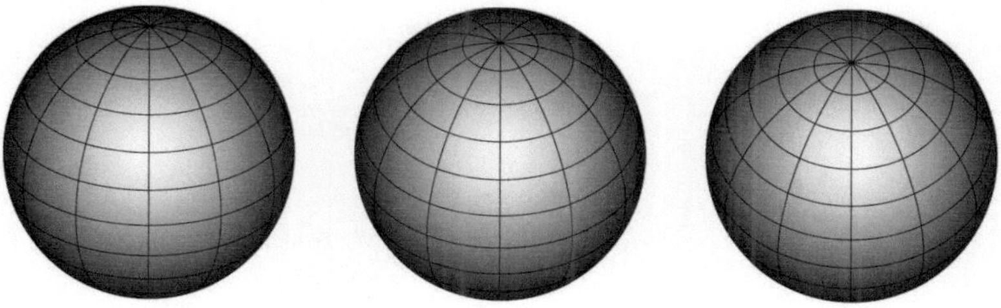

Fig. 2 How the Sun appears with the view angle of 30 degree (*left*), 40 degree (*middle*) or 50 degree (*right*). The *grey scale* is schematically representing the degree of difficulty in obtaining oscillation signals, caused by projection, foreshortening etc.

(Miesch et al. 2007). Most experts agree that the best chance is offered by stereoscopic helioseismology (Gizon et al. 2010). The idea is to use multi vantage points, one given by an out-of-ecliptic spacecraft and others by ground-based telescopes or other satellites (the extreme case is the mission 4π mentioned in the following section), for example to determine travel times along a deeply penetrating, long ray path (Fig. 1). A fully quantitative study of the improvement of having N vantage points distributed in a certain manner is yet to be done, even for the case of $N = 2$. Even if this is done, our knowledge of whatever the feature we are looking for at the tachocline depths is so poor, it is doubtful if one can come up with a quantitative estimate of feasibility. And yet it is clear that this is the only way to improve our chance of detecting and measuring anomalies in the tachocline region.

Additional Scientific Objective: Solar Irradiance Measurement Solar irradiance measurements at various latitudes will, for the first time, enable us to measure the anisotropy of the total solar irradiance. The unexplained low photometric variability of the Sun, another puzzle surrounding the solar activity, may be explained by higher variability at higher latitudes, likely caused by faculae. If it is not the case, then we must conclude that the Sun is a rather atypical star, which will lead to more fundamental questions in astrophysics.

Other Scientific Questions There are other important heliophysics questions that can be addressed from out-of-ecliptic orbits. Even though they may not be part of the main goals, they are often part of why we pursue the main goals. They are therefore added to the list of scientific questions:

- How does magnetic activity shape the structure and evolution of the polar solar corona and how does it affect the Earth?
- How are variations in the solar wind linked to the Sun at all latitudes?
- How are solar energetic particles accelerated and transported in radius and latitude?

2.2 Solar Polar Imager (SPI)/POLARIS

This section on a solar polar imager concept is based on a "white paper" by Liewer et al. (2011) to the 2012 NRC Decadal Survey in Solar and Space Physics (Baker et al. 2012), and a paper by Appourchaux et al. (2009a), who proposed a very similar mission concept under the name POLARIS (POLAR Investigation of the Sun) in response to a call for missions for ESA's Cosmic Vision program. The proposed mission concept uses a solar sail to place a spacecraft in a 0.48-AU circular orbit around the Sun with an inclination of 75°.

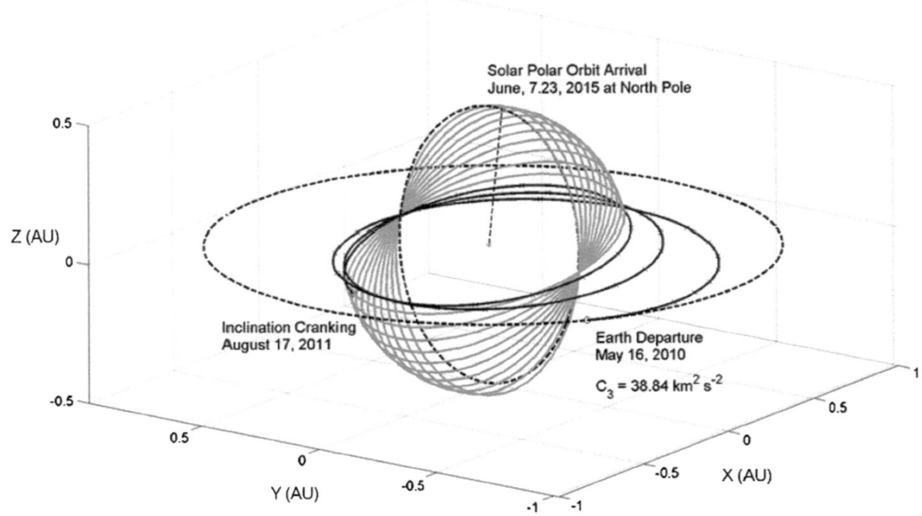

Fig. 3 Orbit of the reference mission in the study of Macdonald et al. (2006)

2.2.1 Mission Profile

To address questions discussed in Section 2.1, SPI/POLARIS will carry a comprehensive payload that includes both remote as well as in situ instruments:

- Dopplergraph and magnetograph imager
- White-light coronagraph
- EUV imager
- UV spectrograph
- Total Solar Irradiance monitor
- Magnetometer
- Solar wind ion composition and electron spectrometer
- Energetic particle package
- Radio and plasma wave package

To be able to measure flows and solar rotation near the poles, the observations need to have a sufficiently high signal to noise (S/N) ratio. To achieve the required S/N ratio, extended periods (many days) of nearly continuous observations of the polar regions are required, which—for an inner heliospheric orbit—translates into requirements for the orbit inclination of more than 65°. The proposed mission profile of SPI/POLARIS foresees the spacecraft to be launched into an elliptical heliocentric orbit. After spiraling down towards the cranking radius, the "cranking phase" utilizing a solar sail to increase the orbit inclination will begin (Fig. 3). It will take just over 5 years to reach an inclination of 90° (Macdonald et al. 2006). For SPI/POLARIS, a radius of 0.48 AU and inclination of 75° was chosen for the final science orbit, because (a) there the radiation pressure is bigger and the "cranking" to higher inclination more efficient, and (b) the orbital period there is exactly 1/3 of Earth's (3:1 resonance). A detailed assessment of a SPI/POLARIS mission using a solar sail of 153 × 153 m square to achieve the chosen orbit is presented in Macdonald et al. (2006). The spacecraft requirements for such an orbit are minimal, because the spacecraft environment is relatively benign compared to, e.g., that of Solar Orbiter. The key technological challenges are related to the solar sail, which requires advanced boom and new thin-film technology. For more details about leading solar sail technologies see Macdonald et al. (2006) and Section 2.4.

2.3 Solar-D

2.3.1 Background

Following the successes of a series of solar observing satellites: Astro-A (*Hinotori*), Solar-A (*Yohkoh*) and Solar-B (*Hinode*), the Japanese solar physics community began discussion of what the next step should be. Two mission plans survived to the stage of serious international discussions. One was the plan that was eventually adopted, and was called Solar-C Plan B until the decision by the community was made. Solar-C (Plan B) aims to answer questions around fundamental plasma processes and phenomena deeply intertwined with these processes, such as chromospheric/coronal heating and local dynamo. A suite of telescopes, covering visual to X-ray, is employed to observe small-scale processes.

The other plan, called Solar-C Plan A, was an out-of-ecliptic mission now tentatively called Solar-D, addressing the issues discussed in Section 2.1. Below we only describe the outline of the mission; more details can be found at http://hinode.nao.ac.jp/SOLAR-C/Documents/Interim2011/SC_plan-a.pdf.

2.3.2 Mission Overview

Aside from the proposed payload, the orbital aspect is of a paramount importance to the Solar-D mission.

Science Strawman Payload The total allocated payload mass is about 130 kg. The suite of instruments of the mission is as follows:

- Helioseismic and Activity Imager
- EUV & X-ray Spectroscopic Imaging Telescopes
- Solar Irradiance Monitor
- Inner Heliospheric Imager (optional)
- In-Situ Instruments (optional)

Orbit Two factors of direct importances in designing the orbit are: (i) inclination to the ecliptic plane, and (ii) distance from the Earth. The inclination obviously is about how high in latitude Solar-D is going to be, or how long it is above a certain latitude. The distance from the Earth is important in guaranteeing a telemetry rate that is compatible with the scientific objectives.

The orbits were required to have the following characteristics in terms of these factors:

- The final 1AU orbit has an inclination larger than 40°
- In the observing phase, the downlink rate is higher than 100 kbps

Ballistic orbits based on swing-bys have some merits and have been discussed, but eventually the options using ion engines for electric propulsion, to increase the orbital inclination, were preferred primarily because of the shorter mission life required.

2.3.3 Programmatic Aspects

Neither Solar-C nor Solar-D has formally been proposed to ISAS/JAXA yet, although the Solar-C Working Group was approved by the agency and it is now preparing a formal proposal. Solar-C will be proposed to be launched in FY2018.

One of the reason that Solar-C Plan A was not chosen as the next Japanese solar mission is in the technical readiness of the μ-20 ion engine, a scaled up version of μ-10 which was used on the asteroid visiting *Hayabusa* (Yoshikawa et al. 2010). For a scientific mission the ion engine needs to be field proven, which is not the case of the μ-20 engine. DESTINY (Demonstration and Experiment of Space Technology for INterplanetary voYage), to be proposed for the ISAS/JAXA small satellite series, is an engineering (as opposed to scientific) mission that can provide such a basis for credibility. If DESTINY can be flown more or less in parallel with Solar-C, then by the time of the Solar-C launch we will be in a position to propose Solar-D for a launch in the latter half of 2020s.

2.4 Solaris

2.4.1 Background and Programmatic Aspects

The *Solaris* mission is a simplified version of the POLARIS mission aiming at very specific scientific objectives. The *Solaris* mission has been proposed to the CNES "Appel à idées" which was issued for their 2014 prospective plan, and to the 2013 L2/L3 call for white papers of ESA. This mission is the result of a study lead in the framework of an ESA-DLR joint working group which aims at defining realistic missions using solar sails. This so-called Solar Sail Mission working group (SSMWG) was established in the Fall of 2011 under the auspices of Malcolm Macdonald. One of the working groups aims at studying a Solar Polar mission based on the use of solar sails. The SSMWG met in November 2011 and June 2012 for discussing the various technology developments required and the possible missions made possible by solar sails.

The current solar sail technology development on which we would like to base a mission is derived from the DLR-ESA Gossamer road map. This is a 3-step approach that will have three missions for enabling the solar sail technology:

- Gossamer-1: launch in 2013, demonstration of the safe deployment of a 5 m $\times$ 5 m solar sail in a 320 km Earth orbit. Documentation of the deployment by at least two on-board cameras
- Gossamer-2: launch in 2014, deployment of a 20 m $\times$ 20 m solar sail in a 500 km Earth orbit. Test of a limited orbit and attitude control. Documentation by at least two on-board wide angle cameras and an additional CubeSat inspector camera. Sail material presumably much thinner than the 7.5 μm Kapton used in Gossamer-1. The scheduled lifetime is about four weeks.
- Gossamer-3: launch in 2015, deployment of a 50 m $\times$ 50 m solar sail in a $>$ 10,000 km Earth orbit. Test of the full orbit and attitude control. Documentation as for Gossamer-2. An acceleration > 0.1 mm s^{-2} together with the sufficiently high initial orbital altitudes will enable the sail craft to leave the Earth gravitational field after 100 days. It is planned to perform a lunar swing-by after about 600 days.

More details about the road map can be found in Geppert et al. (2011). Similar solar-sail developments are under way with the L'Garde's Sunjammer mission due to be launched in the third quarter of 2014. Sunjammer is a mission that will deploy a 35 m $\times$ 35 m solar sail for demonstrating attitude control, executing navigation sequences and flying to L1 and pole sitter positions (www.lgarde.com).

After the successful Gossamer-3 and Sunjammer missions, solar sails will have reached a Technology Readiness Level (TRL) of at least 6; then solar sails will be ready for being baselined for any kind of space mission (scientific or commercial). With this perspective in mind, the SSMWG devised several missions that would use solar sails as an enabling technology for specific mission profiles: *Solaris* is one of them.

2.4.2 Mission Overview

The scientific objectives of *Solaris* have already been discussed in Section 2.1. To address these questions, *Solaris* will carry a minimal payload that includes only remote-sensing instruments:

- Dopplergraph and magnetograph imager
- Total Solar Irradiance monitor

This very minimal payload ensures a very high scientific return for a spacecraft system allowing it to have circular orbit above the solar poles. The so-called Xmas tree syndrome where all the branches of heliophysics are represented has been suppressed. This kind of approach necessarily may not be very wise politically but is technically adapted to the scientific objectives.

The main mission requirements are as follows:

- Data collection for the primary science goals shall begin within five years of launch
- Data collection for the primary science goals shall begin when the payload can observe above ± 60 degrees of solar latitude
- In an orbit passing directly over the solar poles, the pole shall be continuously visible at an observation-zenith angle of less than 30 degrees (TBC) for one full rotation period
- The orbit shall have a very low eccentricity $e < 0.2$ (close to a circular orbit)

These requirements are much more severe than those put on the Solar Orbiter mission which is not suited for long helioseismic observation and solar irradiance measurements mainly due to the lower inclination and to the highly eccentric orbit ($e > 0.8$).

The preliminary mission analysis assumes that the orbit is reached using a two-phase approach: Phase 1 during which a solar sail is deployed and the orbit radius is reduced; Phase 2 during which the orbit is cranked to 90 degrees. Various options have been examined, and Table 1 provides the spacecraft and payload characteristics for one such option (called A^2). With a solar sail of 100 m, a payload mass of 35 kg is enabled within a total launch mass of about 320 kg. Increasing the sail size to 125 m, a payload mass of 60 kg is enabled within a total launch mass of about 500 kg.

2.5 The Mission 4π

The following section is based on the white paper "4PI: A Global Understanding of the Solar Cycle" by Strong et al. (2011) to the 2012 NRC Decadal Survey in Solar and Space Physics (Baker et al. 2012).

4π (Fig. 4) has been proposed as a flagship-class mission (> 1 G$). It would comprise four spacecraft to measure the magnetic field and surface velocity over the entire Sun. Three of the four spacecraft would be on circular orbits in the ecliptic plane within 1 AU, the fourth one would be in a high-inclination orbit to obtain data from the solar polar regions. The data

Table 1 Summary of mission architecture option A^2 mass allowance estimates, including ROM payload allowance estimate

Sail side length (in m)	50	75	100	125
S/C mass range with 7.5 μm film (in kg)	120–107	202–172	306–255	435–360
Platform mass range with 7.5 μm film (in kg)	negative	< 13	23–73	78–154
Payload mass range with 7.5 μm film (in kg)	negative	< 5	5–15	15–30
S/C mass range with 2.5 μm film (in kg)	95–83	146–116	207–156	281–205
Platform mass range with 2.5 μm film (in kg)	negative	39–69	122–172	232–308
Payload mass range with 2.5 μm film (in kg)	negative	10–15	25–35	45–60

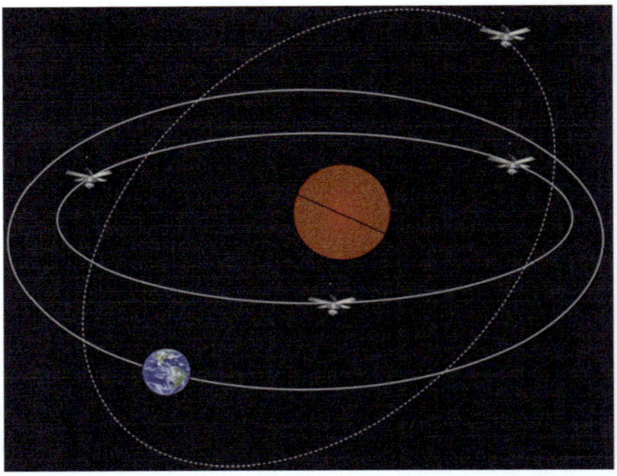

Fig. 4 Illustration of a possible spacecraft configuration for the 4PI mission from Strong et al. (2011)

would be combined to produce full-Sun magnetic field maps and maps of the subsurface dynamics to address the scientific questions discussed in Section 2.1.

A minimum payload complement would include a Doppler/Magnetograph and a coronal imager similar to those on SDO. An enhanced payload would include coronagraphs, irradiance monitors and particle and field detectors.

3 Ecliptic Missions

Here we review two missions that are ecliptic but still stereoscopic.

3.1 SAFARI

This section on the Solar Activity Far Side Investigation (SAFARI) is based on a "white paper" by Velli et al. (2011) to the 2012 NRC Decadal Survey in Solar and Space Physics (Baker et al. 2012). SAFARI (Fig. 5) is a small-mission concept to explore the origins of solar magnetic activity by carrying out observations of the velocity and magnetic fields at the solar surface from a vantage point widely separated from Earth in longitude.

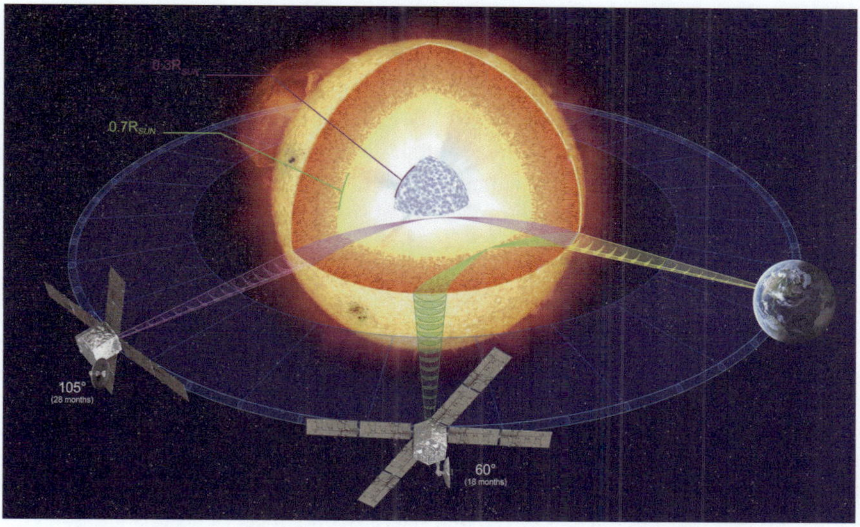

Fig. 5 SAFARI mission concept (courtesy of M. Velli)

3.1.1 Scientific Objectives

The SAFARI science goals are described in Velli et al. (2011). They are similar to the goals of SDO/HMI: To quantify where solar variability originates, how it emerges to the solar surface and develops there. Velli et al. (2011) lists the following top level scientific objectives:

- What are the sources of solar magnetic activity clusters and preferred longitudes?
- What are the structure of flows and solar rotation in the solar interior?
- How do active regions grow and evolve on the Sun and what determines their persistence?

These objectives will be achieved by detecting mass flows and thermal inhomogeneities deep inside the Sun and by measuring the evolution of active regions over time scales of hours to months.

The SAFARI mission concept involves sending a spacecraft into a trailing orbit around the Sun at 1AU, similar to the orbit of the STEREO Behind spacecraft. A drift rate of 45°–60° from Earth per year was suggested. This would allow accomplishing the major science objectives within 2.5 years, with the possibility of an extended mission reaching Sun-Earth opposition in 4 years. The minimum, bare bones SAFARI spacecraft would carry only one instrument, a compact Doppler magnetograph. A simple realization sufficient to accomplish the scientific objectives is a 10-cm aperture telescope (even smaller than MDI) with a dual focal plane magneto-optical filter assembly allowing full Sun images for right and left hand circular polarizations in two wings of the K I D_1 line at 770 nm. Images could be recorded with a 2048 × 2048 pixels detector and, depending on the available telemetry rate, transmitted as binned images (e.g. 512 × 512) at 60 s cadence for the Dopplergrams, and full resolution images with a cadence of 10 min for the magnetograms.

SAFARI consists of two elements: a spacecraft in an orbit trailing the Earth at 1 AU (SAFARI-S), and an Earth-side imaging component (SAFARI-E), that could be either ground based, e.g. operating at the GONG sites, or—if still in orbit—SDO or a successor mission.

3.2 L5 Mission

This section on a mission to the Sun-Earth L5 Lagrangian point is based on a "white paper" by Vourlidas et al. (2011) to the 2012 NRC Decadal Survey in Solar and Space Physics (Baker et al. 2012). The key science driver for an L5 mission is to improve our capabilities for space weather research. The L5 Lagrangian point, located 60° east of the Sun-Earth line, is considered an ideal location for heliophysics and space weather research, as it allows the direct viewing of structures propagating towards Earth, in contrast to the L1 location, and it permits imaging of solar activity and the in-situ sampling of solar wind structures such as corotating interaction regions (CIRs) four to five days before they arrive at Earth. Helioseismology would be only a side aspect of such an L5 mission, the proposed strawman payload which includes an EUV imager, a coronagraph, a heliospheric imager, an off-limb spectrograph, a hard X-ray imager, instruments to measure solar wind ions, electrons and energetic particles, and a magnetograph/Dopplergraph. As with SAFARI, the novel helioseismic measurements would come from the stereoscopic view in combination with helioseismology assets on ground or near Earth. This will allow consideration of new acoustic ray paths that probe the deeper layers of the Sun (stereoscopic local helioseismology). Global helioseismology will benefit from the increased coverage as well, as the reduction of spatial leaks will make it easier to disentangle the various modes.

4 G-Mode Detection

4.1 GOLD

The following subsection differs from the previous sections as it does not describe a new mission concept but an improved instrument concept. We decided to include it because it is currently the only instrument concept that aims at continuing and improving global oscillation measurements of the Sun. GOLD (Global Oscillations of Low-Degree modes; Turck-Chièze et al. 2006; Turck-Chièze et al. 2008 is a compact space version of GOLF-NG (GOLF New Generation), which is an improved version of the GOLF instrument onboard SOHO. GOLF has been designed to detect solar gravity modes that are sensitive to the deep core. While the analysis of ten years of GOLF data has revealed a tantalizing periodic structure in agreement with the period separation predicted by theory for gravity dipole modes (Garcia et al. 2007), "there is currently no undisputed detection of solar g-modes" (Appourchaux et al. 2010). A GOLD-type instrument may finally provide conclusive evidence of the elusive g-modes and may help in answering questions about the dynamics of the deep solar interior such as: Why is the gradient of the solar core rotation so small? Is the axis of the core rotation the same as that of the rest of the Sun? Could we possibly see a direct effect of a fossil magnetic field? Is there any latitudinal dependence of the rotation in the radiative zone?

A prototype instrument of GOLF-NG has been built by the Saclay group. In 2008 and 2010, in collaboration with the IAC, the instrument was deployed at Izana, Tenerife, to take solar measurements. The results from these early campaigns showed that the electronic noise of the detector was too high and a new, CCD-based detector system has been developed.

The key design feature of GOLF-NG (and GOLD) in comparison to the GOLF or BISON instruments, which were designed to measure at a single wavelength position in the red and blue wing in the Na D1 and K line, respectively, is the use of a multi-channel resonant spectrometer, making measurements at 15 positions in the Na D1 line, from the outer wings

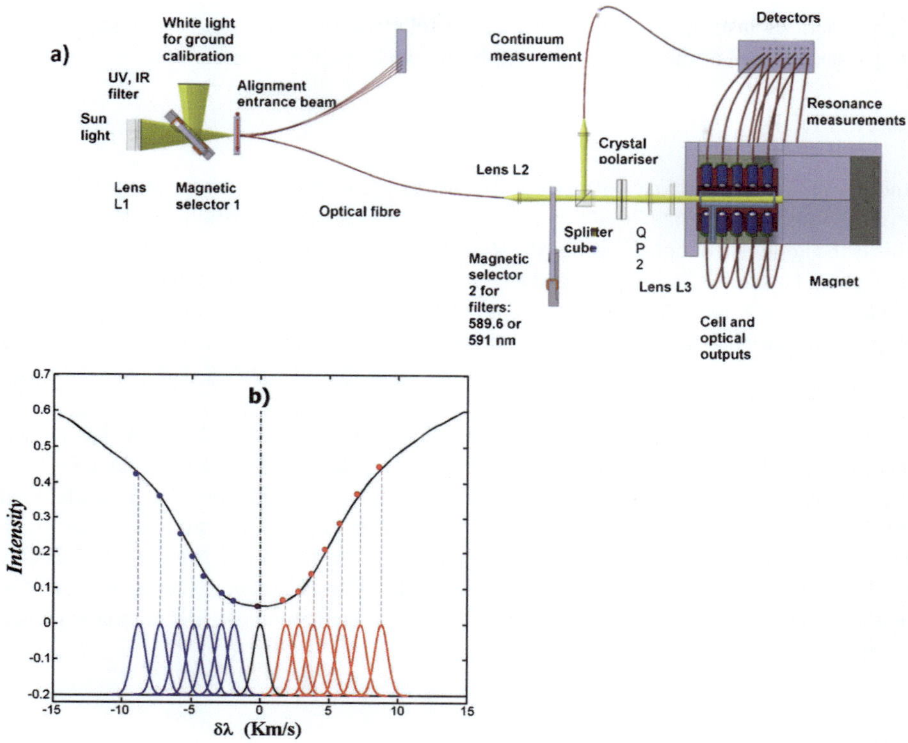

Fig. 6 Schematic view and measurement principle of the GOLF-NG instrument

down to the core of the line (see Fig. 6). The 8 different positions along the cell translate to 8 positions in the solar atmosphere between 200 km and 800 km above the photosphere at which velocity signals can be extracted. The hope is to increase the S/N in the g-mode regime not only by improving the photon statistics, but also because the convective signals (e.g. granulation) change significantly from the low photosphere to the chromosphere which should make it possible to identify and remove the convective noise.

GOLF-NG uses a 31-channel resonant spectrometer, where a small portion of the absorbed solar light is re-emitted after entering the sodium vapour cell and is split into its Zeeman components by means of a static longitudinal magnetic field whose strength varies along the longitudinal axis (see Fig. 6). The cell is placed in a varying permanent magnet between 0 to 12 kG, and the scattered light, resulting from, alternatively, left and right circular polarized incident light, is extracted at 8 different positions along the cell. Four outputs per position, connected to optical fibers by some circular lens, are placed around the cell, thus increasing the total photon counting rate compared to the GOLF instrument by a factor 10, in a total of 31 outputs (Turck-Chièze et al. 2006). A CCD-based detector system has also been developed and realized. It has successfully reached the required noise performances.

A formation flying mission at the L1 point has been proposed to fly GOLD together with many other instruments for space weather and space climate study (Turck-Chièze et al. 2009). The advantages of this concept are the high mass capability of the payload and the high thermal stability. The first platform could be composed of different instruments for deep helioseismology, local helioseismology, and irradiance measurement from middle ultraviolet to infrared. The second platform, partly occulted by the first one, could receive complementary instruments analyzing external solar phenomena and a coronograph with a unique observation of the transition region between photosphere and corona.

4.2 ASTROD

Another proposed technique for detecting g modes is to directly measure the gravitational perturbation they generate, which will be measured as spacetime strain (or deformation). This is the approach adopted for the ASTROD (Astrodynamical Space Test of Relativity using Optical Devices) mission proposed to ESA's Cosmic Vision (Appourchaux et al. 2009b). The strain sensitivity of ASTROD would allow the detection of solar g modes with the lowest predicted amplitudes (Appourchaux and Pallé 2013), with a sensitivity at least 2 orders of magnitude better (Appourchaux et al. 2009b) than that of LISA (Laser Interferometer Space Antenna, Danzmann 2003).

At the time of writing, LISA has not been selected as a mission by ESA, but its science theme ("The gravitational Universe") has been selected as the theme for ESA's L3 mission (see the ESA senior survey committee report issued in October 2013, available at http://sci.esa.int/ssc_report at the time of writing). LISA Pathfinder, a mission for testing in flight the concept of low-frequency gravitational wave detection (McNamara et al. 2008), will be launched in September 2015. As for the ASTROD mission, no space agency has it in their long-term plans. Consequently, without a programmatic window, it is not known when it may fly.

5 Summary

Future space-mission concepts currently discussed in the helioseismology community have been reviewed. Reflecting where the currently unexplored regions are, one popular idea is to observe the Sun from high latitudes (SPI/POLARIS, Solar-D, Solaris and 4π) to explore the polar regions as well as to probe the deep interior using stereoscopic techniques, by combining observations from high latitudes with observations from within the ecliptic plane (or even more vantage points, in the case of 4π). The stereoscopic techniques do not necessarily require an out-of-ecliptic orbit, and some have chosen to stay within the ecliptic plane (SAFARI and L5). For studying the solar core region, detecting g modes and measuring their frequencies precisely would enable simpler approaches than stereoscopic techniques, and a new instrument (GOLD) and a mission (ASTROD) aiming for these are also discussed.

Most of the mission ideas described above are still rather sketchy. We are not aware of any detailed quantitative studies of how much will be gained from a multi-point view of the Sun (for instance from having another Dopplergraph located at L5 with a 60° offset from the Sun-Earth line). We recommend that the helioseismology community initiates such studies in the near future, to be prepared for forthcoming mission opportunities.

References

T. Appourchaux, P.-L. Pallé, in *Fifty Years of Seismology of the Sun and Stars*, ed. by K. Jain, S.C. Tripathy, F. Hill, J.W. Leibacher, A.A. Pevtsov (2013), p. 125 (ASP San Francisco)

T. Appourchaux, P. Liewer, M. Watt, D. Alexander, V. Andretta et al., Exp. Astron. **23**, 1079 (2009a)

T. Appourchaux, R. Burston, Y. Chen, M. Cruise et al., Exp. Astron. **23**, 491 (2009b)

T. Appourchaux et al., Astron. Astrophys. Rev. **18**, 197 (2010)

D. Baker, T. Zurbuchen et al., *Solar and Space Physics: A Science for a Technological Society*. ISBN:978-0-309-16428-3 (2012). Committee on a Decadal Strategy for Solar and Space Physics (Heliophysics), Space Studies Board Aeronautics and Space Engineering Board, Division on Engineering and Physical Sciences, National Research Council
A. Claverie, G.R. Isaak, C.P. McLeod, H.B. van der Raay, Nature **282**, 591 (1979)
S. Couvidat, R.A. García, S. Turck-Chièze et al., Astrophys. J. **597**, L77 (2003)
K. Danzmann (LISA Science Team), Adv. Space Res. **32**, 1233 (2003)
V. Domingo, D. Wyn-Roberts, DISCO—Results of the Phase a Study, in *SCI(82)3* (ESA, Noordwijk, 1982)
V. Domingo, D. Wyn-Roberts, Mem. Soc. Astron. Ital. **55**, 375 (1984)
V. Domingo, B. Fleck, A. Poland, Sol. Phys. **162**, 1 (1995)
T.L. Duvall Jr., A.G. Kosovichev, P.H. Scherrer et al., Sol. Phys. **170**, 63 (1997)
B. Fleck, V. Domingo, A. Poland, Solar Phys. 162(1–2) (1995)
C. Fröhlich, J. Romero, H. Roth, C. Wehrli, B.N. Andersen, T. Appourchaux, V. Domingo et al., Sol. Phys. **162**, 101 (1995)
A.H. Gabriel, G. Grec, J. Charra, J.-M. Robillot, T. Roca Cortés, S. Turck-Chièze et al., Sol. Phys. **162**, 61 (1995)
R.A. Garcia, S. Turck-Chièze, P. Boumier, J.M. Robillot et al., Astron. Astrophys. **442**, 385 (2005)
R.A. Garcia, S. Turck-Chièze, S.J. Jiménez-Reyes, J. Ballot et al., Science **316**, 1591 (2007)
U. Geppert, B. Biering, F. Lura, J. Block, M. Straubel, R. Reinhard, Adv. Space Res. **48**, 1695 (2011)
P.M. Giles, T.L. Duvall Jr., P.H. Scherrer, Nature **390**, 52 (1997)
L. Gizon, A.C. Birch, H.C. Spruit, Annu. Rev. Astron. Astrophys. **48**, 289 (2010)
G. Grec, E. Fossat, M. Pomerantz, Nature **288**, 541 (1980)
D.A. Haber, B.W. Hindman, J. Toomre et al., Astrophys. J. **570**, 855 (2002)
D. Hathaway, L. Rightmire, Science **327**, 1350 (2010)
R. Howe, J. Christensen-Dalsgaard, F. Hill et al., Astrophys. J. **533**, L163 (2000)
M.C.E. Huber, R.M. Bonnet, D.C. Dale, M. Arduini, C. Fröhlich, V. Domingo, G. Whitcomb, ESA Bull. **86**, 25 (1996)
S. Ilonidis, J. Zhao, A.G. Kosovichev, Science **333**, 993 (2011)
A.G. Kosovichev, J. Schou, Astrophys. J. **482**, L207 (1997)
T. Kosugi, K. Matsuzaki, T. Sakao, T. Shimizu, Y. Sone et al., Sol. Phys. **243**, 3 (2007)
P.C. Liewer, D. Alexander, T. Appourchaux, J. Ayon, L. Floyd et al., *A Solar Polar Imager Concept: Observing Solar Activity from a New Perspective* (2011). White Paper to the 2012 NRC Heliophysics Decadal
C. Lindsey, D.C. Braun, Science **287**, 1799 (2000)
B. Löptien, A.C. Birch, L. Gizon, et al. (2014, this issue). doi:10.1007/s11214-014-0065-3
M. Macdonald, G.W. Hughes, C.R. McInnes, A. Lyngvi, P. Falker, A. Atzei, J. Spacecr. Rockets **43**, 960 (2006)
P. McNamara, S. Vitale, K. Danzmann (LISA Pathfinder Science Working Team), Class. Quantum Gravity **25**, 114034 (2008)
M.S. Miesch, P.A. Gilman, M. Dikpati, Astrophys. J. Suppl. Ser. **168**, 337 (2007)
U. Mitra-Kraev, A.G. Kosovichev, T. Sekii, Astron. Astrophys. **481**, L1 (2008)
K. Nagashima, J. Zhao, A.G. Kosovichev, T. Sekii, Astrophys. J. **726**, L17 (2011)
A.A. Norton, J. Pietarila Graham, R.K. Ulrich et al., Sol. Phys. **239**, 69 (2006)
W.D. Pesnell, B.J. Thompson, P.C. Chamberlin, Sol. Phys. **275**, 3 (2012)
P.H. Scherrer, R.S. Bogart, R.I. Bush, J.T. Hoeksema, A.G. Kosovichev, J. Schou et al., Sol. Phys. **162**, 129 (1995)
P.H. Scherrer, J. Schou, R.I. Bush, A.G. Kosovichev, R.S. Bogart et al., Sol. Phys. **275**, 207 (2012)
J. Schou, H.M. Antia, S. Basu et al., Astrophys. J. **505**, 390 (1998)
J. Schou, P.H. Scherrer, R.I. Bush, R. Wachter, S. Couvidat et al., Sol. Phys. **275**, 229 (2012)
T. Sekii, A.G. Kosovichev, J. Zhao, S. Tsuneta et al., Publ. Astron. Soc. Jpn. **59**, S637 (2007)
R. Simoniello, W. Finsterle, D. Salabert, R.A. Garcia et al., Astron. Astrophys. **539**, 135 (2012)
K. Strong, J. Saba, D. Pesnell, J. Luhman, F. Hill, T. Duvall, *4PI: A Global Understanding of the Solar Cycle* (2011). White Paper to the 2012 NRC Heliophysics Decadal
Y. Suematsu, S. Tsuneta, K. Ichimoto, T. Shimizu, M. Otsubo et al., Sol. Phys. **249**, 167 (2008)
S. Tsuneta, K. Ichimoto, Y. Katsukawa, S. Nagata, M. Otsubo et al., Sol. Phys. **249**, 197 (2008)
S. Turck-Chièze, S. Couvidat, A.G. Kosovichev et al., Astrophys. J. **555**, L69 (2001)
S. Turck-Chièze et al., Adv. Space Res. **38**, 1812 (2006)
S. Turck-Chièze et al., Astron. Nachr. **329**, 521 (2008)
S. Turck-Chièze et al., Exp. Astron. **23**, 1017 (2009)
S. Turck-Chièze, S. Couvidat, Rep. Prog. Phys. **74**, 086901 (2011)
F. Vandenbussche, ESA Bull. **97**, 37 (1999)

M. Velli, A. Ruzmaikin, N. Murphy, V. Angelopoulos, P.C. Liewer et al., *SAFARI: Solar Activity Far Side Investigation, White Paper to the 2012 NRC Heliophysics Decadal* (2011)
S.V. Vorontsov, J. Christensen-Dalsgaard, J. Schou et al., Science **296**, 101 (2002)
A. Vourlidas, R. Howard, Y.-K. Ko, D. Biesecker, S. Krucker et al., *Mission to the Sun-Earth L5 Lagrangian Point: An Optimal Platform for Heliophysics & Space Weather Research* (2011). White Paper to the 2012 NRC Heliophysics Decadal
M. Yoshikawa, J. Kawaguchi, H. Yano, in *Lunar and Planetary Institute Science Conference Abstracts* vol. 41, p. 2746 (2010)
J. Zhao, A.G. Kosovichev, T.L. Duvall Jr., Astrophys. J. **557**, 384 (2001)
J. Zhao, A.G. Kosovichev, T. Sekii, Astrophys. J. **708**, 304 (2010)

Space Sci Rev (2015) 196:303–356
DOI 10.1007/s11214-014-0117-8

The Solar-Stellar Connection

A.S. Brun · R.A. García · G. Houdek · D. Nandy · M. Pinsonneault

Received: 16 June 2014 / Accepted: 29 October 2014 / Published online: 26 November 2014

Abstract We discuss how recent advances in observations, theory and numerical simulations have allowed the stellar community to progress in its understanding of stellar convection, rotation and magnetism and to assess the degree to which the Sun and other stars share similar dynamical properties. Ensemble asteroseismology has become a reality with the advent of large time domain studies, especially from space missions. This new capability has provided improved constraints on stellar rotation and activity, over and above that obtained via traditional techniques such as spectropolarimetry or CaII H&K observations. New data and surveys covering large mass and age ranges have provided a wide parameter space to confront theories of stellar magnetism. These new empirical databases are complemented by theoretical advances and improved multi-D simulations of stellar dynamos. We trace these pathways through which a lucid and more detailed picture of magnetohydrodynamics of solar-like stars is beginning to emerge and discuss future prospects.

Updated version.

A.S. Brun (✉) · R.A. García
AIM, CEA/CNRS/University of Paris 7, CEA-Saclay, 91191 Gif-sur-Yvette, France
e-mail: sacha.brun@cea.fr

R.A. García
e-mail: rafael.garcia@cea.fr

G. Houdek
Stellar Astrophysics Centre (SAC), Department of Physics and Astronomy, Aarhus University, Ny Munkegade 120, 8000 Aarhus C, Denmark
e-mail: hg@phys.au.dk

D. Nandy
Center of Excellence in Space Sciences India (CESSI) and Department of Physical Sciences, Indian Institute for Science Education and Research Kolkata, Mohanpur 741252, West Bengal, India
e-mail: dnandi@iiserkol.ac.in

M. Pinsonneault
Department of Astronomy, The Ohio State University, 140 West 18th Avenue, Columbus, OH 43210, USA
e-mail: pinsonneault.1@osu.edu

Keywords Sun · Stars · Asteroseismology · Gyrochronology · Structure · Rotation · Magnetism · Dynamo

1 Solar Dynamics and Magnetism in a Stellar Context

Time-dependent stellar magnetic fields, such as that of the Sun, are generated in their interior through a magnetohydrodynamic (MHD) dynamo mechanism that involves interactions between plasma flows and magnetic fields (e.g. Moffatt 1978; Charbonneau 2010). The generated magnetic fields are manifested at the surface through starspots or polar caps, and contribute to the dynamic activity of the star. This activity governs the structure of stellar coronae, the stellar radiative flux output, and transient but energetic events such as flares (Güdel 2004; Schrijver et al. 2012). Stellar activity can also have a strong impact on the properties of stellar astrospheres which may host planets (Wood 2004). Magnetized solar-like winds extract angular momentum which drives angular momentum in low mass stars. Understanding the origin of stellar magnetism is therefore of crucial importance, and remains an outstanding problem in astrophysics.

The Sun, our nearest star, provides easy access to spatially resolved observations of magnetic activity and thus sets constraints on the dynamo origin of stellar magnetic fields. However, significant uncertainties in our current understanding of the solar dynamo mechanism motivates us to confront our theoretical ideas of magnetic field generation with the wider space parameters available from stellar activity observations (Noyes et al. 1984a; Baliunas et al. 1995; Wright et al. 2004; Giampapa 2005; Böhm-Vitense 2007; Oláh et al. 2009; Donati and Landstreet 2009; Reiners 2012, and references therein). While, in principle, different stars with different properties may host distinct classes of MHD dynamos, a good starting assumption is that a reasonable model of solar and stellar magnetism should be able to reproduce the range of stellar activity taking into account appropriate changes in the internal driving mechanisms. For example, high mass stars have shallow surface convective zones with high turbulent fluxes relative to low mass stars. Stars with different rotation rates and ages will have different magneto-convective properties (usually characterized by varying non-dimensional parameters such as the Reynolds or Rossby numbers). To reduce this problem to a practical exercise, it is also useful to deal with solar-like stars (narrowing the range of spectral type considered), within which, there is substantial variation in the nature of magnetic output (Petit et al. 2008; Marsden et al. 2013). The observation and characterization of stellar convection, rotation and magnetism has undergone a recent boom with the outcome of better ground-based instruments and space-born missions. For instance, until recently chromospheric proxies such as emission in the core of CaII H & K lines and white light modulation were the most commonly used techniques for characterizing stellar magnetism (Wright et al. 2004; Giampapa 2005). The development of very sensitive spectropolarimeter such as ESPaDOnS or NARVAL has helped to further improve our understanding of stellar magnetism by providing detailed magnetic maps and sometimes the surface differential rotation of the star (Donati et al. 1997, 2006; Jardine et al. 1999). With the success of planet transit/asteroseismology missions such as Corot, *Kepler*, the picture can now be refined even further (Garcia et al. 2014). We have now entered an era in which systematic coupled analyses of chromospheric activity proxies, spectropolarimetry maps and asteroseismic data is being performed. This allows us to better characterize magnetism of solar-like stars and its variability on short and relatively long terms (Berdyugina 2005; Petit et al. 2008; Morgenthaler et al. 2011a, 2011b; Morin et al. 2013; Mathur et al. 2013a). The purpose of this review paper is to recap recent progress made, with the aid of asteroseismology, theory and numerical simulations, in our understanding of stellar dynamics and magnetism in solar-like stars. Other proxies, such as the time evolution of surface lithium abundances, can also be useful to constrain stellar dynamics but they are outside the scope of this review (for discussions see Brun et al. 1999; Talon and Charbonnel 2004; Sbordone et al. 2010; Do Nascimento et al. 2010; Spite et al. 2012, and references therein). We begin by discussing recent observational findings with a special emphasis on ensemble asteroseismology and on stellar dynamics, magnetism and cyclic activity. We then present the current theories of stellar rotational history based on 1-D models and of stellar magnetism with an introduction to dynamo theory based on 2-D kinematic dynamo models and 3-D nonlinear turbulent numerical simulations of stellar dynamo. We continue with a discussion on how theoretical models hold up against observations such as scaling laws and rotation-activity relationship (Pizzolato et al. 2003; Wright et al. 2011) and provide a critical assessment of the current understanding. Finally we conclude by proposing possible future developments in asteroseismology and stellar dynamics to better constrain solar-like star rotation and activity.

1.1 New Asteroseismic Diagnostics of Global Stellar Convection, Rotation and Magnetism

Oscillations are intrinsically damped and stochastically excited by the turbulent motion in the outer convective layers of cool stars (e.g. Goldreich and Keeley 1977; Balmforth 1992; Houdek et al. 1999; Chaplin et al. 2005; Belkacem et al. 2008; Houdek 2010). We refer to stars exhibiting such an oscillation pattern as solar-type stars hereafter. These oscillations are also present with higher amplitudes in red-giant stars (e.g. Houdek and Gough 2002; de Ridder et al. 2009; Bedding et al. 2010; Mosser et al. 2010) (and even in mores massive B-type stars as recently shown by Belkacem et al. 2009). The frequency of maximum power scales as $(GM/R^2)T_{\mathrm{eff}}^{-1/2}$ (e.g., Kjeldsen and Bedding 1995); given the wide dynamic range in stellar surface gravities the relevant periods span the range of minutes to weeks.

The magnetic activity cycles in the Sun and other solar-like stars are often the consequence of dynamo processes induced by the interaction of three fundamental mechanisms: convection, rotation, and magnetic fields (see detailed discussion in Sect. 2). Asteroseismic observations place powerful observational constraints on each of these three processes as we will see in Sects. 1.4, 1.5, and 1.6. For example, activity-cycle induced variations of pulsation properties (mode linewidth and amplitudes) in the Sun and in solar-like stars were discussed, for example, by Houdek et al. (2001) and Chaplin et al. (2007).

The long and uninterrupted high-precision photometric observations required for asteroseismology are ideal for characterizing long-term variations such as the ones induced by starspots crossing the visible disk of stars. Indeed, the dimming produced by these spots leads to a measurable modulation in the photometry with a period corresponding to the surface rotation at the active latitudes (e.g. Basri et al. 2011; García et al. 2012, 2014; Lanza 2013; Walkowicz and Basri 2013; Nielsen et al. 2013; Reinhold et al. 2013; McQuillan et al. 2014) in which the spots develop. One must keep in mind that on the Sun faculae and the small-scale field around over-compensate this dimming effect, leading to a net brightening at solar maximum (Spruit 2000; Fröhlich 2012). Moreover their relative amplitude depends on how close to the limb the starspot is and to the stellar inclination angle with respect to the line of sight. Mapping the relative ratio of dimming through star spot vs. brightening through faculae and small-scale field changes as a function of stellar spectral type, age, rotation rate and metallicity is an emerging field, and as of this paper the overall pattern is not firmly established. Some studies have started to address the physical origins of stellar variability (e.g. Shapiro et al. 2014) and found that a transition from net brightening to net dimming should occur around an index of chromospheric activity $\log(R'_{\rm HK})$ of -4.9, a value which is consistent with observations (Lockwood et al. 2007; Hall et al. 2009).

Ground based stellar variability studies have a long history, and they have traditionally focused on young and active stars. The Convection Rotation and planetary Transits (CoRoT) and *Kepler* space missions have contributed continuous and precise data series for large samples, which has had a dramatic impact on the quantity and quality of stellar rotation data. CoRoT and *Kepler* have extended our understanding to large numbers of older and less active field stars (e.g. Mosser et al. 2009; Mathur et al. 2010b, 2011b; Campante et al. 2011). Asteroseismology has also extended our understanding of longer term cycles. In the solar case the number of spots in the Sun changes during the solar magnetic cycle (Hathaway 2010), introducing a modulation of the variance in the photometric observations that can be used as a proxy of the magnetic activity cycle in the Sun (Mathur et al. 2009) and in other solar-like stars (e.g. García et al. 2010; Chaplin et al. 2011a). As the temperature rises from the photosphere to the chromosphere, the cores of very strong lines, such as Ca II H (396.8 nm) and K (393.4 nm) emission lines (Leighton 1959; Wilson 1978; Baliunas et al. 1995), develop emission features whose strength depends on the magnitude of the temperature inversion. However, seismic techniques can go further by probing structural and dynamical changes beneath the photosphere as the magnetic cycles develop. Indeed, since the dawn of helioseismology, several authors noticed that the frequency of the acoustic modes increase near the solar maximum (Woodard and Noyes 1985; Fossat et al. 1987; Pallé et al. 1989; Libbrecht and Woodard 1990; Anguera Gubau et al. 1992). Nowadays, it is possible to measure variations in all the p-mode properties not only in the Sun (e.g. Chaplin et al. 2000; Jiménez-Reyes et al. 2007), but also in other stars such as the F star HD49933 measured by CoRoT, in which the amplitude as well as the frequency shifts of low-degree modes change with time (García et al. 2010).

Asteroseismic measurements can also be used to extract information about convection. The p modes are observed as a power excess at distinct frequencies on top of a continuous Fourier spectrum. This background spectrum is likely due to the presence of convective motions at all scales with granulation playing a key role (e.g. Harvey 1985; Lefebvre et al. 2008), but may also be due to the presence of faculae (e.g. Karoff et al. 2013). Until now, neither classical observations nor one-dimensional analytical convection models could put constraints on granulation. The situation has changed with asteroseismic observations of thousands of stars thanks to the CoRoT (Baglin et al. 2006) and *Kepler* (Borucki et al. 2010) missions. For instance, Mathur et al. (2011a) have determined the timescale and power of granulation for around 1000 *Kepler* red giants at different stages of their evolution going from the red giant branch up to the clump. Moreover, the depth $d_{\rm bcz}$ of the convective zone—which is often the location used to compute a characteristic convection overturning time e.g. $\tau_{\rm conv} \sim d_{\rm bcz}/v_{\rm conv}$, with $v_{\rm conv}$ being a characteristic convective velocity—is a region characterized by a sharp change in stratification, since it changes from (super)-adiabatic to subadiabatic, modulo the extent of convective penetration (e.g., Zahn 1991). Therefore, the sound speed undergoes an abrupt variation producing an acoustic glitch that leaves a fingerprint in the frequencies of the acoustic modes (e.g., Gough and Thompson 1988; Vorontsov 1988; Gough 1990) see also Monteiro et al. (1994), Basu and Antia (1995), Houdek and Gough (2007), Christensen-Dalsgaard et al. (2011), Mazumdar et al. (2014).

1.2 The Link Between Helio- and Asteroseismology: Global Oscillations and 1-D Models

Helioseismology has been providing a wealth of high-quality information about the structure and physics of the solar interior since the first detection of the global five-minute oscillation modes (Claverie et al. 1979). The large number of observable global solar oscillation modes, with spherical degrees up to several thousands, has allowed helioseismologists to infer with high accuracy from the measured oscillation frequency the solar sound speed and density as a function of solar radius. Such helioseismic inversion techniques make use of the fact that the displacement eigenfunctions are related to the adiabatic eigenfrequencies through a variational principle (Chandrasekhar 1963; Lynden-Bell and Ostriker 1967) leading to a linear relation between a small perturbation in the density (and squared sound speed) of a solar model and the corresponding change in the frequencies (e.g. Gough 1978, 1984b). Before the results of the Sudbury Neutrino Observatory (SNO) experiment (see further down), the solar neutrino problem, together with the high-quality helioseismic data collected from space, indicated clear problems in our understanding of either basic physics or the structure of the solar interior. In response, there was a major sustained effort to improve solar models; for a review see e.g. Christensen-Dalsgaard (2004) and references therein. The rather good agreement (<1 %) between the Sun and 1-D solar models computed in the late 1990s (Christensen-Dalsgaard et al. 1996; Brun et al. 1998, 1999) was, however, degraded if the solar photospheric abundances reported by Asplund et al. (2004) were adopted. The problem still holds even after the revision to a slightly higher value of metalicity by Asplund et al. (2009). We refer to Christensen-Dalsgaard and Houdek (2010), Villante et al. (2014) and Chap. 2 (e.g., Basu et al. 2014) of the ISSI book (Thompson et al. 2015) for a recent discussion on helioseismology constraints on metallicity/abundances.

Still, the high-quality helioseismic data has led to many fundamental achievements in solar physics such as inverting the internal solar angular-velocity profile or helping to disentangle the long-standing solar neutrino problem. Seismic inversions have revealed that the non-uniform, surface differential rotation extends all the way to the base of the convective envelope and becomes almost uniform (e.g. solid body rotation) in the radiative interior (e.g. Brown and Morrow 1987; Thompson 1990; Thompson et al. 2003; García et al. 2004; Mathur et al. 2008; Eff-Darwich and Korzennik 2012). The inferred radial interior density and sound-speed profiles also place strong constraints on the standard solar model and consequently on the theory of stellar structure and evolution. Since the neutrino fluxes predicted by solar models could not be explained by standard zero mass neutrino physics, particle physicists had to consider massive neutrinos and their associated flavor oscillations (e.g. Turck-Chièze et al. 2001). Thanks to the SNO experiment measurements (Ahmad et al. 2001) the scientific community has since confirmed, by measuring both charged and neutral currents, that the original deficit of electronic neutrinos is due to their transition to another neutrino flavor as they travel from the solar core to Earth's detectors. Another important issue, which has again become topical very recently by analyzing the high-quality data from the Heliospheric and Magnetic Imager (HMI) on NASA's Solar Dynamics Observatory (SDO), is the solar oblateness (Kuhn et al. 2012) and the consequent distortion of the Sun's gravitational field which may account, in part, for the remaining 0.2 % discrepancy in the precession of Mercury's orbit (e.g. Schou et al. 1998; Gough 2012). Detailed discussions about what we have learnt from helioseismology and how helioseismology can help to address the remaining problems in solar physics can be found in recent reviews by Christensen-Dalsgaard (2004) and Gough (2013) and all the chapters of the ISSI book (Thompson et al. 2015).

That stellar variability provides deep insights into stellar structure has been known since the beginning of the 20th century (e.g. Shapley 1914). In nonradial oscillators such as the Sun, the numerous pulsation modes with distinct frequencies encode detailed information about their interior structure. The Sun oscillates in approximately 10 million distinct oscillation modes and with the help of space-borne instruments we are now able to measure some 6000 individual modes and up to several million modes via so-called ridge fitting. The amplitudes of solar-like oscillations, however, are small compared to those observed in classical pulsators, such as Cepheids or Delta Scuti stars, and their frequencies can also be quite high (5 minutes in the solar case). Therefore only the latest generation of instruments have been able to detect solar-like oscillations in stars other than the Sun.

The first indication of excess acoustic power of solar-like oscillations in other stars with a frequency dependence similar to the Sun was reported by Brown et al. (1991) for the F5 star Procyon A (α CMi), for which the first unambiguous detection of solar-like oscillations was reported by Martić et al. (1999) from the ground, as well as by Schou and Buzasi (2001) using data collected by the WIRE spacecraft. These results were confirmed later by Mosser et al. (2008). The first (unconfirmed) detection of individual peaks in the acoustic power spectrum from high-precision time-resolved spectroscopic observations was published for the G0 star η Boo by Kjeldsen et al. (1995), but it was not until 2003 that an unambiguous confirmation was established by Carrier et al. (2003), Kjeldsen et al. (2003).

High-degree oscillation modes will not be accessible in other stars in the foreseeable future because they require spatially resolved variability data. However, the observable low-degree modes in distant stars still provide rich information about the internal structure and dynamics of a large number of solar-like stars at different evolutionary stages, as demonstrated recently by the results obtained by the CoRoT and *Kepler* missions.

Additional projects, such as the Sun-in-time (Güdel 2007) or the Sun as a star (Giampapa 2005), try to characterize the Sun by comparing its property to solar-like stars at various evolutionary stages. In particular, the understanding of solar rotation, convection, variability and magnetic activity in a stellar context allows to better constraint theory by deriving key scaling laws as a function of age, rotation, chemical abundances and mass of those various physical processes. We defer to Sects. 1.4 to 1.6 for discussions of the relevant observations and to Sect. 2 for the description of the current theoretical understanding.

Returning to asteroseismic diagnostic, even with the limited seismic information of low-degree oscillation modes, some theoretically important properties have become accessible, such as the gross structure of the energy-generating core and the extent to which it is convective, and a large-scale core to envelope contrast in the angular velocity of evolved stars (e.g. Mosser et al. 2012; Deheuvels et al. 2012; Mathur et al. 2012; García 2013, and references therein). Such information will be of crucial importance for checking, and then calibrating, the theory of the structure and evolution of stars.

One of the many interesting applications of utilizing low-degree modes for measuring various aspects of the structure in distance stars is the measurement of rapid variations in the stellar background, brought about by, for example, the rapid variation in the acoustic cut-off frequency at the base of the surface convection zone or the variation in the first adiabatic exponent in the helium ionization zones. The design and analysis of low-degree seismic signatures of the structure of spherical stars is therefore of central importance to asteroseismic diagnosis, and various research groups have been working on improved seismic diagnostic techniques (e.g. Houdek and Gough 2007, and references therein). First applications of low-degree seismic diagnostic techniques have been carried out recently on 19 solar-type stars observed by the *Kepler* mission (Mazumdar et al. 2014).

However our poor understanding of the physics of the stellar surface layers causes a major problem in the analysis of global solar-like oscillation, because the properties of the near-surface layers have a major impact on oscillation frequencies. Typical stellar structure calculations treat the superficial layers with simplified atmosphere models and the stratification of the superadiabatic region by means of a local, time-independent mixing-length approach (e.g. Böhm-Vitense 1958). The effects of the (turbulent) Reynolds stresses in both

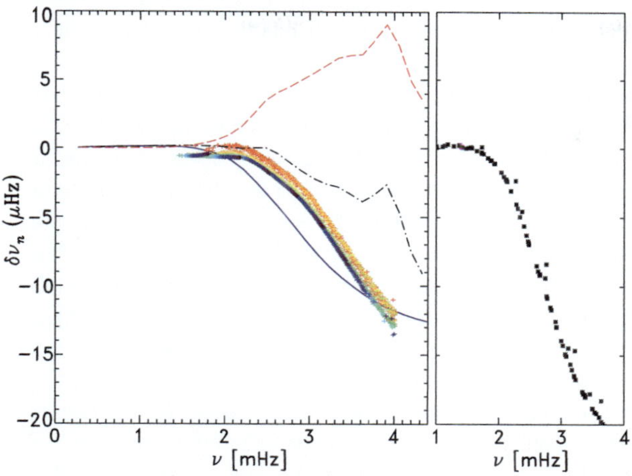

Fig. 1 Frequency residuals between the Sun and solar models as a function of frequency. *Left*: scaled differences between observed GONG frequencies and adiabatically computed frequencies of the "standard" solar Model S. Different colours indicate different degrees of the oscillations modes (adapted from Christensen-Dalsgaard et al. 1996). The *curves* show frequency residuals between models computed with a nonlocal time-dependent and a "standard" mixing-length formulation for convection: The *solid* (*blue*) *curve* shows the adiabatic residuals cause by the Reynolds stress in the mean model. The *dashed* (*red*) *curve* is the frequency shift caused by nonadiabaticity and convection dynamics. The overall frequency shift is plotted by the *dot–dashed curve* (adapted from Houdek 1996). *Right*: Scaled adiabatic frequency differences between a model for which the near-surface layers were represented by a hydrodynamical simulation, and the "standard" solar Model S (adapted from Rosenthal et al. 1995)

the hydrostatic equilibrium and pulsation calculations are essentially always ignored. Linear pulsation calculations typically adopt the adiabatic approximation and ignore the momentum flux perturbations (turbulent pressure fluctuations). However, nonadiabatic effects and the fluctuations of the turbulent fluxes (heat and momentum) do modify the modelled pulsation eigenfunctions and consequently also the oscillation frequencies (e.g. Gough 1984a; Balmforth 1992; Houdek 1996, 2010). For a "standard" solar model, such as Model S (Christensen-Dalsgaard et al. 1996), the frequency differences, scaled with the mode inertia, between the solar and model frequencies are dominated by the near-surface effects and are predominantly a function of frequency alone. Differences between solar frequencies observed with the GONG instruments and adiabatically computed Model S frequencies are illustrated by the symbols (plusses) in the left panel of Fig. 1 (colours indicate different spherical degrees). The increase of the frequency residuals with oscillation frequency depends on the modelling details of the functional form of the acoustic cutoff frequency (as it affects the acoustic potential) with radius in the near-surface layers. In the Sun the acoustic cutoff frequency is about 5.5 mHz.

Sophisticated 3-D hydrodynamical simulations of stellar convection have been used to estimate the effect of turbulent pressure in the equilibrium model (the mean Reynolds stress). For example, Rosenthal et al. (1995) investigated the effect on adiabatic eigenfrequencies of the contribution that the turbulent pressure makes to the mean hydrostatic stratification. They examined a hydrodynamical simulation by Stein and Nordlund (1991) of the outer 2 % by radius of the Sun, matched continuously in sound speed to a model envelope calculated, as in a "standard" solar model, with a local mixing-length formulation. The resulting frequency shifts of adiabatic oscillations between the simulations and the "standard" solar reference model, Model S, are illustrated in the right panel of Fig. 1. The frequency residuals behave similarly to the data (plus symbols) in the left panel of Fig. 1 though with larger shifts at higher oscillation frequencies.

The effects of nonadiabaticity and convection dynamics on the pulsation frequencies were, for example, studied by Balmforth (1992), Rosenthal et al. (1995) and Houdek (1996). In these studies the nonlocal, time-dependent generalization of the mixing-length formulation by Gough (1977a, 1977b) was adopted to model the heat and momentum flux consistently in both the equilibrium envelope model and in the linear nonadiabatic stability analysis. The outcome of such calculations is shown in the left panel of Fig. 1 (Houdek 1996, 2010) by the dot-dashed curve. Similar as for the hydrodynamical simulations the effect of the Reynolds stresses in the mean structure decreases the adiabatic frequencies (solid curve) with a maximum deficit of about 12 μHz, which is much smaller than the deficit of about 24 μHz from the hydrodynamical simulations (symbols in the right panel).

The effects of nonadiabaticity and convection dynamics (dashed curve in the left panel), however, lead to an increase of the mode frequencies by as much as $\sim$9 μHz, nearly cancelling the downshifts from the effect of the Reynolds stress (turbulent pressure) in the mean structure (solid curve in the left panel), leading to the overall effect indicated by the dot-dashed curve in the left panel. If the positive frequency shifts between models computed with and without nonadiabatic effects and convection dynamics (dashed curve in the left panel) are interpreted as the nonadiabatic and momentum flux corrections to the oscillation frequencies then their effects are to bring the frequency residuals of the hydrodynamical simulations (right panel) in better agreement with the data (plus symbols in the left panel of Fig. 1). If we were to understand the reason why the semi-analytical, nonlocal convection model underestimates the effect of the Reynolds stress in the mean model on the adiabatic eigenfrequencies, we would come closer to a self-consistent explanation for the still unsolved problem of reproducing theoretically the observed high-frequency oscillations in the Sun and in sun-like stars. Currently only simple procedures for estimating the near-surface frequency corrections (Kjeldsen et al. 2008) are adopted by the community (e.g. Mathur et al. 2012). This simple procedure is based on the calculations of a model, in which the frequency shifts are scaled by a power-law in frequency (Christensen-Dalsgaard and Gough 1980). A physically more realistic description of the surface effects with reliable applications to solar-like stars is therefore of imminent importance for the whole community to determine more reliable radii and ages in distant stars from the observed mean large frequency separation and small frequency separation, for they depend crucially on the near-surface frequency corrections (e.g. Christensen-Dalsgaard and Gough 1980; Houdek 2010).

1.3 Asteroseismic Scaling Relations

With the advent of space missions, asteroseismologists studying solar-like stars started to apply techniques originally developed for Sun-as-a-star observations to other stars. With only one or two solar-like stars per target field observed by CoRoT (e.g. HD 49933, HD 181420 and HD 181906, Appourchaux et al. 2008; Barban et al. 2009; García et al. 2009, respectively), it was possible to do a full analysis of the power spectrum, determine the individual p-mode frequencies and, after combining the seismic observables with the spectroscopic ones (e.g. Bruntt 2009), perform a full modeling of each target using stellar evolution codes (e.g. Mathur et al. 2013b). Later on, with the study of red giants in the CoRoT exoplanet field and the arrival of *Kepler*, asteroseismologists faced a new challenge: the need to analyze massive samples of thousands of stars with high-quality light curves. The combination of the high data volume and the surface effects above made manual analysis impractical. As a result, seismologists turned to characterizing the pulsation frequency patterns with a measurement of the typical frequency of oscillation and the typical frequency spacing between carefully selected and classified modes. To do this for large sample, automatic pipelines were developed in order to extract these mean seismic observables: $\Delta\nu$ and $\nu_{\rm max}$ (e.g. Huber et al. 2009; Mosser and Appourchaux 2009; Hekker et al. 2010; Mathur et al. 2010a).

The large frequency spacing, $\Delta\nu_\ell(n) = \nu_{n,\ell} - \nu_{n-1,\ell}$, is proportional to the sound travel time in the cavity in which the mode propagates and, therefore, it is proportional to the density inside the star: $\Delta\nu \approx (M/R^3)^{1/2}$. This relation shows a very good agreement with modeling (e.g. Ulrich 1986) and can be scaled from the solar values as (Kjeldsen and Bedding 1995):

$$\Delta\nu \approx \Delta\nu_\odot \left(\frac{M}{M_\odot}\right)^{1/2} \left(\frac{R}{R_\odot}\right)^{-3/2}, \tag{1}$$

in which $\Delta\nu_\odot = 135.1 \pm 0.1$ μHz, as derived by Huber et al. (2011) using 111 VIRGO subseries of 30-day, each spanning from 1996 to 2005 and analyzed in the same way as the *Kepler* data. More recent work (White et al. 2012; Mosser et al. 2013) based on numerical simulations suggests the need for small correction terms.

In 1991, Brown et al. suggested that the frequency at the maximum power of the p-mode bump, $\nu_{\rm max}$, should scale with the acoustic cutoff frequency (see also Belkacem et al. 2011), and following Kjeldsen and Bedding (1995) it could be scaled from the solar values according to:

$$\nu_{\rm max} \approx \nu_{\rm max,\odot} \left(\frac{M}{M_\odot}\right) \left(\frac{R}{R_\odot}\right)^{-2} \left(\frac{T_{\rm eff}}{T_{\rm eff,\odot}}\right)^{-1/2}, \tag{2}$$

in which $T_{\rm eff,\odot} = 5770$ K, and $\nu_{\rm max,\odot} = 3090 \pm 30$ μHz (Huber et al. 2011). This constraint is less strongly motivated by theory and requires empirical guidance.

Therefore, from Eqs. 1 and 2, one can obtain the stellar mass and radius as a function of the seismic variables $\nu_{\rm max}$ and $\Delta\nu$—assuming that $T_{\rm eff}$ is known independently of evolutionary models (e.g. Miglio et al. 2009):

$$R \approx R_\odot \left(\frac{\Delta\nu_\odot}{\Delta\nu}\right)^2 \left(\frac{\nu_{\rm max}}{\nu_{\rm max,\odot}}\right) \left(\frac{T_{\rm eff}}{T_{\rm eff,\odot}}\right)^{1/2}, \tag{3}$$

$$M \approx M_\odot \left(\frac{\Delta\nu_\odot}{\Delta\nu}\right)^4 \left(\frac{\nu_{\rm max}}{\nu_{\rm max,\odot}}\right)^3 \left(\frac{T_{\rm eff}}{T_{\rm eff,\odot}}\right)^{3/2}. \tag{4}$$

It has also recently been shown (e.g. Stello et al. 2009; Hekker et al. 2009, 2011; Mosser et al. 2010) that solar-like oscillations in main-sequence stars follow to a good approximation the following relation:

$$\Delta\nu \approx \Delta\nu_\odot \left(\frac{\nu_{\rm max}}{\nu_{\rm max,\odot}}\right)^b, \tag{5}$$

where b has often values around 0.75. Figure 2 (top) shows the relation $\Delta\nu$ versus $\nu_{\rm max}$ for 1700 stars from the main sequence phase (black diamonds at the right hand side) to the red clump phase (red triangles appearing in a diagonal branch between 20 to 50 μHz in the bottom panel) observed by the *Kepler* satellite. Although the relation appears to be constant for all of these stars, several authors (Mosser et al. 2010; Huber et al. 2010) have suggested that the slope is different between red-giant and main-sequence stars. To enhance such a difference, we have subtracted the luminosity dependence by raising $\nu_{\rm max}$ to the power of

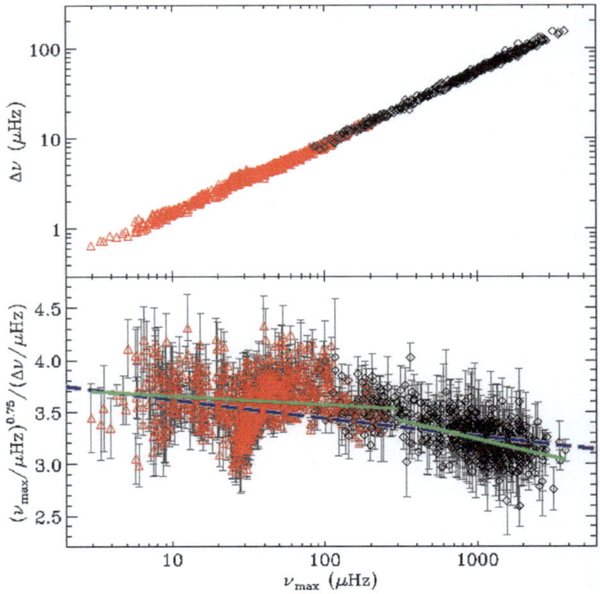

Fig. 2 $\Delta\nu$ versus ν_{max} (*top panel*) for 1700 stars observed by *Kepler* taken from Huber et al. (2011). The *red triangles* show stars observed in long cadence (red giants), while *black diamonds* are stars observed in short cadence (solar-like stars). In the *bottom panel* we have shown the same relation but after removing the luminosity dependence by raising ν_{max} to the power of 0.75. *Green lines* show power law fits to the $\Delta\nu - \nu_{max}$ relation for two different intervals of ν_{max}. The *blue dashed line* shows the relation of Eq. (5) derived using both red giants and main-sequence stars (Stello et al. 2009)

$b = 0.75$. A fit to the residuals below and above $\nu_{max} = 300$ μHz—which roughly marks the transition from low-luminosity red giants to sub-giants results in a steeper slope with increasing ν_{max}. It is important to note that for ν_{max} close to the solar value, for example, the use of a power-law relation calibrated to red-giant stars would lead to an underestimation in $\Delta\nu$ by $\sim$10 %.

Very precise surface-gravity values, (g),—with uncertainties much less than 0.05 dex—can also be inferred directly from the global seismic parameters and from the effective temperature: $g \approx \nu_{max}\sqrt{T_{eff}}$ (e.g. Morel and Miglio 2012; Thygesen et al. 2012; Hekker et al. 2013). This is very important because the seismic $\log g$ values can be used as an independent fixed parameter in the spectroscopic analysis to compute the effective temperatures in an iterative way (Silva Aguirre et al. 2012). Indeed, both parameters are correlated and cannot be determined independently from the spectra. Therefore, the combination of a seismic $\log g$ and a spectroscopic T_{eff} can reduce the uncertainties in both parameters. CoRoT and *Kepler* stars with asteroseismic surface gravities will be used as calibration stars for the GAIA mission (Creevey et al. 2013).

1.4 Stellar Convection and the Background Signal in Oscillation Spectra

Solar-like pulsating stars are those in which their acoustic modes are excited by turbulent motions occurring in their convective outer layers (e.g. Goldreich and Keeley 1977; Balmforth 1992; Houdek et al. 1999; Chaplin et al. 2005; Belkacem et al. 2008). The first observations of granulation are quite old (Herschel 1801), but it is not until 1933 that their turbulent origin was unveiled (Siedentopf 1933).

In solar-like stars, exhibiting outer convective envelopes, the granules evolve with time producing small photometric fluctuations. Therefore, the analysis of stellar light curves in the Fourier domain shows a continuous "background" on top of which the acoustic modes are superimposed (see some examples of main sequence stars, subgiants and red giants in Kallinger et al. 2010; Chaplin et al. 2010; Campante et al. 2011; Ballot et al. 2011; Mathur et al. 2011b, 2013b; Metcalfe et al. 2012).

The first successful attempt to model this continuous background was done by Harvey (1985). He approximated each scale of the convective motions (granulation, supergranulation...) by an exponentially decaying time function corresponding to the autocovariance of its time evolution with two parameters: the characteristic time of the granules and the total brightness fluctuations associated with them (see Mathur et al. 2011a, for a complete review of the different formulations commonly used in asteroseismology). Moreover, it has been demonstrated very recently that the properties of the granulation and the global seismic variables are related (e.g. Chaplin et al. 2011b; Mathur et al. 2011a). It is also important to notice that in some stars another feature is visible close to the p-mode envelope that has been related with the presence of faculae in the star (Karoff et al. 2013, and references therein). Recent theoretical work by Samadi et al. (2013) has also tried to explain the observed relationship between granulation amplitude in the oscillation spectrum and stellar spectral type.

In the context of stellar dynamics, putting observational constraints in the convection inside stars is very important because convection is one of the major ingredients to understand stellar magnetism and dynamo processes. Therefore, precise time domain measurements can offer direct constraints on the granulation characteristics as well as the precise determination of the convective outer envelope. Recently, using *Kepler* data, it was for example possible to calibrate the mixing-length parameter $\alpha_{\rm MLT}$ used in classical 1-D treatment of convection, and to show that the solar value, usually used for generating grids of stellar models (Ekström et al. 2012) is overestimated (Bonaca et al. 2012). Likewise granulation data can put constraints on local (Magic et al. 2013) and global (Miesch et al. 2012) stellar convection simulations, that could in turn help us to characterize stellar oscillations, and convection parameters such as the mixing length (Trampedach and Stein 2011). This important by-product of asteroseismology studies can help to better constrain convective heat transport in 1-D stellar models, and hence to constrain the stellar radius which depends on $\alpha_{\rm MLT}$.

Asteroseismology can also provide important information on the depth of convective envelopes. At the base of the surface convection zone, there is an abrupt change of stratification (going from adiabatic to subadiabatic) which impacts the sound speed and density profiles (see the pionnering work on the base of the solar convection by Christensen-Dalsgaard et al. (1991). By using appropriate seismic diagnostic techniques it is then possible to infer the extent of the convective zone of solar-like stars (Gough 1990; Ballot et al. 2004; Houdek 2004; Houdek and Gough 2007). This approach has now been successfully applied to *Kepler* data for field stars (Mazumdar et al. 2014).

Complementary to the analysis of asteroseismic data, numerical simulations to model and describe the turbulent state of stellar convective envelopes or cores have been pursued. Several groups have undergone a systematic study of the global properties of stellar convections and their associated mean flows (differential rotation and meridional circulation). We refer to Sect. 2 for a more detailed discussion of numerical simulations of rotating and magnetized convective stars.

1.5 Stellar Rotation and Spin Down

All stars rotate, and understanding their rotation history is key to our overall understanding of their dynamics. The angular momentum evolution of stars is an initial value problem, and a wide range of rotation rates are observed in the youngest protostars (see for example Rebull et al. 2006, for the Orion Nebula Cluster). The relatively slow rotation of such stars (Vogel and Kuhi 1981) demands efficient angular momentum loss from the infalling material during the hydrodynamic collapse phase, probably through strong magnetic interactions (Shu et al. 1994). Gaseous accretion disks persist around protostars for up to 12 Myr (Bell et al. 2013), and the exchange of angular momentum between them and their disks can dramatically modify their evolution. In the limit of strong magnetic interaction, star-disk coupling can regulate stellar rotation, preventing protostars from spinning up as they contract (Koenigl 1991; Keppens et al. 1995) but see also (Matt et al. 2012). The early angular momentum evolution of stars is therefore a window into the star and planet formation process. Stars also reach the main sequence with a range of rotation rates. This can cause star-to-star differences in the degree of rotationally induced mixing (see Pinsonneault 1997, for a review). Once on the main sequence, the surface rotation of stars responds both to internal angular momentum transport and to angular momentum loss from magnetized solar-like winds. There are two strong practical applications that result:

(1) We can use the observed angular momentum evolution as a test of stellar physics. Many instabilities and physical processes such as dynamo action, mixing, large scale flows are directly connected to the rotational rate of stars. Stars differ in their thermal structure, rotation rates, and lifetimes; we therefore have natural laboratories for testing physical models. This is especially important for our understanding of solar-like winds, as angular momentum loss driven by them is a central driver of angular momentum evolution in low mass stars.
(2) With calibrated models we can then use rotation as a stellar population diagnostic. Weber and Davis (1967) explained the slow solar rotation as a consequence of angular momentum loss from a magnetized solar-like wind. They predicted $\mathrm{d}J/\mathrm{d}t \sim \Omega^3$, which implies $\Omega \sim t^{-0.5}$ in the asymptotic limit. Skumanich (1972) confirmed this time dependence in rotation, chromospheric activity, and lithium depletion in open cluster stars. Rotation is therefore a potential age indicator, provided both that it can be measured and that the rotation-mass-age relationship can be calibrated.

Both our observational knowledge and the sophistication of our theoretical models have advanced considerably since this pioneering work. We therefore briefly summarize the observational picture (see Fig. 3) before proceeding to the theoretical picture that has emerged. Rotation rates can be inferred spectroscopically from Doppler line broadening. Such measurements, however, suffer from projection effects ($v \sin i$ is what is measured) and become difficult when the rotational broadening becomes comparable to or smaller than other sources, such as microturbulence or thermal broadening. Rotation velocity measurements are therefore primarily used for young and massive stars, and through the 1980s and early 1990s most rotation data was spectroscopic. In a pioneering paper, Stauffer and Hartmann (1987) demonstrated that cool stars in the Pleiades arrived on the main sequence with a wide range of rotation rates, which created three significant problems for simple angular momentum evolution models. Extremely large torques would be predicted for rapid rotators (Pinsonneault et al. 1990), and an unsaturated wind law would not permit the survival of high rotation rates onto the main sequence. Magnetized solar-like winds are too inefficient to spin down young stars to the observed rates for the slowest rotators, requiring an additional mechanism for extracting angular momentum in the pre-main sequence. Finally, once on the main sequence, there is a rapid initial drop in rotation rate, followed by a pause where surface rotation changes little; older stars then resume spinning down at the predicted asymptotic rate. This is evidence for core-envelope decoupling (MacGregor and Brenner 1991; Krishnamurthi et al. 1997) on timescales of tens to hundreds of Myr (see Denissenkov et al. 2010; Gallet and Bouvier 2013, for recent discussions). The observed pattern is therefore richer than a simple power-law relationship, requiring both more sophisticated theory and stronger empirical constraints.

Star spots cause periodic brightness modulations as they transit across the visible disk. Dedicated time domain studies of stars can therefore measure rotation periods. Rotation

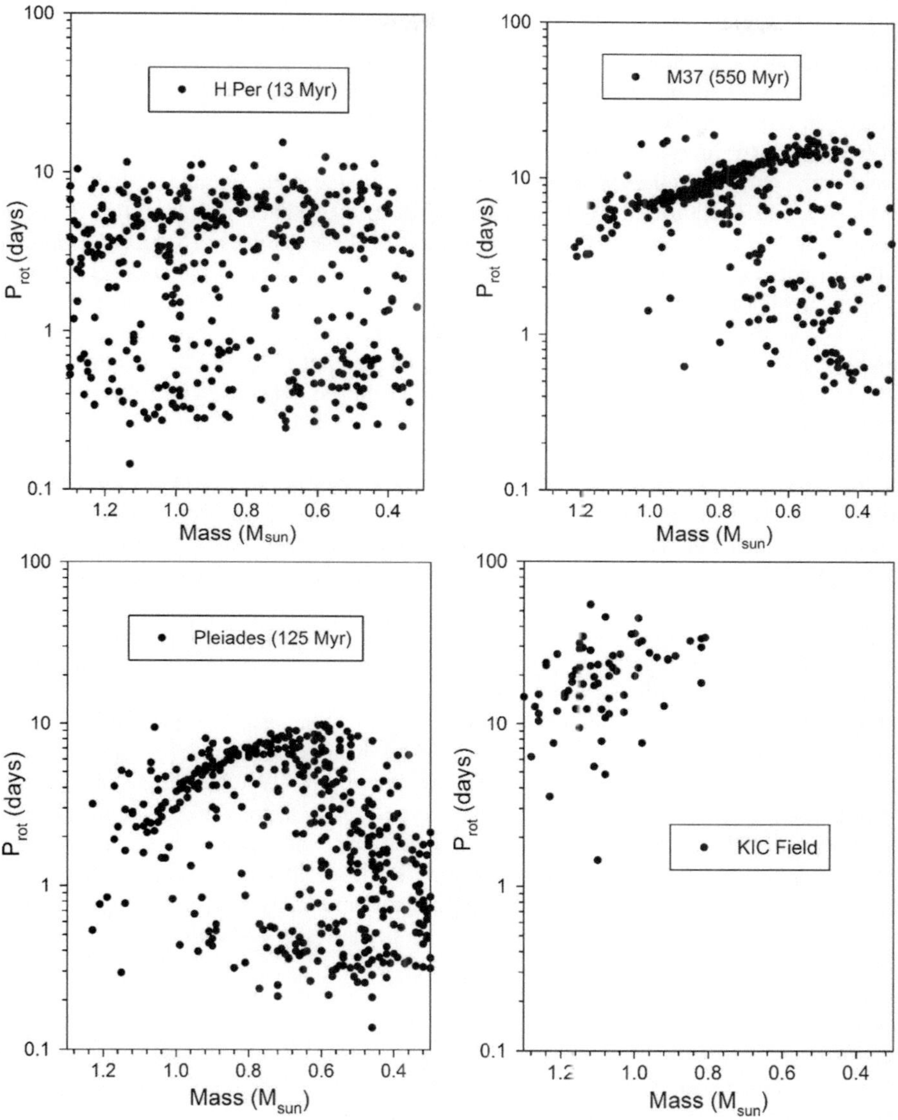

Fig. 3 The observed distribution of rotation rates in stars of different ages. Data for the young protostars in H Per (*upper left*) is taken from Moraux et al. (2013). The data for the 125 Myr old Pleiades system (*lower left*) is taken from the reanalysis by Coker and Pinsonneault (2015, in preparation). M37 data is originally from Hartman et al. (2009) as interpolated onto mass by van Saders and Pinsonneault (2013). Data for old field dwarfs is taken from Garcia et al. (2014)

period measurements do not suffer from inclination effects (although they require a favorable inclination angle, e.g. not pole-on, to produce a signal). There are, however, significant observational selection effects that favor the detection of short period systems, which also tend to be high amplitude. There has been an enormous increase in the quantity and quality of stellar rotation period data. In particular, the advent of large surveys designed to search for extrasolar planets has yielded very large data sets of stellar rotation periods as a natural byproduct for both field stars (Hartman et al. 2011) and open clusters stars. Examples of the latter include the extensive MONITOR program (Aigrain et al. 2007) (see Moraux et al. 2013, for the most recent results), the SuperWASP data for the Hyades and Praesepe (Delorme et al. 2011) and the large (Hartman et al. 2009, 2010) M37 and Pleiades data sets. The *Kepler* and CoRoT missions reach lower amplitudes and can detect stars with longer rotation periods, pushing stellar rotation measurements into the solar regime (Basri et al. 2011; Garcia et al. 2014; Lanza 2013; Walkowicz and Basri 2013; Nielsen et al. 2013; Reinhold et al. 2013; McQuillan et al. 2014). With the aid of these samples, a standard empirical picture of angular momentum evolution has emerged, with the following basic ingredients.

Stars are born with a range of rotation rates, and they would be expected to spin up as they contract and approach the main sequence. Through protostar-disk interactions some stars lose additional angular momentum in the first $\sim$10 Myr, widening the observed distribution. The slow rotator population cannot be generated from the young star forming regions, as noted above, but is already in place in the 13 Myr old system H Per (Moraux et al. 2013). This confirms the hypothesis that protostar-disk interactions are important for stellar angular momentum evolution. The distribution of rotation rates is initially broad and narrows with age, which is a natural consequence of a torque that increases more than linearly with rotation rate. This convergence process is most rapid for solar mass stars and the timescale for spin down increases with decreased mass. This is usually interpreted as evidence for mass-dependent saturation in the angular momentum loss rate, typically parameterized with a Rossby scaling (rotation period relative to the convective overturn timescale). Core-envelope decoupling is required to explain young slow rotators, with a time scale that increases with decreased mass (Denissenkov et al. 2010; Delorme et al. 2011). There is no break or discontinuity in the observed spin down behavior at the fully convective boundary, which is surprising given dynamo model predictions. New data sets in older solar analogs in *Kepler* stars will provide tests of the mass and composition dependence of rotation in old dwarfs, which is currently lacking. Field star studies will need to consider the presence of background populations when modeling angular momentum evolution. For example, angular momentum lost from tidally synchronized binaries is extracted from the orbital rather than the spin angular momentum reservoir, which results in very different angular momentum evolution properties (see Andronov et al. 2006, for a discussion in the context of blue stragglers).

These large data sets have spurred renewed interest in the usage of stellar rotation as an age indicator, or gyrochronology (Barnes 2003). Distinct theoretical scalings and fitting function have been proposed (e.g. Mamajek and Hillenbrand 2008; Barnes and Kim 2010; Epstein and Pinsonneault 2014, for a discussion). There are significant zero-point shifts between these formulations, which will require resolution when compared with empirical data. There are strict lower bounds to the uncertainties in gyrochronology relations even for an ideal theoretical framework. Epstein and Pinsonneault (2014) identified two fundamental limiting effects for the precision of gyrochronology: The intrinsic range of rotation rates at fixed age and the range of surface rotation rates in a given star from latitudinal differential rotation. The former is more important for young and low mass stars, while the latter dominates for older higher mass stars. The net effect is an age uncertainty at the 2 Gyr level for old stars. van Saders and Pinsonneault (2013) extended this to subgiants, and argue for a different gyrochronology relationship for them than for dwarfs. Recently the gyrochronology concept has been adapted to consider magnetic field instead of rotation as a proxy for age: e.g. *magnetochronology* (Vidotto et al. 2014). It is indeed clear (see next section) that stellar magnetic properties vary over the course of stellar evolution, older stars being less active their younger counterparts (Giampapa 2005; Gregory et al. 2012).

The theoretical interpretation, numerical models and multi-D simulations allowing us to study and to understand the stellar age-rotation-activity relationship will be discussed in Sect. 2.

1.6 Stellar Activity & Magnetism

The great advantage of comparing the Sun to other stars is to disentangle what is specific to our star from what is generic to solar-like stars. Systematic observations of rotation and magnetic properties of solar-like stars have revealed interesting trends. For instance surface differential rotation is found to increase with $T_{\rm eff}$; F-star possess a larger latitudinal contrast than K-stars do (Donahue et al. 1996; Barnes et al. 2005; Collier Cameron 2007). Similarly a rotation-activity relationship has been known for quite some time now (Pizzolato et al. 2003). A relation between cycle and rotation periods has also been advocated (Noyes et al. 1984b; Böhm-Vitense 2007).

Meaningful information regarding magnetic cycles underlying stellar activity, or alternatively to rule out the existence of magnetic cycles, require long-term stellar activity observations and their analysis. Direct observations of starspots is a relatively new capability and studies of their behavior over timescales long enough to be relevant for stellar dynamos, do not yet exist. Nonetheless, indirect evidence of magnetic activity cycles (or the lack thereof!) can be obtained by different techniques. Our traditional tools for measuring stellar cycles have been synoptic stellar chromospheric activity observations, complemented with more limited stellar coronal X-ray variability data. Asteroseismology may prove a useful additional set of data.

The most important contribution in the context of long-term stellar activity observations has been the seminal Mount Wilson Observatory (MWO) Calcium (Ca) H+K Project (Wilson 1978; Noyes et al. 1984b; Baliunas et al. 1985, 1995). The emission in these line cores, expressed as a Ca H+K activity index, is understood to be a measure of the magnetically mediated, non-thermal heating of the chromosphere (provided that the basal thermal flux is properly accounted for, e.g. Schrijver et al. 1989; Cuntz et al. 1999). Thus, the long-term variation of the Ca H+K index can be taken as a proxy for variability in stellar magnetic fields (see the review by Hall 2008). The Lowell Observatory's Solar-Stellar Spectrograph program provides complementary observations of Sun-like stars (Hall et al. 2007, 2009; Hall 2008) adding to the knowledge base of the MWO program. An independent measure of stellar magnetic output is the coronal X-ray flux—which originates in the magnetic heating of stellar coronae and which is even more tightly correlated with magnetic flux (Pevtsov et al. 2003). Although stellar X-ray cycle observations are rare, the amplitude of stellar X-ray flux have been observed for multiple stars (Hempelmann et al. 1996; Micela and Marino 2003; Güdel 2004) and this usefully constrains the strength of the dynamo generated magnetic fields. More recent surveys analyzing the activity of solar-like stars are to be found in Wright et al. (2004) and Giampapa et al. (2006). We provide here a brief summary of the important trends and relationships relevant for stellar activity cycles that have been gleaned from these long-term observations.

1.6.1 General Activity Trends with Rotation

First we focus on the scaling behavior of magnetic activity with the stellar rotation rate. Stars that are faster rotators have higher magnetic activity amplitudes than slower rotators. They are usually variable, but do not display regular cycles. Stars with intermediate rotation rates have intermediate magnetic activity output, which sometimes display periodic behavior. Stars which rotate relatively slowly, such as the Sun, are more likely to have regular,

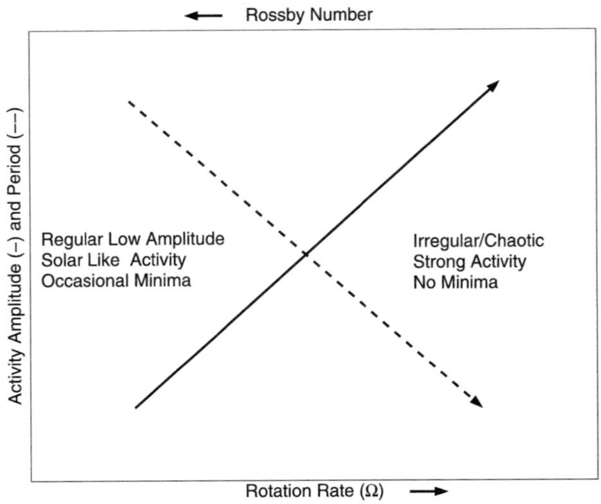

Fig. 4 Stellar activity trends as determined from observations of Ca H+K and X-ray emission from F-G-K stars. The *arrows* in the axis label show the direction of increase of the rotation rate and Rossby number. In general, magnetic cycle amplitude increases while cycle period decreases with increasing rotation rate (i.e., decreasing rotation period). The critical reader is referred to the original observations cited in the text for the exact scaling laws and other subtleties in the data. From Nandy (2004)

periodic cycles, but their activity amplitudes are also relatively smaller. The latter category also sometimes display "flat" (i.e., non-varying) activity with low amplitude, akin to global magnetic minima such as the solar Maunder minimum. We note here that fast rotators also sometimes display a "flat" (non-varying) activity profile albeit with high activity amplitudes—which is not to be confused with global minimum like states.

In general the activity amplitudes increase with rotation rates, but for very fast rotators, the measured amplitudes tend to reach a saturation (Pizzolato et al. 2003; Wright et al. 2011), which in X-rays is of order $\log(L_x/L_{\mathrm{bol}}) \sim -3.2$. For G type stars this saturation is found for rotation rate above 35 $\mathrm{km\,s^{-1}}$, for K type stars at about 10 $\mathrm{km\,s^{-1}}$ and for M dwarfs around 3–4 $\mathrm{km\,s^{-1}}$. How stellar magnetic flux scales with rotation rate is important to understand, since it is telling us how the magnetic field generated by dynamo action inside the stars emerges and imprints the stellar surface. Understanding the apparent saturation phenomenon is also important: for example, it could arise if only the filling factor f reaches a maximum (Reiners 2012). Multi-periodic cycles are also sometimes observed, especially in faster rotators (Oláh et al. 2009). For those stars with cyclic magnetic activity, the systematic analysis of stellar data revealed that for solar type stars there is a good correlation between the cycle and rotation periods of the stars and that correlation is even stronger when using the Rossby number ($Ro = P_{\mathrm{rot}}/\tau$) that takes into account the convection turnover time τ at the base of the stellar convective envelope (Noyes et al. 1984a, 1984b; Soon et al. 1993; Baliunas et al. 1996). As the star rotates faster, its cycle period is found to be shorter. This trend is summarized in Fig. 4. More precisely, Noyes et al. (1984a) found that $P_{\mathrm{cyc}} \propto P_{\mathrm{rot}}^n$, with $n = 1.25 \pm 0.5$. Saar and Brandenburg (1999), Saar (2002) have used an extended stellar sample to argue that there are actually two branches when plotting the cycle period vs. the rotation period of the stars: the primary (starspot) cycle and the Gleissberg, or grand minima type, modulation of stellar activity. For the active branch they found an exponent $n \sim 0.8$ and for the inactive one they inferred $n \sim 1.15$.

1.6.2 Maunder Minima Stars

Many of the stars (roughly 15 %, see Giampapa et al. 2006), studied in the Mount Wilson surveys, displayed activity that were flat (i.e., non-cyclic) and at or below current solar minimum levels. Similar studies with *Kepler* data recently found that among the sampled solar-like seismic stars about 2 % show a flat activity level. One possible interpretation of these results is that these are "Maunder minima" stars where the global magnetic activity has temporarily ceased. Complementary studies of coronal emission also point to the existence of such low-activity stars (Poppenhäger et al. 2009). Baliunas et al. (1995) also provides evidence indicative of a star entering into a Maunder minimum like phase—with its activity variation flattening out accompanied by a reduction in amplitude.

1.6.3 Starspots, Polar Caps and Stellar Butterfly Diagrams

Recent advances in stellar photometry and doppler imaging techniques now allow us to observe starspots and their dynamics. Such observations (reviewed in Berdyugina 2005) show that there is oftentimes a preferred longitude for starspots. Faster rotators tend to have large polar spots which contributes to a strong polar cap, while in general, in very active stars, the spot distribution tends to be more uniform across all latitudes. In some stars, the spots evidently appear at mid-low latitudes and subsequently migrate to higher latitudes, reminiscent of the sunspot migration (although towards low latitude) that generates the solar butterfly diagram. These observations do not span a long period of time and hence detailed studies of stellar butterfly diagrams is still in its infancy. Nevertheless, observations of spot emergence latitudes and migration are expected to provide more stringent constraints in the future for stellar dynamo and flux emergence models.

1.6.4 Asteroseismology and Activity

Ensemble asteroseismology can now add to our developing understanding of stellar activity obtained by other methods, such as spectropolarimetry and CaII H & K studies.

The first contribution of helio and astro-seismology comes through their ability to provide stellar properties and, in particular, stellar ages for field stars that are otherwise difficult to analyze. Our lack of knowledge of stellar ages has been an important limitation in our understanding of the Mt. Wilson survey data, for example. Many of the solar-like stars showing a flat S-index in the original sample analyzed by Baliunas and Jastrow (1990)—and assumed to be in a similar state to the Maunder minimum in the Sun—were found to be more evolved stars (which would be expected to less active) when accurate parallaxes from the Hipparcos mission were taken into account (Wright 2004). Therefore, accurate stellar properties are essential for proper interpretation of their magnetic indexes (see an extended discussion in the review by Judge and Thompson 2012). But seismology can also be employed to directly study stellar activity. Magnetic fields modify the conditions inside the stars and the properties of the modes are changed accordingly. Seismology can therefore provide complementary information to the one obtained by the above mentioned techniques. Moreover, it can for example provide evidences of the existence of subsurface on-going magnetic cycles even during Maunder minimum-like states (Beer et al. 1998) in which no external manifestation of the internal magnetism can be observed. However, the amplitude of the modes decreases for active stars as observed in the Sun (e.g. Jiménez-Reyes et al. 2003), HD 49933 observed by CoRoT (García et al. 2010), and solar-like stars observed by *Kepler* (Chaplin et al. 2011a) difficulting their observation.

The first diagnostic or "proxy for magnetic activity" used in helioseismology was the frequency shift of the acoustic modes, i.e., the displacement of the modes towards higher frequencies as the cycle evolves. These frequency shifts—which follow an inverse mode-mass scaling (Libbrecht and Woodard 1990)—are explained to arise from changes in the outer layers of the Sun along the 11-year magnetic cycle (Schwabe cycle) (Goldreich et al. 1991; Balmforth et al. 1996; Dziembowski and Goode 2004, 2005). Therefore, the high-frequency p modes—having higher upper turning points than the low-frequency p modes—are then more sensitive to the perturbations induced by the magnetic field in the outer part of the Sun. Thus, these high-frequency p modes are observed to have large frequency shifts with the Schwabe cycle, while the low-frequency p modes present little or no variation with it (e.g., Ronan et al. 1994; Chaplin et al. 1998; Gelly et al. 2002; Howe et al. 2008; Jain et al. 2009). A similar behavior has also been found in the F star HD 49933 (Salabert et al. 2011a).

It is now commonly accepted that, apart from the rotational splitting (García et al. 2008; Broomhall et al. 2012), all the rest of the solar low-degree p-modes properties change with time (e.g. Chaplin et al. 2000; Jiménez-Reyes et al. 2003, 2007; Salabert et al. 2011b; Garcia et al. 2013) in a time scale of 11 years following the corresponding surface Schwabe cycle. Another shorter timescale has been recently confirmed using helioseismology. Indeed, the quasi-periodic modulations of 1 to 2 years observed during the last twenty years in the Sun (e.g. Benevolenskaya 1995; Mursula et al. 2003) has then been measured (Fletcher et al. 2010) in the frequency shifts of low-degree modes measured by the Global Oscillation at Low Frequencies (GOLF Gabriel et al. 1995) instrument on board the Solar and Heliospheric Observatory (SoHO Domingo et al. 1995), as well as with the Birmingham Solar Oscillation Network (BiSON Chaplin et al. 1996). Although the origin of this modulation is not well understood yet, the latest results priviledge the hypothesis that the origin of the perturbation is located in the subsurface layers (Simoniello et al. 2013).

The continuous photometric monitoring of hundreds of thousands of stars by CoRoT and *Kepler* provide long baseline stellar variability samples of unprecedented scale and precision. Their duration is also substantial, comprising a period of more than 4 years in the case of *Kepler*; this is long enough to begin to see the effect of activity cycles. The first ensemble analysis *Kepler* light curve variability was done by Basri et al. (2010, 2011). However, because the *Kepler* pipeline was not designed to measure long-period stellar variability they used raw data from only a 90 day quarter, which introduced important limitations in their analysis. Basri et al. (2011) found that the amplitude of variability was in general larger for stars that were clearly periodic than for those that were not. As expected, the largest group of non-periodic variables were giant stars. To improve on the *Kepler* pipeline, which was designed for transit detection and not stellar variability, several groups developed their own correction software (e.g. García et al. 2011, 2014; McQuillan et al. 2014). With these new analysis tools to was possible to detect longer-than-a-month rotation periods as well as to study the magnetic variability of the stars.

Based on broadband photometric variability, a new proxy for stellar magnetic activity has been developed (García et al. 2010; Chaplin et al. 2011a; Mathur et al. 2014b). The time evolution of the variance of the signal appears to be a good indicator of the magnetic activity cycle. It has been successfully tested on the Sun using VIRGO/SPM data recorded during the last 16.5 years (see for more details Garcia et al. 2013). Using the same methodology, Mathur et al. (2014a) studied a subset of 22 *Kepler* F stars with rotation periods of less than 12 days for which p-mode oscillations were reported. Because of the relation between the rotation period of the star and the period of the cycle (e.g. Böhm-Vitense 2007; Jouve et al. 2010), they were expecting to see in some stars cycle-like variations such as the one observed by Metcalfe et al. (2010) in the F8V star *i* Horologii. That star exhibited a magnetic cycle period of 1.6 years for a rotation period of 8.5 days. Mathur et al. (2014a) found that only a small fraction of stars showed a regular cycle-like behavior (see Fig. 5) and there were no clear trends between activity and rotation in the F-star sample. A more systematic study still

Fig. 5 Evolution of the normalized averaged variance of the *Kepler* F star KIC 12009504 for which p-mode oscillations have been measured (in *black*) and smoothed by a boxcar function (160 days wide) in *blue*. A regular cycle of around 800 days is observed (see Mathur et al. 2014a, for further details)

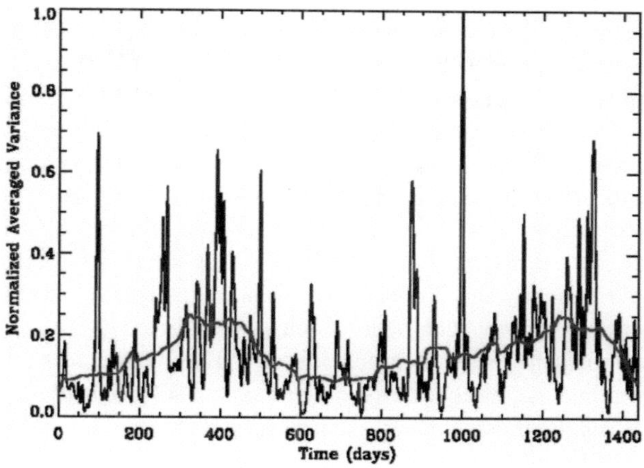

needs to be done to confirm or refute the existence of a general relationship between stellar cycle period and stellar rotation period for these hot stars.

2 Global Models of Stellar Structure, Rotation, and Magnetism

In order to understand, characterize and describe the physical processes at the origin of stellar convection, rotation and magnetism, with the Sun being an archetype of solar-like stars, we must develop a self-consistent, ideally nonlinear, description of their dynamics. Depending on the time scale of interest and given the current computing resources such model can be either 1-D, 2-D or 3-D. In the following section we briefly summarize the recent progress made in modeling the multi-scale dynamics of stars and their success at explaining the observed properties discussed in Sect. 1.

2.1 1-D Stellar Evolution and Core-Envelope Coupling

The angular momentum evolution of a star—its rotational history from its birth until its death—has profound consequences for its structure, evolution, and dynamics. Rotationally induced mixing for low mass stars can induce mild mixing that alters surface abundances (see Pinsonneault 1997 for a review), while for massive stars, which rotate much more rapidly, the impact of mixing is far more dramatic (see Maeder and Meynet 2012, for a recent review). Stellar rotation is also intimately connected with the origin and generation of stellar magnetic fields, and the two together shape wind outflows from stars. Hence it is fundamental to be able to characterize it. The overall picture of angular momentum evolution was discussed in Sect. 1.3. Here we focus on modeling.

Structural evolution in stars will naturally generate internal angular velocity gradients, and in some evolutionary stages these shears can be large. Mass loss from the surface will also carry angular momentum away from the star, providing another powerful driver for radial differential rotation. In convective regions the interaction of rotation, magnetism, and convection will enforce an angular momentum distribution on a convective overturn timescale, much shorter than the natural nuclear evolution timescale. The response of radiatively stable regions is more complex, both because of our limited knowledge of the physics and because of limited empirical constraints on the relevant time scales. Internal redistribution by hydrodynamic and magnetohydrodynamic physical process can both be important, and different classes of models testing different prescriptions have been employed.

Solar-like stars on the main sequence evolves on long time scales; the solar main sequence lifetime is of the order 10 Gyr. Computing such long temporal (secular) evolution with multi-D codes is impractical. One must rely on 1-D stellar evolution models to describe the structure, chemical and rotational evolution of stars. We refer to the recent review by Basu et al. (2014) for a detailed discussion of 1-D stellar evolution codes, their success and difficulty at reproducing the solar internal structure, the influence of chemical composition, and how such helioseismically calibrated 1-D solar models have been extended to compute up-to-date 1-D stellar models. Here we focus on internal angular momentum transport and its consequences for the rotation history of stars. To this end, let's apply the concept of angular momentum exchange and conservation to a star like our Sun.

A useful framework that has been extensively employed in the literature is the two zones model of MacGregor and Brenner (1991). In this model, the solar-like star is treated as a sphere divided into two zones: an inner stably stratified core and a turbulent convective envelope both independently rotating around an axis aligned with the $\hat{\mathbf{e}}_z$ direction, where $\hat{\mathbf{e}}_z$ is the unit vector along the z axis of a 3-D (x, y, z) cartesian system (see Fig. 6 left panel). Each possess their own moment of inertia I and angular velocity Ω; both structural evolution and magnetized winds can generate differential rotation. Their angular momentum is respectively $J_{\rm core} = I_{\rm core}\Omega_{\rm core}$ and $J_{\rm env} = I_{\rm env}\Omega_{\rm env}$. Let's now find the exact quantity of angular momentum exchange ΔJ between the two zones require to have them rotate uniformly at the rotation rate Ω. The initial state is: $J^c_{\rm init} = J_{\rm core}$ and $J^e_{\rm init} = J_{\rm env}$. Applying total angular momentum conservation e.g. $J^c_{\rm init} + J^e_{\rm init} = J^c_{\rm final} + J^e_{\rm final}$, implies that the final state is: $J^c_{\rm final} = J_{\rm core} - \Delta J = I_{\rm core}\Omega$ and $J^e_{\rm final} = J_{\rm env} + \Delta J = I_{\rm env}\Omega$. A simple substitution yields:

$$\Delta J = \frac{I_{\rm env} J_{\rm core} - I_{\rm core} J_{\rm env}}{I_{\rm core} + I_{\rm env}}. \tag{6}$$

If an external torque, such as the one coming from a stellar wind $\tau_{\rm wind}$ is applied to the star, the angular momentum evolution of the convective envelope and radiative interior can be expressed as (MacGregor and Brenner 1991):

$$\frac{dJ_{\rm core}}{dt} = -\frac{\Delta J}{t_{\rm c}} \tag{7}$$

$$\frac{dJ_{\rm env}}{dt} = \frac{\Delta J}{t_{\rm c}} - \frac{J_{\rm env}}{t_{\rm w}}, \tag{8}$$

with $t_{\rm c}$ the coupling time scale between the two zones due to the combined action of turbulence, waves, magnetic fields and viscous stresses and $t_{\rm w}$ the wind braking timescale. These formula further assumes that the convective envelope transmits instantaneously the applied surface torque to the base of the convective envelope.

Such models have been used to explore the relevant coupling time scale between the convective envelope and the radiative interior in solar-like stars over the course of their evolution (MacGregor and Brenner 1991; Keppens et al. 1995; Krishnamurthi et al. 1997; Allain 1998) and (for recent developments see Denissenkov et al. 2010; Bouvier 2013; Gallet and Bouvier 2013; Oglethorpe and Garaud 2013; Zhang and Penev 2014). It is found that a time scale of tens to hundreds of Myr can explain the core-envelope coupling in young open cluster stars (see Fig. 6).

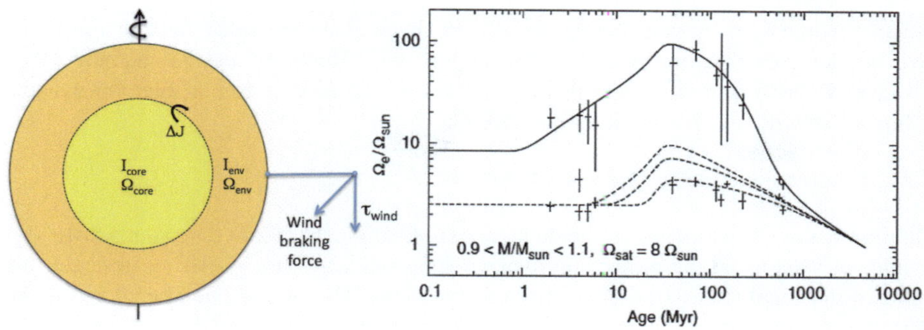

Fig. 6 *Left*: The two layers model of a star, showing the exchange of angular momentum ΔJ between a core and an envelope rotating at different rates and the braking action of a stellar wind. *Right*: Evolutionary track for two layer models showing how the coupling time scale can explain slow and fast rotators (from Denissenkov et al. 2010)

One can identify key angular momentum transport processes which are expected to act inside a stellar convection zone or radiative interior: Reynolds and Maxwell stresses, meridional circulation, viscous processes (usually negligible), large scale magnetic torque, gravity waves (see Gilman et al. 1989; Zahn 1992; Pinsonneault 1997; Mathis and Zahn 2004, 2005; Denissenkov et al. 2010; Mathis 2011, 2013b; Brun et al. 2011; Reiners and Mohanty 2012; Gallet and Bouvier 2013; van Saders and Pinsonneault 2013, and references therein). The amplitude and role of each process varies as a function of time. For example, it is known that young stars are more active than old ones, so internal Maxwell stresses are expected to play a less significant role later in the star's evolution on the main sequence and likewise for the stellar wind torque. Of course when the star undergoes a large structural change due to its secular evolution (as for instance when it goes up the red giant branch) the relative importance of different processes must be re-evaluated.

Kawaler (1988) proposed a general analytical formulation of the applied torque that has been widely used. He adopted the (Mestel 1968) framework and adopted parameters tuned to reproduce Skumanich's law (Skumanich 1972) for the time dependence of surface rotation. Many other studies have since improved upon Kawaler's model by taking into account better observational information or theoretical improvements in order to assess how the magnetic torque applied by a stellar wind depends on stellar parameters and dynamics (rotation, dynamo); see for example the Rossby-scaled saturation threshold adopted by Krishnamurthi et al. (1997), which has been extensively employed subsequently. There has been much recent activity in this area, with new frameworks being proposed by Cranmer and Saar (2011), Matt et al. (2012, 2014), Reiners and Mohanty (2012), van Saders and Pinsonneault (2013), Brown (2014).

Self-consistent solar models including angular momentum transport from hydrodynamic mechanisms were first generated in the 1980s (Endal and Sofia 1981; Pinsonneault et al. 1989). Meridional circulation, hydrodynamic instabilities and gravity waves are the primary mixing and angular momentum transport agents, so these models play an important role in rotational mixing studies (Mathis 2013a; Brun and Mathis 2014). Such models are sometimes coupled with approximations of magnetic effects, such as the Taylor-Spruit mechanism (Spruit 2002). This approach has been used to model massive rotating stars over secular time scales (Palacios et al. 2003; Marques et al. 2013; Meibom et al. 2013) and to generate grids of stellar models used for isochrones and chemical evolution studies (Ekström et al. 2012; Georgy et al. 2013; Lagarde et al. 2012). Some attempts have also been made to model in 2-D the secular evolution of internal angular momentum by considering the role of large scale magnetic field in the radiative interior (see Charbonneau and MacGregor 1993; Ruediger and Kitchatinov 1996; Spada et al. 2010, and references therein). However such models do not solve for dynamo action as we will now discuss in the next section.

2.2 2-D Mean Field Stellar Dynamo Models

On time scales of centuries to millenia it is possible to model in 2-D the magnetohydrodynamics of stars using the mean field dynamo framework by parameterizing or neglecting some of the detailed description of convective motions. This allows one to explore the underlying physics and a wide range of stellar parameter space which we describe below.

2.2.1 The Theoretical Framework

Stellar magnetic cycles involve the generation and conversion of two components of the Sun's magnetic field—the toroidal component and the poloidal component. Under the assumption of axisymmetry, the stellar magnetic field and plasma flow can be expressed as

$$\mathbf{B} = B_\phi \mathbf{e}_\phi + \nabla \times (A\,\mathbf{e}_\phi) \tag{9}$$

$$\mathbf{v} = r\sin(\theta)\Omega\,\hat{\mathbf{e}}_\phi + \mathbf{v}_p \tag{10}$$

In this axisymmetric decomposition, the first term on the R.H.S. of Eq. (9) is the toroidal component and the second term the poloidal component of the magnetic field (of which A is the vector potential). The first term on the R.H.S. of Eq. (10) signifies the ϕ-component of the velocity (i.e., angular velocity Ω), while $\mathbf{v}_p$ denotes the meridional (r–θ) component of the velocity.

Our understanding of solar and stellar cycles is based primarily on 2-D kinematic $\alpha\Omega$ dynamo models and full MHD simulations (cf. §2.2, Brun et al. (2013) (also published as Chap. 5 of the ISSI book (Thompson et al. 2015)) and Weiss 1994; Charbonneau 2010), most of which have been constructed in the solar context. Here we briefly review this understanding. The toroidal component of the magnetic field is produced by stretching of an initial poloidal field by the differential rotation inside stars (the dynamo Ω-effect) and is thought to be stored in a layer stable to convection near the base of the convection zone. Strong toroidal flux tubes rise up due to magnetic buoyancy and erupt as magnetic (star) spots. The toroidal flux must be recycled to poloidal flux for the dynamo cycle to perpetuate and this conversion mechanism (traditionally termed as the dynamo α-effect) could be achieved by various means. The two most popular and widely studied mechanisms are the mean-field α-effect driven by helical turbulence (Parker 1955) and the decay and dispersal of tilted bipolar active regions—i.e., bipolar sunspot or starspot pairs (Babcock 1961; Leighton 1969), hereby the BL mechanism. It is thought that strong toroidal flux tubes, believed to be necessary to produce sunspots, would quench the mean-field α-effect, if there is no segregation of the poloidal and toroidal generation process and thus the BL mechanism must be the dominant mechanism for poloidal field generation in the Sun; however, this debate is far from over and we will return to this later. We display in Fig. 7 a typical representation of a BL like dynamo model.

In the kinematic approach, the plasma velocity is prescribed and the magnetic field evolution equation is solved under the assumption of no direct magnetic feedback on the flows. The magnetic induction equation is given by

$$\frac{\partial \mathbf{B}}{\partial t} = \nabla \times (\mathbf{v} \times \mathbf{B} - \eta \nabla \times \mathbf{B}). \tag{11}$$

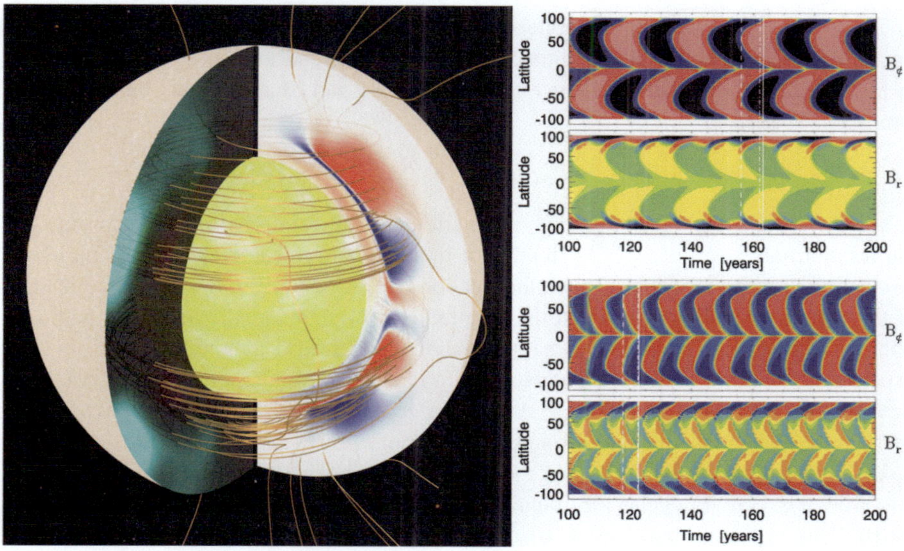

Fig. 7 *Left*: a 2.5D kinematic dynamo simulation depicting a snapshot of the toroidal (right meridional plane) and poloidal field (left meridional place) configurations at a phase close to solar minimum. The simulation shows that the toroidal component of the field is generated and stored deep in the convection zone, from where it buoyantly erupts to form sunspots. The poloidal field is created in the near-surface layers in this particular dynamo model of the Babcock-Leighton type (Nandy et al. 2011). 3-D rendering of field lines (*magenta*) show the expected wound-up field structure in the solar interior. *Right*: Butterfly diagram for 2 models representative of pumping dominated dynamo, rotating at respectively $0.7\Omega_\odot$ and $3.0\Omega_\odot$. We note the equatorward branch, the correct phase relationship between the toroidal and poloidal component of field (shown as *vertical lines*) and the shorter rotation period for the case rotating at three time the solar rate (Do Cao and Brun 2011)

On substituting Eq. (9) and Eq. (10) in the induction equation and separating out the toroidal and poloidal magnetic field components we get the evolution equations for the poloidal and toroidal components, respectively:

$$\frac{\partial A}{\partial t} + \frac{1}{s}\left[\mathbf{v}_p \cdot \nabla(sA)\right] = \eta\left(\nabla^2 - \frac{1}{s^2}\right)A + S_\alpha, \tag{12}$$

$$\frac{\partial B_\phi}{\partial t} + s\left[\mathbf{v}_p \cdot \nabla\left(\frac{B_\phi}{s}\right)\right] + (\nabla \cdot \mathbf{v}_p)B_\phi = \eta\left(\nabla^2 - \frac{1}{s^2}\right)B_\phi + s\left(\left[\nabla \times (A\hat{\mathbf{e}}_\phi)\right] \cdot \nabla\Omega\right) + \frac{1}{s}\frac{\partial(sB'_\phi)}{\partial r}\frac{\partial \eta}{\partial r}. \tag{13}$$

Here $s = r\sin(\theta)$ and η is the turbulent diffusion coefficient. The source term for the poloidal field S_α could denote either the mean-field α-effect (in which case $S_\alpha = \alpha B_\phi$) or the BL source due to the buoyant eruption and flux dispersal of tilted active regions. The flow in the meridional plane $\mathbf{v}_p$ represents meridional circulation ($\mathbf{v}_p = \mathbf{v}_m$, where $\mathbf{v}_m$ denotes the meridional circulation), and in the presence of turbulent pumping (say $\mathbf{v}_\gamma$) it represents the combined effect of both (i.e., $\mathbf{v}_p = \mathbf{v}_m + \mathbf{v}_\gamma$). Appropriate prescriptions for the flows, turbulent diffusion and nature and profile of the poloidal field source along with the boundary conditions, completely describes the kinematic dynamo system defined in Eq. (12) and Eq. (13).

In BL flux transport dynamo models, the cycle period P_{cyc} is found to depend strongly on the meridional circulation amplitude v_0 and its profile and less on the rotation rate Ω_0 or the amplitude of the surface source term s_0 (Dikpati and Charbonneau 1999; Jouve and Brun 2007):

$$P_{\text{cyc}} \propto \Omega_0^{0.05} S_0^{0.07} v_0^{-0.83} \tag{14}$$

As will be seen in the next section, the meridional circulation is found to decrease with the rotation rate as $v_0 \propto \Omega_0^{-0.45}$. This is not intuitive as one could expect the meridional circulation velocity to increases with the rotation rate. A careful study of the ϕ component of the vorticity equation shows that the longitudinal vorticity actually decreases with rotation rate as more and more kinetic energy is being transferred to longitudinal motions at the expense of meridional motions. The fact that in recent 3-D simulations (Ballot et al. 2007; Brown et al. 2008; Matt et al. 2011; Augustson et al. 2012), the meridional circulation is found to weaken as the models is rotated faster implies that standard advection dominated flux transport dynamo models yield the opposite dependency with rotation than the one observed, e.g. activity cycles are found to be longer for faster rotating stars (Jouve et al. 2010). This fact alone requires us to revise our current dynamo paradigm for solar-like stars. One way is to short-circuit the advection path by, for instance, adding more cells in latitude or increasing the radial diffusion as was done in Jouve et al. (2010), Hazra et al. (2014); another is to introduce magnetic turbulent pumping (Mason et al. 2008; Guerrero and de Gouveia Dal Pino 2008; Do Cao and Brun 2011). These large-scale magnetic flux transport processes, e.g. meridional circulation, turbulent diffusion and turbulent (or topological) pumping of magnetic flux are extremely important for dynamo operation because they link the spatially segregated source layers of the dynamo, specifically coupling the near-surface layers to the deep convection zone. The competition between these processes influences the overall behavior of the dynamo significantly (Jouve and Brun 2007; Yeates et al. 2008; Guerrero and de Gouveia Dal Pino 2008; Karak and Nandy 2012; Do Cao and Brun 2011). Meridional circulation is also thought to critically influence the nature of magnetic butterfly diagrams, including their structure, overlap between cycles and cycle periods (Charbonneau and Dikpati 2000; Nandy and Choudhuri 2002; Jouve and Brun 2007; Nandy et al. 2011; Hazra et al. 2014) and also impacts scaling relationships in stellar activity (Nandy 2004; Jouve et al. 2010). However, the exact profile of the meridional flow remains unconstrained in stars. For the Sun, evidence for more than one meridional flow cell are being reported (Haber et al. 2002; Mitra-Kraev and Thompson 2007; Zhao et al. 2013) and there is serious debate about its relative role vis-a-vis other flux transport processes.

For instance new stellar dynamo models including magnetic pumping (Do Cao and Brun 2011) yield the following dependency for the cycle period P_{cyc}:

$$P_{\text{cyc}} \propto v_0^{-0.40} \gamma_{r0}^{-0.30} \gamma_{\theta 0}^{-0.15} \tag{15}$$

where γ_{r0} and $\gamma_{\theta 0}$ represent the amplitude of the radial and latitudinal turbulent pumping terms. We see that the cycle period dependency on the meridional circulation is reduced and that if we want to recover the observational trends, in particular the shorter cycle period (again, we assume that $v_0 \propto \Omega^{-0.45}$), and assuming that γ_r/γ_θ remains constant, the pumping effect should roughly scale as Ω_0^2. Such a scaling may be too extreme and only systematic 3-D numerical simulations will tell us if this is the case or not. Nevertheless, pumping dominated stellar dynamos are a plausible solution to explain stellar observations.

2.2.2 *Stellar Activity and Magnetism: Confronting Mean Field Dynamo Models with Observations*

The general trends in stellar activity observations, in particular the rotation, cycle amplitude, and cycle period relationship is succinctly expressed when confronted with the stellar Rossby number, $R_o = P_\Omega/\tau_c$, the ratio of the rotation period, P_Ω to the convective turn-over-time of the star (Durney and Latour 1978; Baliunas et al. 1996; Montesinos et al. 2001; Jouve et al. 2010). This number can be calculated using a combination of observations with some theoretical ideas about the nature of stellar convection zones, assuming that the convective turn-over-time is $\tau_c = L/V = L^2/\eta$, where L is a characteristic length scale, V a typical convective speed and η the magnetic diffusivity. The latter equality is valid if the magnetic Reynolds number $R_m = VL/\eta$ is taken to be of order unity, an assumption that depends on the properties of stellar convection zones and on the convection zone depth. It can be interpreted as characterizing turbulent mixing of magnetic fields in stellar interiors, while the rotation period plays a role in determining differential rotation. It is well known (both theoretically and observationally) that the differential rotation in the convective envelope of solar-type stars is directly connected to the star's rotation rate Ω_0 (Donahue et al. 1996; Barnes et al. 2005; Ballot et al. 2007; Brown et al. 2008; Küker and Rüdiger 2008; Matt et al. 2011; Augustson et al. 2012). However, the exact scaling n_r (i.e. $\Delta\Omega \propto \Omega_0^{n_r}$) is still a matter of debate among the observers and theoreticians, being sensitive to the observational techniques used and to the modelling approach (see Sect. 2.2). Note that the α-effect used in mean field theory is related to helical turbulence and represents a measure or parameterization of the electromotive mean force (emf) (e.g. Moffatt 1978; Pouquet et al. 1976). It is thus also linked to the rotation rate of the star and the amount of kinetic helicity present in its convective envelope. However in Babcock-Leighton flux transport dynamo models the classical α-effect is replaced by a surface term linked to the tilt of the active regions with respect to the east-west direction (the so-called Joy's law, e.g. Hale et al. 1919; Babcock 1961; Kosovichev and Stenflo 2008; Dasi-Espuig et al. 2010). This tilt is thought to be due to the action of the Coriolis force during the rise of the toroidal structures that emerge as active regions (D'Silva and Choudhuri 1993). Recent 3-D simulations in spherical shells (Jouve and Brun 2009; Fan 2009; Weber et al. 2011, 2013; Pinto and Brun 2013; Jouve et al. 2013) seem to indicate that this is not the only effect responsible for the observed tilt and that the twist and arching of the toroidal structures as well as the continuous action of the surface convection during the emergence influence the resulting tilt. The Rossby number, overall, characterizes stellar convection zones—the zone where the dynamo mechanism generates stellar magnetic fields. Not surprisingly, the Rossby number is simply related through scaling arguments to the dynamo number, $D \sim 1/R_o{}^2$. Thus the Rossby number can be utilized to relate stellar observations to the dynamo mechanism (see e.g., Noyes et al. 1984a, 1984b; Tobias 1998; Montesinos et al. 2001; Nandy 2004).

On can also implement a non linear feed back via the large scale Lorentz force (also called the Malkus-Proctor effect; Malkus and Proctor 1975) on the dynamo field generation. This has the advantage to yield a saturation (quenching) process based on the actual dynamics of the models (Tobias 1997; Moss and Brooke 2000; Bushby 2006, and references therein). These models demonstrate that for low magnetic Prandtl number (e.g. $P_m = \nu/\eta$), such that the viscous time scale is much longer than the magnetic diffusion time scale, the dynamo is irregular, with grand minimum-like behavior and switching of parity between symmetric and antisymmetric dynamo families as observed in the Sun (DeRosa et al. 2012).

From the observational perspective, analysis shows that stars with low Rossby numbers typically exhibit strong activity amplitudes and have irregular cycles, while those stars with relatively higher Rossby numbers tend to have lower activity levels, are more likely to host magnetic cycles and are more likely to be found in grand minima like phases. Since lower Rossby number indicates a higher, underlying dynamo number, and vice-versa, this result is consistent with dynamo theory. Stars which host very efficient dynamos (having high D) are expected to have high activity levels; the non-linear nature of the dynamo mechanism, in conjunction with a high dynamo number also generates irregular cycles because the magnetic feedback is likely stronger. On the other hand, stars with low or moderate dynamo numbers are relatively less active, and are more likely to host regular cyclic behavior with occasional Maunder-like states—such as the Sun. The consistency of stellar observations with the theoretically expected behaviour of a MHD dynamo mechanism leaves little doubt that internal dynamos are the ultimate source of stellar activity.

A closer look at the activity period versus rotation rate relationship yields more stringent constraints on the nature of stellar dynamos. In particular, it is seen that the periods of stellar magnetic cycles decrease with increasing rotation rate (i.e., decreasing rotation period P_{Ω}). As discussed in Sect. 2.1, there is an inter-dependance between dynamo action, magnetic activity levels, rotation and stellar evolution via the torque that magnetized thermally driven stellar winds apply on solar-like stars. The mass dependance of this torque is subject to intense studies (see Bouvier 2013, for a recent discussion). It is understood that solar dynamo models that rely on the BL mechanism for poloidal field generation are critically dependent on flux transport processes such as meridional circulation which couples the source regions for the toroidal and poloidal components of the magnetic field. Such BL dynamo models have been successful in reproducing various features of the solar cycle. Based on such a BL dynamo model, Nandy (2004) has shown that the stellar activity amplitude and period versus rotation rate relationship can only be reproduced if both the rotational shear and meridional flow speed scales positively with rotation rate. While full MHD numerical simulations show that the former holds, they also indicate that meridional circulation speed does not increase with rotate rate (Augustson et al. 2012). Do Cao and Brun (2011) point out in a recent study that the inclusion of turbulent pumping, with the speed of the former increasing with rotation rate, is necessary to recover the stellar scaling laws in the light of these new developments in full MHD simulations. Dubé and Charbonneau (2013) have also computed 2-D kinematic dynamo using ingredients (such as the alpha effect) deduced from 3-D simulations. They show that essential ingredients of the complex 3-D dynamics (such as for instance the cycle period or the sense of propagation of the dynamo wave) can be recovered via such techniques. 3-D kinematic dynamo models or convective B-L models have also recently been developed (Chan et al. 2007; Miesch and Brown 2012; Miesch and Dikpati 2014).

Hence confronting stellar activity observations with diverse simulation techniques is very useful and allows the community to test new ideas and concepts.

We now discuss in more details nonlinear, 3-D MHD simulations of solar convection zone and magnetism.

2.3 3-D Global Models of Stellar Convection and Dynamo Action

With the advent of massively parallel computers it is now possible to attack the difficult problem of stellar dynamics and magnetism with direct 3-D nonlinear numerical simulations. As is frequently the case with multidimensional MHD numerical simulations, two approaches are possible: local high resolution simulations that describe the small scale turbulent motions and local dynamo (Cattaneo 1999; Vögler and Schüssler 2007) and global ones that take into account the correct geometry and the existence of large scale flows in rotating convective zones Glatzmaier (1985a, 1985b); Brun et al. (2004), Miesch and Toomre

(2009). Given the broadness of the topic, here we present the progress made in modeling stellar convection and dynamos with the global approach. These simulations allow us to describe turbulent, rotating and magnetized convection either in a full sphere, a shell or a wedge-like geometry (cf. the recent book by Charbonneau 2013a). Contrary to the kinematic mean field stellar dynamo approach discussed in the previous sections, 3-D simulations use the full set of MHD equations (we refer to these recent references for simpler 2-D models of stellar differential rotation assuming a parametrization of turbulent convection Küker and Rüdiger 2008; Hotta and Yokoyama 2011; Kitchatinov and Olemskoy 2012). These equations couple classical fluid mechanic equations to the induction equation and as such do not prescribe the velocity field but compute it self consistently (cf. Brun et al. 2004; Miesch 2005). This is the main advantage of the global full MHD approach: the influence of rotation and magnetic field on turbulence and convection is computed not parametrized.

Over the last three decades several groups have developed stellar convection and dynamo models in spherical geometry following the pioneering work of P. Gilman and G. Glatzmaier in the early-mid 80's (Gilman and Miller 1981; Glatzmaier and Gilman 1982; Gilman 1983; Glatzmaier 1985a, 1985b). Prominent examples of modern global or wedge-like parallel codes that have in recent years published original studies on stellar convection and dynamo include:

- the Anelastic Spherical Harmonics (ASH) code: Miesch et al. (2000), Brun and Toomre (2002), Ballot et al. (2007), Miesch et al. (2008), Browning (2008), Bessolaz and Brun (2011), Augustson et al. (2012) on surface convection and zonal flows and Brun (2004), Brun et al. (2004, 2011), Browning et al. (2006), Brown et al. (2008, 2010, 2011), Miesch and Brown (2012), Nelson et al. (2013, 2014), Augustson et al. (2013) on solar-like star dynamo action, cycles and magnetic wreaths,
- the Eulag MHD code: Ghizaru et al. (2010), Racine et al. (2011), Charbonneau (2013b) on solar and stellar dynamo and cycles,
- the PENCIL code: Warnecke et al. (2013), Käpylä et al. (2013), Guerrero et al. (2013) on stellar convection, dynamos and cycles,
- the Magics code: Christensen and Aubert (2006); Christensen et al. (2009), Christensen (2010), Gastine et al. (2013, 2014) on convection and dynamos
- K. Chan's code Cai et al. (2011) on core convection

Other codes have been mostly used to model the geodynamo with some applications to the dynamo in M dwarfs.

- Glatzmaier's code: adapted to model Earth's dynamo Glatzmaier and Roberts (1996) or giant planets dynamics Glatzmaier et al. (2009),
- EBE code: Chan et al. (2007),
- Leeds code Hollerbach and Jones (1993), Sreenivasan and Jones (2006); dynamo review by Jones (2011),
- PARODY code: Goudard and Dormy (2008), Morin et al. (2011), Schrinner et al. (2012, 2014) on geo and stellar dynamos and Aubert et al. (2007) on the role of thermal wind in setting mean large scale flows and Aubert et al. (2008) on dynamo
- Busse and Simitev's code: Grote and Busse (2001), Simitev and Busse (2009) on dynamo and bistability
- Aurnou and Heimpel's code: Heimpel et al. (2005), Aurnou et al. (2007) on dynamo in convective envelopes
- Several research groups in Japan: Takehiro and Hayashi (1999), Yoshida and Kageyama (2004), Kimura et al. (2011) on convection or zonal flows or Kageyama and Sato (1997), Nishikawa and Kusano (2008), Masada et al. (2013) on dynamo, solar convection and

magnetism, the most recent studies making use of a novel "ying-yang" grid to map a spherical shell (Kageyama and Sato 2004).

Many of the codes used in the geophysics community were originally developed to model the geodynamo. They therefore rely on the Boussinesq approximation, which is not appropriate for highly stratified and compressible convective envelope as found in solar-like stars. Modelers have been converting these codes to be able to deal with stellar-like stratified flow by using the anelastic approximation (Gough 1969; Glatzmaier 1984; Vasil et al. 2013) and validating their codes with a recent international benchmark (Jones et al. 2011).

Given the intrinsic complexity of computing nonlinear convection and dynamo simulations and the large diversity of stars to model, we first briefly describe purely hydrodynamical models (see also Chap. 3 in the ISSI book (Thompson et al. 2015)). Such models reveal the fundamental ingredients at the origin of large scale mean flow, such as differential rotation and meridional circulation in rotating global convection models. We then follow up by describing the origin of magnetic fields (either cyclic or irregular) through nonlinear dynamo action and the feedback of such fields on the flows—making the link with the 2-D models discussed earlier when necessary.

2.3.1 Convection and Large Scale Flows

Convection is present in stars of all masses on the main sequence. It takes either the form of a convective core in massive ones (greater than about 1.3 $M_\odot$) or of a convective envelope of varying thickness for lows mass stars (from early F down to about M3), with stars less massive than about 0.3 $M_\odot$ being fully convective. Here we will limit our discussion to solar-like stars possessing a deep convective envelope. For a discussion on two other categories (e.g. high mass and very low mass stars) interested readers may refer to (e.g. Browning et al. 2004; Brun et al. 2005; Dobler et al. 2006; Brown et al. 2008; Christensen et al. 2009; Featherstone et al. 2009; Augustson et al. 2011; Cai et al. 2011; Morin et al. 2011; Schrinner et al. 2012, and references therein).

Convection plays a key role in transporting the star's heat by having strong correlations between vertical motions and temperature fluctuations. It is a fundamentally turbulent process in stars as the Rayleigh number (*Ra*, e.g. the ratio between buoyancy driving and dissipation process) is huge, of order 10^{18} in the Sun. This leads to a large range of scales from possibly giant cells on the order of the thickness of the convective layer (e.g. 200 Mm in the Sun) down to granulation size (1 Mm), with intermediate scales such as supergranulation (20–50 Mm) well delineated by the surface magnetic network (Rieutord and Rincon 2010). As the luminosity (energy output) of the star changes so does the amplitude of the convective flow. A simple scaling argument based on mixing length treatment (MLT) gives $v_r \propto (L_*/(\rho_{cz} R_*^2))^{1/3}$, with L_* and R_* respectively the star's luminosity and radius and ρ_{cz} the averaged density of the convective envelope (Kippenhahn et al. 2013). Since the luminosity varies by at least 2 orders of magnitude across the spectral type we are considering (i.e. early F to early M) ($L_* \propto M_*^4$), the stellar radius only by a few ($R_* \propto M_*^{0.9}$), it is expected that the vigor of the convective flows do likewise modulo the effect of density. As can been seen in Fig. 8, 1-D stellar structure models of solar-like stars indicate that the averaged mean density changes significantly in the convective envelope of these stars, in such a way that more massive stars have on average a lower mean density than low mass ones (Matt et al. 2011). This is easily understood as the base of the envelope moves up towards the surface with the thinning of the convective envelope. This yields a large amplitude difference between F and K stars for their characteristic convection velocity. This difference in velocity amplitude is recovered in global numerical simulations of stellar convection. We illustrate

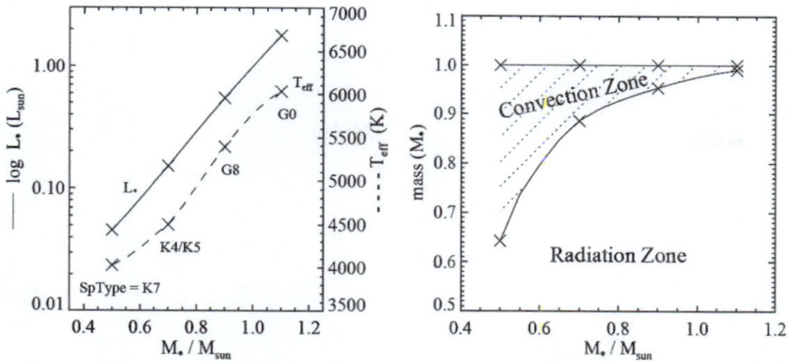

Fig. 8 Stellar luminosity, effective temperature and mass contained in the convective envelope vs. spectral type in solar like stars with the solar metallicity (computed with the CESAM code Morel 1997)

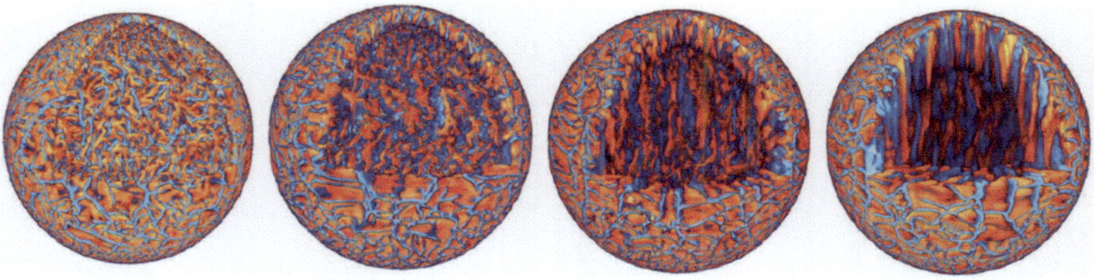

Fig. 9 3-D rendering of turbulent surface convection in solar-like stars with varying convection depths (Bessolaz and Brun 2011). The radial velocity is shown with cold downflows in *blue tones*

an example of such 3-D turbulent convective simulations in Fig. 9, where we clearly see the complex network of downflows surrounding broad upflows.

The turbulent motions within the convective envelopes of rotating stars experience the effect of the Coriolis force to a degree that depends on their fluid Rossby number $Ro = \omega/2\Omega_*$, here defined as the ratio between turbulent and the so-called planetary vorticity (Pedlosky 1982). We note that several definitions of the Rossby number exist in the community (stellar Rossby based on the ratio between convective turnover time and stellar rotation period; convective Rossby $\sqrt{Ra/Ta\,Pr}$; local Rossby number), but it can be shown that they depend almost linearly on one another (see for example Kim and Demarque 1996; Landin et al. 2010; Matt et al. 2014). We here choose to use the fluid Rossby Ro because a clear transition of behavior occurs at $Ro \sim 1$ with the differential rotation becoming prograde or retrograde. Indeed if the stars rotate fast enough such that the large scale motions (larger than the Rossby radius of deformation) feel the continuous action of the Coriolis force as they rise and sink then the redistribution of angular momentum is such that a non uniform rotation profile is established in the convective envelope.

In Fig. 10 we display 4 realizations of the longitudinally averaged differential rotation and meridional circulation computed with the ASH code (Matt et al. 2014, in preparation). These profiles have been obtained by changing either the stellar mass (i.e. aspect ratio and luminosity) or the reference frame rotation rate. We note several trends: As the rotation rate is increased, the differential rotation becomes more and more aligned with the rotation axis (along $\hat{e}_z$) and likewise the meridional circulation posses more and more cells both in latitude and radius. The 1.1 solar mass case rotating at one time solar rate exhibits an anti-solar differential rotation, with slow equator and fast pole. It is indeed the only case with

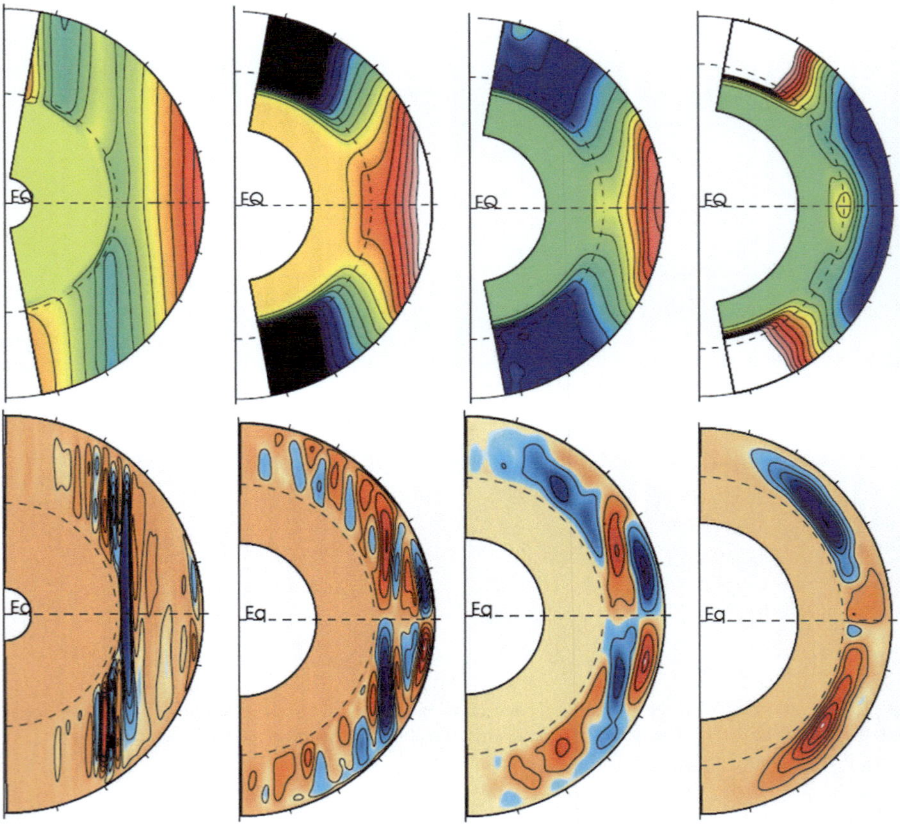

Fig. 10 Differential rotation and meridional circulation realized in 3-D models of solar-like convective envelope coupled nonlinearly (i.e. both zones back react on one another) to a stably stratified interior (Matt et al. 2014, in preparation). 3 masses (0.5, 0.9 and 1.1) $M_\odot$ and 3 rotation rates (solar, 3 and 5 time solar) are being shown. Starting from the left in each row: color contours showing $\Omega(r,\theta)$ and the meridional streamfunction for a 0.5 $M_\odot$ star rotating at 5 $\Omega_\odot$, 0.9 $M_\odot$ star rotating at 3 and 1 $\Omega_\odot$ and 1.1 $M_\odot$ star rotating at $\Omega_\odot$. Prograde rotation is shown in *red/white colors* and retrograde rotation in *dark/blue*. Clockwise meridional flows in *red*

a Rossby number greater than 1 as seen in many recent studies (Bessolaz and Brun 2011; Matt et al. 2011; Guerrero et al. 2013; Gastine et al. 2014; Käpylä et al. 2014). We thus see that the transition between prograde and retrograde differential rotation will not occur at the same rotation rate ($v \sin i$) for each stellar spectral type. We also note the strong shear layer at the base of the convective domain (e.g. the tachocline Spiegel and Zahn 1992) that naturally develops in each modeled stars with a profile closer to the solar case for all the cases with a Rossby number $Ro \sim 0.6$.

These profiles can be understood by considering the transport of angular momentum in a spherical convective shell. When performing this analysis as in Brun and Toomre (2002), Brown et al. (2008), Augustson et al. (2012), we find that the Reynolds stresses are at the origin of the prograde equator for models with a low Rossby number. Through gyroscopic pumping the meridional circulation transport angular momentum to counterbalance the torque applied by the Reynolds stresses (from which we have subtracted the contribution from viscous effect). We refer to Zahn (1992), Mathis and Zahn (2004), McIntyre (2007),

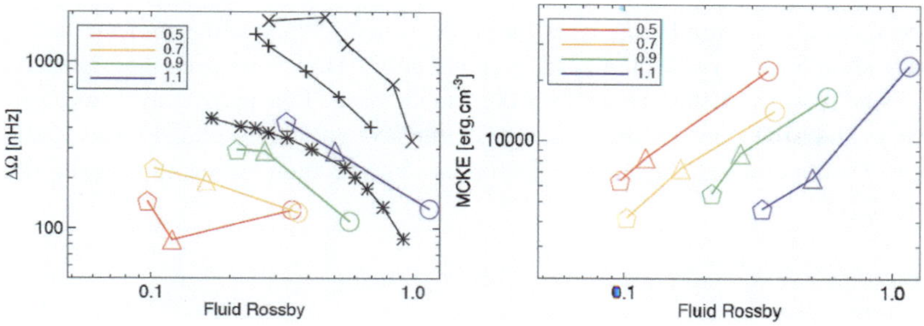

Fig. 11 Differential rotation $\Delta\Omega$ as a function of Rossby number for stellar masses ranging from 0.5 up to 1.3 $M_\odot$. 1 $M_\odot$ cases from Brown et al. (2008), 1.2 and 1.3 $M_\odot$ cases coming from Augustson et al. (2012). Kinetic energy of the axisymmetric meridional circulation as a function of the Rossby number for models ranging from 0.5 to 1.1 $M_\odot$, from Matt et al. (2014, in preparation)

Garaud and Bodenheimer (2010), Brun et al. (2011), Miesch and Hindman (2011) for a more thorough discussion of the role of the meridional circulation in reaching angular momentum balance in rotating stars. The orientation of the iso-contours of Ω from being cylindrical to being more conical (solar-like) at mid latitude is due to the effect of an efficient thermal wind associated with latitudinal entropy variations (with relatively hotter poles and cooler equator) (Miesch et al. 2006). These variations are enhanced further by the presence of a tachocline (Brun et al. 2011).

One important aspect of performing 3-D numerical simulations is that one can obtain general trends on the variations of the main physical processes acting in convective envelope with various parameters such as rotation rate, luminosity, aspect ratio. We focus on scaling laws for the differential rotation and meridional circulation by synthesizing the recently published 3-D stellar convection models computed with the ASH code. The models range from 0.5 up to 1.3 times the solar mass and from 1 to 20 times the solar rotation rate (Brown et al. 2008; Matt et al. 2011; Augustson et al. 2012). In Fig. 11 we show how the latitudinal differential rotation contrast varies as a function of the Rossby number. There is a clear trend for $\Delta\Omega$ to rise with increasing rotation rate. This is in qualitative agreement with direct and indirect observational data that imply larger absolute shear rates in rapid rotators to produce stronger magnetic fields and enhanced angular momentum loss relative to slow rotators. The actual slope seems a bit larger than observed by Collier Cameron (2007) but not so when compared with the work by Donahue et al. (1996). Saar (2011) found that the presence of binaries in the published samples, which can be distinct from single stars because of tidal effects have led to an underestimate of the dependency of $\Delta\Omega$ on rotation rate. Clearly more observational data are required to constrain better the exponent n in the equation $\Delta\Omega \propto \Omega_*^n$. We also clearly see that more massive stars possess larger $\Delta\Omega$ for a given Rossby number. This is in reasonable agreement with the observations published in Barnes et al. (2005). Although, given the paucity of data and the rather large error bars, the exact exponent is uncertain. Data from the *Kepler* mission and in the (future) PLATO mission will be a major asset in this context (Garcia et al. 2014; Rauer et al. 2013).

Another interesting trait is that the meridional circulation is found to decrease in strength as the stars are made to rotate faster. Rapid rotation causes more kinetic energy to go into toroidal motions, which can be anticipated by looking at the equation for gyroscopic pumping (Landin et al. 2010; Augustson et al. 2012). This has a very important consequence: Flux-transport dynamo models as discussed in the previous section can't be easily extended to other stellar spectral type to explain the dependence of their cycle period with their rotation period. Indeed flux-transport models of the advective type depends directly on the meridional circulation amplitude to set the cycle period. Observations indicate that the cycle period should decrease with increasing rotation rate. However, given that the meridional flow is found to weaken, this can only result in an increase of the activity cycle as was shown by Jouve et al. (2010). One needs either to take into account multi-cellular flows as observed also in 3-D simulations (Jouve et al. 2010) or take into account turbulent pumping (Do Cao and Brun 2011).

2.3.2 Nonlinear Dynamo Effect, Magnetic Activity, and Cycles

The observed magnetic activity shows the existence and period of cycles seem to depend on stellar spectral type and rotation rate. An obvious trend is that old (older than the age of the Hyades cluster ~800 Myr), slowly rotating solar-like stars are less active than their younger counterparts. We have also discussed in the previous section that convection and large scale flows change with spectral type. It is certainly not surprising that the associated dynamo action and magnetic activity do so as well.

Several groups have recently published extensive parameter studies of dynamo action in rotating convective spherical shells of various thickness and degree of stratifications. The key results can be summarized as follow: 3-D numerical simulations of dynamo action in solar-like stars have revealed a large range of behavior, from steady dynamo, to irregular and cyclic ones (Brun et al. 2004; Brown et al. 2010, 2011; Racine et al. 2011; Gastine et al. 2012; Schrinner et al. 2012, 2014; Augustson et al. 2013; Käpylä et al. 2013; Nelson et al. 2013, and references therein).

For low mass (M-type) stars, it is seen that a bistability similar to that found for the geodynamo is possible. Observations by Morin et al. (2010) seems to indicate that a strong dipolar state and a multipolar state may coexist for stars having similar stellar parameters. Dynamo models show that around a local Rossby number of 0.1 a weak (multipolar) and strong (dipolar) dynamo branch may coexist (Christensen and Aubert 2006; Morin et al. 2011). Such bistable states are found in a given range of parameters (low stratification, high Pm) and may not actually exist in real stars. It may be the case that these stars are observed in a different phase (low vs. high) of their activity cycle. Other similarity between the geodynamo and low mass stellar dynamos also exist such as the scaling of the magnetic energy with the available heat flux (Christensen et al. 2009). Such scaling tends to break down for higher mass stars for which the tachocline may play a significant role.

Others 3-D simulations of F, G and K stars show that for high rotation rates, large scale magnetic wreaths (see Fig. 13 left panel) are obtained without requiring the presence of a tachocline (Brown et al. 2010, 2011). We show on Fig. 12, four butterfly diagrams (time-latitude plots of the azimuthally averaged toroidal magnetic field near the base of the CZ) realized in such simulations for of a solar-like star rotating at three times the solar rate (Nelson et al. 2013). We remark that as the model is made more turbulent, the steady magnetic wreaths become more time dependent and can lead to cyclic activity (bottom right panel).

In such stars, along with the degree of turbulence, rotation plays an important role in determining the global properties of their magnetism. This is due to a shift in the balance of forces driving the flow between the advection, Coriolis and Lorentz terms. As the rotation rate increases the Lorentz force tends to balance the Coriolis force yielding larger magnetic energy in superequipartion with the kinetic energy of the flow (a direct consequence of a magnetostrophic state; cf. strong scaling below) as in the Earth's iron core. The Elsasser number $\Lambda = B^2/4\pi\bar{\rho}_{cz}\eta\Omega_0$, with $\bar{\rho}_{cz}$ mean density in the convective envelope, η magnetic diffusivity, Ω_0 stellar rotation rate, B a characteristic magnetic field of the CZ, is useful to discuss this balance of terms in the Navier-Stokes (N.V.) equation. Depending on

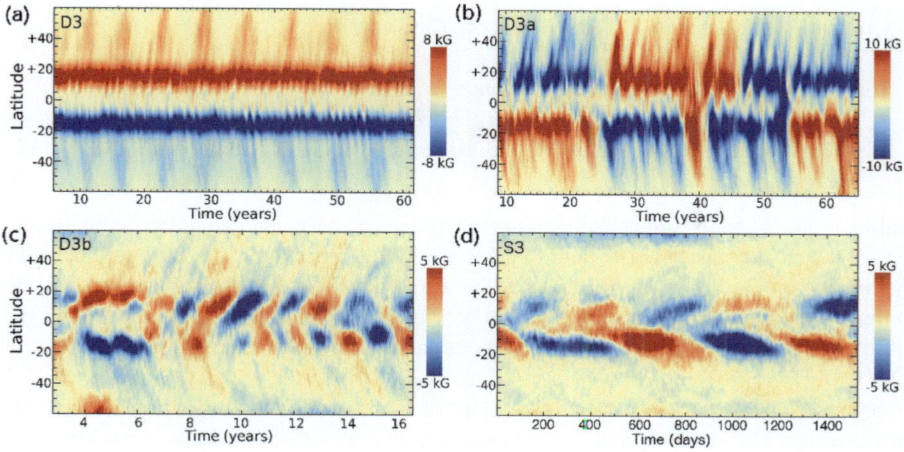

Fig. 12 Magnetic wreaths yielding in turn steady (case D3; *top left*), irregular (case D3a; *top right*) and quasi cyclic (cases D3b & S3; *bottom left* and *right*) magnetic butterfly diagrams (Nelson et al. 2013). *Red tones* correspond to positive toroidal polarity

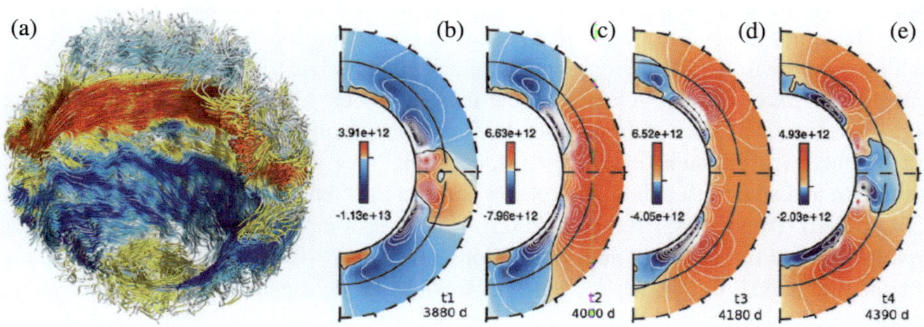

Fig. 13 Magnetic wreaths shown by 3-D rendering of magnetic field lines (Brown et al. 2010). Temporal sequence of meridional cuts showing the mean poloidal field realized in case D5 of Brown et al. (2011) which was extrapolated using potential field approximation to about 1.5 stellar radius. Note the global polarity field reversal of the poloidal field

the amplitude of this number and on the balance assumed in the Navier-Stokes equation, various scaling of the magnetic field amplitude can be expected (Fauve and Pétrélis 2007; Christensen 2010):

- To an order of magnitude, an estimate of the equilibrium magnetic field (assuming ideal gas law) can easily be obtained: $B_{eq} \sim \sqrt{8\pi P_{gas}} \sim \sqrt{\bar{\rho}_{cz}}$, since T_{eff} varies by a factor 2 to 3 between early F and late K stars, whereas $\bar{\rho}_{cz}$ varies by more than a factor of 100 (Matt et al. 2011).

If one assumes for the sake of simplicity that the magnetic Reynolds number $R_m = vd/\eta \sim 1$ such that a characteristic velocity is given by $v \sim \eta/d$,[1] the balance of terms in Navier-Stokes Equation is then given by:

[1] Note however that such choice of characteristic velocity scale is subject to caution in stellar turbulent convective envelopes that usually possess very large Rm.

- Laminar (weak) scaling: Lorentz $\sim$ viscous diffusion
  $\Rightarrow B^2_{\rm weak} \sim \bar{\rho}_{\rm cz} \nu v/d \sim \bar{\rho}_{\rm cz} \nu\eta/d^2$
- Turbulent (equipartition) scaling: Lorentz $\sim$ advection
  $\Rightarrow B^2_{\rm turb} \sim \bar{\rho}_{\rm cz} v^2 \sim \bar{\rho}_{\rm cz}\eta^2/d^2 \Leftrightarrow |B_{\rm weak}| \sim |B_{\rm turb}| P_m^{1/2}$
- Magnetostrophic (strong) scaling (e.g. Elsasser nb $\Lambda \sim 1$): Lorentz $\sim$ Coriolis
  $\Rightarrow B^2_{\rm strong} \sim \bar{\rho}_{\rm cz} \Omega_0 \eta$

with v, d are characteristic velocity and length scales, $P_m = \nu/\eta$ the magnetic Prandtl number. Of course there is an upper limit to the magnitude of the magnetic energy ultimately set by the amount of energy (likely the star's outward energy flux) that can be made available to the dynamo process (Schubert et al. 1996; Christensen et al. 2009). We recall here that dynamo action does not exist for all fluid motions due to its intrinsic 3-D character (Moffatt 1978).

So a possible scenario for stellar dynamos based on heuristic scaling arguments could be the following: Stars rotating at moderate rate (such that their Elsasser number is small), have a level of magnetic energy (or averaged global field strength) that are less than or of the order of the equipartition field given by either the weak or turbulent scalings described above. As stars rotate faster and get closer to a magnetostrophic state with an Elssasser number of order 1 or larger, the formation of large and intense magnetic wreaths starts. The magnetic field becomes more and more dominated by its toroidal component and the magnetic energy becomes larger and larger going above the equipartition value set by the turbulent scaling as it now follows the strong scaling which depends linearly on the rotation rate. The consequence may be the following. As the magnetic energy (or field amplitude) becomes large, the associated Lorentz force starts back reacting strongly on the mean flow. The first consequence is what can be called an "Ω-quenching", e.g. the differential rotation reduces in strength and an almost solid body rotation state in the convective zone (envelope or core) is established (Brun 2004; Brun et al. 2005). Using mean-field classification, one may then hypothesize that the stellar dynamo transitions from an $\alpha - \Omega$ or $\alpha^2 - \Omega$ to being an α^2 dynamo, i.e. helical turbulence is solely responsible for field generation and maintenance, the large scale shear now plays a more marginal role. At that stage what remains of the magnetic wreaths is still unclear, more work must be done. As the rotation is made even faster, quenching of the α effect, due to the large scale magnetic field being more and more intense, occurs naturally as now the Lorentz force back-reacts directly on the helical convective motions and not solely on the large-scale mean flow. Following this discussion, one question naturally crops up which requires detailed investigations: How are these diverse set of plausible dynamo scaling relationships (involving magnetic activity amplitudes) related to stellar activity trends and can the latter impose more stringent constraints on solar and stellar dynamo processes?

As we have seen in Sect. 1.6, the activity of solar-like stars can be assessed by observing their X-ray luminosity L_x. Using such proxy it has been shown that magnetic activity increases with rotation rate and that above a given rotation rate (that depends on the stellar spectral type but is around $R_{\rm os} \sim 0.1$), stellar magnetic activity first saturates (Pizzolato et al. 2003; Wright et al. 2011) and then possibly "over-saturates" for extremely fast rotation rate such as the ones found in young solar-like T-Tauri stars (Feigelson and Montmerle 1999). One can thus simply seek to characterize how the saturation of the dynamo generated field can be linked to the saturation of solar-like star X-ray emissivity. Such a link is not as straightforward to deduce as one could anticipate. Indeed one must also assess how the filling factor f of the magnetic field on the star's surface evolves with stellar parameters not just the field strength. As we have noted the field amplitude does not easily saturate since the Lorentz force can achieve a dynamical balance with various terms whose importance and amplitude vary with stellar parameters. One may conjecture that the first saturation of the X-ray luminosity is due to Ω-quenching and to the physical limitation of the spot coverage (e.g., filling factor f) on the stellar surface and that the second "over-saturation" may be due to "α-quenching", so of the actual field strength (see Gondoin 2012 for an alternative explanation). Currently no global-scale 3-D nonlinear MHD dynamos actually include spot formation and emergence. Global flux emergence simulations have been computed to assess the role of the Coriolis force, Hoop stresses, convective motions, large scale mean flows and background magnetic fields on the evolution and morphology of simplified active regions by varying the properties of idealized global flux ropes (Fan 2009; Jouve and Brun 2009; Weber et al. 2011, 2013; Işık et al. 2011; Jouve et al. 2013; Pinto and Brun 2013) but until very recently, none had self-consistently dealt with the formation of sunspots in a cyclic convective dynamo. So in order to be able to physically characterize the saturation of B and f independently, *spot-dynamos*—wherein dynamos self consistently produce rising omega-loops—must be developed and the parameter space explored systematically.

There has been recent progress in this direction using improved numerical treatment in 3-D simulations of the diffusivities (via dynamical Smagorinsky, slope limited diffusion or implicit large eddy simulation (ILES)). Authors of these studies have been able to lower the diffusivity level low enough to yield more time dependent dynamo solutions possessing both cyclic behavior, regular butterfly diagram and torsional oscillation-like behavior (Käpylä et al. 2011, 2012, 2013; Dubé and Charbonneau 2013; Smolarkiewicz and Charbonneau 2013; Beaudoin et al. 2013; Augustson et al. 2013). Some have even obtained for the first time buoyant magnetic wreaths (Nelson et al. 2011, 2013). In these recent simulations, magnetic wreath-like structures become turbulent and intermittent enough, that many intense bundles of fields reaching 50 kG start to form Ω-loop like structures with statistical properties for rising loops that qualitatively agree with observations (Nelson et al. 2014). We believe that such simulations are the progenitors of future more realistic *spot-dynamos* that will allow us to characterize the link between dynamo action, flux emergence and large scale magnetic topology as a function of stellar parameters.

Finally also note that the torque applied by the magnetized wind (see discussion in Sects. 1.5 and 2.1) depends mostly on the low order fields and that the field topology is expected to change as a function of stellar cycle and overall rotation rate (see Pinto et al. 2011; Reville et al. 2014). Hence assessing the large-scale magnetic topology of stellar magnetic fields and how it changes as a function of stellar parameters is very important as it has a direct bearing on stellar evolution.

3 Future Directions

The characterization of the many facets of stellar dynamics (convection, rotation, magnetism) has made tremendous progress over the last 20 years. We have moved away from a quasi-static view of stars to a time dependent, turbulent and magnetic description of these complex objects. Asteroseismology has significantly contributed to this progress, along with spectropolarimetry and ever improving numerical simulations and models of stellar convection and dynamo action. Still much remains to be done. Many of the assumptions have been deduced by extrapolating our (incomplete) understanding of our star, the Sun, to other stars. The continuous back and forth comparison between the Sun and other stars has been most productive and will continue being so with many projects relying both on ground-based telescopes and space-born missions in the planning. The launch of ESA's M3 Cosmic Vision

PLATO in 2024 (Rauer et al. 2013) will continue the rich tradition of data collection initiated by its predecessors Corot and *Kepler*, helping refine our view of stellar dynamics and magnetism. PLATO has a field of view of more than 2200 square degrees and will observe two different fields for two years. It will combine huge statistics (about 300,000 stars) with high temporal resolution and length of observations to properly extract internal and surface stellar rotations as well as signatures of the existence of stellar cycles. Because one of the PLATO fields will include the *Kepler* field, long-term magnetic variability (of the order of decades) will be potentially unveiled. Fields of shorter observing times of six months (covering in total half of the sky) will add around a million new stars for which oscillations will be measured. This large amount of high-quality seismic data will be a treasure trove for potential new discoveries and will put new constraints on the theories of stellar activity. In the more immediate future, in 2017, the launch of NASA's new planet hunting mission TESS is planned. Although this mission is much smaller in scope than *Kepler*, TESS will observe all stars in the sky down to magnitude 12.5. Apart from the polar regions of the sky (observed for six months), other fields—where the density of stars is higher—will be observed for around 30 days. TESS will therefore contribute significantly to stellar rotation period studies, but it is by design focused on shorter timescale variability studies than PLATO.

Furthermore, other recently launched space missions have already or will soon provide asteroseismic measurements. Apart from the Canadian mission MOST, whose typical 3 weeks observations per target are not necessarily suited for the type of studies we have discussed here, the BRITE constellation of six nano-satellites (http://www.brite-constellation.at) and *Kepler's* K2 missions are also contributing to the generation new constraints. The latter will be particularly important because it provides the ability to employ the new tools of asteroseismology in well-studied star clusters, such as M67, the Pleiades, and the Hyades. K2 will also sample a variety of stellar populations in the Galaxy, important for our understanding of stellar population effects.

In addition to the space-borne instruments, the Danish-led ground-based Stellar Observations Network Group (SONG, Grundahl et al. 2014) will soon provide very high-quality spectroscopic asteroseismic data of distant stars. Asteroseismic data observed in velocity posses a much higher signal to noise ratio compared to the photomoetric data obtained in space, but for much less stars though. Two of the planned eight observing sites, distributed all over the Earth, are already funded, and either in construction (China) or already in operation (Teneriffe). The very high-quality data expected from SONG will allow us to obtain, for example, information from the very outer stellar layers with unprecedented accuracy, information which is much needed to improve our knowledge about the still ill-understood surface-effects (see Sect. 1.2) and the details of the physical mechanisms that drive solar-like oscillations.

The detailed analysis of the legacy star catalog observed by *Kepler*, the exciting and newly planned asteroseismic observation campaigns, the improvements in computation power and, last but not the least, our own improved understanding of the physical processes in stellar interior will not only bring a wealth of new discoveries, but will also bring stellar physics to a new level of sophistication.

On the theoretical side, recent advances based on analytical work and multi-D models of stars of various mass, rotation rates and age, have started paving the way to a direct comparison with asteroseismic and spectropolarimetric data. Simulations coupling the angular evolution and magnetic activity on long time scale are in the making and should help explain the rotation history of stars. Acoustic and gravity modes excitation and propagation are now computed in detail and stellar cycles have now been found in many self-consistent convective dynamo simulations. General trends such as getting higher activity levels and shorter cycle period for faster rotating stars have been recovered. Still the exact dynamo mechanisms at work and resulting magnetic properties (topology, intensity, variability) for any given star are not yet fully elucidated. This comes about because the relative importance, spatial and temporal phasing and nonlinear interactions of the identified processes (e.g. shear in the tachocline, α and ω effects, stability of the magnetic wreaths, role of the meridional flow and turbulent pumping) at the origin of the organisation of the global (large scale) dynamo and the establishment of a cyclic magnetic activity vary. Thus the characterisation of the exact dependence of the stellar dynamo properties to stellar parameters such as mass, rotation/age and stratification is still an evolving understanding. In order to progress further theoretical models and high performance multi-D numerical simulations will benefit from having the following inputs/constraints:

- better constrained variations of the differential rotation in solar-like stars as a function of stellar parameters (mass, rotation/age, metallicity) and its latitudinal and radial profiles
- improved characterization of the $P_{\rm cyc}$ vs. $P_{\rm rot}$ relation
- some assessment of the surface horizontal flows
- convective power and extent of the convective envelopes
- link between star spot numbers and stellar parameters, relative contribution to stellar brightness between dark spot, and bright faculae
- high cadence spectropolarimetric maps over the course of many years hence possibly mapping stellar cycles for many different stars
- calibration with asteroseismology of gyrochronology by studying stellar clusters of various ages

It is our hope that in the next decade or so, these multitude of information will become available and will aid the community in constraining and further developing theories and computational models of solar and stellar activity.

Acknowledgements This work was partly supported by ERC STARS2 207430 grant, ANR Toupies and IDEE grants, FP7 IRSES 269194 and SPACEINN 312844 programs, INSU/PNST. Funding for the Stellar Astrophysics Centre is provided by The Danish National Research Foundation (Grant agreement no.: DNRF106). The research is supported by the ASTERISK project (ASTERoseismic Investigations with SONG and *Kepler*) funded by the European Research Council (Grant agreement no.: 267864). CESSI is supported by the Ministry of Human Resource Development, Government of India and IISER Kolkata. A.S.B and R.A.G. acknowledge support by the CNES via GOLF, CoRoT and Solar Orbiter grants. D.N. acknowledges support from the Ramanujan Fellowship (Department of Science and Technology) and a grant from the (United States) Asian Office of Aerospace Research and Development. A.S. Brun is grateful to both the University of Kyoto, RIMS and Prof. M. Yamada, K. Shibata, H. Isobe & S. Takahiro for their invitation in Fall 2013 and to the University of California at Santa Barbara, KITP and the organisers of the waves and flow program in Spring 2014 where most of this paper was written.

References

Q.R. Ahmad, R.C. Allen, T.C. Andersen, J.D. Anglin, et al., Measurement of the rate of $\nu_e + d -> p + p + e^-$ interactions produced by ^{8}B solar neutrinos at the sudbury neutrino observatory. Phys. Rev. Lett. **87**(7), 071301 (2001). doi:10.1103/PhysRevLett.87.071301

S. Aigrain, S. Hodgkin, J. Irwin, L. Hebb, M. Irwin, F. Favata, E. Moraux, F. Pont, The monitor project: searching for occultations in young open clusters. Mon. Not. R. Astron. Soc. **375**, 29–52 (2007). doi:10.1111/j.1365-2966.2006.11303.x

S. Allain, Modelling the angular momentum evolution of low-mass stars with core-envelope decoupling. Astron. Astrophys. **333**, 629–643 (1998)

N. Andronov, M.H. Pinsonneault, D.M. Terndrup, Mergers of close primordial binaries. Astrophys. J. **646**, 1160–1178 (2006). doi:10.1086/505127

M. Anguera Gubau, P.L. Palle, F. Perez Hernandez, C. Regulo, T. Roca Cortes, The low L solar p-mode spectrum at maximum and minimum solar activity. Astron. Astrophys. **255**, 363–372 (1992)

T. Appourchaux, E. Michel, M. Auvergne, A. Baglin, et al., CoRoT sounds the stars: p-mode parameters of Sun-like oscillations on HD 49933. Astron. Astrophys. **488**, 705–714 (2008). doi:10.1051/0004-6361:200810297

M. Asplund, N. Grevesse, A.J. Sauval, C. Allende Prieto, D. Kiselman, Line formation in solar granulation. IV. [O I], O I and OH lines and the photospheric O abundance. Astron. Astrophys. **417**, 751–768 (2004). doi:10.1051/0004-6361:20034328

M. Asplund, N. Grevesse, A.J. Sauval, P. Scott, The chemical composition of the Sun. Annu. Rev. Astron. Astrophys. **47**, 481–522 (2009). doi:10.1146/annurev.astro.46.060407.145222

J. Aubert, H. Amit, G. Hulot, Detecting thermal boundary control in surface flows from numerical dynamos. Phys. Earth Planet. Inter. **160**, 143–156 (2007). doi:10.1016/j.pepi.2006.11.003

J. Aubert, J. Aurnou, J. Wicht, The magnetic structure of convection-driven numerical dynamos. Geophys. J. Int. **172**, 945–956 (2008). doi:10.1111/j.1365-246X.2007.03693.x

K.C. Augustson, A.S. Brun, J. Toomre, Convection and dynamo action in B stars, in *IAU Symposium*, ed. by N.H. Brummell, A.S. Brun, M.S. Miesch, Y. Ponty. IAU Symposium, vol. 271 (2011), pp. 361–362. doi:10.1017/S1743921311017790

K.C. Augustson, B.P. Brown, A.S. Brun, M.S. Miesch, J. Toomre, Convection and differential rotation in F-type stars. Astrophys. J. **756**, 169 (2012). doi:10.1088/0004-637X/756/2/169

K.C. Augustson, A.S. Brun, J. Toomre, Dynamo action and magnetic cycles in F-type stars. Astrophys. J. **777**, 153 (2013). doi:10.1088/0004-637X/777/2/153

J. Aurnou, M. Heimpel, J. Wicht, The effects of vigorous mixing in a convective model of zonal flow on the ice giants. Icarus **190**, 110–126 (2007). doi:10.1016/j.icarus.2007.02.024

H.W. Babcock, The topology of the Sun's magnetic field and the 22-year cycle. Astrophys. J. **133**, 572 (1961). doi:10.1086/147060

A. Baglin, M. Auvergne, L. Boisnard, T. Lam-Trong, P. Barge, C. Catala, M. Deleuil, E. Michel, W. Weiss, CoRoT: a high precision photometer for stellar evolution and exoplanet finding, in *36th COSPAR Scientific Assembly*. COSPAR, Plenary Meeting, vol. 36 (2006), p. 3749

S. Baliunas, R. Jastrow, Evidence for long-term brightness changes of solar-type stars. Nature **348**, 520–523 (1990). doi:10.1038/348520a0

S.L. Baliunas, J.H. Horne, A. Porter, D.K. Duncan, J. Frazer, H. Lanning, A. Misch, J. Mueller, R.W. Noyes, D. Soyumer, A.H. Vaughan, L. Woodard, Time-series measurements of chromospheric CA II H and K emission in cool stars and the search for differential rotation. Astrophys. J. **294**, 310–325 (1985). doi:10.1086/163299

S.L. Baliunas, R.A. Donahue, W.H. Soon, J.H. Horne, J. Frazer, et al., Chromospheric variations in main-sequence stars. Astrophys. J. **438**, 269–287 (1995). doi:10.1086/175072

S.L. Baliunas, E. Nesme-Ribes, D. Sokoloff, W.H. Soon, A dynamo interpretation of stellar activity cycles. Astrophys. J. **460**, 848 (1996). doi:10.1086/177014

J. Ballot, S. Turck-Chièze, R.A. García, Seismic extraction of the convective extent in solar-like stars. The observational point of view. Astron. Astrophys. **423**, 1051–1061 (2004). doi:10.1051/0004-6361:20035898

J. Ballot, A.S. Brun, S. Turck-Chièze, Simulations of turbulent convection in rotating young solarlike stars: differential rotation and meridional circulation. Astrophys. J. **669**, 1190–1208 (2007). doi:10.1086/521617

J. Ballot, L. Gizon, R. Samadi, G. Vauclair, et al., Accurate p-mode measurements of the G0V metal-rich CoRoT target HD 52265. Astron. Astrophys. **530**, 97 (2011). doi:10.1051/0004-6361/201116547

N.J. Balmforth, Solar pulsational stability. I—Pulsation-mode thermodynamics. Mon. Not. R. Astron. Soc. **255**, 603–649 (1992)

N.J. Balmforth, D.O. Gough, W.J. Merryfield, Structural changes to the Sun through the solar cycle. Mon. Not. R. Astron. Soc. **278**, 437–448 (1996)

C. Barban, S. Deheuvels, F. Baudin, T. Appourchaux, et al., Solar-like oscillations in HD 181420: data analysis of 156 days of CoRoT data. Astron. Astrophys. **506**, 51–56 (2009). doi:10.1051/0004-6361/200911937

S.A. Barnes, On the rotational evolution of solar- and late-type stars, its magnetic origins, and the possibility of stellar gyrochronology. Astrophys. J. **586**, 464–479 (2003). doi:10.1086/367639

S.A. Barnes, Y.-C. Kim, Angular momentum loss from cool stars: an empirical expression and connection to stellar activity. Astrophys. J. **721**, 675–685 (2010). doi:10.1088/0004-637X/721/1/675

J.R. Barnes, A. Collier Cameron, J.-F. Donati, D.J. James, S.C. Marsden, P. Petit, The dependence of differential rotation on temperature and rotation. Mon. Not. R. Astron. Soc. **357**, 1–5 (2005). doi:10.1111/j.1745-3933.2005.08587.x

G. Basri, L.M. Walkowicz, N. Batalha, R.L. Gilliland, et al., Photometric variability in Kepler target stars: The Sun among stars a first look. Astrophys. J. Lett. **713**, 155–159 (2010). doi:10.1088/2041-8205/713/2/L155

G. Basri, L.M. Walkowicz, N. Batalha, R.L. Gilliland, et al., Photometric variability in Kepler target stars. II. An overview of amplitude, periodicity, and rotation in first quarter data. Astron. J. **141**, 20 (2011). doi:10.1088/0004-6256/141/1/20

S. Basu, H.M. Antia, Helium abundance in the solar envelope. Mon. Not. R. Astron. Soc. **276**, 1402–1408 (1995)

S. Basu, N. Grevesse, S. Mathis, S. Turck-Chièze, Understanding the internal chemical composition and physical processes of the solar interior. Space Sci. Rev. (2014). doi:10.1007/s11214-014-0035-9

P. Beaudoin, P. Charbonneau, E. Racine, P.K. Smolarkiewicz, Torsional oscillations in a global solar dynamo. Sol. Phys. **282**, 335–360 (2013). doi:10.1007/s11207-012-0150-2

T.R. Bedding, D. Huber, D. Stello, Y.P. Elsworth, et al., Solar-like oscillations in low-luminosity red giants: first results from Kepler. Astrophys. J. Lett. **713**, 176–181 (2010). doi:10.1088/2041-8205/713/2/L176

J. Beer, S. Tobias, N. Weiss, An active Sun throughout the Maunder minimum. Sol. Phys. **181**, 237–249 (1998)

K. Belkacem, R. Samadi, M.-J. Goupil, M.-A. Dupret, Stochastic excitation of non-radial modes. I. High-angular-degree p modes. Astron. Astrophys. **478**, 163–174 (2008). doi:10.1051/0004-6361:20077775

K. Belkacem, R. Samadi, M. Goupil, L. Lefèvre, et al., Solar-like oscillations in a massive star. Science **324**, 1540 (2009). doi:10.1126/science.1171913

K. Belkacem, M.J. Goupil, M.A. Dupret, R. Samadi, F. Baudin, A. Noels, B. Mosser, The underlying physical meaning of the $\nu_{\max} - \nu_c$ relation. Astron. Astrophys. **530**, 142 (2011). doi:10.1051/0004-6361/201116490

C.P.M. Bell, T. Naylor, N.J. Mayne, R.D. Jeffries, S.P. Littlefair, Pre-main-sequence isochrones—II. Revising star and planet formation time-scales. Mon. Not. R. Astron. Soc. **434**, 806–831 (2013). doi:10.1093/mnras/stt1075

E.E. Benevolenskaya, Double magnetic cycle of solar activity. Sol. Phys. **161**, 1–8 (1995). doi:10.1007/BF00732080

S.V. Berdyugina, Starspots: a key to the stellar dynamo. Living Rev. Sol. Phys. **2**, 8 (2005). doi:10.12942/lrsp-2005-8

N. Bessolaz, A.S. Brun, Hunting for giant cells in deep stellar convective zones using wavelet analysis. Astrophys. J. **728**, 115 (2011). doi:10.1088/0004-637X/728/2/115

E. Böhm-Vitense, Über die Wasserstoffkonvektionszone in Sternen verschiedener Effektivtemperaturen und Leuchtkräfte. Mit 5 Textabbildungen. Z. Astrophys. **46**, 108 (1958)

E. Böhm-Vitense, Chromospheric activity in G and K main-sequence stars, and what it tells us about stellar dynamos. Astrophys. J. **657**, 486–493 (2007). doi:10.1086/510482

A. Bonaca, J.D. Tanner, S. Basu, W.J. Chaplin, et al., Calibrating convective properties of solar-like stars in the Kepler field of view. Astrophys. J. Lett. **755**, 12 (2012). doi:10.1088/2041-8205/755/1/L12

W.J. Borucki, D. Koch, G. Basri, N. Batalha, T. Brown, et al., Kepler planet-detection mission: introduction and first results. Science **327**, 977 (2010). doi:10.1126/science.1185402

J. Bouvier, Observational studies of stellar rotation, in *EAS Publications Series*. EAS Publications Series, vol. 62 (2013), pp. 143–168. doi:10.1051/eas/1362005

A.-M. Broomhall, D. Salabert, W.J. Chaplin, R.A. García, Y. Elsworth, R. Howe, S. Mathur, Misleading variations in estimated rotational frequency splittings of solar p modes: consequences for helioseismology and asteroseismology. Mon. Not. R. Astron. Soc. **422**, 3564–3573 (2012). doi:10.1111/j.1365-2966.2012.20868.x

T.M. Brown, The metastable dynamo model of stellar rotational evolution. Astrophys. J. **789**, 101 (2014). doi:10.1088/0004-637X/789/2/101

T.M. Brown, C.A. Morrow, Depth and latitude dependence of solar rotation. Astrophys. J. Lett. **314**, 21–26 (1987). doi:10.1086/184843

T.M. Brown, R.L. Gilliland, R.W. Noyes, L.W. Ramsey, Detection of possible p-mode oscillations on Procyon. Astrophys. J. **368**, 599–609 (1991). doi:10.1086/169725

B.P. Brown, M.K. Browning, A.S. Brun, M.S. Miesch, J. Toomre, Rapidly rotating suns and active nests of convection. Astrophys. J. **689**, 1354–1372 (2008). doi:10.1086/592397

B.P. Brown, M.K. Browning, A.S. Brun, M.S. Miesch, J. Toomre, Persistent magnetic wreaths in a rapidly rotating Sun. Astrophys. J. **711**, 424–438 (2010). doi:10.1088/0004-637X/711/1/424

B.P. Brown, M.S. Miesch, M.K. Browning, A.S. Brun, J. Toomre, Magnetic cycles in a convective dynamo simulation of a young solar-type star. Astrophys. J. **731**, 69 (2011). doi:10.1088/0004-637X/731/1/69

M.K. Browning, Simulations of dynamo action in fully convective stars. Astrophys. J. **676**, 1262–1280 (2008). doi:10.1086/527432

M.K. Browning, A.S. Brun, J. Toomre, Simulations of core convection in rotating A-type stars: differential rotation and overshooting. Astrophys. J. **601**, 512–529 (2004). doi:10.1086/380198
M.K. Browning, M.S. Miesch, A.S. Brun, J. Toomre, Dynamo action in the solar convection zone and tachocline: pumping and organization of toroidal fields. Astrophys. J. Lett. **648**, 157–160 (2006). doi:10.1086/507869
A.S. Brun, On the interaction between differential rotation and magnetic fields in the Sun. Sol. Phys. **220**, 333–345 (2004). doi:10.1023/B:SOLA.0000031384.75850.68
A.S. Brun, S. Mathis, Angular momentum transport in stars: from short to long time scales. Extraterrestrial seismology book, (2014)
A.S. Brun, J. Toomre, Turbulent convection under the influence of rotation: sustaining a strong differential rotation. Astrophys. J. **570**, 865–885 (2002). doi:10.1086/339228
A.S. Brun, S. Turck-Chièze, P. Morel, Standard solar models in the light of new helioseismic constraints. I. The solar core. Astrophys. J. **506**, 913–925 (1998). doi:10.1086/306271
A.S. Brun, S. Turck-Chièze, J.P. Zahn, Standard solar models in the light of new helioseismic constraints. II. Mixing below the convective zone. Astrophys. J. **525**, 1032–1041 (1999). doi:10.1086/307932
A.S. Brun, M.S. Miesch, J. Toomre, Global-scale turbulent convection and magnetic dynamo action in the solar envelope. Astrophys. J. **614**, 1073–1098 (2004). doi:10.1086/423835
A.S. Brun, M.K. Browning, J. Toomre, Simulations of core convection in rotating A-type stars: magnetic dynamo action. Astrophys. J. **629**, 461–481 (2005). doi:10.1086/430430
A.S. Brun, M.S. Miesch, J. Toomre, Modeling the dynamical coupling of solar convection with the radiative interior. Astrophys. J. **742**, 79 (2011). doi:10.1088/0004-637X/742/2/79
A.S. Brun, M.K. Browning, M. Dikpati, H. Hotta, A. Strugarek, Recent advances on solar global magnetism and variability. Space Sci. Rev. (2013). doi:10.1007/s11214-013-0028-0
H. Bruntt, Accurate fundamental parameters of CoRoT asteroseismic targets. The solar-like stars HD 49933, HD 175726, HD 181420, and HD 181906. Astron. Astrophys. **506**, 235–244 (2009). doi:10.1051/0004-6361/200911925
P.J. Bushby, Zonal flows and grand minima in a solar dynamo model. Mon. Not. R. Astron. Soc. **371**, 772–780 (2006). doi:10.1111/j.1365-2966.2006.10706.x
T. Cai, K.L. Chan, L. Deng, Numerical simulation of core convection by a multi-layer semi-implicit spherical spectral method. J. Comput. Phys. **230**, 8698–8712 (2011). doi:10.1016/j.jcp.2011.08.014
T.L. Campante, R. Handberg, S. Mathur, T. Appourchaux, et al., Asteroseismology from multi-month Kepler photometry: the evolved Sun-like stars KIC 10273246 and KIC 10920273. Astron. Astrophys. **534**, 6 (2011). doi:10.1051/0004-6361/201116620
F. Carrier, F. Bouchy, P. Eggenberger, New seismological results on the G0 Iv bootis, in *Asteroseismology Across the HR Diagram*, ed. by M.J. Thompson, M.S. Cunha, M.J.P.F.G. Monteiro (2003), pp. 315–318
F. Cattaneo, On the Origin of magnetic fields in the quiet photosphere. Astrophys. J. Lett. **515**, 39–42 (1999). doi:10.1086/311962
K.H. Chan, K. Zhang, L. Li, X. Liao, A new generation of convection-driven spherical dynamos using EBE finite element method. Phys. Earth Planet. Inter. **163**, 251–265 (2007). doi:10.1016/j.pepi.2007.04.017
S. Chandrasekhar, Letter to the editor: a general variational principle governing the radial and the non-radial oscillations of gaseous masses. Astrophys. J. **138**, 896 (1963). doi:10.1086/147694
W.J. Chaplin, Y. Elsworth, R. Howe, G.R. Isaak, C.P. McLeod, B.A. Miller, H.B. van der Raay, S.J. Wheeler, R. New, BiSON performance. Sol. Phys. **168**, 1–18 (1996)
W.J. Chaplin, Y. Elsworth, G.R. Isaak, R. Lines, C.P. McLeod, B.A. Miller, R. New, An analysis of solar p-mode frequencies extracted from BiSON data: 1991–1996. Mon. Not. R. Astron. Soc. **300**, 1077–1090 (1998). doi:10.1046/j.1365-8711.1998.01999.x
W.J. Chaplin, Y. Elsworth, G.R. Isaak, B.A. Miller, R. New, Variations in the excitation and damping of low-l solar p modes over the solar activity cycle*. Mon. Not. R. Astron. Soc. **313**, 32–42 (2000). doi:10.1046/j.1365-8711.2000.03176.x
W.J. Chaplin, G. Houdek, Y. Elsworth, D.O. Gough, G.R. Isaak, R. New, On model predictions of the power spectral density of radial solar p modes. Mon. Not. R. Astron. Soc. **360**, 859–868 (2005). doi:10.1111/j.1365-2966.2005.09041.x
W.J. Chaplin, Y. Elsworth, G. Houdek, R. New, On prospects for sounding activity cycles of Sun-like stars with acoustic modes. Mon. Not. R. Astron. Soc. **377**, 17–29 (2007). doi:10.1111/j.1365-2966.2007.11581.x
W.J. Chaplin, T. Appourchaux, Y. Elsworth, R.A. García, G. Houdek, et al., The asteroseismic potential of Kepler: first results for solar-type stars. Astrophys. J. Lett. **713**, 169–175 (2010). doi:10.1088/2041-8205/713/2/L169
W.J. Chaplin, T.R. Bedding, A. Bonanno, A. Broomhall, R.A. García, et al., Evidence for the impact of stellar activity on the detectability of solar-like oscillations observed by Kepler. Astrophys. J. Lett. **732**, 5 (2011a). doi:10.1088/2041-8205/732/1/L5

W.J. Chaplin, H. Kjeldsen, T.R. Bedding, J. Christensen-Dalsgaard, et al., Predicting the detectability of oscillations in solar-type stars observed by Kepler. Astrophys. J. **732**, 54 (2011b). doi:10.1088/0004-637X/732/1/54

P. Charbonneau, Dynamo models of the solar cycle. Living Rev. Sol. Phys. **7**, 3 (2010). doi:10.12942/lrsp-2010-3

P. Charbonneau, Solar and stellar dynamos. Society for Astronomical Sciences Annual Symposium **39** (2013a). doi:10.1007/978-3-642-32093-4

P. Charbonneau, Stellar dynamos. Society for Astronomical Sciences Annual Symposium **39**, 187 (2013b)

P. Charbonneau, M. Dikpati, Stochastic fluctuations in a Babcock-Leighton model of the solar cycle. Astrophys. J. **543**, 1027–1043 (2000). doi:10.1086/317142

P. Charbonneau, K.B. MacGregor, Angular momentum transport in magnetized stellar radiative zones. II. The solar spin-down. Astrophys. J. **417**, 762 (1993). doi:10.1086/173357

U.R. Christensen, Dynamo scaling laws and applications to the planets. Space Sci. Rev. **152**, 565–590 (2010). doi:10.1007/s11214-009-9553-2

U.R. Christensen, J. Aubert, Scaling properties of convection-driven dynamos in rotating spherical shells and application to planetary magnetic fields. Geophys. J. Int. **166**, 97–114 (2006). doi:10.1111/j.1365-246X.2006.03009.x

U.R. Christensen, V. Holzwarth, A. Reiners, Energy flux determines magnetic field strength of planets and stars. Nature **457**, 167–169 (2009). doi:10.1038/nature07626

J. Christensen-Dalsgaard, Physics of solar-like oscillations. Sol. Phys. **220**, 137–168 (2004). doi:10.1023/B:SOLA.0000031392.43227.7d

J. Christensen-Dalsgaard, D.O. Gough, Is the sun helium-deficient. Nature **288**, 544–547 (1980). doi:10.1038/288544a0

J. Christensen-Dalsgaard, G. Houdek, Prospects for asteroseismology. Astrophys. Space Sci. **328**, 51–66 (2010). doi:10.1007/s10509-009-0227-z

J. Christensen-Dalsgaard, D.O. Gough, M.J. Thompson, The depth of the solar convection zone. Astrophys. J. **378**, 413–437 (1991). doi:10.1086/170441

J. Christensen-Dalsgaard, W. Dappen, S.V. Ajukov, E.R. Anderson, H.M. Antia, et al., The current state of solar modeling. Science **272**, 1286–1292 (1996). doi:10.1126/science.272.5266.1286

J. Christensen-Dalsgaard, M.J.P.F.G. Monteiro, M. Rempel, M.J. Thompson, A more realistic representation of overshoot at the base of the solar convective envelope as seen by helioseismology. Mon. Not. R. Astron. Soc. **414**, 1158–1174 (2011). doi:10.1111/j.1365-2966.2011.18460.x

A. Claverie, G.R. Isaak, C.P. McLeod, H.B. van der Raay, T.R. Cortes, Solar structure from global studies of the 5-minute oscillation. Nature **282**, 591–594 (1979). doi:10.1038/282591a0

A. Collier Cameron, Differential rotation on rapidly rotating stars. Astron. Nachr. **328**, 1030 (2007). doi:10.1002/asna.200710880

S.R. Cranmer, S.H. Saar, Testing a predictive theoretical model for the mass loss rates of cool stars. Astrophys. J. **741**, 54 (2011). doi:10.1088/0004-637X/741/1/54

O.L. Creevey, F. Thévenin, S. Basu, W.J. Chaplin, L. Bigot, Y. Elsworth, D. Huber, M.J.P.F.G. Monteiro, A. Serenelli, A large sample of calibration stars for Gaia: log *g* from Kepler and CoRoT. ArXiv e-prints (2013)

M. Cuntz, W. Rammacher, P. Ulmschneider, Z.E. Musielak, S.H. Saar, Two-component theoretical chromosphere models for K dwarfs of different magnetic activity: exploring the Ca II emission-stellar rotation relationship. Astrophys. J. **522**, 1053–1068 (1999). doi:10.1086/307689

M. Dasi-Espuig, S.K. Solanki, N.A. Krivova, R. Cameron, T. Peñuela, Sunspot group tilt angles and the strength of the solar cycle. Astron. Astrophys. **518**, 7 (2010). doi:10.1051/0004-6361/201014301

J. de Ridder, C. Barban, F. Baudin, F. Carrier, et al., Non-radial oscillation modes with long lifetimes in giant stars. Nature **459**, 398–400 (2009). doi:10.1038/nature08022

S. Deheuvels, R.A. García, W.J. Chaplin, S.e. Basu, Seismic evidence for a rapidly rotating core in a lower-giant-branch star observed with Kepler. Astrophys. J. **756**, 19 (2012). doi:10.1088/0004-637X/756/1/19

P. Delorme, A. Collier Cameron, L. Hebb, J. Rostron, T.A. Lister, A.J. Norton, D. Pollacco, R.G. West, Stellar rotation in the Hyades and Praesepe: gyrochronology and braking time-scale. Mon. Not. R. Astron. Soc. **413**, 2218–2234 (2011). doi:10.1111/j.1365-2966.2011.18299.x

P.A. Denissenkov, M. Pinsonneault, D.M. Terndrup, G. Newsham, Angular momentum transport in solar-type stars: testing the timescale for core-envelope coupling. Astrophys. J. **716**, 1269–1287 (2010). doi:10.1088/0004-637X/716/2/1269

M.L. DeRosa, A.S. Brun, J.T. Hoeksema, Solar magnetic field reversals and the role of dynamo families. Astrophys. J. **757**, 96 (2012). doi:10.1088/0004-637X/757/1/96

M. Dikpati, P. Charbonneau, A Babcock-Leighton flux transport dynamo with solar-like differential rotation. Astrophys. J. **518**, 508–520 (1999). doi:10.1086/307269

O. Do Cao, A.S. Brun, Effects of turbulent pumping on stellar activity cycles. Astron. Nachr. **332**, 907 (2011). doi:10.1002/asna.201111623
J.D. Do Nascimento, J.S. da Costa, J.R. de Medeiros, Rotation and lithium abundance of solar-analog stars. Theoretical analysis of observations. Astron. Astrophys. **519**, 101 (2010). doi:10.1051/0004-6361/200811026
W. Dobler, M. Stix, A. Brandenburg, Magnetic field generation in fully convective rotating spheres. Astrophys. J. **638**, 336–347 (2006). doi:10.1086/498634
V. Domingo, B. Fleck, A.I. Poland, The SOHO mission: an overview. Sol. Phys. **162**, 1–37 (1995)
R.A. Donahue, S.H. Saar, S.L. Baliunas, A relationship between mean rotation period in lower main-sequence stars and its observed range. Astrophys. J. **466**, 384 (1996). doi:10.1086/177517
J.-F. Donati, J.D. Landstreet, Magnetic fields of nondegenerate stars. Annu. Rev. Astron. Astrophys. **47**, 333–370 (2009). doi:10.1146/annurev-astro-082708-101833
J.-F. Donati, M. Semel, B.D. Carter, D.E. Rees, A. Collier Cameron, Spectropolarimetric observations of active stars. Mon. Not. R. Astron. Soc. **291**, 658 (1997)
J.-F. Donati, I.D. Howarth, M.M. Jardine, P. Petit, C. Catala, J.D. Landstreet, J.-C. Bouret, E. Alecian, J.R. Barnes, T. Forveille, F. Paletou, N. Manset, The surprising magnetic topology of τ Sco: fossil remnant or dynamo output? Mon. Not. R. Astron. Soc. **370**, 629–644 (2006). doi:10.1111/j.1365-2966.2006.10558.x
S. D'Silva, A.R. Choudhuri, A theoretical model for tilts of bipolar magnetic regions. Astron. Astrophys. **272**, 621 (1993)
C. Dubé, P. Charbonneau, Stellar dynamos and cycles from numerical simulations of convection. Astrophys. J. **775**, 69 (2013). doi:10.1088/0004-637X/775/1/69
B.R. Durney, J. Latour, On the angular momentum loss of late-type stars. Geophys. Astrophys. Fluid Dyn. **9**, 241–255 (1978)
W.A. Dziembowski, P.R. Goode, Helioseismic probing of solar variability: the formalism and simple assessments. Astrophys. J. **600**, 464–479 (2004). doi:10.1086/379708
W.A. Dziembowski, P.R. Goode, Sources of oscillation frequency increase with rising solar activity. Astrophys. J. **625**, 548–555 (2005). doi:10.1086/429712
A. Eff-Darwich, S.G. Korzennik, The dynamics of the solar radiative zone. Sol. Phys., 149 (2012). doi:10.1007/s11207-012-0048-z
S. Ekström, C. Georgy, P. Eggenberger, G. Meynet, N. Mowlavi, A. Wyttenbach, A. Granada, T. Decressin, R. Hirschi, U. Frischknecht, C. Charbonnel, A. Maeder, Grids of stellar models with rotation. I. Models from 0.8 to 120 $M_{\odot}$ at solar metallicity ($Z = 0.014$). Astron. Astrophys. **537**, 146 (2012). doi:10.1051/0004-6361/201117751
A.S. Endal, S. Sofia, Rotation in solar-type stars. I—Evolutionary models for the spin-down of the sun. Astrophys. J. **243**, 625–640 (1981). doi:10.1086/158628
C.R. Epstein, M.H. Pinsonneault, How good a clock is rotation? The stellar rotation-mass-age relationship for old field stars. Astrophys. J. **780**, 159 (2014). doi:10.1088/0004-637X/780/2/159
Y. Fan, Magnetic fields in the solar convection zone. Living Rev. Sol. Phys. **6**, 4 (2009). doi:10.12942/lrsp-2009-4
S. Fauve, F. Pétrélis, Scaling laws of turbulent dynamos. C. R. Phys. **8**, 87–92 (2007). doi:10.1016/j.crhy.2006.12.011
N.A. Featherstone, M.K. Browning, A.S. Brun, J. Toomre, Effects of fossil magnetic fields on convective core dynamos in A-type stars. Astrophys. J. **705**, 1000–1018 (2009). doi:10.1088/0004-637X/705/1/1000
E.D. Feigelson, T. Montmerle, High-energy processes in young stellar objects. Annu. Rev. Astron. Astrophys. **37**, 363–408 (1999). doi:10.1146/annurev.astro.37.1.363
S.T. Fletcher, A. Broomhall, D. Salabert, S. Basu, W.J. Chaplin, Y. Elsworth, R.A. Garcia, R. New, A seismic signature of a second dynamo? Astrophys. J. Lett. **718**, 19–22 (2010). doi:10.1088/2041-8205/718/1/L19
E. Fossat, B. Gelly, G. Grec, M. Pomerantz, Search for solar P-mode frequency changes between 1980 and 1985. Astron. Astrophys. **177**, 47 (1987)
C. Fröhlich, Total solar irradiance observations. Surv. Geophys. **33**, 453–473 (2012). doi:10.1007/s10712-011-9168-5
A.H. Gabriel, G. Grec, J. Charra, J.-M. Robillot, et al., Global oscillations at low frequency from the SOHO mission (GOLF). Sol. Phys. **162**, 61–99 (1995)
F. Gallet, J. Bouvier, Improved angular momentum evolution model for solar-like stars. Astron. Astrophys. **556**, 36 (2013). doi:10.1051/0004-6361/201321302
P. Garaud, P. Bodenheimer, Gyroscopic pumping of large-scale flows in stellar interiors and application to lithium-dip stars. Astrophys. J. **719**, 313–334 (2010). doi:10.1088/0004-637X/719/1/313
R.A. García, Solar-like stars observed by Kepler: an incredible adventure, in *Stellar Pulsations: Impact of New Instrumentation and New Insights*, ed. by J.C. Suárez, R. Garrido, L.A. Balona, J. Christensen-Dalsgaard. Advances in Solid State Physics, vol. 31 (2013), p. 171

R.A. García, T. Corbard, W.J. Chaplin, S. Couvidat, A. Eff-Darwich, S.J. Jiménez-Reyes, S.G. Korzennik, J. Ballot, P. Boumier, E. Fossat, C.J. Henney, R. Howe, M. Lazrek, J. Lochard, P.L. Pallé, S. Turck-Chièze, About the rotation of the solar radiative interior. Sol. Phys. **220**, 269–285 (2004). doi:10.1023/B:SOLA.0000031395.90891.ce

R.A. García, S. Mathur, J. Ballot, A. Eff-Darwich, S.J. Jiménez-Reyes, S.G. Korzennik, Influence of low-degree high-order p-mode splittings on the solar rotation profile. Sol. Phys. **251**, 119–133 (2008). doi:10.1007/s11207-008-9144-5

R.A. García, C. Régulo, R. Samadi, J. Ballot, et al., Solar-like oscillations with low amplitude in the CoRoT target HD 181906. Astron. Astrophys. **506**, 41–50 (2009). doi:10.1051/0004-6361/200911910

R.A. García, S. Mathur, D. Salabert, J. Ballot, C. Régulo, T.S. Metcalfe, A. Baglin, CoRoT reveals a magnetic activity cycle in a sun-like star. Science **329**, 1032 (2010). doi:10.1126/science.1191064

R.A. García, S. Hekker, D. Stello, J. Gutiérrez-Soto, et al., Preparation of Kepler light curves for asteroseismic analyses. Mon. Not. R. Astron. Soc. **414**, 6–10 (2011). doi:10.1111/j.1745-3933.2011.01042.x

R.A. García, T. Ceillier, T.L. Campante, G.R. Davies, S. Mathur, PE11 Team of Kepler WG#1, Fast rotating solar-like stars using asteroseismic datasets, in *Progress in Solar/Stellar Physics with Helio- and Asteroseismology*, ed. by H. Shibahashi, M. Takata, A.E. Lynas-Gray. Astronomical Society of the Pacific Conference Series, vol. 462 (2012), p. 133

R.A. Garcia, D. Salabert, S. Mathur, C. Regulo, J. Ballot, G.R. Davies, A. Jimenez, R. Simoniello, Towards solar activity maximum 24 as seen by GOLF and VIRGO/SPM instruments. ArXiv e-prints (2013)

R.A. Garcia, T. Ceillier, D. Salabert, S. Mathur, J.L. van Saders, M. Pinsonneault, J. Ballot, P.G. Beck, S. Bloemen, T.L. Campante, G.R. Davies, J.-D. do Nascimento Jr., S. Mathis, T.S. Metcalfe, M.B. Nielsen, J.C. Suarez, W.J. Chaplin, A. Jimenez, C. Karoff, Towards asteroseismically calibrated age-rotation-activity relations for Kepler solar-like stars. ArXiv e-prints (2014)

T. Gastine, L. Duarte, J. Wicht, Dipolar versus multipolar dynamos: the influence of the background density stratification. Astron. Astrophys. **546**, 19 (2012). doi:10.1051/0004-6361/201219799

T. Gastine, J. Wicht, J.M. Aurnou, Zonal flow regimes in rotating anelastic spherical shells: An application to giant planets. Icarus **225**, 156–172 (2013). doi:10.1016/j.icarus.2013.02.031

T. Gastine, R.K. Yadav, J. Morin, A. Reiners, J. Wicht, From solar-like to antisolar differential rotation in cool stars. Mon. Not. R. Astron. Soc. **438**, 76–80 (2014). doi:10.1093/mnrasl/slt162

B. Gelly, M. Lazrek, G. Grec, A. Ayad, F.X. Schmider, C. Renaud, D. Salabert, E. Fossat, Solar p-modes from 1979 days of the GOLF experiment. Astron. Astrophys. **394**, 285–297 (2002). doi:10.1051/0004-6361:20021106

C. Georgy, S. Ekström, P. Eggenberger, G. Meynet, L. Haemmerlé, A. Maeder, A. Granada, J.H. Groh, R. Hirschi, N. Mowlavi, N. Yusof, C. Charbonnel, T. Decressin, F. Barblan, Grids of stellar models with rotation. III. Models from 0.8 to 120 $M_\odot$ at a metallicity $Z = 0.002$. Astron. Astrophys. **558**, 103 (2013). doi:10.1051/0004-6361/201322178

M. Ghizaru, P. Charbonneau, P.K. Smolarkiewicz, Magnetic cycles in global large-eddy simulations of solar convection. Astrophys. J. Lett. **715**, 133–137 (2010). doi:10.1088/2041-8205/715/2/L133

M.S. Giampapa, Stellar analogs of solar activity: the Sun in a stellar context, in *Saas-Fee Advanced Course 34: The Sun, Solar Analogs and the Climate*, ed. by J.D. Haigh, M. Lockwood, M.S. Giampapa, I. Rüedi, M. Güdel, W. Schmutz (2005), pp. 307–415

M.S. Giampapa, J.C. Hall, R.R. Radick, S.L. Baliunas, A survey of chromospheric activity in the solar-type stars in the open cluster M67. Astrophys. J. **651**, 444–461 (2006). doi:10.1086/507624

P.A. Gilman, Dynamically consistent nonlinear dynamos driven by convection in a rotating spherical shell. II—Dynamos with cycles and strong feedbacks. Astrophys. J. Suppl. **53**, 243–268 (1983). doi:10.1086/190891

P.A. Gilman, J. Miller, Dynamically consistent nonlinear dynamos driven by convection in a rotating spherical shell. Astrophys. J. Suppl. **46**, 211–238 (1981). doi:10.1086/190743

P.A. Gilman, C.A. Morrow, E.E. Deluca, Angular momentum transport and dynamo action in the sun—Implications of recent oscillation measurements. Astrophys. J. **338**, 528–537 (1989). doi:10.1086/167215

G.A. Glatzmaier, Numerical simulations of stellar convective dynamos. I—The model and method. J. Comput. Phys. **55**, 461–484 (1984). doi:10.1016/0021-9991(84)90033-0

G.A. Glatzmaier, Numerical simulations of stellar convective dynamos. II—Field propagation in the convection zone. Astrophys. J. **291**, 300–307 (1985a). doi:10.1086/163069

G.A. Glatzmaier, Numerical simulations of stellar convective dynamos. III—At the base of the convection zone. Geophys. Astrophys. Fluid Dyn. **31**, 137–150 (1985b). doi:10.1080/03091928508219267

G.A. Glatzmaier, P.A. Gilman, Compressible convection in a rotating spherical shell. V—Induced differential rotation and meridional circulation. Astrophys. J. **256**, 316–330 (1982). doi:10.1086/159909

G.A. Glatzmaier, P.H. Roberts, Rotation and magnetism of Earth's inner core. Science **274**, 1887–1891 (1996). doi:10.1126/science.274.5294.1887

G. Glatzmaier, M. Evonuk, T. Rogers, Differential rotation in giant planets maintained by density-stratified turbulent convection. Geophys. Astrophys. Fluid Dyn. **103**, 31–51 (2009). doi:10.1080/03091920802221245

P. Goldreich, D.A. Keeley, Solar seismology. II—The stochastic excitation of the solar p-modes by turbulent convection. Astrophys. J. **212**, 243–251 (1977). doi:10.1086/155043

P. Goldreich, N. Murray, G. Willette, P. Kumar, Implications of solar p-mode frequency shifts. Astrophys. J. **370**, 752–762 (1991). doi:10.1086/169858

P. Gondoin, Dynamo regime transition among Sun-like stars in M 34. A time evolution model of X-ray activity on the main sequence. Astron. Astrophys. **546**, 117 (2012). doi:10.1051/0004-6361/201219823

L. Goudard, E. Dormy, Relations between the dynamo region geometry and the magnetic behavior of stars and planets. Europhys. Lett. **83**, 59001 (2008). doi:10.1209/0295-5075/83/59001

D.O. Gough, The anelastic approximation for thermal convection. J. Atmos. Sci. **26**, 448–456 (1969). doi:10.1175/1520-0469(1969)026<0448:TAAFTC>2.0.CO;2

D. Gough, Mixing-length theory for pulsating stars. Astrophys. J. **214**, 196–213 (1977a). doi:10.1086/155244

D. Gough, The current state of stellar mixing-length theory, in *Problems of Stellar Convection*, ed. by E.A. Spiegel, J.-P. Zahn. Lecture Notes in Physics, vol. 71 (Springer, Berlin, 1977b), pp. 15–56

D. Gough, New data from solar oscillations. Nature **274**, 739 (1978). doi:10.1038/274739a0

D.O. Gough, Immediate and long-term prospects for helioseismology. Adv. Space Res. **4**, 85–102 (1984a). doi:10.1016/0273-1177(84)90370-3

D.O. Gough, Solar inverse theory, in *Solar Seismology from Space*, vol. 84, ed. by R.K. Ulrich, J. Harvey, E. Rhodes, J. Toomre (1984b), pp. 49–78

D.O. Gough, Comments on helioseismic inference, in *Progress of Seismology of the Sun and Stars*, ed. by Y. Osaki, H. Shibahashi. Lecture Notes in Physics, vol. 367 (Springer, Berlin, 1990), p. 283. doi:10.1007/3-540-53091-6

D. Gough, How oblate is the Sun? Science **337**, 1611 (2012). doi:10.1126/science.1226988

D. Gough, What have we learned from helioseismology, what have we really learned, and what do we aspire to learn? Sol. Phys. **287**, 9–41 (2013). doi:10.1007/s11207-012-0099-1

D.O. Gough, M.J. Thompson, Magnetic perturbations to stellar oscillation eigenfrequencies, in *Advances in Helio- and Asteroseismology*, ed. by J. Christensen-Dalsgaard, S. Frandsen. IAU Symposium, vol. 123 (1988), p. 155

S.G. Gregory, J.-F. Donati, J. Morin, G.A.J. Hussain, N.J. Mayne, L.A. Hillenbrand, M. Jardine, Can we predict the global magnetic topology of a pre-main-sequence star from its position in the Hertzsprung-Russell diagram? Astrophys. J. **755**, 97 (2012). doi:10.1088/0004-637X/755/2/97

E. Grote, F.H. Busse, Dynamics of convection and dynamos in rotating spherical fluid shells. Fluid Dyn. Res. **28**, 349–368 (2001). doi:10.1016/S0169-5983(01)00004-1

F. Grundahl, J. Christensen-Dalsgaard, P.L. Pallé, M.F. Andersen, S. Frandsen, K. Harpsøe, U.G. Jørgensen, H. Kjeldsen, P.K. Rasmussen, J. Skottfelt, A.N. Sørensen, A.T. Hage, Stellar observations network group: the prototype is nearly ready, in *IAU Symposium*, ed. by J.A. Guzik, W.J. Chaplin, G. Handler, A. Pigulski. IAU Symposium, vol. 301 (2014), pp. 69–75. doi:10.1017/S1743921313014117

M. Güdel, X-ray astronomy of stellar coronae. Astron. Astrophys. Rev. **12**, 71–237 (2004). doi:10.1007/s00159-004-0023-2

M. Güdel, The Sun in time: activity and environment. Living Rev. Sol. Phys. **4**, 3 (2007). doi:10.12942/lrsp-2007-3

G. Guerrero, E.M. de Gouveia Dal Pino, Turbulent magnetic pumping in a Babcock-Leighton solar dynamo model. Astron. Astrophys. **485**, 267–273 (2008). doi:10.1051/0004-6361:200809351

G. Guerrero, P.K. Smolarkiewicz, A.G. Kosovichev, N.N. Mansour, Differential rotation in solar-like stars from global simulations. Astrophys. J. **779**, 176 (2013). doi:10.1088/0004-637X/779/2/176

D.A. Haber, B.W. Hindman, J. Toomre, R.S. Bogart, R.M. Larsen, F. Hill, Evolving submerged meridional circulation cells within the upper convection zone revealed by ring-diagram analysis. Astrophys. J. **570**, 855–864 (2002). doi:10.1086/339631

G.E. Hale, F. Ellerman, S.B. Nicholson, A.H. Joy, The magnetic polarity of sun-spots. Astrophys. J. **49**, 153 (1919). doi:10.1086/142452

J.C. Hall, Stellar chromospheric activity. Living Rev. Sol. Phys. **5**, 2 (2008). doi:10.12942/lrsp-2008-2

J.C. Hall, G.W. Lockwood, B.A. Skiff, The activity and variability of the Sun and sun-like stars. I. Synoptic Ca II H and K observations. Astron. J. **133**, 862–881 (2007). doi:10.1086/510356

J.C. Hall, G.W. Henry, G.W. Lockwood, B.A. Skiff, S.H. Saar, The activity and variability of the Sun and sun-like stars. II. Contemporaneous photometry and spectroscopy of bright solar analogs. Astron. J. **138**, 312–322 (2009). doi:10.1088/0004-6256/138/1/312

J.D. Hartman, B.S. Gaudi, M.H. Pinsonneault, K.Z. Stanek, M.J. Holman, B.A. McLeod, S. Meibom, J.A. Barranco, J.S. Kalirai, Deep MMT transit survey of the open cluster M37. III. Stellar rotation at 550 Myr. Astrophys. J. **691**, 342–364 (2009). doi:10.1088/0004-637X/691/1/342

J.D. Hartman, G.Á. Bakos, G. Kovács, R.W. Noyes, A large sample of photometric rotation periods for FGK Pleiades stars. Mon. Not. R. Astron. Soc. **408**, 475–489 (2010). doi:10.1111/j.1365-2966.2010.17147.x
J.D. Hartman, G.Á. Bakos, R.W. Noyes, B. Sipőcz, G. Kovács, T. Mazeh, A. Shporer, A. Pál, A photometric variability survey of field K and M dwarf stars with HATNet. Astron. J. **141**, 166 (2011). doi:10.1088/0004-6256/141/5/166
J.W. Harvey, in *Probing the depths of a star: the study of solar oscillations*, vol. 400, ed. by R.W. Noyes, E.J. Rhodes, JPL (1985), p. 327
D.H. Hathaway, The solar cycle. Living Rev. Sol. Phys. **7**, 1 (2010). doi:10.12942/lrsp-2010-1
G. Hazra, B.B. Karak, A.R. Choudhuri, Is a deep one-cell meridional circulation essential for the flux transport solar dynamo? Astrophys. J. **782**, 93 (2014). doi:10.1088/0004-637X/782/2/93
M.H. Heimpel, J.M. Aurnou, F.M. Al-Shamali, N. Gomez Perez, A numerical study of dynamo action as a function of spherical shell geometry. Earth Planet. Sci. Lett. **236**, 542–557 (2005). doi:10.1016/j.epsl.2005.04.032
S. Hekker, T. Kallinger, F. Baudin, J. De Ridder, C. Barban, F. Carrier, A.P. Hatzes, W.W. Weiss, A. Baglin, Characteristics of solar-like oscillations in red giants observed in the CoRoT exoplanet field. Astron. Astrophys. **506**, 465–469 (2009). doi:10.1051/0004-6361/200911858
S. Hekker, A. Broomhall, W.J. Chaplin, Y.P. Elsworth, S.T. Fletcher, R. New, T. Arentoft, P. Quirion, H. Kjeldsen, The Octave (Birmingham-Sheffield Hallam) automated pipeline for extracting oscillation parameters of solar-like main-sequence stars. Mon. Not. R. Astron. Soc. **402**, 2049–2059 (2010). doi:10.1111/j.1365-2966.2009.16030.x
S. Hekker, Y. Elsworth, J. De Ridder, B. Mosser, R.A. García, et al., Solar-like oscillations in red giants observed with Kepler: comparison of global oscillation parameters from different methods. Astron. Astrophys. **525**, 131 (2011). doi:10.1051/0004-6361/201015185
S. Hekker, Y. Elsworth, B. Mosser, T. Kallinger, S. Basu, W.J. Chaplin, D. Stello, Asteroseismic surface gravity for evolved stars. Astron. Astrophys. **556**, 59 (2013). doi:10.1051/0004-6361/201321630
A. Hempelmann, J.H.M.M. Schmitt, K. Stępień, Coronal X-ray emission of cool stars in relation to chromospheric activity and magnetic cycles. Astron. Astrophys. **305**, 284 (1996)
W. Herschel, Observations tending to investigate the nature of the Sun, in order to find the causes or symptoms of its variable emission of light and heat; with remarks on the use that may possibly be drawn from solar observations. Philos. Trans. R. Soc. Lond. Ser. I **91**, 265–318 (1801)
R. Hollerbach, C.A. Jones, Influence of the Earth's inner core on geomagnetic fluctuations and reversals. Nature **365**, 541–543 (1993). doi:10.1038/365541a0
H. Hotta, T. Yokoyama, Modeling of differential rotation in rapidly rotating solar-type stars. Astrophys. J. **740**, 12 (2011). doi:10.1088/0004-637X/740/1/12
G. Houdek, Pulsation of solar-type stars. Ph.D. Thesis, Formal- und Naturwisseschaftliche Fakult at der Universität Wien (1996)
G. Houdek, Asteroseismic helium abundance determination, in *Equation-of-State and Phase-Transition in Models of Ordinary Astrophysical Matter*, ed. by V. Celebonovic, D. Gough, W. Däppen. American Institute of Physics Conference Series, vol. 731 (2004), pp. 193–207. doi:10.1063/1.1828405
G. Houdek, Stellar turbulence and mode physics. Astrophys. Space Sci. **328**, 237–244 (2010). doi:10.1007/s10509-009-0226-0
G. Houdek, D.O. Gough, Modelling pulsation amplitudes of ξ Hydrae. Mon. Not. R. Astron. Soc. **336**, 65–69 (2002). doi:10.1046/j.1365-8711.2002.06024.x
G. Houdek, D.O. Gough, An asteroseismic signature of helium ionization. Mon. Not. R. Astron. Soc. **375**, 861–880 (2007). doi:10.1111/j.1365-2966.2006.11325.x
G. Houdek, N.J. Balmforth, J. Christensen-Dalsgaard, D.O. Gough, Amplitudes of stochastically excited oscillations in main-sequence stars. Astron. Astrophys. **351**, 582–596 (1999)
G. Houdek, W.J. Chaplin, T. Appourchaux, J. Christensen-Dalsgaard, W. Däppen, Y. Elsworth, D.O. Gough, G.R. Isaak, R. New, M.C. Rabello-Soares, Changes in convective properties over the solar cycle: effect on p-mode damping rates. Mon. Not. R. Astron. Soc. **327**, 483–487 (2001). doi:10.1046/j.1365-8711.2001.04749.x
R. Howe, D.A. Haber, B.W. Hindman, R. Komm, F. Hill, I. Gonzalez Hernandez, Helioseismic Frequency shifts in active regions, in *Subsurface and Atmospheric Influences on Solar Activity*, ed. by R. Howe, R.W. Komm, K.S. Balasubramaniam, G.J.D. Petrie. Astronomical Society of the Pacific Conference Series, vol. 383 (2008), p. 305
D. Huber, D. Stello, T.R. Bedding, W.J. Chaplin, T. Arentoft, P. Quirion, H. Kjeldsen, Automated extraction of oscillation parameters for Kepler observations of solar-type stars. Commun. Asteroseismol. **160**, 74 (2009)
D. Huber, T.R. Bedding, D. Stello, B. Mosser, S. Mathur, et al., Asteroseismology of red giants from the first four months of Kepler data: global oscillation parameters for 800 stars. Astrophys. J. **723**, 1607–1617 (2010). doi:10.1088/0004-637X/723/2/1607

D. Huber, T.R. Bedding, D. Stello, S. Hekker, et al., Testing scaling relations for solar-like oscillations from the main sequence to red giants using Kepler data. Astrophys. J. **743**, 143 (2011). doi:10.1088/0004-637X/743/2/143

E. Işık, D. Schmitt, M. Schüssler, Magnetic flux generation and transport in cool stars. Astron. Astrophys. **528**, 135 (2011). doi:10.1051/0004-6361/201014501

K. Jain, S.C. Tripathy, F. Hill, Solar activity phases and intermediate-degree mode frequencies. Astrophys. J. **695**, 1567–1576 (2009). doi:10.1088/0004-637X/695/2/1567

M. Jardine, J.R. Barnes, J.-F. Donati, A. Collier Cameron, The potential magnetic field of AB Doradus: comparison with Zeeman-Doppler images. Mon. Not. R. Astron. Soc. **305**, 35–39 (1999). doi:10.1046/j.1365-8711.1999.02621.x

S.J. Jiménez-Reyes, R.A. García, A. Jiménez, W.J. Chaplin, Excitation and damping of low-degree solar p-modes during activity cycle 23: analysis of GOLF and VIRGO sun photometer data. Astrophys. J. **595**, 446–457 (2003). doi:10.1086/377304

S.J. Jiménez-Reyes, W.J. Chaplin, Y. Elsworth, R.A. García, R. Howe, H. Socas-Navarro, T. Toutain, On the variation of the peak asymmetry of low-l solar p modes. Astrophys. J. **654**, 1135–1145 (2007). doi:10.1086/509700

C.A. Jones, Planetary magnetic fields and fluid dynamos. Annu. Rev. Fluid Mech. **43**, 583–614 (2011). doi:10.1146/annurev-fluid-122109-160727

C.A. Jones, P. Boronski, A.S. Brun, G.A. Glatzmaier, T. Gastine, M.S. Miesch, J. Wicht, Anelastic convection-driven dynamo benchmarks. Icarus **216**, 120–135 (2011). doi:10.1016/j.icarus.2011.08.014

L. Jouve, A.S. Brun, On the role of meridional flows in flux transport dynamo models. Astron. Astrophys. **474**, 239–250 (2007). doi:10.1051/0004-6361:20077070

L. Jouve, A.S. Brun, Three-dimensional nonlinear evolution of a magnetic flux tube in a spherical shell: influence of turbulent convection and associated mean flows. Astrophys. J. **701**, 1300–1322 (2009). doi:10.1088/0004-637X/701/2/1300

L. Jouve, B.P. Brown, A.S. Brun, Exploring the P_{cyc} vs. P_{rot} relation with flux transport dynamo models of solar-like stars. Astron. Astrophys. **509**, 32 (2010). doi:10.1051/0004-6361/200913103

L. Jouve, A.S. Brun, G. Aulanier, Global dynamics of subsurface solar active regions. Astrophys. J. **762**, 4 (2013). doi:10.1088/0004-637X/762/1/4

P.G. Judge, M.J. Thompson, Solar and stellar activity: diagnostics and indices, in *IAU Symposium*, ed. by C.H. Mandrini, D.F. Webb. IAU Symposium, vol. 286 (2012), pp. 15–26. doi:10.1017/S1743921312004589

A. Kageyama, T. Sato, Generation mechanism of a dipole field by a magnetohydrodynamic dynamo. Phys. Rev. E **55**, 4617–4626 (1997). doi:10.1103/PhysRevE.55.4617

A. Kageyama, T. Sato, "Yin-Yang grid": An overset grid in spherical geometry. Geochem. Geophys. Geosyst. **5**, 9005 (2004). doi:10.1029/2004GC000734

T. Kallinger, B. Mosser, S. Hekker, D. Huber, et al., Asteroseismology of red giants from the first four months of Kepler data: Fundamental stellar parameters. Astron. Astrophys. **522**, 1 (2010). doi:10.1051/0004-6361/201015263

P.J. Käpylä, M.J. Mantere, G. Guerrero, A. Brandenburg, P. Chatterjee, Reynolds stress and heat flux in spherical shell convection. Astron. Astrophys. **531**, 162 (2011). doi:10.1051/0004-6361/201015884

P.J. Käpylä, M.J. Mantere, A. Brandenburg, Cyclic magnetic activity due to turbulent convection in spherical wedge geometry. Astrophys. J. Lett. **755**, 22 (2012). doi:10.1088/2041-8205/755/1/L22

P.J. Käpylä, M.J. Mantere, E. Cole, J. Warnecke, A. Brandenburg, Effects of enhanced stratification on equatorward dynamo wave propagation. Astrophys. J. **778**, 41 (2013). doi:10.1088/0004-637X/778/1/41

P.J. Käpylä, M.J. Mantere, A. Brandenburg, Confirmation of bistable stellar differential rotation profiles. ArXiv e-prints (2014)

B.B. Karak, D. Nandy, Turbulent Pumping of Magnetic Flux Reduces Solar Cycle Memory and thus Impacts Predictability of the Sun's Activity. Astrophys. J. Lett. **761**, 13 (2012). doi:10.1088/2041-8205/761/1/L13

C. Karoff, T.L. Campante, J. Ballot, T. Kallinger, M. Gruberbauer, R.A. Garcia, D.A. Caldwell, J.L. Christiansen, K. Kinemuchi, Observations of intensity fluctuations attributed to granulation and faculae on Sun-like stars from the Kepler mission. ArXiv e-prints (2013)

S.D. Kawaler, Angular momentum loss in low-mass stars. Astrophys. J. **333**, 236–247 (1988). doi:10.1086/166740

R. Keppens, K.B. MacGregor, P. Charbonneau, On the evolution of rotational velocity distributions for solar-type stars. Astron. Astrophys. **294**, 469–487 (1995)

Y.-C. Kim, P. Demarque, The theoretical calculation of the Rossby number and the "nonlocal" convective overturn time for pre-main-sequence and early post-main-sequence stars. Astrophys. J. **457**, 340 (1996). doi:10.1086/176733

K. Kimura, S.-I. Takehiro, M. Yamada, Stability and bifurcation diagram of Boussinesq thermal convection in a moderately rotating spherical shell. Phys. Fluids **23**(7), 074101 (2011). doi:10.1063/1.3602917

R. Kippenhahn, A. Weigert, A. Weiss, Stellar structure and evolution (2013). doi:10.1007/978-3-642-30304-3

L.L. Kitchatinov, S.V. Olemskoy, Differential rotation of main-sequence dwarfs: predicting the dependence on surface temperature and rotation rate. Mon. Not. R. Astron. Soc. **423**, 3344–3351 (2012). doi:10.1111/j.1365-2966.2012.21126.x

H. Kjeldsen, T.R. Bedding, Amplitudes of stellar oscillations: the implications for asteroseismology. Astron. Astrophys. **293**, 87–106 (1995)

H. Kjeldsen, T.R. Bedding, M. Viskum, S. Frandsen, Solarlike oscillations in eta Boo. Astron. J. **109**, 1313–1319 (1995). doi:10.1086/117363

H. Kjeldsen, T.R. Bedding, I.K. Baldry, H. Bruntt, et al., Confirmation of solar-like oscillations in η bootis. Astron. J. **126**, 1483–1488 (2003). doi:10.1086/377315

H. Kjeldsen, T.R. Bedding, J. Christensen-Dalsgaard, Correcting stellar oscillation frequencies for near-surface effects. Astrophys. J. Lett. **683**, 175–178 (2008). doi:10.1086/591667

A. Koenigl, Disk accretion onto magnetic T Tauri stars. Astrophys. J. Lett. **370**, 39–43 (1991). doi:10.1086/185972

A.G. Kosovichev, J.O. Stenflo, Tilt of emerging bipolar magnetic regions on the Sun. Astrophys. J. Lett. **688**, 115–118 (2008). doi:10.1086/595619

A. Krishnamurthi, M.H. Pinsonneault, S. Barnes, S. Sofia, Theoretical models of the angular momentum evolution of solar-type stars. Astrophys. J. **480**, 303 (1997). doi:10.1086/303958

J.R. Kuhn, R. Bush, M. Emilio, I.F. Scholl, The precise solar shape and its variability. Science **337**, 1638 (2012). doi:10.1126/science.1223231

M. Küker, G. Rüdiger, Modelling solar and stellar differential rotation. J. Phys. Conf. Ser. **118**(1), 012029 (2008). doi:10.1088/1742-6596/118/1/012029

N. Lagarde, T. Decressin, C. Charbonnel, P. Eggenberger, S. Ekström, A. Palacios, Thermohaline instability and rotation-induced mixing. III. Grid of stellar models and asymptotic asteroseismic quantities from the pre-main sequence up to the AGB for low- and intermediate-mass stars of various metallicities. Astron. Astrophys. **543**, 108 (2012). doi:10.1051/0004-6361/201118331

N.R. Landin, L.T.S. Mendes, L.P.R. Vaz, Theoretical values of convective turnover times and Rossby numbers for solar-like, pre-main sequence stars. Astron. Astrophys. **510**, 46 (2010). doi:10.1051/0004-6361/200913015

A.F. Lanza, Current state of the modelling of photospheric activity, in *Advances in Solid State Physics*, ed. by J.C. Suárez, R. Garrido, L.A. Balona, J. Christensen-Dalsgaard. Advances in Solid State Physics, vol. 31 (2013), p. 215. doi:10.1007/978-3-642-29630-7

S. Lefebvre, R.A. García, S.J. Jiménez-Reyes, S. Turck-Chièze, S. Mathur, Variations of the solar granulation motions with height using the GOLF/SoHO experiment. Astron. Astrophys. **490**, 1143–1149 (2008). doi:10.1051/0004-6361:200810344

R.B. Leighton, Observations of solar magnetic fields in plage regions. Astrophys. J. **130**, 366 (1959). doi:10.1086/146727

R.B. Leighton, A magneto-kinematic model of the solar cycle. Astrophys. J. **156**, 1 (1969). doi:10.1086/149943

K.G. Libbrecht, M.F. Woodard, Solar-cycle effects on solar oscillation frequencies. Nature **345**, 779–782 (1990). doi:10.1038/345779a0

G.W. Lockwood, B.A. Skiff, G.W. Henry, S. Henry, R.R. Radick, S.L. Baliunas, R.A. Donahue, W. Soon, Patterns of photometric and chromospheric variation among sun-like stars: a 20 year perspective. Astrophys. J. Suppl. **171**, 260–303 (2007). doi:10.1086/516752

D. Lynden-Bell, J.P. Ostriker, On the stability of differentially rotating bodies. Mon. Not. R. Astron. Soc. **136**, 293 (1967)

K.B. MacGregor, M. Brenner, Rotational evolution of solar-type stars. I—Main-sequence evolution. Astrophys. J. **376**, 204–213 (1991). doi:10.1086/170269

A. Maeder, G. Meynet, Rotating massive stars: From first stars to gamma ray bursts. Rev. Mod. Phys. **84**, 25–63 (2012). doi:10.1103/RevModPhys.84.25

Z. Magic, R. Collet, M. Asplund, R. Trampedach, W. Hayek, A. Chiavassa, R.F. Stein, Å. Nordlund, The Stagger-grid: A grid of 3D stellar atmosphere models. I. Methods and general properties. Astron. Astrophys. **557**, 26 (2013). doi:10.1051/0004-6361/201321274

W.V.R. Malkus, M.R.E. Proctor, The macrodynamics of alpha-effect dynamos in rotating fluids. J. Fluid Mech. **67**, 417–443 (1975). doi:10.1017/S0022112075000390

E.E. Mamajek, L.A. Hillenbrand, Improved age estimation for solar-type dwarfs using activity-rotation diagnostics. Astrophys. J. **687**, 1264–1293 (2008). doi:10.1086/591785

J.P. Marques, M.J. Goupil, Y. Lebreton, S. Talon, A. Palacios, K. Belkacem, R.-M. Ouazzani, B. Mosser, A. Moya, P. Morel, B. Pichon, S. Mathis, J.-P. Zahn, S. Turck-Chièze, P.A.P. Nghiem, Seismic diagnostics for transport of angular momentum in stars. I. Rotational splittings from the pre-main sequence to the red-giant branch. Astron. Astrophys. **549**, 74 (2013). doi:10.1051/0004-6361/201220211

S.C. Marsden, P. Petit, S.V. Jeffers, J. Morin, R. Fares, A. Reiners, J.D. do Nascimento Jr., M. Auriere, J. Bouvier, B.D. Carter, C. Catala, B. Dintrans, J.-F. Donati, T. Gastine, M. Jardine, R. Konstantinova-Antova, J. Lanoux, F. Lignieres, A. Morgenthaler, J.C. Ramirez-Velez, S. Theado, V. Van Grootel, the Bcool Collaboration, A Bcool magnetic snapshot survey of solar-type stars. ArXiv e-prints (2013)

M. Martić, J. Schmitt, J.-C. Lebrun, C. Barban, et al., Evidence for global pressure oscillations on Procyon. Astron. Astrophys. **351**, 993–1002 (1999)

Y. Masada, K. Yamada, A. Kageyama, Effects of penetrative convection on solar dynamo. Astrophys. J. **778**, 11 (2013). doi:10.1088/0004-637X/778/1/11

J. Mason, D.W. Hughes, S.M. Tobias, The effects of flux transport on interface dynamos. Mon. Not. R. Astron. Soc. **391**, 467–480 (2008). doi:10.1111/j.1365-2966.2008.13918.x

S. Mathis, Angular momentum transport by regular gravito-inertial waves in stellar radiation zones, in *Lecture Notes in Physics*, ed. by J.-P. Rozelot, C. Neiner, vol. 832 (Springer, Berlin, 2011), p. 275

S. Mathis, Angular momentum transport by internal waves in stellar interiors, in *EAS Publications Series*, vol. 63 (2013a), pp. 269–284. doi:10.1051/eas/1363031

S. Mathis, Transport processes in stellar interiors, in *Lecture Notes in Physics*, ed. by M. Goupil, K. Belkacem, C. Neiner, F. Lignières, J.J. Green, vol. 865 (Springer, Berlin, 2013b), p. 23

S. Mathis, J.-P. Zahn, Transport and mixing in the radiation zones of rotating stars. I. Hydrodynamical processes. Astron. Astrophys. **425**, 229–242 (2004). doi:10.1051/0004-6361:20040278

S. Mathis, J.-P. Zahn, Transport and mixing in the radiation zones of rotating stars. II. Axisymmetric magnetic field. Astron. Astrophys. **440**, 653–666 (2005). doi:10.1051/0004-6361:20052640

S. Mathur, A. Eff-Darwich, R.A. García, S. Turck-Chièze, Sensitivity of helioseismic gravity modes to the dynamics of the solar core. Astron. Astrophys. **484**, 517–522 (2008). doi:10.1051/0004-6361:20078839

S. Mathur, S.J. Jiménez-Reyes, R.A. García, GOLF: a new proxy for solar magnetism, in *Solar-Stellar Dynamos as Revealed by Helio- and Asteroseismology: GONG 2008/SOHO 21*, ed. by M. Dikpati, T. Arentoft, I. González Hernández, C. Lindsey, F. Hill. Astronomical Society of the Pacific Conference Series, vol. 416 (2009), p. 529

S. Mathur, R.A. García, C. Régulo, O.L. Creevey, et al., Determining global parameters of the oscillations of solar-like stars. Astron. Astrophys. **511**, 46 (2010a). doi:10.1051/0004-6361/200913266

S. Mathur, R.A. García, C. Catala, H. Bruntt, et al., The solar-like CoRoT target HD 170987: spectroscopic and seismic observations. Astron. Astrophys. **518**, 53 (2010b). doi:10.1051/0004-6361/201014103

S. Mathur, S. Hekker, R. Trampedach, J. Ballot, et al., Granulation in red giants: observations by the Kepler mission and three-dimensional convection simulations. Astrophys. J. **741**, 119 (2011a). doi:10.1088/0004-637X/741/2/119

S. Mathur, R. Handberg, T.L. Campante, R.A. García, et al., Solar-like oscillations in KIC 11395018 and KIC 11234888 from 8 months of Kepler Data. Astrophys. J. **733**, 95 (2011b). doi:10.1088/0004-637X/733/2/95

S. Mathur, T.S. Metcalfe, M. Woitaszek, H. Bruntt, et al., A uniform asteroseismic analysis of 22 solar-type stars observed by Kepler. Astrophys. J. **749**, 152 (2012). doi:10.1088/0004-637X/749/2/152

S. Mathur, R.A. García, A. Morgenthaler, D. Salabert, P. Petit, J. Ballot, C. Régulo, C. Catala, Constraining magnetic-activity modulations in three solar-like stars observed by CoRoT and NARVAL. Astron. Astrophys. **550**, 32 (2013a). doi:10.1051/0004-6361/201117913

S. Mathur, H. Bruntt, C. Catala, O. Benomar, et al., Study of HD 169392A observed by CoRoT and HARPS. Astron. Astrophys. **549**, 12 (2013b). doi:10.1051/0004-6361/201219678

S. Mathur, R.A. García, J. Ballot, T. Ceillier, D. Salabert, T.S. Metcalfe, C. Régulo, A. Jiménez, S. Bloemen, Magnetic activity of F stars observed by Kepler. Astron. Astrophys. **562**, 124 (2014a). doi:10.1051/0004-6361/201322707

S. Mathur, D. Salabert, R.A. García, T. Ceillier, Photometric magnetic-activity metrics tested with the Sun: application to Kepler M dwarfs. J. Space Weather Space Clim. **4**(27), 15 (2014b). doi:10.1051/swsc/2014011

S.P. Matt, O. Do Cao, B.P. Brown, A.S. Brun, Convection and differential rotation properties of G and K stars computed with the ASH code. Astron. Nachr. **332**, 897 (2011). doi:10.1002/asna.201111624

S.P. Matt, K.B. MacGregor, M.H. Pinsonneault, T.P. Greene, Magnetic braking formulation for sun-like stars: dependence on dipole field strength and rotation rate. Astrophys. J. Lett. **754**, 26 (2012). doi:10.1088/2041-8205/754/2/L26

S. Matt, A.S. Brun, A. Strugarek, O. Do Cao, Differential rotation and overshooting in solar-like stars. Astrophys. J. (2014)

A. Mazumdar, M.J.P.F.G. Monteiro, J. Ballot, H.M. Antia, S. Basu, G. Houdek, S. Mathur, M.S. Cunha, V. Silva Aguirre, R.A. García, D. Salabert, G.A. Verner, J. Christensen-Dalsgaard, T.S. Metcalfe, D.T. Sanderfer, S.E. Seader, J.C. Smith, W.J. Chaplin, Measurement of acoustic glitches in solar-type stars from oscillation frequencies observed by Kepler. Astrophys. J. **782**, 18 (2014). doi:10.1088/0004-637X/782/1/18

M.E. McIntyre, Magnetic confinement and the sharp tachopause, in *The Solar Tachocline*, ed. by D.W. Hughes, R. Rosner, N.O. Weiss (2007), p. 183
A. McQuillan, T. Mazeh, S. Aigrain, Rotation periods of 34,030 Kepler main-sequence stars: the full autocorrelation sample. Astrophys. J. Suppl. **211**, 24 (2014). doi:10.1088/0067-0049/211/2/24
S. Meibom, S.A. Barnes, K. Covey, R.D. Jeffries, S. Matt, J. Morin, A. Palacios, A. Reiners, A. Sicilia-Aguilar, J. Irwin, Angular momentum evolution of cool stars: Toward a synthesis of observations and theory before and after the ZAMS. Astron. Nachr. **334**, 168 (2013). doi:10.1002/asna.201211777
L. Mestel, Magnetic braking by a stellar wind-I. Mon. Not. R. Astron. Soc. **138**, 359 (1968)
T.S. Metcalfe, S. Basu, T.J. Henry, D.R. Soderblom, P.G. Judge, M. Knölker, S. Mathur, M. Rempel, Discovery of a 1.6 year magnetic activity cycle in the exoplanet host star ι Horologii. Astrophys. J. Lett. **723**, 213–217 (2010). doi:10.1088/2041-8205/723/2/L213
T.S. Metcalfe, W.J. Chaplin, T. Appourchaux, R.A. García, et al., Asteroseismology of the solar analogs 16 Cyg A and B from Kepler observations. Astrophys. J. Lett. **748**, 10 (2012). doi:10.1088/2041-8205/748/1/L10
G. Micela, A. Marino, A comparison between the X-ray variable Sun and solar-like main sequence stars. Astron. Astrophys. **404**, 637–643 (2003). doi:10.1051/0004-6361:20030517
M.S. Miesch, Large-scale dynamics of the convection zone and tachocline. Living Rev. Sol. Phys. **2**, 1 (2005). doi:10.12942/lrsp-2005-1
M.S. Miesch, B.P. Brown, Convective Babcock-Leighton dynamo models. Astrophys. J. Lett. **746**, 26 (2012). doi:10.1088/2041-8205/746/2/L26
M.S. Miesch, M. Dikpati, A three-dimensional Babcock-Leighton solar dynamo model. Astrophys. J. Lett. **785**, 8 (2014). doi:10.1088/2041-8205/785/1/L8
M.S. Miesch, B.W. Hindman, Gyroscopic pumping in the solar near-surface shear layer. Astrophys. J. **743**, 79 (2011). doi:10.1088/0004-637X/743/1/79
M.S. Miesch, J. Toomre, Turbulence, magnetism, and shear in stellar interiors. Annu. Rev. Fluid Mech. **41**, 317–345 (2009). doi:10.1146/annurev.fluid.010908.165215
M.S. Miesch, J.R. Elliott, J. Toomre, T.L. Clune, G.A. Glatzmaier, P.A. Gilman, Three-dimensional spherical simulations of solar convection. I. Differential rotation and pattern evolution achieved with laminar and turbulent states. Astrophys. J. **532**, 593–615 (2000). doi:10.1086/308555
M.S. Miesch, A.S. Brun, J. Toomre, Solar differential rotation influenced by latitudinal entropy variations in the tachocline. Astrophys. J. **641**, 618–625 (2006). doi:10.1086/499621
M.S. Miesch, A.S. Brun, M.L. De Rosa, J. Toomre, Structure and evolution of giant cells in global models of solar convection. Astrophys. J. **673**, 557–575 (2008). doi:10.1086/523838
M.S. Miesch, N.A. Featherstone, M. Rempel, R. Trampedach, On the amplitude of convective velocities in the deep solar interior. Astrophys. J. **757**, 128 (2012). doi:10.1088/0004-637X/757/2/128
A. Miglio, J. Montalbán, F. Baudin, P. Eggenberger, A. Noels, S. Hekker, J. De Ridder, W. Weiss, A. Baglin, Probing populations of red giants in the galactic disk with CoRoT. Astron. Astrophys. **503**, 21–24 (2009). doi:10.1051/0004-6361/200912822
U. Mitra-Kraev, M.J. Thompson, Meridional flow profile measurements with SOHO/MDI. Astron. Nachr. **328**, 1009–1012 (2007). doi:10.1002/asna.200710873
H.K. Moffatt, *Magnetic Field Generation in Electrically Conducting Fluids* (University of Bristol, Cambridge, Bristol, 1978)
M.J.P.F.G. Monteiro, J. Christensen-Dalsgaard, M.J. Thompson, Seismic study of overshoot at the base of the solar convective envelope. Astron. Astrophys. **283**, 247–262 (1994)
B. Montesinos, J.H. Thomas, P. Ventura, I. Mazzitelli, A new look at the relationship between activity, dynamo number and Rossby number in late-type stars. Mon. Not. R. Astron. Soc. **326**, 877–884 (2001). doi:10.1046/j.1365-8711.2001.04476.x
E. Moraux, S. Artemenko, J. Bouvier, J. Irwin, M. Ibrahimov, T. Magakian, K. Grankin, E. Nikogossian, C. Cardoso, S. Hodgkin, S. Aigrain, T.A. Movsessian, The Monitor Project: stellar rotation at 13 Myr. I. A photometric monitoring survey of the young open cluster h Persei. Astron. Astrophys. **560**, 13 (2013). doi:10.1051/0004-6361/201321508
P. Morel, CESAM: A code for stellar evolution calculations. Astron. Astrophys. Suppl. **124**, 597–614 (1997). doi:10.1051/aas:1997209
T. Morel, A. Miglio, Assessing the accuracy of the surface gravity determination in late-type stars with solar-like pulsators. Mon. Not. R. Astron. Soc. **419**, 34–38 (2012). doi:10.1111/j.1745-3933.2011.01172.x
A. Morgenthaler, P. Petit, J. Morin, M. Aurière, B. Dintrans, R. Konstantinova-Antova, S. Marsden, Direct observation of magnetic cycles in Sun-like stars. Astron. Nachr. **332**, 866 (2011a). doi:10.1002/asna.201111592
A. Morgenthaler, P. Petit, J. Morin, M. Aurière, B. Dintrans, Long-term magnetic monitoring of 19 sun-like stars, in *SF2A-2011: Proceedings of the Annual meeting of the French Society of Astronomy and Astrophysics*, ed. by G. Alecian, K. Belkacem, R. Samadi, D. Valls-Gabaud (2011b), pp. 497–501

J. Morin, J.-F. Donati, P. Petit, X. Delfosse, T. Forveille, M.M. Jardine, Large-scale magnetic topologies of late M dwarfs. Mon. Not. R. Astron. Soc. **407**, 2269–2286 (2010). doi:10.1111/j.1365-2966.2010.17101.x

J. Morin, E. Dormy, M. Schrinner, J.-F. Donati, Weak- and strong-field dynamos: from the Earth to the stars. Mon. Not. R. Astron. Soc. **418**, 133–137 (2011). doi:10.1111/j.1745-3933.2011.01159.x

J. Morin, M. Jardine, A. Reiners, D. Shulyak, B. Beeck, G. Hallinan, L. Hebb, G. Hussain, S.V. Jeffers, O. Kochukhov, A. Vidotto, L. Walkowicz, Multiple views of magnetism in cool stars. Astron. Nachr. **334**, 48 (2013). doi:10.1002/asna.201211771

D. Moss, J. Brooke, Towards a model for the solar dynamo. Mon. Not. R. Astron. Soc. **315**, 521–533 (2000). doi:10.1046/j.1365-8711.2000.03452.x

B. Mosser, T. Appourchaux, On detecting the large separation in the autocorrelation of stellar oscillation times series. Astron. Astrophys. **508**, 877–887 (2009). doi:10.1051/0004-6361/200912944

B. Mosser, F. Bouchy, M. Martić, T. Appourchaux, et al., Asteroseismology of Procyon with SOPHIE. Astron. Astrophys. **478**, 197–202 (2008). doi:10.1051/0004-6361:20078337

B. Mosser, F. Baudin, A.F. Lanza, J.C. Hulot, C. Catala, A. Baglin, M. Auvergne, Short-lived spots in solar-like stars as observed by CoRoT. Astron. Astrophys. **506**, 245–254 (2009). doi:10.1051/0004-6361/200911942

B. Mosser, K. Belkacem, M. Goupil, A. Miglio, et al., Red-giant seismic properties analyzed with CoRoT. Astron. Astrophys. **517**, 22 (2010). doi:10.1051/0004-6361/201014036

B. Mosser, M.J. Goupil, K. Belkacem, E. Michel, D. Stello, et al., Probing the core structure and evolution of red giants using gravity-dominated mixed modes observed with Kepler. Astron. Astrophys. **540**, 143 (2012). doi:10.1051/0004-6361/201118519

B. Mosser, E. Michel, K. Belkacem, M.J. Goupil, A. Baglin, C. Barban, J. Provost, R. Samadi, M. Auvergne, C. Catala, Asymptotic and measured large frequency separations. Astron. Astrophys. **550**, 126 (2013). doi:10.1051/0004-6361/201220435

K. Mursula, B. Zieger, J.H. Vilppola, Mid-term quasi-periodicities in geomagnetic activity during the last 15 solar cycles: Connection to solar dynamo strength To the memory of Karolen I. Paularena (1957–2001). Sol. Phys. **212**, 201–207 (2003). doi:10.1023/A:1022980029618

D. Nandy, Exploring magnetic activity from the Sun to the stars. Sol. Phys. **224**, 161–169 (2004). doi:10.1007/s11207-005-4990-x

D. Nandy, A.R. Choudhuri, Explaining the latitudinal distribution of sunspots with deep meridional flow. Science **296**, 1671–1673 (2002). doi:10.1126/science.1070955

D. Nandy, A. Muñoz-Jaramillo, P.C.H. Martens, The unusual minimum of sunspot cycle 23 caused by meridional plasma flow variations. Nature **471**, 80–82 (2011). doi:10.1038/nature09786

N.J. Nelson, B.P. Brown, A.S. Brun, M.S. Miesch, J. Toomre, Buoyant magnetic loops in a global dynamo simulation of a young sun. Astrophys. J. Lett. **739**, 38 (2011). doi:10.1088/2041-8205/739/2/L38

N.J. Nelson, B.P. Brown, A.S. Brun, M.S. Miesch, J. Toomre, Magnetic wreaths and cycles in convective dynamos. Astrophys. J. **762**, 73 (2013). doi:10.1088/0004-637X/762/2/73

N.J. Nelson, B.P. Brown, A. Sacha Brun, M.S. Miesch, J. Toomre, Buoyant magnetic loops generated by global convective dynamo action. Sol. Phys. **289**, 441–458 (2014). doi:10.1007/s11207-012-0221-4

M.B. Nielsen, L. Gizon, H. Schunker, C. Karoff, Rotation periods of 12 000 main-sequence Kepler stars: Dependence on stellar spectral type and comparison with v sin i observations. Astron. Astrophys. **557**, 10 (2013). doi:10.1051/0004-6361/201321912

N. Nishikawa, K. Kusano, Simulation study of the symmetry-breaking instability and the dipole field reversal in a rotating spherical shell dynamo. Phys. Plasmas **15**(8), 082903 (2008). doi:10.1063/1.2959120

R.W. Noyes, N.O. Weiss, A.H. Vaughan, The relation between stellar rotation rate and activity cycle periods. Astrophys. J. **287**, 769–773 (1984a). doi:10.1086/162735

R.W. Noyes, L.W. Hartmann, S.L. Baliunas, D.K. Duncan, A.H. Vaughan, Rotation, convection, and magnetic activity in lower main-sequence stars. Astrophys. J. **279**, 763–777 (1984b). doi:10.1086/161945

R.L.F. Oglethorpe, P. Garaud, Spin-down dynamics of magnetized solar-type stars. Astrophys. J. **778**, 166 (2013). doi:10.1088/0004-637X/778/2/166

K. Oláh, Z. Kolláth, T. Granzer, K.G. Strassmeier, A.F. Lanza, S. Järvinen, H. Korhonen, S.L. Baliunas, W. Soon, S. Messina, G. Cutispoto, Multiple and changing cycles of active stars. II. Results. Astron. Astrophys. **501**, 703–713 (2009). doi:10.1051/0004-6361/200811304

A. Palacios, S. Talon, C. Charbonnel, M. Forestini, Rotational mixing in low-mass stars. I Effect of the mu-gradients in main sequence and subgiant Pop I stars. Astron. Astrophys. **399**, 603–616 (2003). doi:10.1051/0004-6361:20021759

P.L. Pallé, C. Régulo, T. Roca Cortés, Solar cycle induced variations of the low L solar acoustic spectrum. Astron. Astrophys. **224**, 253–258 (1989)

E.N. Parker, Hydromagnetic dynamo models. Astrophys. J. **122**, 293 (1955). doi:10.1086/146087

J. Pedlosky, *Geophysical Fluid Dynamics* (1982)

P. Petit, B. Dintrans, S.K. Solanki, J.-F. Donati, M. Aurière, et al., Toroidal versus poloidal magnetic fields in Sun-like stars: a rotation threshold. Mon. Not. R. Astron. Soc. **388**, 80–88 (2008). doi:10.1111/j.1365-2966.2008.13411.x

A.A. Pevtsov, G.H. Fisher, L.W. Acton, D.W. Longcope, C.M. Johns-Krull, C.C. Kankelborg, T.R. Metcalf, The relationship between X-Ray radiance and magnetic flux. Astrophys. J. **598**, 1387–1391 (2003). doi:10.1086/378944

M. Pinsonneault, Mixing in stars. Annu. Rev. Astron. Astrophys. **35**, 557–605 (1997). doi:10.1146/annurev.astro.35.1.557

M.H. Pinsonneault, S.D. Kawaler, S. Sofia, P. Demarque, Evolutionary models of the rotating sun. Astrophys. J. **338**, 424–452 (1989). doi:10.1086/167210

M.H. Pinsonneault, S.D. Kawaler, P. Demarque, Rotation of low-mass stars—A new probe of stellar evolution. Astrophys. J. Suppl. **74**, 501–550 (1990). doi:10.1086/191507

R.F. Pinto, A.S. Brun, Flux emergence in a magnetized convection zone. Astrophys. J. **772**, 55 (2013). doi:10.1088/0004-637X/772/1/55

R.F. Pinto, A.S. Brun, L. Jouve, R. Grappin, Coupling the solar dynamo and the corona: wind properties, mass, and momentum losses during an activity cycle. Astrophys. J. **737**, 72 (2011). doi:10.1088/0004-637X/737/2/72

N. Pizzolato, A. Maggio, G. Micela, S. Sciortino, P. Ventura, The stellar activity-rotation relationship revisited: Dependence of saturated and non-saturated X-ray emission regimes on stellar mass for late-type dwarfs. Astron. Astrophys. **397**, 147–157 (2003). doi:10.1051/0004-6361:20021560

K. Poppenhäger, J. Robrade, J.H.M.M. Schmitt, J.C. Hall, 51 Pegasi—a planet-bearing Maunder minimum candidate. Astron. Astrophys. **508**, 1417–1421 (2009). doi:10.1051/0004-6361/200912945

A. Pouquet, U. Frisch, J. Leorat, Strong MHD helical turbulence and the nonlinear dynamo effect. J. Fluid Mech. **77**, 321–354 (1976). doi:10.1017/S0022112076002140

É. Racine, P. Charbonneau, M. Ghizaru, A. Bouchat, P.K. Smolarkiewicz, On the mode of dynamo action in a global large-eddy simulation of solar convection. Astrophys. J. **735**, 46 (2011). doi:10.1088/0004-637X/735/1/46

H. Rauer, C. Catala, C. Aerts, T. Appourchaux, W. Benz, A. Brandeker, J. Christensen-Dalsgaard, M. Deleuil, L. Gizon, M.-J. Goupil, M. Güdel, E. Janot-Pacheco, M. Mas-Hesse, I. Pagano, G. Piotto, D. Pollacco, N.C. Santos, A. Smith, J.-C., Suárez, R. Szabó, S. Udry, V. Adibekyan, Y. Alibert, J.-M. Almenara, P. Amaro-Seoane, M. Ammler-von Eiff, M. Asplund, E. Antonello, W. Ball, S. Barnes, F. Baudin, K. Belkacem, M. Bergemann, G. Bihain, A.C. Birch, X. Bonfils, I. Boisse, A.S. Bonomo, F. Borsa, I.M. Brandão, E. Brocato, S. Brun, M. Burleigh, R. Burston, J. Cabrera, S. Cassisi, W. Chaplin, S. Charpinet, C. Chiappini, R.P. Church, S. Csizmadia, M. Cunha, M. Damasso, M.B. Davies, H.J. Deeg, R.F. Díaz, S. Dreizler, C. Dreyer, P. Eggenberger, D. Ehrenreich, P. Eigmüller, A. Erikson, R. Farmer, S. Feltzing, F. de Oliveira Fialho, P. Figueira, T. Forveille, M. Fridlund, R.A. García, P. Giommi, G. Giuffrida, M. Godolt, J. Gomes da Silva, T. Granzer, J.L. Grenfell, A. Grotsch-Noels, E. Günther, C.A. Haswell, A.P. Hatzes, G. Hébrard, S. Hekker, R. Helled, K. Heng, J.M. Jenkins, A. Johansen, M.L. Khodachenko, K.G. Kislyakova, W. Kley, U. Kolb, N. Krivova, F. Kupka, H. Lammer, A.F. Lanza, Y. Lebreton, D. Magrin, P. Marcos-Arenal, P.M. Marrese, J.P. Marques, J. Martins, S. Mathis, S. Mathur, S. Messina, A. Miglio, J. Montalban, M. Montalto, M.J.P.F.G. Monteiro, H. Moradi, E. Moravveji, C. Mordasini, T. Morel, A. Mortier, V. Nascimbeni, R.P. Nelson, M.B. Nielsen, L. Noack, A.J. Norton, A. Ofir, M. Oshagh, R.-M. Ouazzani, P. Pápics, V.C. Parro, P. Petit, B. Plez, E. Poretti, A. Quirrenbach, R. Ragazzoni, G. Raimondo, M. Rainer, D.R. Reese, R. Redmer, S. Reffert, B. Rojas-Ayala, I.W. Roxburgh, S. Salmon, A. Santerne, J. Schneider, J. Schou, S. Schuh, H. Schunker, A. Silva-Valio, R. Silvotti, I. Skillen, I. Snellen, F. Sohl, S.G. Sousa, A. Sozzetti, D. Stello, K.G. Strassmeier, M. Švanda, G.M. Szabó, A. Tkachenko, D. Valencia, V. van Grootel, S.D. Vauclair, P. Ventura, F.W. Wagner, N.A. Walton, J. Weingrill, S.C. Werner, P.J. Wheatley, K. Zwintz, The PLATO 2.0 Mission. ArXiv e-prints (2013)

L.M. Rebull, J.R. Stauffer, S.T. Megeath, J.L. Hora, L. Hartmann, A correlation between pre-main-sequence stellar rotation rates and IRAC excesses in Orion. Astrophys. J. **646**, 297–303 (2006). doi:10.1086/504865

A. Reiners, Observations of cool-star magnetic fields. Living Rev. Sol. Phys. **9**, 1 (2012). doi:10.12942/lrsp-2012-1

A. Reiners, S. Mohanty, Radius-dependent angular momentum evolution in low-mass stars. I. Astrophys. J. **746**, 43 (2012). doi:10.1088/0004-637X/746/1/43

T. Reinhold, A. Reiners, G. Basri, Rotation and differential rotation of active Kepler stars. ArXiv e-prints (2013)

V. Reville, A.S. Brun, S. Matt, A. Strugarek, R. Pinto, The effect of magnetic topology on thermally-driven wind: towards a general formulation of the braking law. Astrophys. J. (2014, in press)

M. Rieutord, F. Rincon, The Sun's supergranulation. Living Rev. Sol. Phys. **7**, 2 (2010). doi:10.12942/lrsp-2010-2

R.S. Ronan, K. Cadora, B.J. Labonte, Solar cycle changes in the high frequency spectrum. Sol. Phys. **150**, 389–392 (1994). doi:10.1007/BF00712900

C.S. Rosenthal, J. Christensen-Dalsgaard, G. Houdek, M.J.P.F.G. Monteiro, A. Nordlund, R. Trampedach, Seismology of the solar surface regions, in *Helioseismology*. ESA Special Publication, vol. 376 (1995), pp. 459–464

G. Ruediger, L.L. Kitchatinov, The internal solar rotation in its spin-down history. Astrophys. J. **466**, 1078 (1996). doi:10.1086/177577

S. Saar, Stellar dynamos: scaling laws and coronal connections, in *Stellar Coronae in the Chandra and XMM-NEWTON Era*, ed. by F. Favata, J.J. Drake. Astronomical Society of the Pacific Conference Series, vol. 277 (2002), p. 311

S.H. Saar, Starspots, cycles, and magnetic fields, in *IAU Symposium*, ed. by D. Prasad Choudhary, K.G. Strassmeier. IAU Symposium, vol. 273 (2011), pp. 61–67. doi:10.1017/S1743921311015018

S.H. Saar, A. Brandenburg, Time evolution of the magnetic activity cycle period. II. Results for an expanded stellar sample. Astrophys. J. **524**, 295–310 (1999). doi:10.1086/307794

D. Salabert, C. Régulo, J. Ballot, R.A. García, S. Mathur, About the p-mode frequency shifts in HD 49933. Astron. Astrophys. **530**, 127 (2011a). doi:10.1051/0004-6361/201116633

D. Salabert, R.A. García, P.L. Pallé, A. Jiménez, Global p-mode oscillations throughout the complete solar cycle 23 and the beginning of cycle 24. J. Phys. Conf. Ser. **271**(1), 012030 (2011b). doi:10.1088/1742-6596/271/1/012030

R. Samadi, K. Belkacem, H.-G. Ludwig, E. Caffau, T.L. Campante, et al., Stellar granulation as seen in disk-integrated intensity. II. Theoretical scaling relations compared with observations. Astron. Astrophys. **559**, 40 (2013). doi:10.1051/0004-6361/201220817

L. Sbordone, P. Bonifacio, E. Caffau, H.-G. Ludwig, N.T. Behara, J.I. González Hernández, M. Steffen, R. Cayrel, B. Freytag, C. van't Veer, P. Molaro, B. Plez, T. Sivarani, M. Spite, F. Spite, T.C. Beers, N. Christlieb, P. François, V. Hill, The metal-poor end of the Spite plateau. I. Stellar parameters, metallicities, and lithium abundances. Astron. Astrophys. **522**, 26 (2010). doi:10.1051/0004-6361/200913282

J. Schou, D.L. Buzasi, Observations of P-modes in α cen, in *SOHO 10/GONG 2000 Workshop: Helio- and Asteroseismology at the Dawn of the Millennium*, ed. by A. Wilson, P.L. Pallé. ESA Special Publication, vol. 464 (2001), pp. 391–394

J. Schou, H.M. Antia, S. Basu, R.S. Bogart, R.I. Bush, et al., Helioseismic studies of differential rotation in the solar envelope by the solar oscillations investigation using the Michelson doppler imager. Astrophys. J. **505**, 390–417 (1998). doi:10.1086/306146

C.J. Schrijver, A.K. Dobson, R.R. Radick, The magnetic, basal, and radiative-equilibrium components in Mount Wilson CA II H+K fluxes. Astrophys. J. **341**, 1035–1044 (1989). doi:10.1086/167561

C.J. Schrijver, J. Beer, U. Baltensperger, E.W. Cliver, M. Güdel, et al., Estimating the frequency of extremely energetic solar events, based on solar, stellar, lunar, and terrestrial records. J. Geophys. Res. (Space Phys.) **117**, 8103 (2012). doi:10.1029/2012JA017706

M. Schrinner, L. Petitdemange, E. Dormy, Dipole collapse and dynamo waves in global direct numerical simulations. Astrophys. J. **752**, 121 (2012). doi:10.1088/0004-637X/752/2/121

M. Schrinner, L. Petitdemange, R. Raynaud, E. Dormy, Topology and field strength in spherical, anelastic dynamo simulations. Astron. Astrophys. **564**, 78 (2014). doi:10.1051/0004-6361/201322801

G. Schubert, K. Zhang, M.G. Kivelson, J.D. Anderson, The magnetic field and internal structure of Ganymede. Nature **384**, 544–545 (1996). doi:10.1038/384544a0

A.I. Shapiro, S.K. Solanki, N.A. Krivova, W.K. Schmutz, W.T. Ball, R. Knaack, E.V. Rozanov, Y.C. Unruh, The variability of Sun-like stars: reproducing observed photometric trends. ArXiv e-prints (2014)

H. Shapley, On the nature and cause of Cepheid variation. Astrophys. J. **40**, 448 (1914). doi:10.1086/142137

F. Shu, J. Najita, E. Ostriker, F. Wilkin, S. Ruden, S. Lizano, Magnetocentrifugally driven flows from young stars and disks. 1: A generalized model. Astrophys. J. **429**, 781–796 (1994). doi:10.1086/174363

H. Siedentopf, Konvektion in Sternatmosphären. Astron. Nachr. **247**, 297 (1933)

V. Silva Aguirre, L. Casagrande, S. Basu, T.L. Campante, et al., Verifying asteroseismically determined parameters of Kepler stars using hipparcos parallaxes: self-consistent stellar properties and distances. ArXiv e-prints (2012)

R.D. Simitev, F.H. Busse, Bistability and hysteresis of dipolar dynamos generated by turbulent convection in rotating spherical shells. Europhys. Lett. **85**, 19001 (2009). doi:10.1209/0295-5075/85/19001

R. Simoniello, K. Jain, S.C. Tripathy, S. Turck-Chièze, C. Baldner, W. Finsterle, F. Hill, M. Roth, The quasi-biennial periodicity as a window on the solar magnetic dynamo configuration. Astrophys. J. **765**, 100 (2013). doi:10.1088/0004-637X/765/2/100

A. Skumanich, Time scales for CA II emission decay, rotational braking, and lithium depletion. Astrophys. J. **171**, 565 (1972). doi:10.1086/151310

P.K. Smolarkiewicz, P. Charbonneau, EULAG, a computational model for multiscale flows: An MHD extension. J. Comput. Phys. **236**, 608–623 (2013). doi:10.1016/j.jcp.2012.11.008

W.H. Soon, S.L. Baliunas, Q. Zhang, An interpretation of cycle periods of stellar chromospheric activity. Astrophys. J. Lett. **414**, 33–36 (1993). doi:10.1086/186989

F. Spada, A.C. Lanzafame, A.F. Lanza, A semi-analytic approach to angular momentum transport in stellar radiative interiors. Mon. Not. R. Astron. Soc. **404**, 641–660 (2010). doi:10.1111/j.1365-2966.2010.16325.x

E.A. Spiegel, J.-P. Zahn, The solar tachocline. Astron. Astrophys. **265**, 106–114 (1992)

M. Spite, F. Spite, P. Bonifacio, The cosmic lithium problem. An observer's perspective. Mem. Soc. Astron. Ital. Suppl. **22**, 9 (2012)

H. Spruit, Theory of solar irradiance variations. Space Sci. Rev. **94**, 113–126 (2000)

H.C. Spruit, Dynamo action by differential rotation in a stably stratified stellar interior. Astron. Astrophys. **381**, 923–932 (2002). doi:10.1051/0004-6361:20011465

B. Sreenivasan, C.A. Jones, The role of inertia in the evolution of spherical dynamos. Geophys. J. Int. **164**, 467–476 (2006). doi:10.1111/j.1365-246X.2005.02845.x

J.R. Stauffer, L.W. Hartmann, The distribution of rotational velocities for low-mass stars in the Pleiades. Astrophys. J. **318**, 337–355 (1987). doi:10.1086/165371

R.F. Stein, Å. Nordlund, Convection and its influence on oscillations, in *Challenges to Theories of the Structure of Moderate-Mass Stars*, ed. by D. Gough, J. Toomre. Lecture Notes in Physics, vol. 388 (Springer, Berlin, 1991), p. 195

D. Stello, W.J. Chaplin, S. Basu, Y. Elsworth, T.R. Bedding, The relation between $\Delta\nu$ and ν_{max} for solar-like oscillations. Mon. Not. R. Astron. Soc. **400**, 80–84 (2009). doi:10.1111/j.1745-3933.2009.00767.x

S.-I. Takehiro, Y.-Y. Hayashi, Mean zonal flows excited by critical thermal convection in rotating spherical shells. Geophys. Astrophys. Fluid Dyn. **90**, 43–77 (1999). doi:10.1080/03091929908203692

S. Talon, C. Charbonnel, Angular momentum transport by internal gravity waves. II. Pop II stars from the Li plateau to the horizontal branch. Astron. Astrophys. **418**, 1051–1060 (2004). doi:10.1051/0004-6361:20034334

M.J. Thompson, A new inversion of solar rotational splitting data. Sol. Phys. **125**, 1–12 (1990). doi:10.1007/BF00154773

M.J. Thompson, J. Christensen-Dalsgaard, M.S. Miesch, J. Toomre, The internal rotation of the Sun. Annu. Rev. Astron. Astrophys. **41**, 599–643 (2003). doi:10.1146/annurev.astro.41.011802.094848

M.J. Thompson, A.S. Brun, J.L. Culhane, L. Gizon, M. Roth, T. Sekii, *Helioseismology and Dynamics of the Solar Interior*. Space Science Review Series (Springer, Berlin, 2015)

A.O. Thygesen, S. Frandsen, H. Bruntt, T. Kallinger, et al., Atmospheric parameters of 82 red giants in the Kepler field. Astron. Astrophys. **543**, 160 (2012). doi:10.1051/0004-6361/201219237

S.M. Tobias, The solar cycle: parity interactions and amplitude modulation. Astron. Astrophys. **322**, 1007–1017 (1997)

S.M. Tobias, Relating stellar cycle periods to dynamo calculations. Mon. Not. R. Astron. Soc. **296**, 653–661 (1998). doi:10.1046/j.1365-8711.1998.01412.x

R. Trampedach, R.F. Stein, The mass mixing length in convective stellar envelopes. Astrophys. J. **731**, 78 (2011). doi:10.1088/0004-637X/731/2/78

S. Turck-Chièze, S. Couvidat, A.G. Kosovichev, A.H. Gabriel, et al., Solar neutrino emission deduced from a seismic model. Astrophys. J. Lett. **555**, 69–73 (2001). doi:10.1086/321726

R.K. Ulrich, Determination of stellar ages from asteroseismology. Astrophys. J. Lett. **306**, 37–40 (1986). doi:10.1086/184700

J.L. van Saders, M.H. Pinsonneault, Fast star, slow star; old star, young star: subgiant rotation as a population and stellar physics diagnostic. Astrophys. J. **776**, 67 (2013). doi:10.1088/0004-637X/776/2/67

G.M. Vasil, D. Lecoanet, B.P. Brown, T.S. Wood, E.G. Zweibel, Energy conservation and gravity waves in sound-proof treatments of stellar interiors. II. Lagrangian constrained analysis. Astrophys. J. **773**, 169 (2013). doi:10.1088/0004-637X/773/2/169

A.A. Vidotto, S.G. Gregory, M. Jardine, J.-F. Donati, P. Petit, J. Morin, C.P. Folsom, J. Bouvier, A.C. Cameron, G. Hussain, S. Marsden, I.A. Waite, R. Fares, S. Jeffers, J.D. do Nascimento Jr., Stellar magnetism: empirical trends with age and rotation. ArXiv e-prints (2014)

F.L. Villante, A.M. Serenelli, F. Delahaye, M.H. Pinsonneault, The chemical composition of the sun from helioseismic and solar neutrino data. Astrophys. J. **787**, 13 (2014). doi:10.1088/0004-637X/787/1/13

S.N. Vogel, L.V. Kuhi, Rotational velocities of pre-main-sequence stars. Astrophys. J. **245**, 960–976 (1981). doi:10.1086/158872

A. Vögler, M. Schüssler, A solar surface dynamo. Astron. Astrophys. **465**, 43–46 (2007). doi:10.1051/0004-6361:20077253

S.V. Vorontsov, A search of the effects of magnetic field in the solar 5-minute oscillations, in *Advances in Helio- and Asteroseismology*, ed. by J. Christensen-Dalsgaard, S. Frandsen. IAU Symposium, vol. 123 (1988), p. 151

L.M. Walkowicz, G.S. Basri, Rotation periods, variability properties and ages for Kepler exoplanet candidate host stars. Mon. Not. R. Astron. Soc. **436**, 1883–1895 (2013). doi:10.1093/mnras/stt1700

J. Warnecke, P.J. Käpylä, M.J. Mantere, A. Brandenburg, Spoke-like differential rotation in a convective dynamo with a coronal envelope. Astrophys. J. **778**, 141 (2013). doi:10.1088/0004-637X/778/2/141

E.J. Weber, L. Davis Jr., The angular momentum of the solar wind. Astrophys. J. **148**, 217–227 (1967). doi:10.1086/149138

M.A. Weber, Y. Fan, M.S. Miesch, The rise of active region flux tubes in the turbulent solar convective envelope. Astrophys. J. **741**, 11 (2011). doi:10.1088/0004-637X/741/1/11

M.A. Weber, Y. Fan, M.S. Miesch, Comparing simulations of rising flux tubes through the solar convection zone with observations of solar active regions: constraining the dynamo field strength. Sol. Phys. **287**, 239–263 (2013). doi:10.1007/s11207-012-0093-7

N.O. Weiss, Solar and stellar dynamos, in *Lectures on Solar and Planetary Dynamos*, ed. by M.R.E. Proctor, A.D. Gilbert (1994), p. 59

T.R. White, T.R. Bedding, M. Gruberbauer, O. Benomar, D. Stello, T. Appourchaux, W.J. Chaplin, J. Christensen-Dalsgaard, Y.P. Elsworth, R.A. García, S. Hekker, D. Huber, H. Kjeldsen, B. Mosser, K. Kinemuchi, F. Mullally, M. Still, Solving the mode identification problem in asteroseismology of F stars observed with Kepler. Astrophys. J. Lett. **751**, 36 (2012). doi:10.1088/2041-8205/751/2/L36

O.C. Wilson, Chromospheric variations in main-sequence stars. Astrophys. J. **226**, 379–396 (1978). doi:10.1086/156618

B.E. Wood, Astrospheres and solar-like stellar winds. Living Rev. Sol. Phys. **1**, 2 (2004). doi:10.12942/lrsp-2004-2

M.F. Woodard, R.W. Noyes, Change of solar oscillation eigenfrequencies with the solar cycle. Nature **318**, 449–450 (1985). doi:10.1038/318449a0

J.T. Wright, Do we know of any Maunder minimum stars? Astron. J. **128**, 1273–1278 (2004). doi:10.1086/423221

J.T. Wright, G.W. Marcy, R.P. Butler, S.S. Vogt, Chromospheric Ca II emission in nearby F, G, K, and M stars. Astrophys. J. Suppl. **152**, 261–295 (2004). doi:10.1086/386283

N.J. Wright, J.J. Drake, E.E. Mamajek, G.W. Henry, The stellar-activity-rotation relationship and the evolution of stellar dynamos. Astrophys. J. **743**, 48 (2011). doi:10.1088/0004-637X/743/1/48

A.R. Yeates, D. Nandy, D.H. Mackay, Exploring the physical basis of solar cycle predictions: flux transport dynamics and persistence of memory in advection-versus diffusion-dominated solar convection zones. Astrophys. J. **673**, 544–556 (2008). doi:10.1086/524352

M. Yoshida, A. Kageyama, Application of the Yin-Yang grid to a thermal convection of a Boussinesq fluid with infinite Prandtl number in a three-dimensional spherical shell. Geophys. Res. Lett. **31**, 12609 (2004). doi:10.1029/2004GL019970

J.-P. Zahn, Convective penetration in stellar interiors. Astron. Astrophys. **252**, 179–188 (1991)

J.-P. Zahn, Circulation and turbulence in rotating stars. Astron. Astrophys. **265**, 115–132 (1992)

M. Zhang, K. Penev, Stars get dizzy after lunch. Astrophys. J. **787**, 131 (2014). doi:10.1088/0004-637X/787/2/131

J. Zhao, R.S. Bogart, A.G. Kosovichev, T.L. Duvall Jr., T. Hartlep, Detection of equatorward meridional flow and evidence of double-cell meridional circulation inside the Sun. Astrophys. J. Lett. **774**, 29 (2013). doi:10.1088/2041-8205/774/2/L29

MIX
Papier aus verantwortungsvollen Quellen
Paper from responsible sources
FSC® C105338

If you have any concerns about our products,
you can contact us on
ProductSafety@springernature.com

In case Publisher is established outside the EU,
the EU authorized representative is:
Springer Nature Customer Service Center GmbH
Europaplatz 3, 69115 Heidelberg, Germany

Printed by Libri Plureos GmbH
in Hamburg, Germany